Theoretical Fluid Dynamics

Theoretical and Mathematical Physics

This series, founded in 1975 and formerly entitled (until 2005) *Texts and Monographs in Physics* (TMP), publishes high-level monographs in theoretical and mathematical physics. The change of title to *Theoretical and Mathematical Physics* (TMP) signals that the series is a suitable publication platform for both the mathematical and the theoretical physicist. The wider scope of the series is reflected by the composition of the editorial board, comprising both physicists and mathematicians.

The books, written in a didactic style and containing a certain amount of elementary background material, bridge the gap between advanced textbooks and research monographs. They can thus serve as a basis for advanced studies, not only for lectures and seminars at graduate level, but also for scientists entering a field of research.

More information about this series at http://www.springer.com/series/720

Achim Feldmeier

Theoretical Fluid Dynamics

 Springer

Achim Feldmeier
Institut für Physik und Astronomie
Universität Potsdam
Potsdam, Germany

ISSN 1864-5879 ISSN 1864-5887 (electronic)
Theoretical and Mathematical Physics
ISBN 978-3-030-31024-0 ISBN 978-3-030-31022-6 (eBook)
https://doi.org/10.1007/978-3-030-31022-6

This Springer imprint is published by the registered company Springer Nature Switzerland AG
The registered company address is: Gewerbestrasse 11, 6330 Cham, Switzerland

For Gudrun

Preface

This book deals with calculational aspects of theoretical fluid dynamics. The emphasis is on calculations and derivations, not on formal proofs.

Roughly, the first half of the book is on exact solutions of stationary fluid equations in two dimensions, often including eddies, and the second half is on waves and instabilities. A more detailed description of the contents is given in the introduction.

I have tried to make some relatively recent research literature available to readers who have not yet specialized in fluid dynamics. Still, the material is somewhat advanced since there are lengthy calculations and a bulk of technicalities that are usually not found in more introductory texts.

There is almost nothing in this book on turbulence, boundary layers, rotating fluids, the Taylor columns, the Couette flow, solitary and cnoidal waves, the Orr–Sommerfeld equation, and wave coupling. Each of these deserves dedicated volumes on their own, and I felt not familiar enough with these topics to write about them.

I thank all the students who have attended my courses in fluid mechanics at the University of Potsdam over the last 20 years, where some of the material in this book was presented first. This book owes very much to the careful reading and critical comments of Prof. Kolumban Hutter from the ETH Zürich, to whom I express my cordial thanks. I thank the library of Technical University Berlin for making available in their *Freihandmagazin* many scientific journals that are not yet digitalized or to which I had no online access. Finally, I thank Springer-Verlag, especially Lisa Scalone and Manopriya Saravanan, for the efforts put into the production of this book. And I thank my wife for her constant interest in the book and for her patience during the years I worked on it.

For confused and mislead explanations, overlooked original work, missing recent references, inconsistencies, and plain errors, please contact me at afeld@uni-potsdam.de.

Berlin, Germany

Achim Feldmeier

Acknowledgements

The following persons, publishers, companies, institutions, and societies are gratefully acknowledged for granting the permission to use copyrighted figures and photographs:

Prof. Douglass Auld
Figure 11.1
Figure 11.3

Prof. Christopher E. Brennen
Figure 4.4—Reproduced from Brennen (1970)

Caltech thesis library
Figure 5.19—Reproduced from Konrad (1976)

Cambridge University Press: Journal of Fluid Mechanics
Figure 5.12—Reproduced from Abernathy and Kronauer (1962)
Figure 5.18—Reproduced from Baker and Shelley (1990)
Figure 5.23—Reproduced from Walker et al. (1987)
Figure 5.26—Reproduced from Moffatt (1969)
Figure 8.15—Reproduced from Longuet-Higgins and Fox (1978)

Centre d'études acadiennes
Figure 7.18

Elsevier: Physics Letters
Figure 5.21—Reproduced from Joyce and Montgomery (1972)

Encyclopaedia Britannica
Figure 7.5—Reproduced from Darwin (1902)

Google Earth
Figure 8.4

Prof. Jon Ove Hagen, Prof. Bjørn Gjevik
Figure 7.4

Achim Hering
Figure 11.4

Hydrotechnik Lübeck GmbH (Werner Evers)
Figure 8.9

International Journal of Fluid Dynamics
Figure 5.9—Reproduced from Lo et al. (2000)

London Mathematical Society: Proceedings of the London Mathematical Society
Figure 7.10—Reproduced from Taylor (1922)

Prof. Marc McCaughrean
Figure 3.16

Sandra Mihm
Figure 5.2

NASA
Figure 3.2
Figure 3.11
Figure 3.12
Figure 4.2
Figure 5.3
Figure 7.7
Figure 7.11
Figure 10.1

Arnold Price
Figure 7.17

Royal Society of London: Proceedings of the Royal Society and *Philosophical Transactions of the Royal Society*
Figure 5.11—Reproduced from Rosenhead (1931)
Figure 7.6—Reproduced from Proudman and Doodson (1924)
Figure 8.19—Reproduced from Akers et al. (2014)
Figure 11.2—Reproduced from Kolsky et al. (1949)

Prof. Satoshi Sakai
Figure 10.2

Meinhard Spitzer
Figure 8.16

Springer Nature: Nature
Figure 5.10—Reproduced from Zhang et al. (2000)

U. S. Army Corps of Engineers, Tulsa District
Figure 3.1

Wikipedia (see https://creativecommons.org/licenses/by-sa/2.0/)
Figure 3.10
Figure 4.3
Figure 5.4
Figure 8.11
Figure 11.9

Prof. Richard Wolfson
Figure 11.11—Reproduced from Wolfson and Holzer (1982)

YouTube
Figure 5.1

Berlin, Germany   Achim Feldmeier

Useful Readings

Abernathy, F.H., and R.E. Kronauer. 1962. The formation of vortex streets. *Journal of Fluid Mechanics* 13: 1.

Akers, B.F., D.M. Ambrose, and J.D. Wright. 2014. Gravity perturbed Crapper waves. *Proceedings of the Royal Society A* 470: 20130526.

Baker, G.R., and M.J. Shelley. 1990. On the connection between thin vortex layers and vortex sheets. *Journal of Fluid Mechanics* 215: 161.

Brennen, C.E. 1970. Cavity surface wave patterns and general appearance. *Journal of Fluid Mechanics* 44: 33.

Darwin, G.H. 1902. Tides. *Encyclopaedia Britannica*, 9th ed. https://www.1902encyclopedia.com/T/TID/tides.html.

Joyce, G., and D. Montgomery. 1972. Simulation of the 'negative temperature' instability for line vortices. *Physics Letters* 39A: 371.

Kolsky, H., J.P. Lewis, M.T. Sampson, A.C. Shearman, and C.I. Snow. 1949. Splashes from underwater explosions. *Proceedings of the Royal Society of London A* 196: 379.

Konrad, J.H. 1976. *An experimental investigation of mixing in two-dimensional turbulent shear flows with applications to diffusion-limited chemical reactions.* Internal Report CIT-8-PU. Pasadena: California Institute of Technology.

Lo, S.H., P.R. Voke, and N.J. Rockliff. 2000. Three-dimensional vortices of a spatially developing plane jet. *International Journal of Fluid Dynamics* 4, Article 1.

Longuet-Higgins, M.S., and M.J.H. Fox. 1978. Theory of the almost-highest wave. Part 2. Matching and analytic extension. *Journal of Fluid Mechanics* 85: 769.

Moffatt, H.K. 1969. The degree of knottedness of tangled vortex lines. *Journal of Fluid Mechanics* 35: 117.

Proudman, J., and A.T. Doodson. 1924. The principal constituent of the tides of the North Sea. *Philosophical Transactions of the Royal Society of London A* 224: 185.

Rosenhead, L. 1931. The formation of vortices from a surface of discontinuity. *Proceedings of the Royal Society of London A* 134: 170.

Taylor, G.I. 1922. Tidal oscillations in gulfs and rectangular basins. *Proceedings of the London Mathematical Society, Series 2*, 20: 148.

Walker, J.D.A., C.R. Smith, A.W. Cerra, and T.L. Doligalski. 1987. The impact of a vortex ring on a wall. *Journal of Fluid Mechanics* 181: 99.

Wolfson, R.L.T., and T.E. Holzer. 1982. Intrinsic stellar mass flux and steady stellar winds. *Astrophysical Journal* 255: 610.

Zhang, J., S. Childress, A. Libchaber, and M. Shelley. 2000. Flexible filaments in a flowing soap film as a model for one-dimensional flags in a two-dimensional wind. *Nature* 408: 835.

Contents

Chapter 1
Introduction

A fluid is an unstructured substance that fills a macroscopic volume in space continuously. There may be some order on the smallest, molecular scale, but there is no long-range order. Fluids can flow freely, i.e., tiny forces can lead to macroscopic motion, as, for example, in oceanic tides. Fluids are almost incompressible (they maintain their density under applied pressure) and show only small shear forces under internal friction. Fluids with strong shear and viscoelastic fluids showing properties of elastic materials are not considered in this book. Typical examples of fluids are liquids and, to a good approximation, gases.

A fluid is described by a scalar field $\rho(t, \vec{x})$ for the continuous mass distribution, and a vector field $\vec{u}(t, \vec{x})$ for the fluid velocity. Both ρ and $\vec{u}$ are assumed to be at least once continuously differentiable in the whole fluid body, except possibly on certain surfaces on which discontinuities in ρ and $\vec{u}$ or their spatial derivatives may occur.

A fluid parcel is an infinitesimal volume dV containing very many molecules. An infinitesimal volume dV has a well-defined location, in accordance with the introduction of smooth fields $\rho(t, \vec{x})$ and $\vec{u}(t, \vec{x})$. The continuum limit of introducing differentiable ρ and $\vec{u}$ and localized fluid parcels means that the infinitesimals of calculus, dx, etc., are far larger than the microscopic lengthscale of molecular distances. Since a fluid parcel contains many molecules, a thermal pressure $p(t, \vec{x})$ and temperature $T(t, \vec{x})$ can be defined. If, however, the continuum limit cannot be taken, one deals with *kinetic theory* or plasma physics. Instead of fields $\rho(t, \vec{x})$ and $\vec{u}(t, \vec{x})$, a distribution function $f(t, \vec{x}, \vec{u})$ is considered then, defined in the six-dimensional phase space of molecules or ions having location $\vec{x}$ and (arbitrary) velocity $\vec{u}$.

The general mathematical description of fluids is developed in Chap. 2, including the continuum limit of a particle ensemble. We discuss the intrinsic stress forces acting in fluids. The viscous force depends on a rank-two tensor, $\nabla \otimes \vec{u}$, for which an expression in arbitrary orthogonal coordinates is derived. Fluid pressure p is shown to act as Lagrange function in an Euler–Lagrange formalism for fluids. The chapter closes with an introduction to exterior differentiation and a derivation of the

© Springer Nature Switzerland AG 2019
A. Feldmeier, *Theoretical Fluid Dynamics*, Theoretical and Mathematical Physics,
https://doi.org/10.1007/978-3-030-31022-6_1

converse Poincaré lemma. This gives the condition under which $\nabla \times \vec{u} = 0$ implies $\vec{u} = \nabla \phi$, where the latter is central to Chap. 3.

Fluids are almost incompressible under applied pressures (and cannot be expanded by underpressure). In the idealized case that the fluid has constant density, the velocity field $\vec{u}$ is *divergence-free* or *solenoidal*, $\nabla \cdot \vec{u} = 0$. We assume solenoidal $\vec{u}$ throughout the book, except in the last two chapters on sound, internal gravity waves, and shocks.

If $\nabla \cdot \vec{u} = 0$, one may ask for vector fields that are also *irrotational*, $\nabla \times \vec{u} = 0$. In Chap. 3, besides $\nabla \cdot \vec{u} = \nabla \times \vec{u} = 0$, we assume that the fluid is inviscid (vanishing viscosity) and the flow is stationary, i.e., no fluid field depends on time. Furthermore, the flow is restricted to be two dimensional in a plane, which means that the fluid is stratified into plane layers with $\vec{u} = u_x(x, y)\,\hat{x} + u_y(x, y)\,\hat{y}$, with no dependence of $\vec{u}$ on z, and there is no velocity component u_z in the z-direction. For such a velocity field, $\nabla \cdot \vec{u} = 0$ becomes $u_x + v_y = 0$ and $\nabla \times \vec{u} = 0$ becomes $u_y - v_x = 0$. Defining functions $\phi(x, y)$ and $\psi(x, y)$ with $u = \phi_x = \psi_y$ and $v = \phi_y = -\psi_x$, one obtains from this the Laplace equation $\Delta \phi = \Delta \psi = 0$. Thus, ϕ and ψ are harmonic functions, and $\Phi = \phi + i\,\psi$ (see Betz 1948 for a similar symbol for the complex speed potential) is an analytic complex function that obeys the Cauchy–Riemann equations, $\phi_x = \psi_y$ and $\phi_y = -\psi_x$. Therefore, every analytic function is the complex potential Φ of a solenoidal, irrotational, and stationary flow of an inviscid fluid in the plane! There is a wealth of methods of complex function theory available for these flows in the complex plane. They are developed in Chap. 3, including a derivation of the Schwarz–Christoffel formula and a discussion of some elementary facts about the Riemann surfaces.

Elementary methods of complex function theory are applied in Chap. 4 on vortices. The theory of lift is presented, i.e., of forces normal to the mean (horizontal) flow direction. We give an extensive analysis of the stability of the Kármán vortex street, both for first- and second-order perturbations. For most of the book, vanishing or small viscosity is assumed. In Chap. 4, however, two examples of eddy flows of viscous fluids are included. The first is the corner eddy analyzed by Yih (1959). The second is a gas ring around a massive central object, which spreads under the influence of viscosity. It is shown that, as $t \to \infty$, all matter falls to the central object and all angular momentum is transported to infinity by no matter.

Chapter 5 applies the Schwarz–Christoffel formula and the Levi-Civita method from complex function theory to fluid jets and to a variety of wakes and cavities in flows. Highlights are Kolscher's cusped cavity (1940) as a counterexample for a claim by Brillouin and the re-entrant jet cavity on the two Riemann sheets. The latter example suggests a periodically alternating solution, which shows for the first and second half of the period the solution from the first and second Riemann sheet, respectively.

After these chapters on stationary flows, the next level of complexity are flows with waves and instabilities, i.e., with periodic time dependence and exponential or power-law growth of perturbations. Chapter 6 addresses the most important fluid instability, the Kelvin–Helmholtz instability of shear flows, including a generalization obtained by Bjerknes. This instability is responsible for the formation of macroscopic eddies in flows, from laboratory to terrestrial and even cosmic scales. It came as a surprise when Moore showed in 1979 that an infinitesimally thin shear layer is an inadequate model for Kelvin–Helmholtz instability leading to macroscopic eddies. Moore finds that singularities form in the almost flat layer before it rolls up into macroscopic vortices. Also, in this chapter, we discuss, using methods of statistical mechanics, why large eddies must occur in turbulent flows, and what are the topological invariants of interwoven, closed vortex loops.

The rest of the book, almost its second half, addresses waves. The short Chap. 7 introduces the phase and group speed, and discusses a somewhat pathological example of a Green's function involving both wave propagation and unstable growth.

Chapters 8 and 9 treat shallow water waves and free surface waves, respectively, which are the leading examples of fluid waves. Wave equations and dispersion relations for both types of waves are derived. For shallow water waves, the wavelength is much larger than the water depth, and the whole body of fluid oscillates essentially in the horizontal direction. For free surface waves, on the other hand, the wavelength is much smaller than the water depth, and, as their name indicates, these waves are essentially a surface phenomenon and one is primarily interested in the wave shape.

Chapter 8 addresses tidal waves and nonlinear bores, i.e., hydraulic jumps. Retardation effects are derived in an estuary with wave reflection, and it is shown that waves propagating toward a coast do not necessarily break, contrary to some claims in the literature. An important paper by Crease (1956) is discussed on waves that hit a long, rotating obstacle, like a landmass, and propagate behind it. The derivation uses the Wiener–Hopf method (1931) that is applied in advanced flow analyses.

In Chap. 9, the Green's function for free surface waves is derived. This solves a famous problem, viz., the motion of a water surface after being hit by a stone. In a well-known argument, Stokes stated that waves of highest elevation, nowadays called the *Stokes waves*, should have a cusp with interior angle of 120 degrees at their crest. For any further increase in wave height, spilling and breaking should set in at this cusp. Another central section of the chapter is the derivation of the Crapper wave (1957), one of the few known exact nonlinear solutions of a wave equation. The Crapper waves are capillary waves, and they suggest a mechanism for the immersion of air into seawater.

While certain properties of the Stokes wave are readily derived from phenomenological arguments, a rigorous existence proof was only found in 1961 by Krasovskii. Since the Stokes wave has maximum elevation, and thus is a strongly nonlinear solution far away from infinitesimal, linear waves, the proof is difficult and uses nonlinear functional analysis. We do not present this proof, but only demonstrate the existence of weakly nonlinear waves close to linear waves. The proof presented here is due to Lichtenstein, and applies the theory of linear and nonlinear integral equations, which is briefly reviewed.

Finally, the last two chapters deal with waves in compressible media, both fluids and gases. In Chap. 11, sound and internal gravity waves (buoyancy waves) are treated, with an application from helioseismology. If a gas is suddenly compressed, e.g., by pushing a piston into a gas-filled tube, a new phenomenon occurs, so-called *shock fronts*. In these, the fluid density and speed have macroscopic jumps on the molecular lengthscale, corresponding to discontinuities in the continuum limit.

Chapter 12 gives a brief introduction to this topic. Different types of transonic flows are considered, e.g., the solar wind and nozzle flow, and the jump conditions at shocks are derived. Shocks are nonlinear waves, and thus a central question is, how their propagation differs from linear sound waves. This is best understood by using the theory of characteristics, which is briefly introduced.

References

Betz, A. 1948. *Konforme Abbildung*. Berlin: Springer.

Crapper, G.D. 1957. An exact solution for progressive capillary waves of arbitrary amplitude. *Journal of Fluid Mechanics* 2: 532.

Crease, J. 1956. Long waves on a rotating Earth in the presence of a semi-infinite barrier. *Journal of Fluid Mechanics* 1: 86.

Kolscher, M. 1940. Unstetige Strömungen mit endlichem Totwasser. *Luftfahrtforschung* (Berlin-Adlershof) vol. 17, Lfg. 5, München: Oldenbourg.

Krasovskii, Yu.P. 1961. On the theory of steady-state waves of finite amplitude. *U.S.S.R. Computational Mathematics and Mathematical Physics* 11: 996.

Moore, D.W. 1979. The spontaneous appearance of a singularity in the shape of an evolving vortex sheet. *Proceedings of the Royal Society of London A* 365: 105.

Wiener, N., and E. Hopf. 1931. Über eine Klasse singulärer Integralgleichungen. *Sitzungsberichte der preussischen Akademie der Wissenschaften, Physikalisch-Mathematische Klasse* XXXI: 696.

Yih, C.-S. 1959. Two solutions for inviscid rotational flow with corner eddies. *Journal of Fluid Mechanics* 5: 36.

Chapter 2
Description of Fluids

2.1 The Euler and Lagrange Picture

In fluid mechanics, infinitesimal masses dM, called *fluid parcels*, occupy infinitesimal volumes dV that have well-defined positions $\vec{r}(t)$ in some fixed frame of reference, similar to point masses in classical point mechanics. The fluid parcels fill some connected volume in space homogeneously (tearing fluids are not treated here), where space is the Euclidean vector space $\mathbb{R}^3$. To distinguish between different fluid parcels, a second variable $\vec{a}$ is introduced in $\vec{r}(t, \vec{a})$ that identifies parcels uniquely, the so-called *Lagrange marker*. Often $\vec{a}$ is chosen to be the initial position $\vec{r}_0$ of the parcel at $t = 0$.

In a major abstraction step, a mass density field $\rho(t, \vec{x}) = dM/dV$ is introduced for the fluid, where a position vector $\vec{x}$ denotes an arbitrary space point. Once $\rho(t, \vec{x})$ is defined this way, any explicit reference to individual fluid parcels vanishes and ρ is a time-dependent *scalar field* on $\mathbb{R}^3$. Similarly, the velocity $\dot{\vec{r}}(t, \vec{a})$ of fluid parcels in the chosen reference frame is used to introduce a time-dependent vector field $\vec{u}(t, \vec{x})$ for the flow velocity at space points $\vec{x}$. The velocity field $\vec{u}$ at t and $\vec{x}$ is identified with the velocity $\dot{\vec{r}}$ of the parcel that passes through $\vec{x}$ at t,

$$\vec{u}(t, \vec{x} = \vec{r}(t, \vec{a})) = \dot{\vec{r}}(t, \vec{a}). \tag{2.1}$$

Let $\vec{a} = \vec{r}_0$ be the initial position of the parcel at $t = 0$. If $\vec{r}(t, \vec{r}_0)$ can be inverted to $\vec{r}_0(t, \vec{r})$, then (2.1) becomes

$$\vec{u}(t, \vec{x}) = \dot{\vec{r}}(t, \vec{r}_0(t, \vec{x})), \tag{2.2}$$

The original version of this chapter was revised: Belated correction from author has been incorporated. The correction to this chapter is available at https://doi.org/10.1007/978-3-030-31022-6_13.

A. Feldmeier, *Theoretical Fluid Dynamics*, Theoretical and Mathematical Physics, https://doi.org/10.1007/978-3-030-31022-6_2

where an $\vec{r}$ on the right side is replaced by $\vec{x}$. For (2.2), see Herglotz (1985), p. 24 and p. 33, after his Eqs. (1.31) and (1.61). The introduction of the fields ρ and $\vec{u}$ is especially fitting for time-independent or *stationary* flows, where $\rho(\vec{x})$ and $\vec{u}(\vec{x})$ are independent of t, i.e., $\partial/\partial t \equiv 0$.

The *Euler picture* (apparently introduced by d'Alembert, see Truesdell 1966, p. 10) is a description of the fluid in terms of $\rho(t, \vec{x})$, $\vec{u}(t, \vec{x})$, and a pressure field $p(t, \vec{x})$. By contrast, the *Lagrange picture* (introduced by Euler) is a description in terms of parcel locations $\vec{r}(t, \vec{a})$, obtained from velocities $\dot{\vec{r}}(t, \vec{a})$, and other parcel properties, like density and pressure. The Lagrange picture directly resembles Newtonian mechanics, as one is interested in trajectories of individual point masses. The Euler picture, by contrast, is a field theory for fluids, like Maxwell's theory is for electromagnetism. Its minimum requirement, for fluids with constant density ρ, is one vector field $\vec{u}(t, \vec{x})$.

Often, the thermal pressure p in a flow is a function of density and temperature T, $p = p(\rho, T)$. If p is a function of density alone, $p = p(\rho)$, as is the case in flows with constant temperature or if a polytropic law $p \sim \rho^\kappa$ holds with some empirical polytropic index κ (which is usually not identical to the adiabatic exponent γ), the flow is termed *barotropic*. Temperature is calculated from the ideal gas law $p/\rho = k_{\mathrm{B}} T/\mu$ with Boltzmann's constant k_{B} and mean molecular mass μ, or from another equation of state $T = T(\rho, p)$. These equations of state for $p(\rho, T)$ and $T(\rho, p)$ belong to thermodynamics and not to fluid dynamics proper.

The fields ρ, $\vec{u}$, p, and T are defined at each space point $\vec{x}$, and assumed to be continuously differentiable ('smooth'). Occasionally, this will be relaxed to *piecewise* continuous differentiability. The fields ρ and $\vec{u}$ can only be smooth if the number of molecules that make up a fluid parcel is very large and changes slowly with $\vec{x}$. Furthermore, p and T are defined by averaging over Maxwellian velocity distributions, and thus also require very large particle numbers in dV. Thus, the infinitesimals dx, dy, dz that define the volume $dV = dx\,dy\,dz$ of the fluid parcel are far above the microscopic, molecular scale.

2.2 The Lagrange Derivative

Let $q = q(t, \vec{r}(t))$ be any scalar property (density, pressure, temperature, etc.) of a parcel with trajectory $\vec{r}(t)$. The total time derivative of q is

$$\frac{dq}{dt} = \frac{\partial q}{\partial t} + \dot{\vec{r}} \cdot \nabla q. \tag{2.3}$$

With the identification $\vec{u} = \dot{\vec{r}}$, see (2.1), this becomes

$$\frac{dq}{dt} = \frac{\partial q}{\partial t} + \vec{u} \cdot \nabla q. \tag{2.4}$$

If instead of the scalar field q a vector field is considered, most notably the parcel velocity $\vec{u}$ itself, one has

$$\frac{d\vec{u}}{dt} = \frac{\partial\vec{u}}{\partial t} + \vec{u} \cdot (\nabla \otimes \vec{u}) = \frac{\partial\vec{u}}{\partial t} + (\vec{u} \cdot \nabla)\,\vec{u}, \tag{2.5}$$

where $\otimes$ is the tensor product and $\nabla \otimes \vec{u}$ is the *vector gradient* of $\vec{u}$, i.e., a tensor of rank 2. In Cartesian coordinates,

$$\nabla \otimes \vec{u} = (\partial_i u_j), \tag{2.6}$$

with a 3×3 matrix on the right side, see (2.12). The last equation sign in (2.5) holds by definition of the tensor product. The functions $q(t, \vec{r}(t))$ and $\vec{u}(t, \vec{r}(t))$ in (2.4) and (2.5) refer to the parcel at $\vec{r}(t)$. Let $\vec{x}$ be the space point at which the parcel with trajectory $\vec{r}(t)$ is at time t, and let $q(t, \vec{x})$ and $\vec{u}(t, \vec{x})$ be fields in the Euler picture. (The same names q and $\vec{u}$ are used in the Euler and Lagrange picture.) Then the *Lagrange* derivative D/Dt of the scalar field q is defined by

$$\frac{D}{Dt}\,q(t, \vec{x}) = \frac{d}{dt}\,q(t, \vec{r}(t)), \tag{2.7}$$

and correspondingly for the vector field $\vec{u}$. From (2.4) and (2.5),

$$\frac{D}{Dt} = \frac{\partial}{\partial t} + \vec{u} \cdot \nabla. \tag{2.8}$$

Often, the Lagrange derivative is directly introduced as total time derivative, by

$$\frac{d}{dt}\,q(t, \vec{x} = \vec{r}(t)) = \frac{\partial q}{\partial t} + \dot{\vec{r}} \cdot \nabla q = \frac{\partial q}{\partial t} + \vec{u} \cdot \nabla q. \tag{2.9}$$

In the following, d/dt will be used as symbol for the Lagrange derivative. For a *stationary* flow, $\partial\vec{u}/\partial t = 0$, thus

$$\frac{d\vec{u}}{dt} = \vec{u} \cdot (\nabla \otimes \vec{u}). \tag{2.10}$$

In Cartesian coordinates, with $\vec{u} = (u, v, w)$ (thus *not* $u = |\vec{u}|$), one has

$$\frac{du}{dt} = uu_x + vu_y + wu_z,$$
$$\frac{dv}{dt} = uv_x + vv_y + wv_z,$$
$$\frac{dw}{dt} = uw_x + vw_y + ww_z, \tag{2.11}$$

and therefore the velocity gradient in Cartesian coordinates is

$$\nabla \otimes \vec{u} = \begin{pmatrix} u_x & v_x & w_x \\ u_y & v_y & w_y \\ u_z & v_z & w_z \end{pmatrix}. \tag{2.12}$$

Note that this matrix is not symmetric. The velocity gradient in cylindrical coordinates is derived in Sect. 2.13, and in general orthogonal coordinates in Sect. 2.14.

2.3   Conservation Laws

If in a fluid certain quantities like mass, momentum, angular momentum, or energy that are attached to fluid parcels are conserved, then *conservation laws* of a general form hold. The subsequent addition of source and sink terms to these equations leads to general *balance laws*. Let $Q(t)$ be any conserved quantity, which can be a tensor field of rank 0, 1, 2, or higher, and let $q(t, \vec{x})$ be its spatial density of the same rank, i. e., the amount dQ of Q in an infinitesimal volume dV. Then $Q_V = \int_V dV\, q$ is the amount of Q in a volume V with boundary ∂V. The word volume and the symbol V shall refer to a space region with a given shape as well as to its volume content, i. e., to a point set and to its measure. Furthermore, in order to apply integral theorems, all volumes are assumed to have a piecewise differentiable boundary, as, for example, in a cube. Since Q is conserved, Q_V can only change by inflow and outflow of parcels through ∂V. For an area element $d\vec{a} = \hat{n}\, da$ of ∂V, the normal $\hat{n}$ shall point away from V. Let $\vec{u}(t, \vec{x})$ be the fluid velocity field in the chosen reference frame. From elementary geometry, an amount $q\, d\vec{a} \cdot \vec{u}\, dt$ of Q flows through the area da in a time interval dt. The rate of change of Q_V is therefore

$$\begin{aligned} \frac{d}{dt} \int_V dV\, q &= -\oint_{\partial V} (d\vec{a} \cdot \vec{u})\, q \\ &\overset{(a)}{=} -\oint_{\partial V} d\vec{a} \cdot (\vec{u} \otimes q) \\ &\overset{(b)}{=} -\int_V dV\, \nabla \cdot (\vec{u} \otimes q), \end{aligned} \tag{2.13}$$

where $\otimes$ is again the tensor product. The minus sign on the right side accounts for $\hat{n}$ pointing away from V, since Q_V drops if $\hat{n} \cdot \vec{u} > 0$. In (a), the associative law $(d\vec{a} \cdot \vec{u})\, q = d\vec{a} \cdot (\vec{u} \otimes q)$ for the tensor product is used. In (b), Gauss' divergence theorem for tensors is applied, which is derived in Eq. (2.58). We state here once and for all the requirements in Gauss' theorem: V is a compact subset of $\mathbb{R}^3$, ∂V is orientable and piecewise smooth. The components of $\vec{u} \otimes q$ have continuous first partial derivatives on an open set U that includes V, $V \subset U$ (i. e., one can differentiate $\vec{u} \otimes q$ across ∂V).

Let the integration domain V be constant in time (no change of shape, size, position, or orientation of V) in the chosen frame of reference of $\vec{x}$ and $\vec{u}$. Then the time derivative on the left side of (2.13) acts on q only. Furthermore, all positions $\vec{x}$ in the volume integral refer to fixed locations in V, thus a partial time derivative acts on $\mathsf{q}(t, \vec{x})$, giving

$$\int_V dV \left[\frac{\partial \mathsf{q}}{\partial t} + \nabla \cdot (\vec{u} \otimes \mathsf{q}) \right] = 0. \tag{2.14}$$

Since the volume V is arbitrary, one can conclude that

$$\boxed{\frac{\partial \mathsf{q}}{\partial t} + \nabla \cdot (\vec{u} \otimes \mathsf{q}) = 0} \tag{2.15}$$

The precise reasoning in this step is as follows, see Milne-Thomson (1968), p. 73: $\int_V dV\, f = 0$ implies $V^{-1} \int_V dV\, f = 0$. Thus, for a bounded function f,

$$0 = \lim_{V \to 0} \frac{1}{V} \int_V dV\, f(\vec{x}) = \frac{1}{V} f(\vec{x}_0)\, V = f(\vec{x}_0), \tag{2.16}$$

where $\vec{x}_0$ is a point included in all the volumes V considered in the limit. Such a sequence of smaller and smaller V exists for any $\vec{x}$, thus $f(\vec{x}) = 0$ everywhere in V.

If mass is conserved and q is the scalar density field ρ of rank 0, then (2.15) becomes the *continuity equation*,

$$\boxed{\frac{\partial \rho}{\partial t} + \nabla \cdot (\rho \vec{u}) = 0} \tag{2.17}$$

Assuming for the moment that no external or internal forces act on fluid parcels, the total fluid momentum is conserved and can change within a fixed volume V by in- and outflow of parcels only. Setting $\mathsf{q} = \rho \vec{u}$, with momentum density $\rho \vec{u}$ of rank 1, (2.15) becomes the equation of momentum conservation,

$$\boxed{\frac{\partial (\rho \vec{u})}{\partial t} + \nabla \cdot (\vec{u} \otimes \rho \vec{u}) = 0} \tag{2.18}$$

Force densities as sources of $\rho \vec{u}$ will be added later to the right side of (2.18).

In numerical calculations, it is often advantageous to have the points of the volume V, instead of being fixed, moving with an arbitrary, predescribed velocity $\vec{w}(t, \vec{x})$ with respect to the frame in which also $\vec{u}(t, \vec{x})$ is measured. This is especially relevant if the fluid (or rather a gas) undergoes strong spatial expansion (e. g., in a stellar explosion) or contraction (e. g., in the collapse of a gas cloud to a star), and the spatial numerical grid on which all scalar, vector, and tensor fields are finitely discretized shall roughly follow the fluid motion in order to achieve good numerical resolution at all times. One starts again with the continuity equation (2.17) in the fixed frame of reference.

To transform this to an equation in the local reference frames moving along with grid points (with reference to which the fields ρ, $\rho\vec{u}$, etc., are defined and discretized), the *relative* velocity $\vec{u} - \vec{w}$ between fluid and grid is introduced in (2.17),

$$\frac{\partial \rho}{\partial t} + \nabla \cdot (\rho(\vec{u} - \vec{w})) + \nabla \cdot (\rho\vec{w}) = 0. \tag{2.19}$$

Applying the product rule in the last term gives

$$\frac{\partial \rho}{\partial t} + \nabla \cdot (\rho(\vec{u} - \vec{w})) + \rho\nabla \cdot \vec{w} + \vec{w} \cdot \nabla\rho = 0. \tag{2.20}$$

The partial time derivative in a frame moving with $\vec{w}(t, \vec{x})$ is, as an operator equation (cf. the foregoing section on the Lagrange derivative),

$$\left.\frac{\partial}{\partial t}\right|_g = \frac{\partial}{\partial t} + \vec{w} \cdot \nabla, \tag{2.21}$$

where a subscript 'g' refers to the moving grid frame. Thus (2.20) becomes

$$\left.\frac{\partial \rho}{\partial t}\right|_g + \nabla \cdot (\rho(\vec{u} - \vec{w})) + \rho\nabla \cdot \vec{w} = 0, \tag{2.22}$$

which is Eq. (14) in Norman and Winkler (1986). The first two terms correspond to (2.17), now in a frame moving with velocity $\vec{w}$. The last term on the left, $\rho\nabla \cdot \vec{w}$, is new and is termed *grid compression* term by Norman and Winkler. Its origin is similar to that of apparent forces in non-inertial frames in classical point mechanics. The derivation of this term in Norman and Winkler (1986) and Winkler et al. (1984) differs slightly from the present one, and starts directly from the integral form of the conservation laws. The momentum equation on a moving grid is also readily obtained.

2.4   Divergence-Free Vector Field

To a high degree of accuracy, it can often be assumed that fluid parcels of fixed mass dM do not change the amount dV of their volume during motion (still, the shape, etc., of the parcel may alter). The change of dV within an infinitesimal time is determined by the motion of its boundary with velocity $\vec{u}$,

$$\frac{d}{dt}(dV) = \oint_{\partial(dV)} d\vec{a} \cdot \vec{u}. \tag{2.23}$$

The *divergence* of a vector field $\vec{u}$ is defined by

$$\nabla \cdot \vec{u} = \frac{1}{dV} \oint_{\partial(dV)} d\vec{a} \cdot \vec{u}. \tag{2.24}$$

Therefore, (2.23) becomes

$$\boxed{\frac{d}{dt}(dV) = dV \, \nabla \cdot \vec{u}} \tag{2.25}$$

and if dV is constant during motion, the important equation

$$\boxed{\nabla \cdot \vec{u} = 0} \tag{2.26}$$

follows. A vector field $\vec{u}$ with $\nabla \cdot \vec{u} = 0$ is called *divergence-free* or *solenoidal*. The two terms are used synonymously in the following.

Let $\rho = \rho(p, T)$, without dependence on further thermodynamic variables. In an *incompressible* fluid, the density of fluid parcels does not respond to pressure changes at constant temperature ('the degree of dependence on pressure measures the compressibility,' Proudman 1953, p. 43), $(\partial\rho/\partial p)|_T = 0$, and the equation of state simplifies to $\rho = \rho(T)$. If the parcel (or the fluid as a whole) has also constant temperature in time, then its density is constant and the Lagrange derivative of ρ vanishes,

$$\frac{d\rho}{dt} = \frac{\partial\rho}{\partial t} + \vec{u} \cdot \nabla\rho = 0. \tag{2.27}$$

The continuity equation (2.17) from mass conservation is

$$\frac{\partial\rho}{\partial t} + \nabla \cdot (\rho\vec{u}) = 0. \tag{2.28}$$

Applying the product rule in the second term on the left and using (2.27) gives again $\nabla \cdot \vec{u} = 0$.

Equation (2.26) can also be derived without using the abstract definition of divergence or Gauss' theorem, see Lamb (1932), p. 4. For this, let x, y, z be Cartesian coordinates and u, v, w the velocity components. Let $dV(0) = dx \, dy \, dz$ be an infinitesimal fluid cube at $t = 0$. The two corner points of a cube edge dx move relative to each other at speeds

$$\frac{\partial u}{\partial x} dx, \qquad \frac{\partial v}{\partial x} dx, \qquad \frac{\partial w}{\partial x} dx \tag{2.29}$$

apart in the x-, y-, and z-directions, respectively, and correspondingly for the edges dy and dz. After a time dt, the fluid cube is stretched to a parallelepiped. Its volume is given by the scalar triple product,

$$dV(dt) = \begin{vmatrix} (1 + u_x \, dt) \, dx & v_x \, dt \, dx & w_x \, dt \, dx \\ u_y \, dt \, dy & (1 + v_y \, dt) \, dy & w_y \, dt \, dy \\ u_z \, dt \, dz & v_z \, dt \, dz & (1 + w_z \, dt) \, dz \end{vmatrix}. \tag{2.30}$$

In lowest order of the differentials, this gives

$$dV(dt) = \left[1 + dt \left(\frac{\partial u}{\partial x} + \frac{\partial v}{\partial y} + \frac{\partial w}{\partial z} \right) \right] dV(0). \tag{2.31}$$

Demanding that the parcel maintains its volume, $dV(dt) = dV(0)$, it follows that

$$\frac{\partial u}{\partial x} + \frac{\partial v}{\partial y} + \frac{\partial w}{\partial z} = 0, \tag{2.32}$$

which is $\nabla \cdot \vec{u} = 0$ in Cartesian coordinates.

2.5 Fluid Boundaries

In this section, it is shown that in frictionless fluids, fluid parcels cannot leave or enter fluid boundaries; parcels in the boundary remain staying there. The proof uses only continuity of fluid motion. The presentation follows Meyer (1971, p. 2) and Lichtenstein (1929, p. 120).

We use the topological concepts of *open* and *closed* sets. A set $U \subset \mathbb{R}^3$ is *connected* if U is *not* the union of two disjoint, non-empty sets (this still allows for 'holes' in U). A connected open subset $U \subset \mathbb{R}^3$ is called a *domain*. The word 'domain' is thus a mathematical term; the word 'region' remains colloquial. A subset $U \subset \mathbb{R}^3$ is a topological *continuum* if U is bounded, open, and connected. In a mathematical sense, fluids are physical objects that occupy continua. An excellent introduction to the mathematical theory of the continuum via Dedekind cuts having neither jumps nor gaps is given in Sierpinski (1958), Chap. XI.

Let $\bar{U}$ be the closure and $\overset{\circ}{U}$ the open interior of $U \subset \mathbb{R}^3$. A fluid occupies $\overset{\circ}{U}$, and boundary conditions are specified on $\partial U = \bar{U} - \overset{\circ}{U}$. (Lichtenstein actually defines that the fluid occupies $\bar{U}$, not just $\overset{\circ}{U}$.) Fluid *motion* is described by a map

$$H_t : \bar{U}_0 \to \bar{U}_t, \tag{2.33}$$

where $\overset{\circ}{U}_0$ and $\overset{\circ}{U}_t$ are the domains occupied by fluid at times 0 and t. The function H_t is assumed to be one-to-one, and both H_t and H_t^{-1} shall be continuous. This requires zero viscosity (inviscid fluid) and zero heat conductivity: dissipative processes are not time reversible, thus H_t is not one-to-one then. Furthermore, fluid volumes are not allowed to merge or to fracture (e. g., droplet formation), since subtle issues occur for tearing surfaces and noncontinuous $\vec{u}(t, \vec{x})$, see Dussan (1976). The set of all functions H_t forms a transformation *group*. Group multiplication, the group inverse, and the neutral element $\mathbb{1}$ are given by

$$H_s \circ H_t = H_{s+t},$$
$$H_t^{-1} = H_{-t},$$
$$\mathbb{1} = H_0. \tag{2.34}$$

Let the fluid occupy an open set $U_0 \subset \mathbb{R}^3$ at $t = 0$, and let $U_t = H_t(U_0)$. For a continuous function, pre-images of open sets are open by definition of continuity. Thus, as $H_t^{-1} : U_t \to U_0$ is continuous, U_t is open since U_0 is open. Furthermore, since H_t is one-to-one, $H_t(\bar{U}_0) = \bar{U}_t$ in (2.33). Using elementary identities, this implies[1]

$$H_t^{-1}(\partial U_t) = H_t^{-1}(\bar{U}_t \backslash U_t)$$
$$\stackrel{(a)}{=} H_t^{-1}(\bar{U}_t) \backslash H_t^{-1}(U_t)$$
$$= \bar{U}_0 \backslash U_0 = \partial U_0. \tag{2.35}$$

Here (a) holds since for arbitrary sets A and B, one has $f^{-1}(A \backslash B) = f^{-1}(A) \backslash f^{-1}(B)$ if $B \subset A$, and the latter condition is fulfilled in the present case, $U_t \subset \bar{U}_t$. Therefore, under the above assumptions on H_t,

$$\partial U_t = H_t(\partial U_0), \tag{2.36}$$

and H_t maps boundaries to boundaries (Lichtenstein 1929, p. 120): thus a parcel cannot reach the boundary of a fluid from its interior. Or, if the fluid occupies a closed domain, its boundary consists always of the same fluid parcels.

These considerations can be extended to the following, much stronger theorem.

Theorem *Let U be a domain with boundary ∂U and let f be a continuous one-to-one function on U. Then the image of the boundary is the boundary of the image, $f(\partial U) = \partial f(U)$.*

In this theorem, only f is assumed to be continuous, whereas above it was assumed that both H_t and H_t^{-1} are continuous. The present theorem is a consequence (see Markushevich 1965, p. 99) of a deep theorem, the invariance of domain theorem (cf. Appendix A), which states that the image of a domain under a continuous, one-to-one function is again a domain. The proof of the latter theorem (ca. four pages) for the two-dimensional xy-plane is given in Markushevich (1965), p. 94.

A domain U is *simply connected* if every closed curve *within* it can be contracted continuously to a point without leaving the domain, see Sect. 2.21, i.e., U has no holes. Then a related theorem is the following.

Theorem *If U is a simply connected domain and f and f^{-1} are continuous and one-to-one, then $f(U)$ is simply connected.*

For the proof, see again Markushevich (1965), p. 100.

---

[1] I thank M. Bukenberger for pointing out an error in an earlier version of this argument.

Fig. 2.1 Continuous
stretching of a fluid
boundary by immersion of a
solid body. After Meyer
(1971), Fig. 2.2

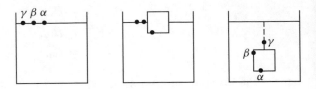

This theorem has an important consequence for inviscid fluids. In the left panel of Fig. 2.1, a thin fluid film is at rest in a container (i.e., there is essentially no third dimension normal to the plane of the paper). In the right panel, a solid body is fully immersed in the fluid over the whole thickness of this fluid film. In the left panel, the fluid domain is simply connected, but it is not simply connected in the right panel: a closed curve around the solid body cannot be contracted to a point without crossing the body. Assuming that both the mapping from the left to the right panel and its inverse are one-to-one and continuous, this would violate the last theorem.

This does not happen due to the fact that points like γ of the original fluid boundary in the left panel come to lie within the fluid in the right panel, establishing a boundary interface there, indicated in the figure by a dashed line. A closed curve within the fluid cannot cross this dashed line, a boundary, thus no closed curve can be drawn around the immersed body, and no contradiction occurs. Put differently, when the points from the dashed line are removed, the fluid domain can be continuously transformed, via bending and stretching, to a rectangle without a hole.

Fluid interfaces like this, across which fluid parcels cannot pass, can be thought of as the origin of fluid wakes; these wakes are treated in Chap. 5.

2.6 Phase Space Fluid

As a nontrivial example of a solenoidal vector field, consider a Hamiltonian mechanical system with n degrees of freedom (no friction). Its motion is described in a $2n$-dimensional phase space of canonical coordinates q_i and momenta p_i. The q_i, p_i are understood as Cartesian coordinates. The state of the system at any time is described by a single point in this phase space, and the time evolution of the system gives a trajectory $q_i(t)$, $p_i(t)$ (in the case of one degree of freedom, a curve $q(t)$, $p(t)$ or, after elimination of t, $p(q)$ in the qp-plane). If a homogeneous 'cloud' of $N \gg 1$ points in phase space is considered, where neighboring points correspond to slightly different initial values $q_i(0)$ and $p_i(0)$ of the system, one obtains a bundle of (non-crossing) trajectories, which, in the continuum limit $N \to \infty$, describes the motion of a fluid filling a phase space volume. The Hamiltonian equations of motion lead to Liouville's theorem, which states that this phase space volume is constant in time, i.e., a constant of motion. This means that the Hamiltonian speed $\vec{\Phi}$ (defined below) in phase space is divergence-free.

In order to demonstrate this, let q_1 to q_n be canonical coordinates, and p_1 to p_n the corresponding canonical momenta. The q_i, p_i are Cartesian coordinates of a $2n$-dimensional phase space. One chooses phase space points (at random) so that (i) neighboring points have infinitesimal distance, (ii) all points lie on different trajectories, and (iii) the point density $\rho = dN/dV$ is spatially constant. Here dN is the number of points in the infinitesimal phase space volume

$$dV = \prod_{i=1}^{n} dq_i \, dp_i. \tag{2.37}$$

In the continuum limit $dN \gg 1$ (1 is here an infinitesimal number), these points define a fluid in phase space. The evolution of any given set of initial values can be followed through all times $t < \infty$, i.e., the number of points N selected at $t = 0$ is conserved during motion. A phase space *speed* $\vec{\Phi}$ is defined by the $2n$-tuple (cf. Boltzmann's kinetic gas theory)

$$\vec{\Phi} = (\dot{q}_1, \ldots, \dot{q}_n, \dot{p}_1, \ldots, \dot{p}_n). \tag{2.38}$$

The $\dot{q}_i$ and $\dot{p}_i$ determine the rate of change of dV, and thus are indeed adequate speeds to describe changes in V. The system shall obey the Hamilton equations, with $i = 1, \ldots, n$,

$$\dot{q}_i = \frac{\partial H}{\partial p_i}, \qquad \dot{p}_i = -\frac{\partial H}{\partial q_i}. \tag{2.39}$$

Then

$$\nabla \cdot \vec{\Phi} = \sum_i \frac{\partial \dot{q}_i}{\partial q_i} + \sum_i \frac{\partial \dot{p}_i}{\partial p_i}$$

$$= \sum_i \frac{\partial^2 H}{\partial q_i \, \partial p_i} - \sum_i \frac{\partial^2 H}{\partial p_i \, \partial q_i} = 0. \tag{2.40}$$

Thus, the velocity field $\vec{\Phi}$ of the phase space fluid is solenoidal. Correspondingly, the phase space volume filled by the fluid is constant in time. This is known as Liouville's theorem.

2.7 Moving Fluid Line

As an application of the velocity gradient tensor, we consider the motion of an arbitrary curve (obtained e.g., by staining the fluid with ink) along with the fluid. Let $d\vec{l}$ be an infinitesimal segment of the curve; $d\vec{l}$ consists of parcels that move with velocity $\vec{u}$. To first differential order, $d\vec{l}$ remains straight. Thus, $d\vec{l}$ experiences only translation, rotation, and dilation.

Fig. 2.2 Motion of an
infinitesimal line d$\vec{l}$ with the
fluid

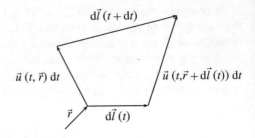

In Fig. 2.2, from vector addition,

$$\vec{u}(t, \vec{r})\, dt + d\vec{l}(t + dt) = d\vec{l}(t) + \vec{u}(t, \vec{r} + d\vec{l}(t))\, dt, \qquad (2.41)$$

or, after rearranging terms,

$$\frac{d\vec{l}(t + dt) - d\vec{l}(t)}{dt} = \vec{u}(t, \vec{r} + d\vec{l}(t)) - \vec{u}(t, \vec{r}). \qquad (2.42)$$

The left side of this equation is the time derivative of d$\vec{l}$. The last two terms on the right side describe the change of $\vec{u}$ over a distance d$\vec{l}$. By definition of the vector gradient, this is d$\vec{l} \cdot \nabla \otimes \vec{u}$ (or, in Cartesian coordinates, $\sum_i dl_i \partial_i u_j$). Therefore,

$$\boxed{\dot{\overrightarrow{dl}} = d\vec{l} \cdot \nabla \otimes \vec{u}} \qquad (2.43)$$

This equation can be integrated (formally), by introducing a time evolution operator U. Discretize time according to $dt = \epsilon$, $t = n\epsilon$ with $n \in \mathbb{N}_0$, and for the discrete timestep $\epsilon \to 0$ is understood at the end of the calculation. This gives a sequence of segments $d\vec{l}_0 = d\vec{l}(0)$ up to $d\vec{l}_n = d\vec{l}(t)$ and $d\vec{l}_{n+1} = d\vec{l}(t + \epsilon)$. Taking the transposed of (2.43), and introducing the tensor $\mathsf{T}_n = (\nabla \otimes \vec{u}(t, \vec{r}))^\tau$, gives

$$d\vec{l}_{n+1} = d\vec{l}_n + \epsilon\, \mathsf{T}_n \cdot d\vec{l}_n. \qquad (2.44)$$

To simplify notation, we do not distinguish between a vector (a column) and its transposed (a row) in $\mathbb{R}^3$. Inserting on the right

$$d\vec{l}_n = d\vec{l}_{n-1} + \epsilon\, \mathsf{T}_{n-1} \cdot d\vec{l}_{n-1}, \qquad (2.45)$$

and iterating this procedure until reaching $n - n$ gives

$$
d\vec{l}_{n+1} = \left(\mathbb{1} + \epsilon \sum_{i=0}^{n} \mathsf{T}_i + \epsilon^2 \sum_{i=1}^{n} \sum_{j=0}^{i-1} \mathsf{T}_i \cdot \mathsf{T}_j + \epsilon^3 \sum_{i=2}^{n} \sum_{j=1}^{i-1} \sum_{k=0}^{j-1} \mathsf{T}_i \cdot \mathsf{T}_j \cdot \mathsf{T}_k \right.
$$

$$
\left. + \cdots + \epsilon^{n+1} \mathsf{T}_n \cdot \mathsf{T}_{n-1} \ldots \mathsf{T}_1 \cdot \mathsf{T}_0 \right) \cdot d\vec{l}_0
$$

$$(2.46)$$

after a short calculation. Note that there is no summation left in the last term in the bracket. Taking $\epsilon \to 0$ and replacing sums by integrals, a linear relation is obtained,

$$
d\vec{l}(t) = \mathsf{U}(t) \cdot d\vec{l}(0),
\tag{2.47}
$$

with a time evolution operator (see Cocke 1969, Eqs. 2 and 3)

$$
\mathsf{U}(t) = \mathbb{1} + \int_0^t dt'\, \mathsf{T}(t') + \int_0^t dt'\, \mathsf{T}(t') \cdot \int_0^{t'} dt''\, \mathsf{T}(t'')
$$

$$
+ \int_0^t dt'\, \mathsf{T}(t') \cdot \int_0^{t'} dt''\, \mathsf{T}(t'') \cdot \int_0^{t''} dt'''\, \mathsf{T}(t''') + \cdots
\tag{2.48}
$$

2.8 Internal Fluid Stress

After these sections on the kinematics of fluids, we turn to the forces that are responsible for fluid motion. The conceptually simplest forces are body or volume forces. Here, the same force acts on each molecule in a fluid parcel, and the acceleration of the parcel is $d\vec{u}/dt = \vec{g}$. In most of the following, $\vec{g}$ is the gravitational acceleration.

Internal fluid forces, by contrast, are obtained by averaging over molecular interactions, in order to find a description for the interaction between fluid parcels. In the following, microscopic molecular forces are assumed to have short range, so that the macroscopic fluid force is of *contact* type, i.e., acts only on the common surface element of neighboring fluid parcels. Each fluid parcel exerts then *stresses* on neighboring parcels. The term *pressure* refers to normal stress and *shear* to tangential stress.

The force from normal pressure on an infinitesimal surface $d\vec{a}$ is $p\, d\vec{a}$. Euler was the first to realize that pressure acting normal on the side faces of a fluid parcel is *isotropic*, and thus can be described by a scalar field. The classical proof is found in Lamb (1932), p. 2 and Milne-Thomson (1968), p. 8.

An expression is now derived for the total surface force acting on a fluid parcel, including shear. The parcel shall have the shape of an infinitesimal tetrahedron ('pyramid') with four side faces. The maximum linear extent of the tetrahedron in any direction shall be ϵ (the minimum linear extent shall be of the same order). Then the volume forces, e.g., gravity, scale with $g\, dM \sim \epsilon^3$, the surface forces with $p\, da \sim \epsilon^2$, thus volume forces can be neglected compared to surface forces for

$\epsilon \rightarrow 0$. Surface forces alone establish force equilibrium on the parcel then, i. e., the sum of the surface forces vanishes,

$$d\vec{F}(d\vec{a}_1) + d\vec{F}(d\vec{a}_2) + d\vec{F}(d\vec{a}_3) + d\vec{F}(d\vec{a}_4) = 0, \qquad (2.49)$$

where $d\vec{a}_i$ are the area vectors of the faces of the tetrahedron, which is closed, thus

$$d\vec{a}_1 + d\vec{a}_2 + d\vec{a}_3 + d\vec{a}_4 = 0. \qquad (2.50)$$

From (2.49),
$$-d\vec{F}(d\vec{a}_4) = d\vec{F}(d\vec{a}_1) + d\vec{F}(d\vec{a}_2) + d\vec{F}(d\vec{a}_3), \qquad (2.51)$$

and from (2.50),
$$d\vec{F}(-d\vec{a}_4) = d\vec{F}(d\vec{a}_1 + d\vec{a}_2 + d\vec{a}_3). \qquad (2.52)$$

Using Newton's third law, $d\vec{F}(-d\vec{a}_i) = -d\vec{F}(d\vec{a}_i)$, gives

$$d\vec{F}(d\vec{a}_1 + d\vec{a}_2 + d\vec{a}_3) = d\vec{F}(d\vec{a}_1) + d\vec{F}(d\vec{a}_2) + d\vec{F}(d\vec{a}_3). \qquad (2.53)$$

The force exerted by neighboring fluid parcels is thus a linear function of the $d\vec{a}$. For this argument, see Milne-Thomson (1968), p. 630. A linear function mapping a vector (here $d\vec{a}$) to a vector (here $d\vec{F}$) is by definition a rank-two tensor. Since for linear functions a distributive law holds similar to that for multiplication, the action of a tensor on its vector argument is written as a so-called inner product, which in Cartesian coordinates becomes the usual scalar product of a matrix with a vector. Therefore, introducing the *stress tensor* T,

$$d\vec{F}(d\vec{a}) = d\vec{a} \cdot \mathsf{T}. \qquad (2.54)$$

This equation is called Cauchy's *fundamental lemma*, see Truesdell (1966), p. 29. Notice that (2.54) is often written as $d\vec{F} = \tilde{\mathsf{T}} \cdot d\vec{a}$ in the literature. Then $\tilde{\mathsf{T}} = \mathsf{T}^\tau$, the transposed stress tensor. If the stress tensor is not symmetric, $\mathsf{T} \neq \mathsf{T}^\tau$, one has to distinguish clearly whether $d\vec{a}$ is on the left or right side of T. We will always use form (2.54), given that a surface integral over $d\vec{a}$ is taken. Obviously, for non-symmetric T, the tensor divergence introduced below, $\nabla \cdot \mathsf{T} = \sum_i \partial_i T_{ij}$ in Cartesian coordinates, is different from $\nabla \cdot \mathsf{T}^\tau = \sum_i \partial_i T_{ji} \equiv \sum_j \partial_j T_{ij}$ (either i or j can serve as free vector index). The total stress force acting on a parcel with volume dV is

$$\boxed{d\vec{F} = \oint_{\partial dV} d\vec{a} \cdot \mathsf{T}} \qquad (2.55)$$

Let $d\vec{F} = dM\,\vec{g}_s$ (subscript 's' for surface forces) be the force exerted on a fluid parcel of mass dM by its surroundings,

$$dM\, \vec{g}_s = \oint_{\partial dV} d\vec{a} \cdot \mathsf{T}. \tag{2.56}$$

Then

$$\rho \vec{g}_s = \frac{1}{dV} \oint_{\partial dV} d\vec{a} \cdot \mathsf{T} = \nabla \cdot \mathsf{T}. \tag{2.57}$$

The last equation is the *definition* of tensor divergence. It leads directly to Gauss' theorem for tensors,

$$\boxed{\int_V dV\, \nabla \cdot \mathsf{T} = \oint_{\partial V} d\vec{a} \cdot \mathsf{T}} \tag{2.58}$$

The proof idea in Cartesian coordinates is as usual: the volume integral on the left side of the equation is replaced by a Riemann sum over small cubes ΔV_i (or tetrahedrons). For each cube, the definition of tensor divergence (2.57) is applied, giving a Riemann sum of surface integrals over the cube faces. In the interior of V, each face belongs to two neighboring cubes. The two surface vectors from neighboring cubes on their common side face have opposite signs. Thus, interior faces cancel in the Riemann sum (if T is continuous). Only the contribution from cube faces at the boundary ∂V remains, which have no neighboring cube further out. This is the Riemann sum for the surface integral on the right side of (2.58).

Jänich (2001), p. 122 remarks that this is not suited for a general proof of the Gauss or Stokes theorem, since 'decomposing the whole manifold into a grid of cells [...] is demanding,' especially in presence of coordinate singularities. He still sees a merit in this proof idea since it reduces the integral 'theorem at the intuitive level to a truism.' For a traditional proof, see Weatherburn (1924), p. 22 and p. 126. For a modern proof using differential forms, see Bott and Tu (1982), p. 31.

The simplest assumption for T is

$$\mathsf{T}(t, x, y, z) = -p(t, x, y, z)\, \mathbb{1}, \tag{2.59}$$

where the minus sign accounts for the parcel moving toward regions of lower pressure. Writing the rank-two identity tensor with dyadic vector products as $\mathbb{1} = \hat{x} \otimes \hat{x} + \hat{y} \otimes \hat{y} + \hat{z} \otimes \hat{z}$, the equation

$$\nabla \cdot (p\mathbb{1}) = \nabla p \cdot \mathbb{1} = \nabla p \tag{2.60}$$

results. Fluids with $\mathsf{T} = -p\mathbb{1}$ show no shear forces from friction. They obey the so-called *Euler* equation of motion, which is the fundamental dynamical equation of fluids,

$$\boxed{\frac{\partial \vec{u}}{\partial t} + (\vec{u} \cdot \nabla)\, \vec{u} = -\frac{\nabla p}{\rho}} \tag{2.61}$$

To see that this agrees with the standard definition of pressure as a force per area, consider a rectangular fluid parcel located at (x, y, z) and spanned by the vectors $dx\,\hat{x}, dy\,\hat{y}$, and $dz\,\hat{z}$. The acceleration in x-direction is given by the excess pressure force in this direction,

$$\rho\,dx\,dy\,dz\,\frac{du}{dx} = \left(p\left(x - \tfrac{1}{2}dx, y, z\right) - p\left(x + \tfrac{1}{2}dx, y, z\right)\right)dy\,dz, \qquad (2.62)$$

or

$$g_x = -\frac{1}{\rho}\frac{\partial p}{\partial x}. \qquad (2.63)$$

Similar equations hold for the y- and z-directions. In vector notation, thus, the acceleration due to normal pressure is indeed

$$\frac{d\vec{u}}{dt} = -\frac{\nabla p}{\rho}. \qquad (2.64)$$

The pressure p is taken to be the *thermodynamic* pressure (see Truesdell 1952, p. 164), e.g., for an ideal gas. Historically, see Love (1901), an important theorem was that

$$dp = \rho(X\,dx + Y\,dy + Z\,dz) \qquad (2.65)$$

is the total differential of a scalar field p (X, Y, Z are forces per mass).

Next, tangential stresses, i.e., shear forces are considered. Together with normal stresses, they enter the equation of motion in the form

$$\boxed{\frac{\partial \vec{u}}{\partial t} + (\vec{u} \cdot \nabla)\,\vec{u} = \frac{1}{\rho}\,\nabla \cdot \mathsf{T}} \qquad (2.66)$$

What information can be obtained about T without going into detailed empirical models of shear forces? We show that if the stresses on a fluid parcel vanish, then $\nabla \cdot \mathsf{T} = 0$. If, furthermore, the *couple stress* (a torque per unit area) on the parcel vanishes, then T is also symmetric. Figure 2.3 shows the shear forces in the xy-plane (no normal pressure, and assuming $T_{xy} = T_{yx} = 1$) that act on a rectangular parcel with volume $dV = dx\,dy\,dz$. The yz- and zx-planes look correspondingly.

The force on a fluid parcel from internal stress (pressure and shear) is given by (2.55)

$$d\vec{F} = \oint_{\partial dV} d\vec{a} \cdot \mathsf{T}. \qquad (2.67)$$

Fig. 2.3 Shear force vectors (bold arrows) in the xy-plane

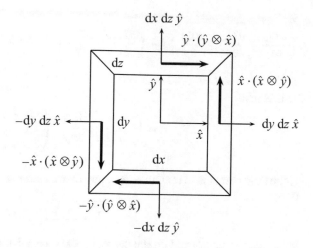

In Cartesian coordinates, T becomes

$$
\begin{aligned}
\mathsf{T} = {} & T_{xx}\,\hat{x}\otimes\hat{x} + T_{xy}\,\hat{x}\otimes\hat{y} + T_{xz}\,\hat{x}\otimes\hat{z} \\
& + T_{yx}\,\hat{y}\otimes\hat{x} + T_{yy}\,\hat{y}\otimes\hat{y} + T_{yz}\,\hat{y}\otimes\hat{z} \\
& + T_{zx}\,\hat{z}\otimes\hat{x} + T_{zy}\,\hat{z}\otimes\hat{y} + T_{zz}\,\hat{z}\otimes\hat{z}.
\end{aligned}
\tag{2.68}
$$

From Fig. 2.3 (and from similar figures for the yz- and zx-planes),

$$
\begin{aligned}
\mathrm{d}\vec{F} = {} & \\
\mathrm{d}y\,\mathrm{d}z\,\hat{x}\cdot{} & \left\{[\hat{x}\otimes\hat{x}\,T_{xx} + \hat{x}\otimes\hat{y}\,T_{xy} + \hat{x}\otimes\hat{z}\,T_{xz}](x+\tfrac{1}{2}\mathrm{d}x,y,z) - [\ldots](x-\tfrac{1}{2}\mathrm{d}x,y,z)\right\} \\
+\mathrm{d}x\,\mathrm{d}z\,\hat{y}\cdot{} & \left\{[\hat{y}\otimes\hat{x}\,T_{yx} + \hat{y}\otimes\hat{y}\,T_{yy} + \hat{y}\otimes\hat{z}\,T_{yz}](x,y+\tfrac{1}{2}\mathrm{d}y,z) - [\ldots](x,y-\tfrac{1}{2}\mathrm{d}y,z)\right\} \\
+\mathrm{d}x\,\mathrm{d}y\,\hat{z}\cdot{} & \left\{[\hat{z}\otimes\hat{x}\,T_{zx} + \hat{z}\otimes\hat{y}\,T_{zy} + \hat{z}\otimes\hat{z}\,T_{zz}](x,y,z+\tfrac{1}{2}\mathrm{d}z) - [\ldots](x,y,z-\tfrac{1}{2}\mathrm{d}z)\right\},
\end{aligned}
\tag{2.69}
$$

or, with $\mathrm{d}V = \mathrm{d}x\,\mathrm{d}y\,\mathrm{d}z$,

$$
\begin{aligned}
\frac{\mathrm{d}\vec{F}}{\mathrm{d}V} = {} & \left\{[\hat{x}\,T_{xx} + \hat{y}\,T_{xy} + \hat{z}\,T_{xz}](x+\tfrac{1}{2}\mathrm{d}x,y,z) - [\ldots](x-\tfrac{1}{2}\mathrm{d}x,y,z)\right\}/\mathrm{d}x \\
& + \left\{[\hat{x}\,T_{yx} + \hat{y}\,T_{yy} + \hat{z}\,T_{yz}](x,y+\tfrac{1}{2}\mathrm{d}y,z) - [\ldots](x,y-\tfrac{1}{2}\mathrm{d}y,z)\right\}/\mathrm{d}y \\
& + \left\{[\hat{x}\,T_{zx} + \hat{y}\,T_{zy} + \hat{z}\,T_{zz}](x,y,z+\tfrac{1}{2}\mathrm{d}z) - [\ldots](x,y,z-\tfrac{1}{2}\mathrm{d}z)\right\}/\mathrm{d}z.
\end{aligned}
\tag{2.70}
$$

With the definition $\mathrm{d}V\,\nabla\cdot\mathsf{T} = \oint_{\partial\mathrm{d}V}\mathrm{d}\vec{a}\cdot\mathsf{T}$, the divergence of a tensor of rank 2 in Cartesian coordinates can be read off,

$\nabla \cdot \mathsf{T} =$

$$\left(\frac{\partial T_{xx}}{\partial x} + \frac{\partial T_{yx}}{\partial y} + \frac{\partial T_{zx}}{\partial z} \right) \hat{x} + \left(\frac{\partial T_{xy}}{\partial x} + \frac{\partial T_{yy}}{\partial y} + \frac{\partial T_{zy}}{\partial z} \right) \hat{y} + \left(\frac{\partial T_{xz}}{\partial x} + \frac{\partial T_{yz}}{\partial y} + \frac{\partial T_{zz}}{\partial z} \right) \hat{z}.$$

$$(2.71)$$

This is just

$$\nabla \cdot \mathsf{T} = \left(\frac{\partial}{\partial x}, \frac{\partial}{\partial y}, \frac{\partial}{\partial z} \right) \cdot \begin{pmatrix} T_{xx} & T_{xy} & T_{xz} \\ T_{yx} & T_{yy} & T_{yz} \\ T_{zx} & T_{zy} & T_{zz.} \end{pmatrix} \tag{2.72}$$

If the fluid parcel is to remain at rest, the internal stress force must vanish, $\mathrm{d}\vec{F}/\mathrm{d}V = 0$, or

$$\boxed{\nabla \cdot \mathsf{T} = 0} \tag{2.73}$$

Next, the torque exerted on the parcel by the internal stress force (2.67) is

$$\mathrm{d}\vec{M} = \oint \vec{r} \times (\mathrm{d}\vec{a} \cdot \mathsf{T}). \tag{2.74}$$

Note that $\vec{r}$ has to be taken inside the integral. Internal couple stresses would make fluid parcels spin. In order that this does not happen, $\mathrm{d}\vec{M}/\mathrm{d}V = 0$ must hold. The following calculation is elementary but lengthy, and is written down on this one occasion. Considering first the x-direction of the parcel's cube, then the y- and z-directions, one obtains for vanishing couple stress,

$$
\begin{aligned}
&\overset{I_+,P_+}{((x+\tfrac{1}{2}\mathrm{d}x)\hat{x}} + \overset{A_+,p_+}{y\hat{y}} + \overset{a_+,i_+}{z\hat{z}}) \times \mathrm{d}y\,\mathrm{d}z\,\hat{x} \cdot [\overset{A_+,a_+}{\hat{x} \otimes \hat{x}\, T_{xx}} + \overset{I_+,i_+}{\hat{x} \otimes \hat{y}\, T_{xy}} + \overset{P_+,p_+}{\hat{x} \otimes \hat{z}\, T_{xz}}](x+\tfrac{1}{2}\mathrm{d}x, y, z) \\
&\overset{I_-,P_-}{-((x-\tfrac{1}{2}\mathrm{d}x)\hat{x}} + \overset{A_-,p_-}{y\hat{y}} + \overset{a_-,i_-}{z\hat{z}}) \times \mathrm{d}y\,\mathrm{d}z\,\hat{x} \cdot [\overset{A_-,a_-}{\hat{x} \otimes \hat{x}\, T_{xx}} + \overset{I_-,i_-}{\hat{x} \otimes \hat{y}\, T_{xy}} + \overset{P_-,p_-}{\hat{x} \otimes \hat{z}\, T_{xz}}](x-\tfrac{1}{2}\mathrm{d}x, y, z) \\
&\overset{J_+,Q_+}{+(x\hat{x}} + \overset{B_+,q_+}{(y+\tfrac{1}{2}\mathrm{d}y)\hat{y}} + \overset{b_+,j_+}{z\hat{z}}) \times \mathrm{d}x\,\mathrm{d}z\,\hat{y} \cdot [\overset{B_+,b_+}{\hat{y} \otimes \hat{x}\, T_{yx}} + \overset{J_+,j_+}{\hat{y} \otimes \hat{y}\, T_{yy}} + \overset{Q_+,q_+}{\hat{y} \otimes \hat{z}\, T_{yz}}](x, y+\tfrac{1}{2}\mathrm{d}y, z) \\
&\overset{J_-,Q_-}{-(x\hat{x}} + \overset{B_-,q_-}{(y-\tfrac{1}{2}\mathrm{d}y)\hat{y}} + \overset{b_-,j_-}{z\hat{z}}) \times \mathrm{d}x\,\mathrm{d}z\,\hat{y} \cdot [\overset{B_-,b_-}{\hat{y} \otimes \hat{x}\, T_{yx}} + \overset{J_-,j_-}{\hat{y} \otimes \hat{y}\, T_{yy}} + \overset{Q_-,q_-}{\hat{y} \otimes \hat{z}\, T_{yz}}](x, y-\tfrac{1}{2}\mathrm{d}y, z) \\
&\overset{K_+,R_+}{+(x\hat{x}} + \overset{C_+,r_+}{y\hat{y}} + \overset{c_+,k_+}{(z+\tfrac{1}{2}\mathrm{d}z)\hat{z}}) \times \mathrm{d}x\,\mathrm{d}y\,\hat{y} \cdot [\overset{C_+,c_+}{\hat{z} \otimes \hat{x}\, T_{zx}} + \overset{K_+,k_+}{\hat{z} \otimes \hat{y}\, T_{zy}} + \overset{R_+,r_+}{\hat{z} \otimes \hat{z}\, T_{zz}}](x, y, z+\tfrac{1}{2}\mathrm{d}z) \\
&\overset{K_-,R_-}{-(x\hat{x}} + \overset{C_-,r_-}{y\hat{y}} + \overset{c_-,k_-}{(z-\tfrac{1}{2}\mathrm{d}z)\hat{z}}) \times \mathrm{d}x\,\mathrm{d}y\,\hat{y} \cdot [\overset{C_-,c_-}{\hat{z} \otimes \hat{x}\, T_{zx}} + \overset{K_-,k_-}{\hat{z} \otimes \hat{y}\, T_{zy}} + \overset{R_-,r_-}{\hat{z} \otimes \hat{z}\, T_{zz}}](x, y, z-\tfrac{1}{2}\mathrm{d}z) \\
&= 0. \tag{2.75}
\end{aligned}
$$

The overbars serve to identify the terms: the three columns of the stress tensor are named $(a, b, c)^\tau$, $(i, j, k)^\tau$, and $(p, q, r)^\tau$, and the same again with capital letters. Subscripts $+$ and $-$ refer to locations $x + \frac{1}{2}\,\mathrm{d}x$ and $x - \frac{1}{2}\,\mathrm{d}x$, and similarly for y and z. Rearranging terms and supplementing 'missing' differentials gives

$$
\begin{aligned}
&\hat{x}\left[y\,\frac{\partial \overbrace{T_{xz}}^{p_\pm}}{\partial x}+\frac{\partial}{\partial y}\overbrace{(yT_{yz})}^{q_\pm}+y\,\frac{\partial \overbrace{T_{zz}}^{r_\pm}}{\partial z}-z\,\frac{\partial \overbrace{T_{xy}}^{i_\pm}}{\partial x}-z\,\frac{\partial \overbrace{T_{yy}}^{j_\pm}}{\partial y}-\frac{\partial}{\partial z}\overbrace{(zT_{zy})}^{k_\pm}\right]\\
&\hat{y}\left[z\,\frac{\partial \overbrace{T_{xx}}^{a_\pm}}{\partial x}+z\,\frac{\partial \overbrace{T_{yx}}^{b_\pm}}{\partial y}+\frac{\partial}{\partial y}\overbrace{(zT_{zx})}^{c_\pm}-z\,\frac{\partial \overbrace{T_{xy}}^{P_\pm}}{\partial x}-z\,\frac{\partial \overbrace{T_{yy}}^{Q_\pm}}{\partial y}-\frac{\partial}{\partial z}\overbrace{(zT_{zy})}^{R_\pm}\right]\\
&\hat{z}\left[\frac{\partial}{\partial x}\overbrace{(xT_{xy})}^{I_\pm}+x\,\frac{\partial \overbrace{T_{yy}}^{J_\pm}}{\partial y}+x\,\frac{\partial \overbrace{T_{zy}}^{K_\pm}}{\partial z}-y\,\frac{\partial \overbrace{T_{xx}}^{A_\pm}}{\partial x}-\frac{\partial}{\partial y}\overbrace{(yT_{yx})}^{B_\pm}-y\,\frac{\partial \overbrace{T_{zx}}^{C_\pm}}{\partial z}\right]=0.
\end{aligned}
\tag{2.76}
$$

Using the Cartesian representation (2.71) for $\nabla\cdot\mathsf{T}=0$, this reduces to

$$
\hat{x}\,[T_{yz}-T_{zy}]+\hat{y}\,[T_{zx}-T_{xz}]+\hat{z}\,[T_{xy}-T_{yx}]=0. \tag{2.77}
$$

Since $\hat{x}$, $\hat{y}$, $\hat{z}$ are linearly independent, T must be *symmetric*,

$$
\begin{aligned}
T_{xy}&=T_{yx},\\
T_{yz}&=T_{zy},\\
T_{zx}&=T_{xz}.
\end{aligned}
\tag{2.78}
$$

This calculation can be simplified by using coordinate names x_i with $i=1,2,3$, tensor components T_{ij}, and the ϵ-tensor for the cross product. Starting again with vanishing force and force couple,

$$
\begin{aligned}
\frac{\mathrm{d}\vec{F}}{\mathrm{d}V}&=\frac{1}{\mathrm{d}V}\oint_{\partial\mathrm{d}V}\mathrm{d}\vec{a}\cdot\mathsf{T}=0,\\
\frac{\mathrm{d}\vec{M}}{\mathrm{d}V}&=\frac{1}{\mathrm{d}V}\oint_{\partial\mathrm{d}V}\vec{r}\times(\mathrm{d}\vec{a}\cdot\mathsf{T})=0,
\end{aligned}
\tag{2.79}
$$

these become, with $\mathrm{d}\vec{a}=\mathrm{d}a\,\hat{n}$ and using the summation convention, i. e., performing a sum over indices that appear twice in a term,

$$
\begin{aligned}
&\frac{1}{\mathrm{d}V}\oint_{\partial\mathrm{d}V}\mathrm{d}a\;n_k\,T_{kl}=0,\\
&\frac{1}{\mathrm{d}V}\oint_{\partial\mathrm{d}V}\mathrm{d}a\;\epsilon_{ijl}\,x_j\,n_k\,T_{kl}=\frac{1}{\mathrm{d}V}\oint_{\partial\mathrm{d}V}\mathrm{d}a\;n_k\,(T_{kl}\,\epsilon_{ijl}\,x_j)=0.
\end{aligned}
\tag{2.80}
$$

The last equation holds since the *numbers* ϵ_{ijl}, x_j, n_k, and T_{kl} commute. The object in brackets corresponds to a rank-two tensor R_{ki}. Thus, with the definition (2.57) of tensor divergence and using (2.71) in the form $(\nabla\cdot\mathsf{T})_j=\partial_i\,T_{ij}$, one has, with the abbreviation $\partial_i=\partial/\partial x_i$,

$$\partial_k T_{kl} = 0, \tag{2.81}$$

$$\partial_k (\epsilon_{ijl}\, x_j\, T_{kl}) = 0. \tag{2.82}$$

The first line is again (2.73). Using $\partial_i x_j = \delta_{ij}$ and that the derivative of ϵ_{ijl} is zero, the second line yields

$$\epsilon_{ikl}\, T_{kl} + \epsilon_{ijl}\, x_j\, \partial_k T_{kl} = 0. \tag{2.83}$$

With (2.81), this becomes

$$\epsilon_{ikl}\, T_{kl} = 0. \tag{2.84}$$

Since ϵ_{ikl} is antisymmetric with respect to k and l, the latter equation can only hold, in general, if T_{kl} is symmetric,

$$T_{kl} = T_{lk}, \tag{2.85}$$

which is again (2.78). For this derivation, see Prager (1961), p. 49. For alternative derivations, see Weatherburn (1924), p. 142, Milne-Thomson (1968), p. 631, and Batchelor (1967), p. 11. Already Boltzmann has pointed out that for fluids in motion, the symmetry of T is a postulate. Continua for which this postulate holds are called the Boltzmann or Cauchy–Boltzmann continua, as opposed to the Cosserat continua, for which fluid parcels respond to stresses from force couples. For a review of this theory of the brothers Cosserat, see Schaefer (1967).

The archetypical shear force in fluids is due to friction, i.e., the resistance of fluid parcels to relative tangential motion. The modern theory of fluid friction was developed by Stokes (1845). Let u_i with $i = 1, 2, 3$ be the Cartesian components of the velocity vector. Newtonian friction is proportional to the velocity difference between sliding layers. The classical viscosity hypothesis is that the stress tensor T due to friction has the Cartesian form

$$T_{ij} = -p\,\delta_{ij} + 2\mu\, S_{ij}, \tag{2.86}$$

where μ is the *dynamic* viscosity, or *first* or *shear* coefficient of viscosity, and S_{ij} is the representation of the *rate-of-strain* tensor,

$$S_{ij} = \frac{1}{2}\left(\frac{\partial u_i}{\partial x_j} + \frac{\partial u_j}{\partial x_i}\right). \tag{2.87}$$

Details on strain can be found in Weatherburn (1924), pp. 132–140. Note that besides shearing $\partial_i u_j$ with $i \neq j$, the stress tensor also includes compression and expansion $\partial_i u_i$. The *Stokes hypothesis* is that *no* viscous energy dissipation is caused by compression or expansion. This can be proved from kinetic theory for monatomic gases, see the book by Truesdell and Muncaster (1980). For diatomic gases, etc., the Stokes hypothesis does not hold in general, since rotational degrees of freedom of the molecules can be excited. To employ the Stokes hypothesis, a further term is added to (2.86),

$$T_{ij} = -p\,\delta_{ij} + 2\mu\,S_{ij} + \lambda\,\partial_k u_k\,\delta_{ij}, \tag{2.88}$$

where the summation convention is used. Taking the trace T_{ii}, i.e., the sum of the diagonal elements of T_{ij}, using $\delta_{ii} = 3$, gives (after division by three)

$$\frac{1}{3}\,T_{ii} = -p + \zeta\,\partial_i u_i, \tag{2.89}$$

with the *second* or *volume* or *bulk* viscosity (coefficient)

$$\zeta = \frac{2}{3}\,\mu + \lambda, \tag{2.90}$$

which measures the viscosity related to compression or expansion of a fluid (see Rosenhead 1954 and subsequent contributions in this volume). The Stokes hypothesis is that $\zeta = 0$, or

$$\lambda = -\frac{2}{3}\,\mu. \tag{2.91}$$

For an in-depth discussion of the Stokes relation (2.91), see Truesdell (1952), p. 228. Why is a *symmetrical* strain tensor $\mathbf{S}$ assumed in the shear tensor $\mathbf{T}$? To answer this, a new fundamental vector field is introduced, the *vorticity* $\vec{\omega}$, by

$$\vec{\omega} = \nabla \times \vec{u}. \tag{2.92}$$

In Cartesian coordinates,

$$\omega_k = \epsilon_{kmn}\partial_m u_n = \frac{1}{2}\,\epsilon_{kmn}(\partial_m u_n - \partial_n u_m), \tag{2.93}$$

where the latter equation holds since $\epsilon_{kmn} = -\epsilon_{knm}$: the Levi-Civita symbol is defined by $\epsilon_{ijk} = 1$ (resp. -1) if (i, j, k) is an even (resp. odd) permutation of the tuple $(1, 2, 3)$, $\epsilon_{ijk} = 0$ else. One ϵ symbol 'calls' for another, since then a simple identity without ϵ holds. Consider thus

$$\begin{aligned}
\epsilon_{ijk}\,\omega_k &= \frac{1}{2}\,\epsilon_{ijk}\,\epsilon_{kmn}(\partial_m u_n - \partial_n u_m) \\
&= \frac{1}{2}\,\epsilon_{kij}\,\epsilon_{kmn}(\partial_m u_n - \partial_n u_m) \\
&= \frac{1}{2}(\delta_{im}\delta_{jn} - \delta_{in}\delta_{jm})(\partial_m u_n - \partial_n u_m) \\
&= \partial_i u_j - \partial_j u_i.
\end{aligned} \tag{2.94}$$

The velocity difference between two fluid parcels at a distance $d\vec{r}$ is (see Fig. 2.2)

$$d\vec{u} = d\vec{r} \cdot (\nabla \otimes \vec{u}). \tag{2.95}$$

A part of this change is due to relative motion, and is linked to shearing and friction. However, the two neighboring fluid parcels may also partake in some global rotation, say in a large eddy, and thus have different velocity vectors, but without shearing and friction. The part of $\partial_j u_i$ that contributes to friction is separated from the part that does not. In Cartesian coordinates, (2.95) becomes

$$
\begin{aligned}
du_i &= dr_j\, \partial_j\, u_i \\
&= \frac{1}{2}\, dr_j\, [(\partial_j u_i + \partial_i u_j) + (\partial_j u_i - \partial_i u_j)] \\
&= dr_j\, S_{ji} + \frac{1}{2}\, dr_j\, (\partial_j u_i - \partial_i u_j)] \\
&= dr_j\, S_{ji} + \frac{1}{2}\, dr_j\, \epsilon_{jik}\, \omega_k \\
&= dr_j\, S_{ji} + \frac{1}{2}\, \epsilon_{ijk}\, \omega_j\, dr_k.
\end{aligned}
\tag{2.96}
$$

The last term gives an acceleration

$$
\frac{d\vec{u}}{dt} = \frac{1}{2}\, \vec{\omega} \times \vec{u}.
\tag{2.97}
$$

This corresponds to Euler's equation $\dot{\vec{r}} = \vec{\Omega} \times \vec{r}$ for the change of a rotating vector $\vec{r}$ with footpoint on the axis of rotation. Thus, (2.97) describes rotation of the fluid parcels as a whole with angular velocity $\vec{\Omega} = \vec{\omega}/2$. It would still be present if the fluid were frozen (solid-body rotation; see Milne-Thomson 1968, p. 48). This term was not included here since it does not contribute to friction. The shear tensor including normal pressure and lateral shear due to friction is thus

$$
\boxed{\mathsf{T} = -\left(p + \frac{2}{3}\mu\, \nabla \cdot \vec{u}\right)\mathbb{1} + 2\mu\, \mathsf{S}}
\tag{2.98}
$$

Let μ be a constant, and introduce the *kinematic viscosity* $\nu = \mu/\rho$, with units m²/s. According to (2.57), the force per mass due to shear viscosity is then, in Cartesian coordinates,

$$
\begin{aligned}
g_i &= \frac{1}{\rho}\, \partial_j T_{ji} \\
&= -\frac{\partial_i p}{\rho} - \frac{2}{3}\, \nu\, \partial_i \partial_j u_j + \nu\, \partial_j(\partial_j u_i + \partial_i u_j) \\
&= -\frac{\partial_i p}{\rho} + \nu\, \partial_j\, \partial_j u_i + \frac{1}{3}\, \nu\, \partial_i\, \partial_j u_j.
\end{aligned}
\tag{2.99}
$$

In coordinate-free notation, $\partial_j \, \partial_j = \nabla \cdot \nabla = \Delta$ is the Laplace operator or Laplacian ('div grad'), and the Navier–Stokes equation for fluids with Newtonian friction is obtained as

$$\frac{\partial \vec{u}}{\partial t} + (\vec{u} \cdot \nabla)\vec{u} = -\frac{\nabla p}{\rho} + \nu \, \Delta \vec{u} + \frac{1}{3} \nu \, \nabla(\nabla \cdot \vec{u}) \qquad (2.100)$$

Note that the assumption of Newtonian friction, i.e., friction that depends linearly on velocity gradients, is quite severe. More refined models of small-scale interaction from modern *rheology* allow to study flows of a wide variety of viscoelastic fluids from glues to glaciers. We do not go into these topics in the present book.

Let us briefly return to (2.96), which shall express now the change of $\vec{u}$ across a *single* fluid parcel. Again, $\frac{1}{2} \, \epsilon_{ijk} \, \omega_j \, dr_k$ describes a solid-body rotation of the parcel, and $S_{ij} \, dr_j$ is the strain on the parcel. The symmetric matrix S_{ij} can be diagonalized, which gives, with $i = 1, 2, 3$ and not applying the summation convention,

$$du_i' = \alpha_i \, dx_i', \qquad (2.101)$$

where a prime refers to the Cartesian coordinate system of the three perpendicular eigenvectors of S_{ij}, and α_i are the eigenvalues of S_{ij}. Pure strain corresponds to squeezing and stretching the parcel along these directions x_i'. For a cubic fluid parcel with volume $dV = dx_1' \, dx_2' \, dx_3'$, the volume change is, if (i, j, k) is an even permutation of $(1, 2, 3)$,

$$\delta(dV)/\delta t = \sum_{i=1}^{3} dx_i' \, dx_j' \, du_k' = (\alpha_1 + \alpha_2 + \alpha_3) \, dV. \qquad (2.102)$$

The du_i' are the deformation rates of the parcel in the directions x_i'. If the parcel maintains its volume, then

$$\alpha_1 + \alpha_2 + \alpha_3 = 0. \qquad (2.103)$$

This means that the parcel contracts in at least one direction and expands in at least one other direction. If the fluid parcel is a sphere, the strain deforms it to an ellipsoid with its principal axes along x_i'. Lighthill (1986) gives an instructive example of a decomposition of a *shearing* motion into pure strain and rotation. For this, let $\vec{u} = (2y, 0, 0)$ be the shearing motion in normalized units, with the horizontal speed growing in vertical direction. The vorticity is $\vec{\omega} = (0, 0, -2)$ and the rate-of-strain matrix is

$$S_{ij} = \begin{pmatrix} 0 & 1 & 0 \\ 1 & 0 & 0 \\ 0 & 0 & 0 \end{pmatrix}. \qquad (2.104)$$

This is brought into diagonal form by a matrix L_{ij} that describes rotation (of a vector) by 45 degrees in the xy-plane,

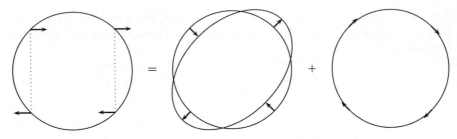

Fig. 2.4 Decomposition of shearing motion acting on a sphere (left) into pure strain (middle) and rotation (right). Adapted from Lighthill (1986), p. 51

$$L_{ij} = \begin{pmatrix} c & c & 0 \\ -c & c & 0 \\ 0 & 0 & 1 \end{pmatrix}, \tag{2.105}$$

with $c = 1/\sqrt{2}$. Indeed,

$$(LSL^{-1})_{ij} = \begin{pmatrix} 1 & 0 & 0 \\ 0 & -1 & 0 \\ 0 & 0 & 0 \end{pmatrix}. \tag{2.106}$$

The corresponding decomposition of the shear motion into pure strain and rotation is shown in Fig. 2.4.

We close with some remarks on the inviscid limit $\nu \to 0$. The Navier–Stokes equation for $\nu > 0$ is of second order in spatial differentials, whereas the Euler equation for $\nu = 0$ is of first order. Thus, the number of boundary conditions differs for $\nu > 0$ and $\nu = 0$. As a consequence, the limit $\nu \to 0$ is not trivial. The boundary condition for the speed normal to a wall is always $u_\perp = 0$. For $\nu = 0$, then, the tangential speed $u_\parallel$ is arbitrary (but is continuous with respect to the inner tangential fluid speed). For $\nu > 0$ instead, one usually assumes $u_\parallel = 0$: any tangential motion at a wall is suppressed by wall friction. Is $u_\parallel = 0$ for $\nu \to 0$ consistent with arbitrary $u_\parallel$ for $\nu = 0$? Prandtl introduced the concept of a *boundary layer* near walls, which should be thin for small ν. At its fluid-facing side, $u_\parallel$ is arbitrary, and at its wall facing side, $u_\parallel = 0$. Thus, the boundary layer separates the two distinct cases $\nu > 0$ and $\nu = 0$. Furthermore, Case (1961) proved that under certain conditions solutions of the Euler and Navier–Stokes equations agree for $\nu \to 0$. This depends on the limit $\nu \to 0$ being interchangeable with a Laplace transform.

2.9 Fluid Equations from Kinetic Theory

On different occasions, we referred to the molecules that make up a fluid parcel. In this section, the fluid equations are derived by taking statistical averages over kinetic phase space equations for molecules. The Boltzmann collision term requires some special attention, but gives no contribution in the end, see Huang (1963), Chap. 5. For this reason, the collision term is neglected here from the beginning and thus only the simple case of collisionless systems is considered. Examples of these are rarefied ion plasmas or large star clusters. To be specific, we refer to 'ions' in the following.

Consider six-dimensional phase space with coordinates (x_i, v_i) where $i = 1, 2, 3$. Note that x_i and v_i are independent, $\partial v_i / \partial x_j = 0$. (To avoid reference to physical speeds v_i, some books introduce general Cartesian coordinates x_1 to x_6 in phase space.) All ions shall have the same mass m. The ion acceleration is assumed to be due to conservative forces. Then along ion trajectories $(x_i(t), v_i(t))$ in phase space,

$$
\begin{aligned}
dx_i/dt &= v_i, \\
dv_i/dt &= -\partial V(x_1, x_2, x_3)/\partial x_i.
\end{aligned}
\tag{2.107}
$$

The ion mass m was absorbed here in the potential V. The phase space *distribution function* f is defined by

$$
dN = f(t, x_1, x_2, x_3, v_1, v_2, v_3) \, dx_1 \, dx_2 \, dx_3 \, dv_1 \, dv_2 \, dv_3,
\tag{2.108}
$$

where dN is the number of ions that is at time t in the infinitesimal 6-D cube

$$
[x_1, x_1 + dx_1] \times \cdots \times [v_3, v_3 + dv_3].
\tag{2.109}
$$

Ions leave this cube at speed $\dot{x}_i = v_i$ in the x_i-direction and at 'speed' $\dot{v}_i = -\partial V/\partial x_i$ in the v_i-direction. The phase space speed (2.38) is thus

$$
\vec{\Phi} = (v_1, v_2, v_3, -\partial V/\partial x_1, -\partial V/\partial x_2, -\partial V/\partial x_3).
\tag{2.110}
$$

Without collisions, the ion number in a phase space volume changes by in- and outflow only. Therefore, a phase space continuity equation holds (the cube is assumed to be fixed in phase space),

$$
\frac{\partial f}{\partial t} + \nabla \cdot (\vec{\Phi} f) = 0.
\tag{2.111}
$$

The divergence operator in Cartesian phase space coordinates is

$$
\nabla = \left(\frac{\partial}{\partial x_1}, \frac{\partial}{\partial x_2}, \frac{\partial}{\partial x_3}, \frac{\partial}{\partial v_1}, \frac{\partial}{\partial v_2}, \frac{\partial}{\partial v_3} \right).
\tag{2.112}
$$

This gives

$$
\frac{\partial f}{\partial t} + \sum_{i=1}^{3} v_i \frac{\partial f}{\partial x_i} - \sum_{i=1}^{3} \frac{\partial V}{\partial x_i} \frac{\partial f}{\partial v_i} = 0.
\tag{2.113}
$$

The Boltzmann collision term would appear on the right side. Integration over velocity space $(d^3 v = dv_1 \, dv_2 \, dv_3)$ gives

$$\frac{\partial}{\partial t} \int d^3v \, f + \sum_{i=1}^{3} \frac{\partial}{\partial x_i} \int d^3v \, v_i \, f - \sum_{i=1}^{3} \frac{\partial V}{\partial x_i} \int d^3v \, \frac{\partial f}{\partial v_i} = 0, \qquad (2.114)$$

where $\partial v_i / \partial x_j = 0$ was used. The last term on the left side is equal to

$$-\frac{\partial V}{\partial x_1} \iint dv_2 \, dv_3 \, \left[f(x_i, v_1 \to \infty, v_2, v_3) - f(x_i, v_1 \to -\infty, v_2, v_3) \right] + \cdots$$
$$(2.115)$$

The dots indicate terms that include f at $v_2 \to \pm\infty$ and $v_3 \to \pm\infty$. Since there are no ions with infinite speed, all these terms are zero. Define now *macroscopic* quantities ρ and u_i by

$$\rho = m \int d^3v \, f,$$

$$\rho u_i = m \int d^3v \, v_i \, f,$$

$$\rho w_{ij} = m \int d^3v \, v_i v_j \, f. \qquad (2.116)$$

Then (2.114) becomes, with the last term set to zero,

$$\frac{\partial \rho}{\partial t} + \sum_{i=1}^{3} \frac{\partial (\rho u_i)}{\partial x_i} = 0. \qquad (2.117)$$

This is the continuity equation for the mass density $\rho(t, x_1, x_2, x_3)$. Next, multiplying (2.113) with v_j and integrating over velocity space gives

$$\frac{\partial}{\partial t} \int d^3v \, v_j \, f + \sum_{i=1}^{3} \frac{\partial}{\partial x_i} \int d^3v \, v_i v_j \, f - \sum_{i=1}^{3} \frac{\partial V}{\partial x_i} \int d^3v \, v_j \, \frac{\partial f}{\partial v_i} = 0, \quad (2.118)$$

where $\partial v_j / \partial t = 0$ was used for the (dummy) integration variable v_j. Integration by parts in the last term gives

$$-\int d^3v \, v_j \, \frac{\partial f}{\partial v_i} = \int d^3v \, \frac{\partial v_j}{\partial v_i} \, f = \delta_{ij} \int d^3v \, f = \delta_{ij} \rho / m. \qquad (2.119)$$

In the first equation, it was used again that no ion has infinite speed. Thus, (2.118) becomes

$$\frac{\partial (\rho u_j)}{\partial t} + \sum_{i=1}^{3} \frac{\partial (\rho w_{ij})}{\partial x_i} + \rho \frac{\partial V}{\partial x_j} = 0. \qquad (2.120)$$

Let a tensor T with Cartesian components T_{ij} be defined by

$$\rho w_{ij} = \rho u_i u_j - T_{ij}. \qquad (2.121)$$

Subtracting the continuity equation multiplied by u_j from (2.120) gives

$$\frac{\partial u_j}{\partial t} + \sum_{i=1}^{3} u_i \frac{\partial u_j}{\partial x_i} = -\frac{\partial V}{\partial x_j} + \frac{1}{\rho} \sum_{i=1}^{3} \frac{\partial T_{ij}}{\partial x_i}, \qquad (2.122)$$

which has the form (2.66) including external forces. This equation was first derived by Maxwell. Later, Jeans applied it to stellar dynamics, and it is therefore called the Jeans equation. Huang (1963) derives Newtonian friction from definitions (2.116).

2.10 Streamlines and Pathlines

Two different families of curves in the fluid flow are introduced now. First, *pathlines* are the trajectories of individual fluid parcels in course of time. However, the Euler picture deals primarily with the velocity field $\vec{u}(t, \vec{x})$, and not with fluid parcels. Let then t_0 be an arbitrary but fixed time. *Streamlines* are defined as those curves that have, at this t_0, in any point along the curve the same slope as the velocity field $\vec{u}$ in this point. That is, the tangent to a streamline agrees in every point with the direction of the velocity field. With an increment (dx, dy, dz) along a streamline and $\vec{u} = (u, v, w)$, this means (see Batchelor 1967, p. 72)

$$\frac{dx}{u(t_0, \vec{x})} = \frac{dy}{v(t_0, \vec{x})} = \frac{dz}{w(t_0, \vec{x})}. \qquad (2.123)$$

The streamline is said to be an instantaneous integral curve of the velocity field $\vec{u}$. Streamlines and pathlines agree only for stationary flows, where ρ, $\vec{u}$, and all other fluid fields do not depend on time. In time-dependent flows, by contrast, streamlines and pathlines are usually different, as demonstrated in Fig. 2.5 for the velocity field of a two-dimensional flow in the xy-plane.

The figure includes a t-axis to show the temporal change of $\vec{u}(t, \vec{x})$, and thus of streamlines η. The pathline $(x(t), y(t))$ of a parcel gives a curve (or graph) $\xi = (t, x(t), y(t))$ with respect to the xyt-axes. At any time t, there is one single streamline η that crosses ξ at this t. This family of streamlines η_t (with family parameter t) forms a smooth surface Ξ in xyt-space. If $\vec{u}(\vec{x})$ is stationary, all these streamlines are identical. In this case, the locations $(x(t), y(t))$ of the pathline lie on one and the same streamline η, and streamlines and pathline agree. Put differently, the projection of ξ to the xy-plane agrees with the single streamline.

To illustrate the difference between pathlines and streamlines in nonstationary flows, consider a lawn sprinkler. Water streams out of a straight tube pointing in the polar $\hat{r}$-direction. The tube rotates about the vertical z-axis at constant angular speed. Assume that water leaves the tube in the form of single droplets.

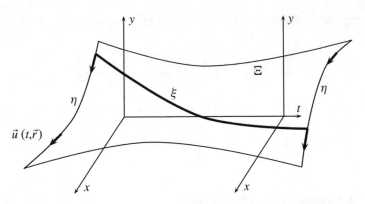

Fig. 2.5 Streamlines η at different times and pathline ξ

Inertial frame. A droplet is force-free after it has left the tube. Thus, droplets move nearly radially outward in straight lines, and the *pathlines* are straight lines in nearly $\hat{r}$-direction. A photograph (fixed time t_0!) of the sprinkler from above shows the *streamlines:* one sees droplets that left the sprinkler at different times. The earlier they have left, the farther away from the rotation axis they are. The connection of the droplets, the streamline, is a spiral.

Corotating frame. This is a frame fixed with respect to the rotating sprinkler; the origin is the rotation axis. Then the flow is stationary, and pathlines and streamlines agree. The Coriolis acceleration bends parcel trajectories against the rotation direction of the sprinkler. Thus, the pathlines are spirals, and so again are the streamlines. Note in this example that the streamlines, not the pathlines, are invariant.

As a second example, consider the *solar wind*. Charged plasma particles, mostly protons, electrons, and helium nuclei, stream away nearly radially from the hot corona of the rotating Sun. In an inertial frame, pathlines are straight and streamlines are spirals. The solar magnetic field is 'frozen-in' into the ion plasma. This means the magnetic field lines are aligned with the streamlines. Therefore, the magnetic field $\vec{B}$ also shows a spiral pattern. In the non-inertial, corotating frame, pathlines and streamlines agree with magnetic field lines. The fluid velocity $\vec{u}$ and the magnetic field lines in both the inertial and corotating frames are shown in Fig. 2.6.

2.11   Vortex Line, Vortex Tube, and Line Vortex

Streamlines were defined as curves tangential to the velocity field $\vec{u}$ at any fixed moment in time ('integral curve'). *Vortex lines* are defined, for each moment in time, as curves tangential to the *vorticity* field $\nabla \times \vec{u}$ in each point. That is, the unit tangent vector to the vortex line at any point and the unit vorticity vector at this point agree. Note that neither the magnitude nor the direction of $\nabla \times \vec{u}$ is generally constant along

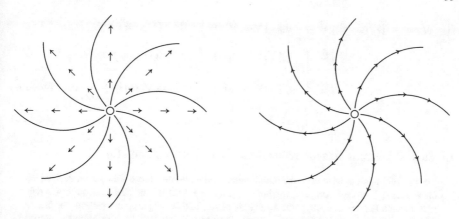

Fig. 2.6 Fluid velocity $\vec{u}$ (arrows) and magnetic field lines (full lines) in the solar wind. Left: inertial frame, right: corotating frame. Adapted from Hundhausen (1972), p. 133

Fig. 2.7 Circulation theorem for a vortex tube

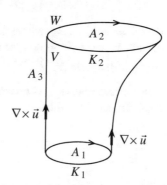

a vortex line. Let K_1 be a differentiable, closed curve in the fluid that is nowhere tangent to $\nabla \times \vec{u}$. Consider the set of all vortex lines that cross K_1 once, see Fig. 2.7. These curves form a surface W of tubular form, called *vortex tube*.

Let K_2 be another closed curve that defines the same vortex tube as K_1, see the Fig. 2.7. K_1 and K_2 shall neither touch nor cross. The vortex lines passing through K_1 and K_2 establish a one-to-one mapping F of the two curves. An increment $dl_1 > 0$ of the curve parameter of K_1 shall be mapped by this F to an increment $dl_2 > 0$ on K_2, i.e., the two curves shall have the same directional sense.

Theorem (Kelvin) *The circulation Γ is the same for K_1 and K_2,*

$$\Gamma_1 = \oint_{K_1} d\vec{l} \cdot \vec{u} = \oint_{K_2} d\vec{l} \cdot \vec{u} = \Gamma_2. \tag{2.124}$$

Proof Let A_1 and A_2 be tube cross sections with $K_1 = \partial A_1$ and $K_2 = \partial A_2$. The curves K_1 and K_2 delimit an area A_3 on the tube W. The three surfaces A_1, A_2, and A_3 form the closed boundary of a volume V inside W. Let $d\vec{a}$ on A_1 and A_3 point away from V. Then $d\vec{a}$ of A_2 points into V. Thus

$$\oint_{K_1} d\vec{l} \cdot \vec{u} - \oint_{K_2} d\vec{l} \cdot \vec{u} \overset{\text{(a)}}{=} \int_{A_1} d\vec{a} \cdot (\nabla \times \vec{u}) - \int_{A_2} d\vec{a} \cdot (\nabla \times \vec{u})$$

$$\overset{\text{(b)}}{=} \int_{A_1} d\vec{a} \cdot (\nabla \times \vec{u}) + \int_{A_2} d\vec{a}' \cdot (\nabla \times \vec{u})$$

$$\overset{\text{(c)}}{=} \int_{A_1} d\vec{a} \cdot (\nabla \times \vec{u}) + \int_{A_2} d\vec{a}' \cdot (\nabla \times \vec{u}) + \int_{A_3} d\vec{a} \cdot (\nabla \times \vec{u})$$

$$\overset{\text{(d)}}{=} \int_V dV \, \nabla \cdot (\nabla \times \vec{u}) = 0. \tag{2.125}$$

(a) Stokes' theorem. Its requirements are stated here once and for all:

Let $V \subset \mathbb{R}^3$ be an open subset. All components of $\vec{u}$ have continuous first partial derivatives on V. Let $A \subset V$ be a compact, oriented, regular two-dimensional surface (roughly, a regular surface has (i) at each point a tangent plane, and local parameterizations or charts $x(u, v), y(u, v), z(u, v)$ that are (ii) homeomorphisms and (iii) have continuous partial derivatives of all orders). The boundary ∂A of A is a non-empty, simple (no self-intersection), closed, piecewise-smooth curve with positive orientation (the enclosed area lies to the left). Then $\int_A d\vec{a} \cdot \nabla \times \vec{u} = \oint_{\partial A} d\vec{l} \cdot \vec{u}$.

(b) $d\vec{a}' = -d\vec{a}$ on A_2, where $d\vec{a}'$ points away from V.
(c) $d\vec{a} \cdot \nabla \times \vec{u} = 0$ on A_3 since $d\vec{a}$ and $\nabla \times \vec{u} = 0$ are orthogonal.
(d) $A_1 \cup A_2 \cup A_3 = \partial V$ and using Gauss' theorem.

Since circulation Γ is constant for a vortex tube, Γ can be used as a measure for the strength of the tube,

$$\Gamma = \oint_{C=\partial A} d\vec{l} \cdot \vec{u} = \int_A d\vec{a} \cdot \vec{\omega}, \tag{2.126}$$

where the differentiable curve $C = \partial A$ is the boundary of a cross section A of the tube.

Imagine that a vortex tube of small cross section is shrunk everywhere to infinitesimal cross section, while ω increases in such a way that a finite limit

$$\Gamma = \lim_{da \to 0} \, d\vec{a} \cdot \vec{\omega} \equiv \kappa \tag{2.127}$$

exists. This defines a *line vortex* or *vortex filament* (Saffman and Baker 1979), to be distinguished from a vortex line. Since $\vec{\omega} \to \infty$ on a line vortex, the latter is a singular curve of the vorticity field $\nabla \times \vec{u}$. The number $\kappa = d\vec{a} \cdot \vec{\omega}$ is the *strength* of the line vortex or simply the vortex strength. Here $d\vec{a}$ can be any arbitrarily tilted cross section through the infinitesimal tube (as long as $d\vec{a} \cdot \vec{\omega}$ does not change its sign): the projection $d\vec{a} \cdot \vec{\omega}$ is a geometrical invariant.

Consider an arbitrary, connected volume V in the fluid. Then again

$$0 = \int_V dV \, \nabla \cdot \vec{\omega} = \oint_{\partial V} d\vec{a} \cdot \vec{\omega}. \tag{2.128}$$

The latter integral has an immediate geometrical interpretation: let V be fibrated by line vortices of unit strength $d\vec{a} \cdot \vec{\omega} = 1$. The line vortices may be open or closed ('vortex rings'). Parts of V may have $\vec{\omega} = 0$, and thus have no vortex lines at all; this is of no concern in the following argument. Since $d\vec{a}$ points away from V everywhere, (2.128) counts the number of unit line vortices that leave V (where $d\vec{a} \cdot \vec{\omega} > 0$) minus the number of line vortices that enter V (where $d\vec{a} \cdot \vec{\omega} < 0$). This difference is zero, and there are thus as many line vortices that leave V as there are vortices that enter V. This would still allow for a line vortex l_1 to end at some point in V, and another vortex l_2 of the same strength to start in another point. However, this is impossible since V is arbitrary, and one can make ∂V cross the gap between the endpoint of one and the start point of another line vortex, thus violating (2.128). By taking V to be the full fluid body, one arrives at the conclusion: a line vortex is either closed (a ring) or extends from one fluid boundary to another. For this argument, see Kotschin et al. (1954), p. 30. Still, this argument cannot count as a mathematical proof, since certain potential 'anomalies' of the line vortex were not accounted for. Thus, the line vortex could split into two at zeros of $\vec{\omega}$ or the line vortex could wander around forever, see Chorin and Marsden (1990), p. 28 and Saffman (1995), p. 9.

Vortices can be 'injected' to fluids via friction at boundaries. Thus, eddies are frequently found behind solid obstacles in fluids.

2.12 Vortex Sheet

In this section, the fluid shall be inviscid and the velocity field solenoidal.

A two-dimensional vortex sheet is a curve in the plane in which the tangential component of the velocity field is discontinuous. Such a field is assumed irrotational elsewhere. In other terms, the vorticity is concentrated in a curve as a delta function. (Benedetto and Pulvirenti 2019, p. 1041)

Furthermore, the curve defining the vortex sheet is assumed to be differentiable and the normal velocity is continuous when crossing the vortex sheet. A vortex sheet is the cut of a (Euclidean or complex) plane with an array of straight-line vortices that lie parallel to each other and form a smooth surface; the cut is performed normal to the extension direction of the vortices. Thus, in three dimensions a vortex sheet is a smooth surface (straight in one direction) in which $|\nabla \times \vec{u}| \to \infty$ everywhere, and $\nabla \times \vec{u} = 0$ outside the sheet. Note that the surface of a *vortex tube* consists of vortex lines with finite $\vec{\omega}$, whereas a *vortex sheet* consists of line vortices with infinite $\vec{\omega}$, which causes the finite jump in tangential speed.

The velocity field of a vortex sheet is similar to the magnetic field of a *current sheet* in electrodynamics, where the electric current flows in one preferential direction in a thin, essentially two-dimensional plasma layer. The current sheet can be thought of as an assembly of parallel electric wires in this surface. When crossing from one side of a current sheet to the other, a jump occurs in the component of the magnetic field parallel to the sheet, as for the tangential velocity component in a vortex sheet. Current sheets occur, for example, in the solar corona.

To derive some basic properties of vortex sheets, we consider a simple shear flow $\vec{u} = (u, 0)$, with tangential fluid speed $u(y) \sim \tanh(y/\epsilon)$ in the x-direction (with $\epsilon > 0$ a small number), i.e.,

$$u(y) = -\frac{\Delta u}{2} \frac{e^{y/\epsilon} - e^{-y/\epsilon}}{e^{y/\epsilon} + e^{-y/\epsilon}}, \qquad (2.129)$$

with Δu an arbitrary constant. For $y \gg \epsilon$, this converges to $u = -\Delta u/2$, and for $y \ll -\epsilon$ to $u = \Delta u/2$ (implying positive vorticity), and thus the jump in horizontal speed between $y \to -\infty$ and $y \to \infty$ is Δu. The velocity field $(u, 0)$ with u from (2.129) is trivially solenoidal, $\partial u/\partial x + \partial v/\partial y = 0$. We are primarily interested in $y > \epsilon$ in (2.129), since the latter law cannot describe the 'microstructure' of the vortex sheet where the essentially circular velocity field of individual vortices making up the sheet becomes visible, thus u and v are of the same order, instead of $v = 0$ here. The vorticity of the velocity field $(u, 0)$ with u from (2.129) is

$$\vec{\omega} = \nabla \times \vec{u} = -\frac{\partial u}{\partial y} \hat{z} = \frac{2\Delta u}{\epsilon} \frac{1}{\left(e^{y/\epsilon} + e^{-y/\epsilon}\right)} \hat{z}. \qquad (2.130)$$

This is proportional to ϵ^{-1} in the layer $-\epsilon < y < \epsilon$ and drops quickly to zero for $|y| > \epsilon$. The circulation Γ over a distance Δx along the x-axis is, if the integration path is a rectangle with two sides below and above the x-axis at the same distance $|y|$, lying parallel to it, and with two sides normal to the x-axis,

$$\Gamma(\Delta x, y) = \oint d\vec{l} \cdot \vec{u} = 2 |u(y)| \, \Delta x, \qquad (2.131)$$

where both sides parallel to the x-axis give positive contribution to Γ because of the sign choice of u, and the two sides normal to the x-axis give no contribution to Γ since $v = 0$.

Consider now the limit $\epsilon \to 0$ or $|y| \gg \epsilon$ for any finite y, which defines a vortex sheet. Setting $\Gamma(x = 0) = 0$, one has then

$$\boxed{\Gamma(x) = \Delta u \, x} \qquad (2.132)$$

This linear proportionality $\Gamma \sim x$ establishes a one-to-one correspondence between the position x of a fluid parcel in the vortex sheet and the circulation $\Gamma(x)$ between the origin and x. Since the initial position of a fluid parcel at time $t = 0$ is a standard Lagrange marker for parcels, so is here, along the vortex sheet, the circulation $\Gamma(x)$. This will be used in the discussion of Moore's singularity for a perturbed vortex sheet in Sect. 6.8.

The vorticity (2.130) attains its maximum at $y = 0$, where, with $\omega = |\vec{\omega}|$,

$$\omega = \frac{\Delta u}{h} \qquad (2.133)$$

Here $h = 2\epsilon$ is the (typical) *thickness* of the vortex sheet. Equation (2.133) characterizes any vortex sheet and is not specific to the velocity law (2.129). Moreover, this equation can be used to define the limit $\omega \to \infty$ for the sheet more precisely, as

$$\Delta u = \lim_{h \to 0} h \omega, \qquad (2.134)$$

where Δu is a finite, constant number. Imagining the vortex sheet to be build up of individual line vortices that lie parallel to each other, each vortex has a diameter $h = 2\epsilon$ in vertical direction. The vortex has the same diameter h in the x-direction, and thus covers a square area $da = 4\epsilon^2$ (or a disk area $\pi\epsilon^2$). Then (2.127) together with (2.132) gives the vortex strength,

$$\Gamma(h) = \Delta u\, h = \omega h^2 \approx \kappa, \qquad (2.135)$$

or

$$\Delta u = \lim_{h \to 0} \frac{\kappa}{h}. \qquad (2.136)$$

Therefore, the vortex strength κ of individual line vortices making up the vortex sheet goes to zero together with the thickness h of the sheet.

A vortex sheet can roll up to form a cylindrical surface with infinite vorticity. One can use a new name *cylindrical vortex sheet* for this, see Batchelor (1967), p. 98, or retain the name vortex tube, assuming now also $\omega \to \infty$ (as stated above, vortex tubes have usually finite vorticity), see Tanaka and Kida (1993). For further discussions on vortex sheets, see Batchelor (1967), p. 98, Milne-Thomson (1968), p. 374, and Kotschin et al. (1954), p. 182.

2.13 Vector Gradient in Cylindrical Coordinates

Since the rank-two tensor $\nabla \otimes \vec{u}$ is of fundamental interest, we derive in this section its representation in cylindrical coordinates. The general representation in arbitrary orthogonal coordinates is derived in the next section. To start with, if $\nabla \otimes \vec{u}$ is multiplied with $\vec{u}$ from the left, one can avoid working with $\nabla \otimes \vec{u}$ by using the identity

$$\vec{u} \cdot (\nabla \otimes \vec{u}) = (\vec{u} \cdot \nabla)\vec{u} = \frac{1}{2}\nabla(\vec{u} \cdot \vec{u}) + (\nabla \times \vec{u}) \times \vec{u}. \qquad (2.137)$$

Often, however, an explicit expression for $\nabla \otimes \vec{u}$ is needed. Let $r\varphi z$ be cylindrical coordinates. The unit vectors $\hat{e}_r$ and $\hat{e}_\varphi$ are functions of φ alone, thus

$$\partial_\varphi \hat{e}_r = \hat{e}_\varphi,$$
$$\partial_\varphi \hat{e}_\varphi = -\hat{e}_r. \tag{2.138}$$

In general, however, the derivative of a unit vector is not necessarily a unit vector, see (2.152). Let $\vec{u} = u_r\,\hat{e}_r + u_\varphi\,\hat{e}_\varphi + u_z\,\hat{e}_z$ and $\vec{v} = v_r\,\hat{e}_r + v_\varphi\,\hat{e}_\varphi + v_z\,\hat{e}_z$ be arbitrary fields. The gradient in cylindrical coordinates is given by

$$\nabla = \hat{e}_r\partial_r + r^{-1}\hat{e}_\varphi\partial_\varphi + \hat{e}_z\partial_z, \tag{2.139}$$

and thus

$$\vec{u}\cdot\nabla = u_r\partial_r + r^{-1}u_\varphi\partial_\varphi + u_z\partial_z. \tag{2.140}$$

With the product rule for derivatives,

$$(\vec{u}\cdot\nabla)\vec{v} = \hat{e}_r(\vec{u}\cdot\nabla)v_r + \hat{e}_\varphi(\vec{u}\cdot\nabla)v_\varphi + \hat{e}_z(\vec{u}\cdot\nabla)v_z$$
$$+ v_r(\vec{u}\cdot\nabla)\hat{e}_r + v_\varphi(\vec{u}\cdot\nabla)\hat{e}_\varphi + v_z(\vec{u}\cdot\nabla)\hat{e}_z. \tag{2.141}$$

The last line gives, with constant $\hat{e}_z$,

$$v_r(\vec{u}\cdot\nabla)\hat{e}_r + v_\varphi(\vec{u}\cdot\nabla)\hat{e}_\varphi + v_z(\vec{u}\cdot\nabla)\hat{e}_z$$
$$= v_r\big(u_r\partial_r + r^{-1}u_\varphi\partial_\varphi + u_z\partial_z\big)\hat{e}_r + v_\varphi\big(u_r\partial_r + r^{-1}u_\varphi\partial_\varphi + u_z\partial_z\big)\hat{e}_\varphi$$
$$= \frac{u_\varphi v_r}{r}\,\partial_\varphi\hat{e}_r + \frac{u_\varphi v_\varphi}{r}\,\partial_\varphi\hat{e}_\varphi$$
$$= -\frac{u_\varphi v_\varphi}{r}\,\hat{e}_r + \frac{u_\varphi v_r}{r}\,\hat{e}_\varphi. \tag{2.142}$$

Therefore, in cylindrical coordinates,

$$(\vec{u}\cdot\nabla)\vec{v} = \hat{e}_r\left[(\vec{u}\cdot\nabla)v_r - \frac{u_\varphi v_\varphi}{r}\right] + \hat{e}_\varphi\left[(\vec{u}\cdot\nabla)v_\varphi + \frac{u_\varphi v_r}{r}\right] + \hat{e}_z(\vec{u}\cdot\nabla)v_z. \tag{2.143}$$

On the right, $\vec{u}\cdot\nabla$ acts on functions v_r, v_φ, v_z, and (2.139) can be used. As an application, consider a force-free fluid parcel,

$$\frac{d\vec{u}(t,\vec{r})}{dt} = \frac{\partial\vec{u}}{\partial t} + \vec{u}\cdot\nabla\vec{u} = 0. \tag{2.144}$$

Then from (2.143),

$$\frac{\partial u_r}{\partial t} + u_r \frac{\partial u_r}{\partial r} + \frac{u_\varphi}{r} \frac{\partial u_r}{\partial \varphi} + u_z \frac{\partial u_r}{\partial z} - \frac{u_\varphi^2}{r} = 0,$$

$$\frac{\partial u_\varphi}{\partial t} + u_r \frac{\partial u_\varphi}{\partial r} + \frac{u_\varphi}{r} \frac{\partial u_\varphi}{\partial \varphi} + u_z \frac{\partial u_\varphi}{\partial z} + \frac{u_r u_\varphi}{r} = 0, \qquad (2.145)$$

$$\frac{\partial u_z}{\partial t} + u_r \frac{\partial u_z}{\partial r} + \frac{u_\varphi}{r} \frac{\partial u_z}{\partial \varphi} + u_z \frac{\partial u_z}{\partial z} = 0.$$

One recognizes here the terms for centrifugal $(-u_\varphi^2/r)$ and Coriolis acceleration $(u_r u_\varphi/r)$. A familiar factor of two is missing in the latter, since Coriolis acceleration occurs here as first time derivative of speed, but in point mechanics as second time derivative of location.

2.14 Vector Gradient in Orthogonal Coordinates

The above expressions $-u_\varphi^2/r$ and $u_r u_\varphi/r$ are called *curvature terms*. Already in spherical coordinates they become slightly unintuitive. The easiest calculation of these terms proceeds via *metrical factors*. Metrical factors turn changes in coordinate variables into length changes. Let q_1, q_2, q_3 be orthogonal curvilinear coordinates, i.e., $\partial q_i / \partial q_j = \delta_{ij}$. Let $\mathrm{d}l_i$ be the length change corresponding to a coordinate change $\mathrm{d}q_i$ (all indices run from 1 to 3),

$$\mathrm{d}l_i = h_i(q_1, q_2, q_3) \, \mathrm{d}q_i. \qquad (2.146)$$

For non-orthogonal q_i, one has instead

$$\mathrm{d}l_i = \sum_{j=1}^{3} h_{ij}(q_1, q_2, q_3) \, \mathrm{d}q_j, \qquad (2.147)$$

with matrix h_{ij} of metrical factors. The analysis is restricted in the following to orthogonal coordinates (2.146). First, (2.138) is generalized: how does $\hat{e}_i$ change with q_j? Consider an arbitrary coordinate pair q_i, q_j. Figure 2.8 shows an infinitesimal element of the coordinate mesh.

Fig. 2.8 Cell of coordinate mesh of the coordinate pair q_i, q_j

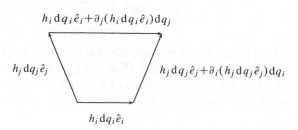

Let again $\partial_i = \partial/\partial q_i$. Vector addition in Fig. 2.8 gives

$$\partial_i(h_j\,dq_j\hat{e}_j)dq_i = \partial_j(h_i\,dq_i\hat{e}_i)dq_j. \tag{2.148}$$

(This argument is taken from Weatherburn 1924, p. 14.) For orthogonal coordinates and $i \neq j$,

$$\partial_i(dq_j) = 0. \tag{2.149}$$

Equation (2.148) becomes, for $i \neq j$,

$$\hat{e}_j\partial_i h_j + h_j\,\partial_i\hat{e}_j = \hat{e}_i\partial_j h_i + h_i\,\partial_j\hat{e}_i. \tag{2.150}$$

There is no summation over i or j here! Since $\hat{e}_i \cdot \hat{e}_j = 0$ for $i \neq j$, this can only hold if

$$\hat{e}_j\partial_i h_j = h_i\,\partial_j\hat{e}_i,$$
$$\hat{e}_i\partial_j h_i = h_j\,\partial_i\hat{e}_j, \tag{2.151}$$

see also Batchelor (1967, p. 598) for this and the following. In (2.151), the same equation appears twice, just with altered indices, and one arrives at the conclusion (no summation)

$$\boxed{\partial_i\hat{e}_j = \frac{\partial_j h_i}{h_j}\,\hat{e}_i \quad (i \neq j)} \tag{2.152}$$

These comprise six equations. Let now (i, j, k) be an even permutation of $(1, 2, 3)$. Then

$$\begin{aligned}\partial_k\hat{e}_k &= \partial_k(\hat{e}_i \times \hat{e}_j)\\ &= \partial_k\hat{e}_i \times \hat{e}_j + \hat{e}_i \times \partial_k\hat{e}_j\\ &= h_i^{-1}\,\partial_i h_k\,\hat{e}_k \times \hat{e}_j + h_j^{-1}\,\partial_j h_k\,\hat{e}_i \times \hat{e}_k\\ &= -h_i^{-1}\,\partial_i h_k\,\hat{e}_j \times \hat{e}_k - h_j^{-1}\,\partial_j h_k\,\hat{e}_k \times \hat{e}_i. \end{aligned} \tag{2.153}$$

If $\hat{e}_1, \hat{e}_2, \hat{e}_3$ form a right-handed system, then (no summation over i, j, or k),

$$\boxed{\partial_k\hat{e}_k = -\frac{\partial_i h_k}{h_i}\,\hat{e}_i - \frac{\partial_j h_k}{h_j}\,\hat{e}_j} \tag{2.154}$$

These form three more equations, for $k = 1, 2, 3$. Their right sides do not change under the replacement $i \leftrightarrow j$. Thus, (i, j, k) does actually not need to be an even permutation of $(1, 2, 3)$. Equations (2.152) and (2.154) are the looked-for generalization of (2.138). The vector gradient $\nabla \otimes \vec{u}$ is determined from its definition (2.10),

$$\mathrm{d}\vec{r} \cdot (\nabla \otimes \vec{u}) = (\mathrm{d}\vec{r} \cdot \nabla)\vec{u}. \tag{2.155}$$

With

$$\mathrm{d}\vec{r} = \sum_i h_i \mathrm{d}q_i \hat{e}_i, \qquad \nabla = \sum_i h_i^{-1}\hat{e}_i \partial_i, \tag{2.156}$$

and the Cartesian representation of the dyadic product,

$$\nabla \otimes \vec{u} = \sum_{i,j} (\nabla \otimes \vec{u})_{ij}\, \hat{e}_i \otimes \hat{e}_j, \tag{2.157}$$

Equation (2.155) becomes (with $\hat{e}_i \cdot \hat{e}_i \otimes \hat{e}_j = \hat{e}_j$)

$$\sum_{i,j} h_i \, \mathrm{d}q_i \, (\nabla \otimes \vec{u})_{ij} \, \hat{e}_j$$

$$= \sum_{i,j} \mathrm{d}q_i \, \partial_i (u_j \hat{e}_j)$$

$$= \sum_{i,j} \mathrm{d}q_i \, (\partial_i u_j) \, \hat{e}_j + \sum_{i,j} \mathrm{d}q_i \, u_j \, \partial_i \hat{e}_j$$

$$= \sum_{i,j} \mathrm{d}q_i \, (\partial_i u_j) \, \hat{e}_j + \sum_{\substack{i,j \\ i \neq j}} \mathrm{d}q_i \, u_j \, \partial_i \hat{e}_j + \sum_k \mathrm{d}q_k \, u_k \, \partial_k \hat{e}_k$$

$$\overset{(a)}{=} \sum_{i,j} \mathrm{d}q_i \, (\partial_i u_j) \, \hat{e}_j + \sum_{\substack{i,j \\ i \neq j}} \mathrm{d}q_i \, u_j \, \frac{\partial_j h_i}{h_j}\hat{e}_i - \sum_k \mathrm{d}q_k \, u_k \sum_{j \neq k} \frac{\partial_j h_k}{h_j}\, \hat{e}_j$$

$$\overset{(b)}{=} \sum_{i,j} \mathrm{d}q_i \, (\partial_i u_j) \, \hat{e}_j + \sum_{\substack{i,k \\ i \neq k}} \mathrm{d}q_i \, u_k \, \frac{\partial_k h_i}{h_k}\, \hat{e}_i - \sum_{\substack{i,j \\ i \neq j}} \mathrm{d}q_i \, u_i \, \frac{\partial_j h_i}{h_j}\, \hat{e}_j. \tag{2.158}$$

(a) Equations (2.152) and (2.154)
(b) Rename $k \to i$ in the last term. For clarity, $j \to k$ in the middle term.

Let (a_{ij}) be a 3×3 matrix. If for arbitrary $\mathrm{d}q_1, \mathrm{d}q_2, \mathrm{d}q_3$ the equation

$$\sum_{i,j} a_{ij} \, \mathrm{d}q_i \, \hat{e}_j = 0 \tag{2.159}$$

holds, then all $a_{ij} = 0$, for $i, j = 1, 2, 3$. Namely, choose $\mathrm{d}q_1 \neq 0$ and $\mathrm{d}q_2 = \mathrm{d}q_3 = 0$. Then $a_{11}\hat{e}_1 + a_{12}\hat{e}_2 + a_{13}\hat{e}_3 = 0$, implying $a_{11} = a_{12} = a_{13} = 0$. Repeat this with $\mathrm{d}q_2 \neq 0$ and $\mathrm{d}q_1 = \mathrm{d}q_3 = 0$, and then with $\mathrm{d}q_3 \neq 0$ and $\mathrm{d}q_1 = \mathrm{d}q_2 = 0$. Therefore, in (2.158), the coefficients of the $\mathrm{d}q_i \, \hat{e}_j$ must be identical. Consider first the $\mathrm{d}q_i \, \hat{e}_j$ with $i \neq j$. The middle sum term in (2.158) does not contribute because of $\mathrm{d}q_i \, \hat{e}_i$, and therefore

$$(\nabla \otimes \vec{u})_{ij} = \frac{1}{h_i}\, \partial_i u_j - \frac{u_i}{h_i\, h_j}\, \partial_j h_i \qquad (i \neq j) \tag{2.160}$$

Next, identify the coefficients of $dq_i\, \hat{e}_i$. Equation (2.158) gives here

$$\sum_i h_i\, dq_i\, (\nabla \otimes \vec{u})_{ii}\, \hat{e}_i = \sum_i dq_i\, (\partial_i u_i)\, \hat{e}_i + \sum_{\substack{i,k \\ i \neq k}} dq_i\, u_k\, \frac{\partial_k h_i}{h_k}\, \hat{e}_i. \tag{2.161}$$

Notice that the middle term in (2.158) contributes to $i = j$. Since $\hat{e}_i$ are linearly independent, their coefficients must agree, giving

$$(\nabla \otimes \vec{u})_{ii} = \frac{1}{h_i}\, \partial_i u_i + \sum_{k \neq i} \frac{u_k}{h_i\, h_k}\, \partial_k h_i \tag{2.162}$$

Equations (2.160) and (2.162) give the components of $\nabla \otimes \vec{u}$ in arbitrary orthogonal coordinates. We specify two of the elements explicitly,

$$(\nabla \otimes \vec{u})_{11} = \frac{1}{h_1}\partial_1 u_1 + \frac{u_2}{h_1\, h_2}\, \partial_2 h_1 + \frac{u_3}{h_1\, h_3}\, \partial_3 h_1,$$

$$(\nabla \otimes \vec{u})_{12} = \frac{1}{h_1}\, \partial_1 u_2 - \frac{u_1}{h_1\, h_2}\, \partial_2 h_1. \tag{2.163}$$

In cylindrical coordinates,

$$q_1 = r, \qquad q_2 = \varphi, \qquad q_3 = z,$$

$$h_1 = 1, \qquad h_2 = r, \qquad h_3 = 1. \tag{2.164}$$

This gives

$$(\nabla \otimes \vec{u})_{ij} = \begin{pmatrix} \partial u_r/\partial r & \partial u_\varphi/\partial r & \partial u_z/\partial r \\ \partial u_r/(r\partial\varphi) - u_\varphi/r & \partial u_\varphi/(r\partial\varphi) + u_r/r & \partial u_z/(r\partial\varphi) \\ \partial u_r/\partial z & \partial u_\varphi/\partial z & \partial u_z/\partial z \end{pmatrix}. \tag{2.165}$$

For $d\vec{u}/dt = \partial\vec{u}/\partial t + \vec{u} \cdot \nabla\vec{u}$, Eq. (2.145) is obtained again. Alternatively to the method presented here, $\nabla \otimes \vec{u}$ can be calculated using the Christoffel symbols. This method is shorter, but also more abstract, see Stone and Norman (1992). The latter paper gives a couple of other formulas from tensor analysis. Frequently, the divergence of a rank-two tensor is encountered. The following expression for this divergence is taken from Truesdell and Noll (1965). Let T be a *symmetric* tensor of rank 2. Then in cylindrical coordinates,

$$(\nabla \cdot \mathsf{T})_r = \partial_r \, T_{rr} + \frac{1}{r} \partial_\varphi \, T_{r\varphi} + \partial_z \, T_{rz} + \frac{T_{rr} - T_{\varphi\varphi}}{r},$$

$$(\nabla \cdot \mathsf{T})_\varphi = \partial_r \, T_{r\varphi} + \frac{1}{r} \partial_\varphi \, T_{\varphi\varphi} + \partial_z \, T_{\varphi z} + \frac{2T_{r\varphi}}{r}, \qquad (2.166)$$

$$(\nabla \cdot \mathsf{T})_z = \partial_r \, T_{rz} + \frac{1}{r} \partial_\varphi \, T_{\varphi z} + \partial_z \, T_{zz} + \frac{T_{rz}}{r}.$$

Stone and Norman (1992) give the general expression for $\nabla \cdot \mathsf{T}$.

2.15   Vorticity Equation

In certain cases, pressure p can be eliminated from the Euler equation. In this section, it is assumed that, for some function H (to be distinguished from the Hamiltonian H),

$$\frac{\nabla p}{\rho} = \nabla \mathsf{H}. \qquad (2.167)$$

This is trivially true for $\rho = \mathrm{const}$, with $\mathsf{H} = p/\rho$. In general, the equation holds for *barotropic* flows where by definition ρ is a function of p alone, $\rho = \rho(p)$. This is proved in a lemma.

Lemma *Let $\rho(\vec{r})$ and $p(\vec{r})$ be two scalar fields in $\mathbb{R}^3$ (time t is allowed as a further variable). Let $\rho = \rho(p)$ be a one-to-one function, and let*

$$\mathsf{H}(\vec{r}) = \int_{\vec{r}_0}^{\vec{r}} \frac{\mathrm{d} p(\vec{r}')}{\rho(p(\vec{r}'))}, \qquad (2.168)$$

where the integral is a line integral along a curve K from $\vec{r}_0$ to $\vec{r}$, with arbitrary but fixed $\vec{r}_0$. Then H is a scalar field $\mathsf{H}(\vec{r})$ independent of K. In physical terms, H is the enthalpy (cf. Chorin and Marsden 1990, p. 13). Especially,

$$\boxed{\frac{\nabla p}{\rho(p)} = \nabla \mathsf{H}} \qquad (2.169)$$

Proof It is shown first that $\mathsf{H}(\vec{r})$ is a scalar field, independent of the curve K from $\vec{r}_0$ to $\vec{r}$. Because of $\rho = \rho(p)$, the scalar fields $\rho(\vec{r})$ and $p(\vec{r})$ have the same level surfaces (by shape, not by value). Thus, $\mathrm{d} p$ is constant everywhere between the surfaces $\rho = \rho_0$ and $\rho = \rho_0 + \mathrm{d}\rho$. On closed curves starting and ending at $\vec{r}_0$, each level surface of ρ is crossed an even number of times. Contributions $\mathrm{d} p/\rho$ to the integral in (2.168) cancel pairwise, because of $+\mathrm{d} p$ and $-\mathrm{d} p$ on crossing the level surface in opposite directions. Thus, $\oint \mathrm{d} p/\rho = 0$, which means path independence of $\int \mathrm{d} p/\rho$ between two points $\vec{r}_0$ and $\vec{r}$ on the closed curve. Second, to show (2.169), one considers the change dH in H when the point $\vec{r}$ is shifted by an infinitesimal amount $d\vec{l}$. Then

$$d\vec{l} \cdot \nabla H = dH = \int_{\vec{r}}^{\vec{r}+d\vec{l}} \frac{dp}{\rho} = \frac{p(\vec{r}+d\vec{l}) - p(\vec{r})}{\rho(\vec{r})} = d\vec{l} \cdot \frac{\nabla p}{\rho}, \qquad (2.170)$$

where the integral was replaced by a single Riemann sum term, since $d\vec{l}$ is infinitesimal. Since $d\vec{l}$ is arbitrary, $\nabla H = \rho^{-1} \nabla p$ follows. $\qquad\qquad\qquad \square$

The Euler equation therefore becomes, for $\rho = \rho(p)$,

$$\frac{\partial \vec{u}}{\partial t} + (\vec{u} \cdot \nabla)\,\vec{u} = -\nabla(H + V), \qquad (2.171)$$

with V the potential per unit mass of conservative forces ($\vec{g} = -\nabla V$). The advection term is rewritten using the identity (2.137),

$$\frac{\partial \vec{u}}{\partial t} + \frac{1}{2}\,\nabla(\vec{u} \cdot \vec{u}) - \vec{u} \times (\nabla \times \vec{u}) = -\nabla(H + V). \qquad (2.172)$$

Taking the curl on both sides, and introducing vorticity $\vec{\omega} = \nabla \times \vec{u}$, this becomes, after exchanging differentials and using $\nabla \times \nabla = 0$,

$$\frac{\partial \vec{\omega}}{\partial t} + \nabla \times (\vec{\omega} \times \vec{u}) = 0. \qquad (2.173)$$

This is a purely kinematic equation, without pressure. Using the identity

$$\nabla \times (\vec{a} \times \vec{b}) = \vec{a}(\nabla \cdot \vec{b}) - \vec{b}(\nabla \cdot \vec{a}) - (\vec{a} \cdot \nabla)\vec{b} + (\vec{b} \cdot \nabla)\vec{a}, \qquad (2.174)$$

this becomes, with div curl $\equiv 0$,

$$\frac{\partial \vec{\omega}}{\partial t} + (\vec{u} \cdot \nabla)\,\vec{\omega} + \vec{\omega}\,(\nabla \cdot \vec{u}) = (\vec{\omega} \cdot \nabla)\,\vec{u}. \qquad (2.175)$$

The first two terms are the Lagrange derivatives of $\vec{\omega}$. The third term is transformed using the continuity equation,

$$\frac{\partial \rho}{\partial t} + \vec{u} \cdot \nabla \rho + \rho\,\nabla \cdot \vec{u} = 0 \qquad (2.176)$$

or

$$\frac{d\rho}{dt} = -\rho\,\nabla \cdot \vec{u}. \qquad (2.177)$$

Inserting this in (2.175) and dividing the emerging equation by ρ gives

$$\frac{1}{\rho}\frac{d\vec{\omega}}{dt} - \frac{\vec{\omega}}{\rho^2}\frac{d\rho}{dt} = \frac{\vec{\omega}}{\rho} \cdot \nabla \vec{u}. \tag{2.178}$$

Using the product rule for derivatives for d/dt, the left side can be simplified, resulting in

$$\boxed{\frac{d}{dt}\frac{\vec{\omega}}{\rho} = \frac{\vec{\omega}}{\rho} \cdot \nabla \vec{u}} \tag{2.179}$$

This is the *vorticity equation*, first derived by Helmholtz.

The motion is restricted now to be two dimensional (2-D) in a plane with polar coordinates r and φ. The coordinate z is normal to the plane, and $\partial/\partial z = 0$ for all fluid fields. The only nonvanishing component of the vorticity is

$$\omega_z = \frac{1}{r}\frac{\partial(ru_\varphi)}{\partial r} - \frac{1}{r}\frac{\partial(u_r)}{\partial \varphi}. \tag{2.180}$$

This follows directly from the definition of curl,

$$d\vec{a} \cdot \nabla \times \vec{u} = \oint_{\partial d\vec{a}} d\vec{l} \cdot \vec{u}. \tag{2.181}$$

The vorticity equation now becomes, since $\vec{\omega}$ has only a z-component,

$$\frac{d}{dt}\frac{\vec{\omega}}{\rho} = \frac{\omega_z}{\rho}\frac{\partial \vec{u}}{\partial z}. \tag{2.182}$$

Since, however, the motion is in the plane, $\vec{u}$ does not depend on z, and thus

$$\frac{d}{dt}\frac{\vec{\omega}}{\rho} = 0. \tag{2.183}$$

Vorticity per unit mass is thus conserved in 2-D motion of an inviscid fluid in a plane. Finally, one can further simplify the expression (2.180) for 2-D vorticity by reinterpreting r and φ as natural curvilinear coordinates along a streamline C. At any point of C, let r be the radius of curvature of C and $r\,d\varphi$ the infinitesimal arclength along C. The unit vector $\hat{\varphi}$ shall be tangent to C and $\hat{r}$ normal to C. By definition of streamlines, $\vec{u}$ is tangent to C, $\vec{u} = u\hat{\varphi}$ (with $u = |\vec{u}|$). Then $u_\varphi = u$ and $u_r = 0$ in (2.180), and thus

$$\omega_z = \frac{\partial u}{\partial r} + \frac{u}{r}. \tag{2.184}$$

2.16 Velocity from Vorticity

Conceptually, the velocity $\vec{u}(t, \vec{x})$ is the primary vector field characterizing a fluid, and determines the vorticity $\vec{\omega} = \nabla \times \vec{u}$. With the introduction of line vortices, the inverse approach becomes meaningful too, to specify a vorticity distribution $\vec{\omega}$ via some geometrical arrangement of discrete, one-dimensional line vortices, e.g., a vortex street, vortex sheet, or vortex tube, and determine its velocity field $\vec{u}$. In this section, the general formula to find $\vec{u}$ for given $\vec{\omega}$ is derived.

Friction shall be negligible and $\nabla \cdot \vec{u} = 0$ is assumed. Furthermore, $\vec{u}$ shall vanish at infinity. Therefore,

$$\nabla \cdot \vec{u} = 0, \tag{2.185}$$

$$\nabla \times \vec{u} = \vec{\omega}, \tag{2.186}$$

for given vorticity $\vec{\omega}$. The velocity field $\vec{u}$ is written as the sum of a curl and a gradient,

$$\vec{u} = \nabla \times \vec{A}' + \nabla \phi. \tag{2.187}$$

Then (2.185) is fulfilled, if the velocity potential ϕ obeys

$$\nabla \cdot \nabla \phi = \Delta \phi = 0. \tag{2.188}$$

This will be assumed in the following. Equation (2.186) gives

$$\vec{\omega} = \nabla \times \vec{u} = \nabla \times \nabla \times \vec{A}' = \nabla(\nabla \cdot \vec{A}') - \Delta \vec{A}'. \tag{2.189}$$

The velocity $\vec{u}$ in (2.187) is left unchanged if, for some scalar function Λ,

$$\vec{A}' \to \vec{A} = \vec{A}' + \nabla \Lambda, \tag{2.190}$$

since $\nabla \times \nabla \Lambda = 0$. Choose Λ so that

$$\nabla \cdot \vec{A} = 0. \tag{2.191}$$

Such a Λ indeed exists, as is easily shown: taking the divergence of (2.190) gives

$$\Delta \Lambda = -\nabla \cdot \vec{A}'. \tag{2.192}$$

This is a Poisson equation (with a known scalar field $-\nabla \cdot \vec{A}'$ on the right side), to which the usual existence theorems for solutions of differential equations with appropriate boundary conditions apply. Equation (2.187) now becomes

$$\vec{u} = \nabla \times \vec{A} + \nabla \phi, \tag{2.193}$$

where $\Delta\phi = 0$, and $\vec{A}$ obeys

$$\Delta\vec{A} = -\vec{\omega}. \tag{2.194}$$

This is the vector Poisson equation. A Green's function G is introduced by

$$\Delta G(\vec{r} - \vec{r}') = -\delta(\vec{r} - \vec{r}'). \tag{2.195}$$

Here $\vec{r}'$ is an auxiliary variable, and the Laplacian Δ refers to $\vec{r}$ alone. If G is known, $\vec{A}(\vec{r})$ can be calculated by direct integration (we drop time t),

$$\vec{A}(\vec{r}) = \int_V d^3r'\, G(\vec{r} - \vec{r}')\, \vec{\omega}(\vec{r}'). \tag{2.196}$$

It is assumed here that $\vec{A}$, $\vec{\omega}$, and their derivatives vanish outside the finite volume V and on its boundary ∂V. If this is not the case, surface integrals over $\vec{A}$ and its normal derivative on ∂V will appear in (2.196) besides the volume integral (Green's third integral theorem). Equation (2.196) is easily checked for correctness by applying the Laplacian Δ on both sides.

In electrostatics, $\Delta\phi(\vec{r}) = -\delta(\vec{r})$ is the Poisson equation for a point charge. The potential for the point charge located in the origin is well known, $\phi \sim r^{-1}$. Including normalization, the Green's function is then

$$G(\vec{r}) = \frac{1}{4\pi r}. \tag{2.197}$$

(For the genitive in 'the Green's function,' see Wright 2006 and Prosperetti 2011, p. 39.) This is also easily checked,

$$\begin{aligned}
\int_V dV\, \Delta G &= \frac{1}{4\pi} \int_V dV\, \nabla \cdot \nabla r^{-1} \\
&= \frac{1}{4\pi} \oint_{\partial V} d\vec{a} \cdot \nabla r^{-1} \\
&= -\frac{1}{4\pi} \oint_{\partial V} d\vec{a} \cdot \frac{\hat{r}}{r^2} \\
&= -\frac{1}{4\pi} \oint_{\partial V} d\Omega.
\end{aligned} \tag{2.198}$$

Here $d\Omega$ is the solid angle covered by $d\vec{a}$, as seen along $\hat{r}$ from a distance r. If the origin $\vec{r} = 0$ lies inside V, as is assumed here, the integral in (2.198) is 4π. This is even true when the volume is non-convex, and straight rays from a point inside V cross the boundary ∂V more than once, due to pairwise compensation of $d\Omega$ on pieces of the boundary that face each other. This is discussed in many textbooks on vector analysis and elementary differential geometry. With $\vec{r} = 0$ lying inside V, therefore

$$\int_V dV\, \Delta G(\vec{r}) = -1 = -\int_V dV\, \delta(\vec{r}), \tag{2.199}$$

which, since V is arbitrary, gives Eq. (2.195). Therefore

$$\vec{A}(\vec{r}) = \frac{1}{4\pi} \int_V d^3r' \, \frac{\vec{\omega}(\vec{r}')}{|\vec{r} - \vec{r}'|}. \tag{2.200}$$

Equation (2.193) becomes, dropping again the time coordinate,

$$\vec{u}(\vec{r}) = \frac{1}{4\pi} \int_V d^3r' \, \nabla \times \left(\frac{\vec{\omega}(\vec{r}')}{|\vec{r} - \vec{r}'|} \right) + \nabla \phi,$$

$$\stackrel{(a)}{=} \frac{1}{4\pi} \int_V d^3r' \, \nabla \left(\frac{1}{|\vec{r} - \vec{r}'|} \right) \times \vec{\omega}(\vec{r}') + \nabla \phi,$$

$$= -\frac{1}{4\pi} \int_V d^3r' \, \frac{\vec{r} - \vec{r}'}{|\vec{r} - \vec{r}'|^3} \times \vec{\omega}(\vec{r}') + \nabla \phi. \tag{2.201}$$

In (a) it was used that for an arbitrary scalar field $f(\vec{r})$ and an arbitrary constant vector $\hat{n}$, $\nabla \times (f(\vec{r})\hat{n}) = \nabla f \times \hat{n}$.

Our results in this section resemble well-known formulas from magnetostatics: Replace $\vec{\omega}$ by the current density $\vec{j}$ and $\vec{u}$ by magnetic induction $\vec{B}$. Then (2.200) holds for the vector potential $\vec{A}$ in the Coulomb gauge, and (2.201), for $\phi = 0$, is the Biot–Savart law (both for $\mu_0 = 1$).

2.17 The Bernoulli Equation

The Bernoulli equation is the law of energy conservation along a streamline. It can be derived from spatial integration of the Euler equation. One should be aware, however, that the Bernoulli equation is older than the Euler equation. A three-dimensional, stationary, barotropic flow of an inviscid fluid is assumed. For barotropic flow, $\rho^{-1} \nabla p = \nabla H$ (see the lemma in Sect. 2.15). The stationary Euler equation is written in the form (2.172),

$$\nabla \left(\frac{1}{2} u^2 + H + V \right) = \vec{u} \times \vec{\omega}, \tag{2.202}$$

where $u^2 = \vec{u} \cdot \vec{u}$. Let $d\vec{s}$ be the arclength along a certain streamline (the flow is stationary, and thus streamlines do not change in time). Then along this streamline, by definition of the gradient operator,

$$d_s \left(\frac{1}{2} u^2 + H + V \right) = d\vec{s} \cdot \nabla \left(\frac{1}{2} u^2 + H + V \right) = d\vec{s} \cdot (\vec{u} \times \vec{\omega}) = 0. \tag{2.203}$$

The last equation holds since $d\vec{s}$ and $\vec{u}$ are parallel by the definition of streamlines. Therefore

$$\frac{1}{2}u^2 + H + V = C_s,\tag{2.204}$$

where C_s is the Bernoulli constant along the streamline under consideration. For $\rho = $ const, this equation takes the form

$$\frac{1}{2}u^2 + \frac{p}{\rho} + V = C_s.\tag{2.205}$$

If the flow is irrotational, $\vec{\omega} = 0$, Eq. (2.202) gives directly

$$\frac{1}{2}u^2 + H + V = C,\tag{2.206}$$

where C is now a constant throughout the fluid, not just along a given streamline. Another form of the Bernoulli equation for irrotational flows, $\nabla \times \vec{u} = 0$, is obtained by writing $\vec{u} = \nabla\phi$. Then the Euler equation for nonstationary flows is

$$(\nabla\phi)_t + \frac{1}{2}\nabla(\nabla\phi \cdot \nabla\phi) + \nabla(H + V) = 0.\tag{2.207}$$

After interchanging differentials, this gives

$$\phi_t + \frac{1}{2}\nabla\phi \cdot \nabla\phi + H + V = C(t),\tag{2.208}$$

where $C(t)$ depends on time but not on spatial variables, i.e., is constant throughout the fluid. Since one is interested primarily in the velocity field $\vec{u} = \nabla\phi$, and since $\nabla C(t) = 0$, one can introduce a new potential $\tilde{\phi} = \phi - \int^t dt'\, C(t')$ to obtain zero on the right side of (2.208).

Stationary, 2-D flows in a plane are a central topic in the following chapters; we therefore derive here the special form the Bernoulli equation takes then. Let r be the radius of curvature at any point of the streamline, and ds be the increase in arclength along this streamline. The tangential and normal accelerations with respect to the streamline are given by

$$u\,\frac{\partial u}{\partial s} \qquad \text{and} \qquad -\frac{u^2}{r}.\tag{2.209}$$

The first expression is obvious and the second expression is the centrifugal acceleration. The equations of motion are then, for stationary motion,

$$u \frac{\partial u}{\partial s} = -\frac{1}{\rho} \frac{\partial p}{\partial s} - \frac{\partial V}{\partial s}, \tag{2.210}$$

$$-\frac{u^2}{r} = -\frac{1}{\rho} \frac{\partial p}{\partial r} - \frac{\partial V}{\partial r}. \tag{2.211}$$

For $\rho = \text{const}$, this can be written as

$$\frac{\partial}{\partial s} \left(\frac{p}{\rho} + \frac{1}{2} u^2 + V \right) = 0, \tag{2.212}$$

$$\frac{\partial}{\partial r} \left(\frac{p}{\rho} + \frac{1}{2} u^2 + V \right) = \frac{1}{2} \frac{\partial u^2}{\partial r} + \frac{u^2}{r} = \left(\frac{\partial u}{\partial r} + \frac{u}{r} \right) u \overset{(2.184)}{=} u\omega_z \tag{2.213}$$

(compare with Milne-Thomson 1968, p. 109), or

$$\frac{p}{\rho} + \frac{1}{2} u^2 + V = C_s, \tag{2.214}$$

$$\frac{\partial C_s}{\partial r} = u\omega_z, \tag{2.215}$$

where the Bernoulli constant C_s is again constant along the streamline. The second equation gives the change of C_s *across* the streamlines. For irrotational motion, C_s is again a global constant C of the flow.

2.18   The Euler–Lagrange Equations for Fluids

In point mechanics and in (quantum) field theory, the equations of motion are often derived from the Euler–Lagrange equations (themselves obtained from a variational principle), using a Lagrange function L that is specific to the given theory. It comes therefore as a surprise that the Euler–Lagrange formalism (or, equivalently, the Hamilton formalism) is usually not applied in fluid dynamics, and one starts directly with the Euler equation for fluids and the relevant boundary conditions. One reason is that *constraints*, which are so important in point mechanics and in field theories, are usually not important in fluid dynamics. An application of the Euler–Lagrange formalism to fluids is given in the next section, and demonstrates its power. In the present section, the Lagrange density for inviscid fluids is derived. Quite astonishingly, this is given by the fluid pressure. We follow closely a seminal paper by Clebsch (1859). To obtain the standard sign for the Hamiltonian H, the variable p (resp. q) is used where Clebsch uses q (resp. p).

The (generally time-dependent) flow shall have a divergence-free velocity field, $\nabla \cdot \vec{u} = 0$, but have nonvanishing vorticity, $\nabla \times \vec{u} \neq 0$. Clebsch's central idea is that any three-dimensional vector can be written, introducing three independent scalar functions ϕ, p, q, as

$$\vec{u} = \nabla\phi + p\nabla q, \tag{2.216}$$

see also Eq. (2.49) in Jackiw (2002), p. 13. This is to be distinguished from the usual Helmholtz decomposition of a vector field into a sum of a vector field without curl and a vector field without divergence. The so-called Clebsch parameterization (2.216) is of importance in higher field-theoretical aspects of fluid dynamics, see especially Sect. 2.5 in Jackiw (2002). Cartesian coordinates x_i are introduced, with indices running from 1 to 3, and the summation convention is used. Thus

$$\boxed{u_i = \frac{\partial \phi}{\partial x_i} + p \, \frac{\partial q}{\partial x_i}} \tag{2.217}$$

The q and p will be interpreted as canonical variables, that is, canonical (or generalized) coordinates and momenta. They depend on space and time,

$$q = q(t, x_1, x_2, x_3), \qquad\qquad p = p(t, x_1, x_2, x_3). \tag{2.218}$$

Since the name p is thus already in use, we write the fluid pressure in this section as P. Then

$$\nabla \times \vec{u} = \nabla \times (\nabla \phi + p \nabla q) = \nabla p \times \nabla q. \tag{2.219}$$

Similarly,

$$0 = \nabla \cdot \vec{u} = \nabla \cdot (\nabla \phi + p \nabla q) = \Delta \phi + p \Delta q + \nabla p \cdot \nabla q. \tag{2.220}$$

Clebsch (1859) shows how ϕ, p, q can be obtained for a given vorticity field $\vec{\omega} = \nabla \times \vec{u}$, where $\nabla \cdot \vec{u} = 0$. The dynamical equations for an inviscid fluid are

$$\frac{\partial u_i}{\partial x_i} = 0,$$

$$\frac{\partial u_i}{\partial t} + u_j \frac{\partial u_i}{\partial x_j} = -\frac{\partial P}{\partial x_i}. \tag{2.221}$$

To these, Clebsch's formula (2.217) is added, as well as the identification $u_i = dx_i/dt$ of the flow velocity field $\vec{u}(t, \vec{x})$ at $\vec{x}$ with the velocity of the fluid parcel located at $\vec{x}$. Variations δx_i in fluid parcel position are considered, but there is no need to introduce variations in time, and therefore $\delta t = 0$. Multiplying (2.221) with δx_i, and summing over i and j,

$$\left(\frac{\partial u_i}{\partial t} + u_j \frac{\partial u_i}{\partial x_j} \right) \delta x_i = -\delta P, \tag{2.222}$$

where $\delta P(t, x_1, x_2, x_3) = (\partial P/\partial x_i) \, \delta x_i$. With these preparations, one finds (with $u_i u_i$ abbreviating $\sum_i u_i u_i$, etc.):

$$-\delta\left(P + \frac{1}{2}u_i u_i\right)$$

$$= \frac{\partial u_i}{\partial t}\,\delta x_i + \frac{\partial u_i}{\partial x_j}\,u_j\,\delta x_i - \frac{\partial u_i}{\partial x_j}\,u_i\,\delta x_j$$

$$= \frac{\partial u_i}{\partial t}\,\delta x_i + \frac{1}{2}\left[\frac{\partial u_i}{\partial x_j}\,u_j\,\delta x_i - \frac{\partial u_i}{\partial x_j}\,u_i\,\delta x_j + \frac{\partial u_j}{\partial x_i}\,u_i\,\delta x_j - \frac{\partial u_j}{\partial x_i}\,u_j\,\delta x_i\right]$$

$$= \frac{\partial u_i}{\partial t}\,\delta x_i + \frac{1}{2}\left(\frac{\partial u_i}{\partial x_j} - \frac{\partial u_j}{\partial x_i}\right)(u_j\,\delta x_i - u_i\,\delta x_j)$$

$$= \frac{\partial u_i}{\partial t}\,\delta x_i + \frac{1}{2}\left(\frac{\partial p}{\partial x_j}\frac{\partial q}{\partial x_i} - \frac{\partial p}{\partial x_i}\frac{\partial q}{\partial x_j}\right)(u_j\,\delta x_i - u_i\,\delta x_j)$$

$$= \frac{\partial u_i}{\partial t}\,\delta x_i + u_i\frac{\partial p}{\partial x_i}\frac{\partial q}{\partial x_j}\,\delta x_j - u_i\frac{\partial q}{\partial x_i}\frac{\partial p}{\partial x_j}\,\delta x_j$$

$$= \frac{\partial u_i}{\partial t}\,\delta x_i + \frac{\partial p}{\partial x_i}\frac{\mathrm{d}x_i}{\mathrm{d}t}\,\delta q - \frac{\partial q}{\partial x_i}\frac{\mathrm{d}x_i}{\mathrm{d}t}\,\delta p$$

$$= \frac{\partial u_i}{\partial t}\,\delta x_i + \left(\frac{\mathrm{d}p}{\mathrm{d}t} - \frac{\partial p}{\partial t}\right)\delta q - \left(\frac{\mathrm{d}q}{\mathrm{d}t} - \frac{\partial q}{\partial t}\right)\delta p$$

$$= \frac{\partial}{\partial x_i}\left(\frac{\partial\phi}{\partial t}\right)\delta x_i + p\frac{\partial}{\partial x_i}\left(\frac{\partial q}{\partial t}\right)\delta x_i + \frac{\partial p}{\partial t}\frac{\partial q}{\partial x_i}\,\delta x_i + (\ldots)\,\delta q - (\ldots)\,\delta p$$

$$= \delta\frac{\partial\phi}{\partial t} + p\,\delta\frac{\partial q}{\partial t} + \frac{\mathrm{d}p}{\mathrm{d}t}\,\delta q - \left(\frac{\mathrm{d}q}{\mathrm{d}t} - \frac{\partial q}{\partial t}\right)\delta p$$

$$= \delta\left[\frac{\partial\phi}{\partial t} + p\frac{\partial q}{\partial t}\right] + \frac{\mathrm{d}p}{\mathrm{d}t}\,\delta q - \frac{\mathrm{d}q}{\mathrm{d}t}\,\delta p. \tag{2.223}$$

Here (2.217) was used twice, and dummy indices were renamed. Therefore

$$\delta\left[P + \frac{1}{2}u_i u_i + \frac{\partial\phi}{\partial t} + p\frac{\partial q}{\partial t}\right] = -\frac{\mathrm{d}p}{\mathrm{d}t}\,\delta q + \frac{\mathrm{d}q}{\mathrm{d}t}\,\delta p. \tag{2.224}$$

This equation suggests to also consider variations δq, δp instead of $\delta x_1, \delta x_2, \delta x_3$. Define a function H that will manifest itself as Hamiltonian of the fluid by

$$H(t, q, p) = P + \frac{1}{2}u_i u_i + \frac{\partial\phi}{\partial t} + p\frac{\partial q}{\partial t}. \tag{2.225}$$

Then, using (2.224) and $\delta t = 0$,

$$\delta H = \frac{\partial H}{\partial q}\,\delta q + \frac{\partial H}{\partial p}\,\delta p = -\frac{\mathrm{d}p}{\mathrm{d}t}\,\delta q + \frac{\mathrm{d}q}{\mathrm{d}t}\,\delta p, \tag{2.226}$$

and one arrives indeed at the Hamilton equations

$$\frac{dq}{dt} = \frac{\partial H}{\partial p},$$

$$\frac{dp}{dt} = -\frac{\partial H}{\partial q}. \tag{2.227}$$

These are supplemented by the Clebsch parameterization (2.217) and by $\partial u_i/\partial x_i = 0$. From (2.225),

$$P = -\frac{\partial \phi}{\partial t} - p\frac{\partial q}{\partial t} + H - \frac{1}{2}u_i u_i$$

$$= -\frac{\partial \phi}{\partial t} - p\frac{\partial q}{\partial t} + H - \frac{1}{2}\left(\frac{\partial \phi}{\partial x_i} + p\frac{\partial q}{\partial x_i}\right)\left(\frac{\partial \phi}{\partial x_i} + p\frac{\partial q}{\partial x_i}\right). \tag{2.228}$$

Read P now as

$$P = P\left(t, q, \frac{\partial q}{\partial t}, \frac{\partial q}{\partial x_i}, p\right), \tag{2.229}$$

without terms in $\partial p/\partial t$ and $\partial p/\partial x_i$. Furthermore, let

$$\boxed{\Pi = \int dt \ dx_1 \ dx_2 \ dx_3 \ P} \tag{2.230}$$

Then the variational problem

$$\delta \Pi = 0 \tag{2.231}$$

leads to the Euler–Lagrange equations

$$0 = \frac{\partial P}{\partial p}, \tag{2.232}$$

$$0 = \frac{\partial P}{\partial q} - \frac{\partial}{\partial t}\frac{\partial P}{\partial(\partial q/\partial t)} - \frac{\partial}{\partial x_i}\frac{\partial P}{\partial(\partial q/\partial x_i)}. \tag{2.233}$$

Equation (2.232) is trivial, and (2.233) is derived at the end of this section. Inserting P from (2.228) in (2.232) gives

$$0 = -\frac{\partial q}{\partial t} + \frac{\partial H}{\partial p} - \left(\frac{\partial \phi}{\partial x_i} + p\frac{\partial q}{\partial x_i}\right)\frac{\partial q}{\partial x_i}$$

$$= \frac{\partial H}{\partial p} - \frac{\partial q}{\partial t} - \frac{\partial q}{\partial x_i}\frac{dx_i}{dt}$$

$$= \frac{\partial H}{\partial p} - \frac{dq}{dt}. \tag{2.234}$$

Inserting P from (2.228) in (2.233) gives

$$0 = \frac{\partial H}{\partial q} + \frac{\partial p}{\partial t} + \left(\frac{\partial \phi}{\partial x_i} + p \frac{\partial q}{\partial x_i} \right) \frac{\partial p}{\partial x_i}$$

$$= \frac{\partial H}{\partial q} + \frac{\partial p}{\partial t} + \frac{\partial p}{\partial x_i} \frac{dx_i}{dt}$$

$$= \frac{\partial H}{\partial q} + \frac{dp}{dt}. \tag{2.235}$$

Equations (2.234) and (2.235) are again the Hamilton equations (2.227), as expected. We have thus found that Π in (2.230) is an 'action,' and P a Lagrange density. To proceed, a canonical transformation to $\bar{H} = 0$ is performed (the Hamilton–Jacobi formalism). We briefly recapitulate from point mechanics with a Lagrange function L the basic equations for a generating function S of this transformation,

$$\delta \int dt\, L = \delta \int dt\, (p\dot{q} - H) \stackrel{!}{=}$$

$$\delta \int dt\, \bar{L} = \delta \int dt\, (\bar{p}\dot{\bar{q}} - \bar{H}) + \delta \int dt\, \frac{dS}{dt}. \tag{2.236}$$

The last term is zero if variations at the boundaries of integration vanish. Assuming that $S = S(t, q, \bar{q})$ gives

$$p\dot{q} - H - \bar{p}\dot{\bar{q}} + \bar{H} = \frac{\partial S}{\partial q} \dot{q} + \frac{\partial S}{\partial \bar{q}} \dot{\bar{q}} + \frac{\partial S}{\partial t} \tag{2.237}$$

or

$$p = \frac{\partial S}{\partial q}, \qquad \bar{p} = -\frac{\partial S}{\partial \bar{q}}, \qquad \bar{H} = H + \frac{\partial S}{\partial t}. \tag{2.238}$$

Let $\bar{H} = 0$, and a new function s is introduced by

$$s(t, x_1, x_2, x_3) = S(t, q, \bar{q}). \tag{2.239}$$

This gives

$$\frac{\partial s}{\partial x_i} = \frac{\partial S}{\partial q} \frac{\partial q}{\partial x_i} + \frac{\partial S}{\partial \bar{q}} \frac{\partial \bar{q}}{\partial x_i}, \tag{2.240}$$

$$\frac{\partial s}{\partial t} = \frac{\partial S}{\partial t} + \frac{\partial S}{\partial q} \frac{\partial q}{\partial t} + \frac{\partial S}{\partial \bar{q}} \frac{\partial \bar{q}}{\partial t}. \tag{2.241}$$

With this follows

$$u_i = \frac{\partial \phi}{\partial x_i} + p \frac{\partial q}{\partial x_i}$$

$$= \frac{\partial \phi}{\partial x_i} + \frac{\partial S}{\partial q} \frac{\partial q}{\partial x_i}$$

$$= \frac{\partial \phi}{\partial x_i} + \frac{\partial s}{\partial x_i} - \frac{\partial S}{\partial \bar{q}} \frac{\partial \bar{q}}{\partial x_i}$$

$$= \frac{\partial \bar{\phi}}{\partial x_i} + \bar{p} \frac{\partial \bar{q}}{\partial x_i} = \bar{u}_i, \tag{2.242}$$

where $\bar{\phi} = \phi + s$. Similarly,

$$P = -\frac{\partial \phi}{\partial t} - p \frac{\partial q}{\partial t} + H - \frac{1}{2} u_i u_i$$

$$= -\frac{\partial \phi}{\partial t} - \frac{\partial S}{\partial q} \frac{\partial q}{\partial t} - \frac{\partial S}{\partial t} - \frac{1}{2} \bar{u}_i \bar{u}_i$$

$$= -\frac{\partial \phi}{\partial t} - \frac{\partial s}{\partial t} + \frac{\partial S}{\partial \bar{q}} \frac{\partial \bar{q}}{\partial t} - \frac{1}{2} \bar{u}_i \bar{u}_i$$

$$= -\frac{\partial \bar{\phi}}{\partial t} - \bar{p} \frac{\partial \bar{q}}{\partial t} - \frac{1}{2} \bar{u}_i \bar{u}_i = \bar{P}. \tag{2.243}$$

Thus, $u_i = \bar{u}_i$ and $P = \bar{P}$ are invariant under the above transformation. The Hamilton equations derived from $\delta \int dt \, dx_1 \, dx_2 \, dx_3 \, \bar{P} = 0$ are then

$$\frac{d\bar{q}}{dt} = \frac{d\bar{p}}{dt} = 0, \tag{2.244}$$

as is familiar from point mechanics (and is the very idea of the Hamilton–Jacobi formalism: to have the canonical variables as constants of motion.) Finally, again as a reminder from point mechanics, we derive the Euler–Lagrange equations (2.233) from the variational principle. Standard symbols are used here that translate directly to those in (2.233). For reasons of brevity, only a single x-coordinate is considered. The generalization to three spatial coordinates is obvious.

$$0 \overset{(a)}{=} \delta \int dt \, dx \;\; L(t, x, f(t, x), f_t(t, x), f_x(t, x))$$

$$\overset{(b)}{=} \int dt \, dx \left(\frac{\partial L}{\partial f} \delta f + \frac{\partial L}{\partial f_t} \delta f_t + \frac{\partial L}{\partial f_x} \delta f_x \right)$$

$$\overset{(c)}{=} \int dt \, dx \left(\frac{\partial L}{\partial f} \delta f + \frac{\partial L}{\partial f_t} (\delta f)_t + \frac{\partial L}{\partial f_x} (\delta f)_x \right)$$

$$\overset{(d)}{=} \int dt \, dx \left(\frac{\partial L}{\partial f} - \frac{\partial}{\partial t} \frac{\partial L}{\partial f_t} - \frac{\partial}{\partial x} \frac{\partial L}{\partial f_x} \right) \delta f \;\; + \;\; \text{b. t.} \tag{2.245}$$

(a) Abbreviations $f_t = \partial f / \partial t$ and $f_x = \partial f / \partial x$.

(b) No variations in independent variables t and x.

(c) Interchangeability $\delta(\partial f/\partial t) = \partial(\delta f)/\partial t$.

(d) Integration by parts with respect to t and x. Boundary terms 'b. t.' shall vanish.

Similar results to those of Clebsch were found by Hargreaves (1908), and Friedrichs (1934) succeeds to derive boundary conditions from a variational principle.

2.19 Water Waves from the Euler–Lagrange Equations

We apply the formalism developed in the last section to derive water wave equations from a Lagrangian $L = -p$, following closely a paper by Luke (1967). In this approach, the kinematic and dynamic boundary conditions at the free water surface (and at the ground) follow directly from a calculation using the Euler–Lagrange formalism, whereas the more standard derivation of these boundary conditions requires physical and geometrical insight. Thus, the Euler–Lagrange formalism shows its intended strength: to replace geometrical considerations by analytical calculations (see, e. g., its historically first application in Bernoulli's brachistochrone problem).

The fluid shall be inviscid, and the velocity field $\vec{u}$ divergence-free, irrotational, and two dimensional in the xy-plane, with $u(t, x, y)$ and $v(t, x, y)$ having no dependence on z. The fluid body shall lie between height $y = 0$, which is the ground, and $y = h(t, x)$, which is the local height of a free water wave: a periodic elevation and suppression of the water surface. To obtain free surface waves, gravitational acceleration g is included, taken here to act in the $-y$-direction. The density is assumed to be constant and is normalized to $\rho = 1$. Then the Euler equations are

$$u_t + uu_x + vu_y = -p_x,$$
$$v_t + uv_x + vv_y = -p_y - g. \tag{2.246}$$

For irrotational flows, a velocity potential $\phi(t, x, y)$ can be introduced by

$$u = \phi_x, \qquad\qquad v = \phi_y. \tag{2.247}$$

Equation (2.246) then becomes

$$\phi_{xt} + \phi_x\phi_{xx} + \phi_y\phi_{xy} + p_x = 0, \tag{2.248}$$
$$\phi_{yt} + \phi_x\phi_{yx} + \phi_y\phi_{yy} + p_y = -g. \tag{2.249}$$

Both equations can be integrated once to give the Bernoulli equation. With integration 'constants' gy and 0, (2.248) and (2.249) imply

$$\phi_t + \frac{1}{2}\phi_x^2 + \frac{1}{2}\phi_y^2 + p + gy = 0. \tag{2.250}$$

As the relevant Lagrange function, negative pressure is chosen, $L = -p$, or

$$L(t, x, y, \phi(t, x, y), \phi_t(t, x, y), \phi_x(t, x, y), \phi_y(t, x, y))$$
$$= \phi_t + \frac{1}{2}\phi_x^2 + \frac{1}{2}\phi_y^2 + gy. \tag{2.251}$$

The minus sign in $L = -p$ gives a positive Hamiltonian in (2.269). The variational principle according to Clebsch (1859) is

$$0 = \delta \int_{t_1}^{t_2} dt \int_{x_1}^{x_2} dx \int_0^{h(t,x)} dy \ L(t, x, y). \tag{2.252}$$

Luke's new idea is to include here the height $y = h(t, x)$ of the free water surface above ground, at position x (there is no z-coordinate), and to allow for variations $\delta\phi$ and δh of the functions ϕ and h. To first order in variations δ,

$$\delta \int_0^{h(t,x)} dy \ L = \int_0^{h+\delta h} dy \ (L + \delta L) - \int_0^h dy \ L$$
$$= \int_h^{h+\delta h} dy \ L + \int_0^h dy \ \delta L = \delta h \ L|_{y=h} + \int_0^h dy \ \delta L, \tag{2.253}$$

so that (2.252) gives

$$0 = \iint dt \ dx \ \left[\delta h(t, x) \ L|_{y=h(t,x)} + \int_0^{h(t,x)} dy \ \delta L \right]. \tag{2.254}$$

The following boundary conditions are applied:

$$0 = \delta\phi(t_1, x, y) = \delta\phi(t_2, x, y) = \delta\phi(t, x_1, y) = \delta\phi(t, x_2, y),$$
$$0 = \delta h(t_1, x) = \delta h(t_2, x) = \delta h(t, x_1) = \delta h(t, x_2). \tag{2.255}$$

The second term in (2.254) is calculated by careful integration by parts, using (2.251),

$$\iiint dt \ dx \ dy \ \delta \left(\phi_t + \frac{1}{2}\phi_x^2 + \frac{1}{2}\phi_y^2 + gy \right)$$
$$= \iiint dt \ dx \ dy \ (\delta\phi_t + \phi_x \ \delta\phi_x + \phi_y \ \delta\phi_y)$$
$$= \iiint dt \ dx \ dy \ \left(\frac{\partial}{\partial t} + \phi_x \frac{\partial}{\partial x} + \phi_y \frac{\partial}{\partial y} \right) \delta\phi$$
$$= \int_{x_1}^{x_2} dx \int_{t_1}^{t_2} dt \int_0^{h(t,x)} dy \ \frac{\partial}{\partial t} \ \delta\phi$$

$$+ \int_{t_1}^{t_2} dt \int_{x_1}^{x_2} dx \int_0^{h(t,x)} dy \left[\frac{\partial}{\partial x} (\phi_x \, \delta\phi) - \phi_{xx} \, \delta\phi \right]$$

$$+ \int_{t_1}^{t_2} dt \int_{x_1}^{x_2} dx \int_0^{h(t,x)} dy \left[\frac{\partial}{\partial y} (\phi_y \, \delta\phi) - \phi_{yy} \, \delta\phi \right]$$

$$\overset{(a)}{=} - \int_{t_1}^{t_2} dt \int_{x_1}^{x_2} dx \int_0^{h(t,x)} dy \, (\phi_{xx} + \phi_{yy}) \, \delta\phi$$

$$+ \int_{x_1}^{x_2} dx \int_{t_1}^{t_2} dt \left[\frac{\partial}{\partial t} \left(\int_0^{h(t,x)} dy \, \delta\phi \right) - \frac{\partial}{\partial T} \int_0^{h(T,x)} dy \, \delta\phi(t, x, y) \Big|_{T=t} \right]$$

$$+ \int_{t_1}^{t_2} dt \int_{x_1}^{x_2} dx \left[\frac{\partial}{\partial x} \left(\int_0^{h(t,x)} dy \, \phi_x \delta\phi \right) - \frac{\partial}{\partial X} \int_0^{h(t,X)} dy \, (\phi_x \delta\phi)(t, x, y) \Big|_{X=x} \right]$$

$$+ \int_{t_1}^{t_2} dt \int_{x_1}^{x_2} dx \left[(\phi_y \, \delta\phi)(t, x, h(t, x)) - (\phi_y \, \delta\phi)(t, x, 0) \right]$$

$$\overset{(b)}{=} - \int_{t_1}^{t_2} dt \int_{x_1}^{x_2} dx \int_0^{h(t,x)} dy \, (\phi_{xx} + \phi_{yy}) \, \delta\phi$$

$$+ \int_{x_1}^{x_2} dx \left[\int_0^{h(t_2,x)} dy \, \delta\phi(t_2, x, y) - \int_0^{h(t_1,x)} dy \, \delta\phi(t_1, x, y) \right]$$

$$+ \int_{t_1}^{t_2} dt \left[\int_0^{h(t,x_2)} dy \, (\phi_x \, \delta\phi)(t, x_2, y) - \int_0^{h(t,x_1)} dy \, (\phi_x \, \delta\phi)(t, x_1, y) \right]$$

$$- \int_{x_1}^{x_2} dx \int_{t_1}^{t_2} dt \, \frac{\partial h(t,x)}{\partial t} \frac{d}{dh(t,x)} \int_0^{h(t,x)} dy \, \delta\phi(t, x, y)$$

$$- \int_{t_1}^{t_2} dt \int_{x_1}^{x_2} dx \, \frac{\partial h(t,x)}{\partial x} \frac{d}{dh(t,x)} \int_0^{h(t,x)} dy \, (\phi_x \, \delta\phi)(t, x, y)$$

$$+ \int_{t_1}^{t_2} dt \int_{x_1}^{x_2} dx \left[(\phi_y \, \delta\phi)(t, x, h(t, x)) - (\phi_y \, \delta\phi)(t, x, 0) \right]$$

$$\overset{(c)}{=} - \int_{t_1}^{t_2} dt \int_{x_1}^{x_2} dx \int_0^{h(t,x)} dy \, (\phi_{xx} + \phi_{yy}) \, \delta\phi(t, x, y)$$

$$+ \int_{t_1}^{t_2} dt \int_{x_1}^{x_2} dx \left[(-h_t - h_x \, \phi_x(t, x, h) + \phi_y(t, x, h)) \, \delta\phi(t, x, h) \right.$$

$$\left. - (\phi_y \, \delta\phi)(t, x, 0) \right]. \tag{2.256}$$

The first three equations use formal differentiation rules for variations δ. Note that $\delta y = 0$.

(a) New variable names T, X are used to limit the scope of differentiation: differentiation acts only on t and x of the upper bound of integration.

(b) Lines 2 and 3 vanish due to the boundary conditions (2.255). In line 4,

$$\frac{d}{dT} \int_0^{h(T,x)} dy\, \delta\phi \bigg|_{T=t} = \frac{\partial h(t,x)}{\partial t} \frac{d}{dh} \int_0^h dy\, \delta\phi \qquad (2.257)$$

was used, and a similar formula holds for $\frac{d}{dX}$ in line 5.

(c) Abbreviate $h(t,x)$ as h, and use $h_t = \partial h/\partial t$ and $h_x = \partial h/\partial x$.

Consider now (2.254) together with (2.256). There are four variations,

$$\delta h(t,x), \qquad \delta\phi(t,x,y), \qquad \delta\phi(t,x,0), \qquad \delta\phi(t,x,h), \qquad (2.258)$$

and consider the following four cases, where $0 < y < h$,

$$\delta h = \delta\phi(0) = \delta\phi(h) = 0,$$
$$\delta h = \delta\phi(y) = \delta\phi(h) = 0,$$
$$\delta h = \delta\phi(0) = \delta\phi(y) = 0,$$
$$\delta\phi(y) = \delta\phi(0) = \delta\phi(h) = 0, \qquad (2.259)$$

omitting the obvious arguments t and x. Variations not appearing in these four cases are assumed to be arbitrary and linearly independent (Luke 1967). Thus, their prefactors in (2.254) and (2.256) have to vanish. This leads, in the same sequence as the four cases above, to the equations

$$\phi_{xx} + \phi_{yy} = 0, \qquad \text{for } 0 < y < h, \qquad (2.260)$$
$$\phi_y = 0, \qquad \text{for } y = 0, \qquad (2.261)$$
$$-h_t - h_x\,\phi_x + \phi_y = 0, \qquad \text{for } y = h(t,x), \qquad (2.262)$$
$$\phi_t + \frac{1}{2}\phi_x^2 + \frac{1}{2}\phi_y^2 + gy = 0, \qquad \text{for } y = h(t,x). \qquad (2.263)$$

From top to bottom, these are
- the Laplace equation,
- the kinematic boundary condition at the bottom $y = 0$,
- the kinematic boundary condition at the free surface $y = h(t,x)$, and
- the dynamic boundary condition at the free surface $y = h(t,x)$.

Together, they impose the system of free surface wave equations. We do not discuss them further here, since they will be (or were) derived from direct physical arguments at other places:
- Equation (2.260) is derived as (3.5).
- Equation (2.261) is clear: no speed component orthogonal to the ground.
- Equation (2.262) is derived as (8.12).
- Equation (2.263) was derived as (2.208); see also (9.72).

To summarize the results found so far, a Lagrange function

$$\hat{L}(t, x) = - \int_0^{h(t,x)} dy \left(\phi_t + \frac{1}{2}\phi_x^2 + \frac{1}{2}\phi_y^2 + gy \right) \tag{2.264}$$

was introduced, and the free surface wave equations follow from a variational principle,

$$0 = \delta \int_{t_1}^{t_2} dt \int_{x_1}^{x_2} dx \ \hat{L}(t, x). \tag{2.265}$$

The novelty of Luke's approach is expressed by Whitham (1967):

> A [...] variational principle for the full equations of water waves, including the free-surface conditions, did not seem to be known. It was also thought that the y-dependence, with the wave propagation occurring only in the (x, t) dependence, may require crucial changes in the general approach.

The basic idea of this approach, however, appeared already in Bateman (1944). Whitham (1965) studies dispersive waves using a Lagrangian. The water wave equations can also be derived using the Hamilton equations. This was first done by Broer (1974) and Miles (1977): let $h(t, x)$ be again the water height measured from ground, and let

$$\hat{\phi}(t, x) = \phi(t, x, h(t, x)) \tag{2.266}$$

be the velocity potential at the free water surface. Miles (1977) performs a Legendre transformation of the Lagrangian in (2.264),

$$\hat{L}(t, x) = h_t(t, x)\, \hat{\phi}(t, x) - \hat{H}(t, x). \tag{2.267}$$

To find $\hat{H}$, write

$$\hat{L}(t, x) = - \int_0^{h(t,x)} dy \left(\phi_t + \frac{1}{2}\phi_x^2 + \frac{1}{2}\phi_y^2 + gy \right)$$

$$= -\frac{\partial}{\partial t} \int_0^{h(t,x)} dy\, \phi + h_t(t, x)\, \hat{\phi}(t, x) - \frac{1}{2} \int_0^{h(t,x)} dy\, (\phi_x^2 + \phi_y^2) - \frac{1}{2} gh^2(t, x). \tag{2.268}$$

The first term is a time derivative and can be omitted since it has no influence on the equations of motion. Comparison with (2.267) then gives

$$\hat{H}(t, x) = \frac{1}{2} \int_0^{h(t,x)} dy\, (\phi_x^2(t, x, y) + \phi_y^2(t, x, y)) + \frac{1}{2} gh^2(t, x). \tag{2.269}$$

The Hamiltonian $\hat{H}$ is thus the energy density of the fluid, as expected. In (2.267), h and $\hat{\phi}$ appear as canonical variables. Indeed, Miles (1977, Eq. 2.3) gives the Hamilton equations in terms of these variables h and $\hat{\phi}$. The physical reasons for this are discussed in Sect. 3 of Miles' paper.

2.20 Stretching in an Isotropic Random Velocity Field

Fluid turbulence is not dealt with in this book, since the subject is highly specialized and mathematically involved. Yet, there is an important general theorem with a simple proof, which can be discussed here. Cocke (1969) and Orszag (1970) show, using statistical methods: If the flow velocity field $\vec{u}(t, \vec{r})$ is random and isotropic, then distances between fluid parcels become on average *larger* with time. This stretching (and subsequent folding) of parcels may be fundamental to turbulence, as is suggested in the following quote.

> Batchelor (1952) argued that a random flow should, on the average, separate irreversibly the vortices of an initially compact small volume moving with the fluid. Incompressibility then implies that the surfaces of the volume must be drawn towards each other, producing an extended thin structure out of the initially compact blob. [...] The stretching mechanism has led a number of authors to conjecture that the small-scale structure should consist typically of extensive thin sheets and ribbons of vorticity, drawn out by the straining action of their own shear fields. (Kraichnan 1974, p. 312)

We follow here the presentation given by Cocke (1969) and Orszag (1970). Let dy_i with $i = 1, 2, 3$ be Cartesian components of an infinitesimal line element in the fluid at $t = 0$, and let dx_i be the components of the same line element at a later time $t > 0$. As in (2.47), let

$$dx_i = U_{ij}\, dy_j, \tag{2.270}$$

where the summation convention is understood. Consider a fluid parcel that at $t = 0$ is a cube with length ϵ of its edges. Then its volume at time $t > 0$ is, from (2.270),

$$dV(t) = \epsilon^3 \det U. \tag{2.271}$$

In a solenoidal velocity field $\vec{u}$, the parcel volume is constant, so

$$dV(t) = dV(0) = \epsilon^3, \tag{2.272}$$

and therefore

$$\det U = 1. \tag{2.273}$$

For the distance-squared between two material points,

$$dx_i\, dx_i = U_{ij}\, U_{ik}\, dy_j\, dy_k = W_{jk}\, dy_j\, dy_k. \tag{2.274}$$

Since $dy_j\, dy_k$ is symmetric, W_{jk} must be symmetric. By a standard theorem, a symmetric matrix can be diagonalized. Denote the eigenvalues of W by w_1, w_2, w_3, and the eigenvectors by $\hat{a}_1, \hat{a}_2, \hat{a}_3$. By the same theorem, $\hat{a}_1, \hat{a}_2, \hat{a}_3$ form an orthogonal triad. With $\det U = 1$, also $\det W = \det U^2 = 1$. Since determinants are invariant under diagonalization (namely, $\det W = \det S^{-1} W S$), it follows that

$$\det W = w_1 \, w_2 \, w_3 = 1. \tag{2.275}$$

The symmetric matrix W has six independent coefficients. Let w_1, w_2, w_3 be continuous random variables that assume values following some probability distribution. Let the remaining random variables be the angles $\alpha_1, \alpha_2, \alpha_3$. The three independent α_i rotate the triad $\hat{e}_1, \hat{e}_2, \hat{e}_3$ of the initial cube into the eigenvector triad $\hat{a}_1, \hat{a}_2, \hat{a}_3$, e. g., by rotation in the xy-, yz-, and zx-planes, respectively. Write (2.274) as $\mathrm{d}x_i \, \mathrm{d}x_i = W'_{ij} \, \mathrm{d}y'_i \, \mathrm{d}y'_j = w_1 \mathrm{d}y'^2_1 + w_2 \mathrm{d}y'^2_2 + w_3 \mathrm{d}y'^2_3$ in the system $\hat{a}_1, \hat{a}_2, \hat{a}_3$. Let $\hat{a}_3$ be the polar axis of a spherical coordinate system. Then

$$(\mathrm{d}y'_1, \mathrm{d}y'_2, \mathrm{d}y'_3) = \sqrt{\mathrm{d}y_i \, \mathrm{d}y_i} \, (\sin\vartheta \cos\varphi, \sin\vartheta \sin\varphi, \cos\vartheta), \tag{2.276}$$

since the vector length is invariant under rotation. Therefore,

$$\mathrm{d}x_i \, \mathrm{d}x_i = (w_1 \, \sin^2\vartheta \, \cos^2\varphi + w_2 \, \sin^2\vartheta \, \sin^2\varphi + w_3 \, \cos^2\vartheta) \, \mathrm{d}y_i \, \mathrm{d}y_i. \tag{2.277}$$

Since $\mathrm{d}x_i \, \mathrm{d}x_i > 0$ and $\mathrm{d}y_i \, \mathrm{d}y_i > 0$, it follows
  from $\vartheta = \pi/2$ and $\varphi = 0$ that $w_1 > 0$;
  from $\vartheta = \pi/2$ and $\varphi = \pi/2$ that $w_2 > 0$;
  from $\vartheta = 0$ that $w_3 > 0$.
An average is performed over the angles ϑ and φ. If turbulence is isotropic, w_1, w_2, w_3 do not depend on the direction angles ϑ and φ. Furthermore, for all three averages over the full unit sphere,

$$\frac{1}{4\pi} \int_0^\pi \mathrm{d}\vartheta \, \sin\vartheta \, \cos^2\vartheta \int_0^{2\pi} \mathrm{d}\varphi = \frac{1}{4\pi} \int_0^\pi \mathrm{d}\vartheta \, \sin^3\vartheta \int_0^{2\pi} \mathrm{d}\varphi \, \sin^2\varphi$$

$$= \frac{1}{4\pi} \int_0^\pi \mathrm{d}\vartheta \, \sin^3\vartheta \int_0^{2\pi} \mathrm{d}\varphi \, \cos^2\varphi = \frac{1}{3}. \tag{2.278}$$

Therefore, with brackets $\langle . \rangle$ indicating (ensemble) angle averages,

$$\langle \mathrm{d}x_i \, \mathrm{d}x_i \rangle = \frac{1}{3} \, \langle w_1 + w_2 + w_3 \rangle \, \mathrm{d}y_i \, \mathrm{d}y_i. \tag{2.279}$$

If $w_i > 0$ for all i, as is here the case, the following general formula holds:

$$\frac{1}{3} \, (w_1 + w_2 + w_3) \geq (w_1 w_2 w_3)^{1/3}. \tag{2.280}$$

The proof is simple and can be found in standard textbooks. The equality in (2.280) holds only if $w_1 = w_2 = w_3 = 1$. This is a single point in the space of the random variables w_1, w_2, w_3. This single point $w_1 = w_2 = w_3$ can be dismissed as statistically exceptional. Using $w_1 \, w_2 \, w_3 = 1$, it follows that

$$\boxed{\langle \mathrm{d}x_i \, \mathrm{d}x_i \rangle > \mathrm{d}y_i \, \mathrm{d}y_i = 3\epsilon^2} \tag{2.281}$$

We have thus found that the *material* space diagonal, i.e., a line made up of fluid parcels, in an infinitesimal fluid cube gets *stretched* in time for almost all realizations of isotropic turbulence. The above proof can easily be carried over to stretching of infinitesimal material surface elements made up of fluid parcels. The movement of material lines and surfaces can easily be followed in fluids by straining them, e.g., with ink.

2.21 The Converse Poincaré Lemma

For any twice continuously differentiable scalar field ϕ and vector field $\vec{A}$ in $\mathbb{R}^3$, the well-known identities

$$\nabla \times (\nabla \phi) = 0,$$
$$\nabla \cdot (\nabla \times \vec{A}) = 0 \tag{2.282}$$

hold. In fluid dynamics (and generally in field theory), one often wants to write a vector field with vanishing curl as gradient of a scalar field (as for path independent, conservative forces in mechanics), and a vector field with vanishing divergence as curl of another vector field. It is shown in this section that the implications

$$\nabla \times \vec{u} = 0 \quad \rightarrow \quad \vec{u} = \nabla \phi,$$
$$\nabla \cdot \vec{u} = 0 \quad \rightarrow \quad \vec{u} = \nabla \times \vec{A} \tag{2.283}$$

hold indeed if the spatial region occupied by fluid is *contractible*. This is the case for $\mathbb{R}^2$ and $\mathbb{R}^3$, but not for the two-sphere S^2. With the usual identification of points $(x, y) \in \mathbb{R}^2$ and points $x + iy \in \mathbb{C}$, this leads to the important conclusion that a complex potential Φ can be introduced for irrotational flows in the complex plane, which is central to Chap. 2.

The proof of (2.283) for contractible spaces or domains is somewhat intricate and uses exterior differential calculus. To justify this effort, we also derive in the following the simple and unique form of the Gauss and Stokes integral theorems in the exterior calculus of Cartan. The exterior differential $\underline{d}$ defined below allows to write the above formulas in a very suggestive form: Eq. (2.282) then becomes

$$\underline{d}\,\underline{d}\,\omega = 0, \tag{2.284}$$

where ω is a differential form. Equation (2.284) is called *Poincaré's lemma*, and is a direct consequence of the definition of $\underline{d}$. A differential form ω is called *closed* if $\underline{d}\,\omega = 0$, and it is called *exact* if $\omega = \underline{d}\,\alpha$. Poincaré's lemma states that every exact form is closed. Equation (2.283) now becomes

$$\underline{d}\alpha = 0 \quad \rightarrow \quad \alpha = \underline{d}\,\omega. \tag{2.285}$$

This implication is called the *converse* Poincaré lemma. We derive in the following the conditions under which (2.285) holds, starting with some topological definitions. *Maps* $f, g : X \to Y$ are continuous functions from a topological space X to a topological space Y. A *homotopy* from f to g is a map (thus continuous)

$$\psi : [0, 1] \times X \to Y, \tag{2.286}$$

with

$$\begin{aligned} \psi(0, x) &= f(x), \\ \psi(1, x) &= g(x). \end{aligned} \tag{2.287}$$

The functions f and g are then called *homotopic*, $f \sim g$. An *open path* is a map $f : [0, 1] \to Y$ and a *closed path* is a map $f : S_1 \to Y$, with S_1 the unit circle. An open path homotopy is thus a map $\psi : [0, 1] \times [0, 1] \to Y$. A domain is called *path connected* if any two points can be connected by a path. Instead of 'domain,' one can also read 'space' here and in the following. A domain D is *simply connected* if it is path-connected (there are no detached domains), and any closed path C can be contracted continuously to a point without leaving D. This means the closed path C is homotopic to a point, where the homotopy between a closed path and a point is a map

$$\psi : [0, 1] \times S_1 \to Y. \tag{2.288}$$

This is equivalent to saying that any two paths with identical start- and endpoints can be continuously deformed into one another ('there are no holes in D'), i.e., are homotopic. Note that $[0, 1] \times S_1 = D$ in (2.288) is the unit disk in the plane, thus $\psi : D \to Y$. A domain V is *contractible* if, for all $\vec{x} \in V$ and a constant point $\vec{a} \in V$,

$$f : \vec{x} \to \vec{x} \quad \sim \quad g : \vec{x} \to \vec{a}, \tag{2.289}$$

i.e., if the identity map is homotopic to a constant map. Contractible spaces are homotopy equivalent to a one-point space. All $\mathbb{R}^m$ and S_∞ are contractible, but no sphere S_n is contractible. For S_1, the intuition behind this is as follows: any two points on the circle are connected by two different paths in S_1, which however cannot be continuously transformed into one another within S_1. Furthermore, any contractible domain or space is simply connected. For when the domain is contracted to a point, so are all the paths within it. Conversely, not every simply connected domain is contractible. Namely, on a sphere S_n, any closed path can be deformed to a point. Thus, spheres S_n are simply connected but, as remarked above, they are not contractible. Even in $\mathbb{R}^2$, there are simply connected, not contractible domains, e.g., the so-called Polish circle, but they carry little physical relevance. The main theorem of this section can now be stated.

Theorem (The converse Poincaré lemma) *On a contractible open domain, every smooth closed form is exact.*

The following proof is adapted from Flanders (1963), and uses only a minimum amount of exterior differential calculus. Consider a connected, open set $U \subset \mathbb{R}^3$, i.e., a domain. Infinitely differentiable functions $f : U \to \mathbb{R}$ are termed 0-forms, $\omega^{(0)} = f$. Let f_1, f_2, f_3 be infinitely differentiable functions. Then in $\mathbb{R}^3$, 1-, 2-, and 3-forms are expressions of the form

$$\omega^{(1)} = f_1 \, \underline{d}x_1 + f_2 \, \underline{d}x_2 + f_3 \, \underline{d}x_3,$$
$$\omega^{(2)} = f_1 \, \underline{d}x_2 \, \underline{d}x_3 + f_2 \, \underline{d}x_3 \, \underline{d}x_1 + f_3 \, \underline{d}x_1 \, \underline{d}x_2,$$
$$\omega^{(3)} = f \, \underline{d}x_1 \, \underline{d}x_2 \, \underline{d}x_3, \tag{2.290}$$

introducing a new differentiation symbol $\underline{d}$ and a new product of such differentials. We first define monomials, then their products, and finally $\underline{d}$. A *p-monomial* is an expression of the form $f(x_1 \ldots, x_n) \, \underline{d}x_{i_1} \ldots \underline{d}x_{i_p}$. For $\alpha \in \mathbb{R}$ let

$$\alpha(f \, \underline{d}x_{i_1} \ldots \underline{d}x_{i_p}) = (\alpha f) \, \underline{d}x_{i_1} \ldots \underline{d}x_{i_p},$$
$$f \, \underline{d}x_{i_1} \ldots \underline{d}x_{i_p} + g \, \underline{d}x_{i_1} \ldots \underline{d}x_{i_p} = (f + g) \, \underline{d}x_{i_1} \ldots \underline{d}x_{i_p}. \tag{2.291}$$

Then *p*-monomials form a basis of the vector space $F^p(U)$ of *p*-forms on U. The *wedge product* of monomials shall obey

$$f \, \underline{d}x_{i_1} \ldots \underline{d}x_{i_p} \, g \, \underline{d}x_{j_1} \ldots \underline{d}x_{j_q} = fg \, \underline{d}x_{i_1} \ldots \underline{d}x_{i_p} \, \underline{d}x_{j_1} \ldots \underline{d}x_{j_q}. \tag{2.292}$$

The wedge product shall be alternating,

$$\underline{d}x_i \, \underline{d}x_j = -\underline{d}x_j \, \underline{d}x_i, \tag{2.293}$$

thus $\underline{d}x_i \, \underline{d}x_i = 0$. Finally, one demands associativity and distributivity of the wedge product. The product $\underline{d}x_i \, \underline{d}x_j$ is often written $dx_i \wedge dx_j$, with a new symbol $\wedge$ for the wedge product. Clearly, it suffices in the following to prove all theorems for monomials only. *Exterior differentiation* $\underline{d}$ is defined to be a linear operation (with $\alpha, \beta \in \mathbb{R}$ and monomials ω, τ),

$$\underline{d}(\alpha\omega + \beta\tau) = \alpha \, \underline{d}\omega + \beta \, \underline{d}\tau, \tag{2.294}$$

and it maps *p*-forms to $(p + 1)$-forms, according to the rule

$$\boxed{\underline{d}(f \, \underline{d}x_{i_1} \ldots \underline{d}x_{i_p}) = \sum_{i=1}^{n} \frac{\partial f(x_1, \ldots, x_n)}{\partial x_i} \, \underline{d}x_i \, \underline{d}x_{i_1} \ldots \underline{d}x_{i_p}} \tag{2.295}$$

The individual terms in the sum are different from zero only if $i \neq i_1, \ldots, i \neq i_p$. The exterior differential $\underline{d} f$ of a 0-form is the total differential of f. From (2.295), then

$$\underline{d}(f \, \underline{d}x_{i_1} \ldots \underline{d}x_{i_p}) = \underline{d}f \, \underline{d}x_{i_1} \ldots \underline{d}x_{i_p}. \tag{2.296}$$

This $\underline{d}$ is the ultimate abstract extension of the usual gradient, curl and divergence of vector calculus. (Bott and Tu 1982, p. 14)

Note again that for monomials, $\underline{d}x_i \, \underline{d}x_i = 0$, but for forms, in general, $\omega \omega \neq 0$. For example, let $\omega = \underline{d}x_1 \, \underline{d}x_2 + \underline{d}x_3 \, \underline{d}x_4$. Then $\omega \omega = 2 \underline{d}x_1 \, \underline{d}x_2 \, \underline{d}x_3 \, \underline{d}x_4$. From now on, products of differentials are even permutations of $(1, \ldots, n)$.

Theorem (The Poincaré lemma) *For any differential form ω,*

$$\underline{d}(\underline{d}\omega) = 0. \tag{2.297}$$

Proof It suffices to consider monomials $\omega = f \, \underline{d}x_{i_1} \ldots \underline{d}x_{i_p}$, including $p = 0$ and $\omega = f$. Abbreviate $f_i = \partial f / \partial x_i$ *only in the next three lines*, and let $\tau = \underline{d}x_{i_1} \ldots \underline{d}x_{i_p}$. Then

$$\underline{d}(\underline{d}f) = \sum_{i,j} f_{ij} \, \underline{d}x_i \, \underline{d}x_j \, \tau$$

$$= \frac{1}{4} \sum_{i,j} (f_{ij} + f_{ji})(\underline{d}x_i \, \underline{d}x_j - \underline{d}x_j \, \underline{d}x_i) \, \tau$$

$$= \frac{1}{4} \sum_{i,j} (f_{ij} \, \underline{d}x_i \, \underline{d}x_j - f_{ji} \, \underline{d}x_j \, \underline{d}x_i + f_{ji} \, \underline{d}x_i \, \underline{d}x_j - f_{ij} \, \underline{d}x_j \, \underline{d}x_i) \, \tau$$

$$= 0, \tag{2.298}$$

after renaming the dummy indices $i \leftrightarrow j$ in the second and fourth terms.

Theorem *For monomials ω and τ, exterior differentiation $\underline{d}$ obeys the product rule,*

$$\underline{d}(\omega\tau) = (\underline{d}\omega)\tau + (-1)^p \omega \, \underline{d}\tau, \tag{2.299}$$

where p is the degree of ω, i. e., the number of $\underline{d}$ in ω.

Proof
$$\underline{d}(\omega\tau) = \underline{d}(fg) \, \underline{d}x_{i_1} \ldots \underline{d}x_{i_p} \, \underline{d}x_{j_1} \ldots \underline{d}x_{j_q}$$

$$= [(\underline{d}f)g + f \, \underline{d}g] \, \underline{d}x_{i_1} \ldots \underline{d}x_{i_p} \, \underline{d}x_{j_1} \ldots \underline{d}x_{j_q}$$

$$= (\underline{d}f) \, \underline{d}x_{i_1} \ldots \underline{d}x_{i_p} \, g \, \underline{d}x_{j_1} \ldots \underline{d}x_{j_q} + (-1)^p f \, \underline{d}x_{i_1} \ldots \underline{d}x_{i_p} \, \underline{d}g \, \underline{d}x_{j_1} \ldots \underline{d}x_{j_q}$$

$$= (\underline{d}\omega)\tau + (-1)^p \omega \, \underline{d}\tau. \tag{2.300}$$

In the second line, the standard product rule for derivatives was used. For the forms (2.290) in $\mathbb{R}^3$, exterior differentiation gives

$$\underline{d}\omega^{(0)} = \frac{\partial f}{\partial x_1}\, \underline{d}x_1 + \frac{\partial f}{\partial x_2}\, \underline{d}x_2 + \frac{\partial f}{\partial x_3}\, \underline{d}x_3,$$

$$\underline{d}\omega^{(1)} = \left(\frac{\partial f_3}{\partial x_2} - \frac{\partial f_2}{\partial x_3}\right)\underline{d}x_2\, \underline{d}x_3 + \left(\frac{\partial f_1}{\partial x_3} - \frac{\partial f_3}{\partial x_1}\right)\underline{d}x_3\, \underline{d}x_1$$

$$+ \left(\frac{\partial f_2}{\partial x_1} - \frac{\partial f_1}{\partial x_2}\right)\underline{d}x_1\, \underline{d}x_2,$$

$$\underline{d}\omega^{(2)} = \left(\frac{\partial f_1}{\partial x_1} + \frac{\partial f_2}{\partial x_2} + \frac{\partial f_3}{\partial x_3}\right)\underline{d}x_1\, \underline{d}x_2\, \underline{d}x_3,$$

$$\underline{d}\omega^{(3)} = 0. \tag{2.301}$$

Note that f, f_1, f_2, f_3 are four different functions. The last line holds since out of the four differentials $\underline{d}x_i$, only three can be different. Thus, one can identify, with $\vec{f} = (f_1, f_2, f_3)$,

$$\underline{d}\omega^{(0)} \equiv \nabla f \cdot \vec{dl},$$

$$\underline{d}\omega^{(1)} \equiv (\nabla \times \vec{f}) \cdot \vec{da},$$

$$\underline{d}\omega^{(2)} \equiv \nabla \cdot \vec{f}\, dV, \tag{2.302}$$

(we write $\equiv$ instead of $=$, since $\underline{d}x_i$ appears on the left side and dx_i on the right side), where

$$\vec{dl} \equiv (\underline{d}x_1,\ \underline{d}x_2,\ \underline{d}x_3),$$

$$\vec{da} \equiv (\underline{d}x_2\, \underline{d}x_3,\ \underline{d}x_3\, \underline{d}x_1,\ \underline{d}x_1\, \underline{d}x_2),$$

$$dV \equiv \underline{d}x_1\, \underline{d}x_2\, \underline{d}x_3. \tag{2.303}$$

The gradient, Stokes, and Gauss theorems then give, respectively,

$$\int_C \underline{d}\omega^{(0)} = \oint_{\partial C} \omega^{(0)},$$

$$\int_A \underline{d}\omega^{(1)} = \oint_{\partial A} \omega^{(1)},$$

$$\int_V \underline{d}\omega^{(2)} = \oint_{\partial V} \omega^{(2)}, \tag{2.304}$$

where for a curve C with endpoints $\vec{a}$ and $\vec{c}$, let

$$\oint_{\partial C} \omega^{(0)} = \omega^{(0)}(\vec{c}) - \omega^{(0)}(\vec{a}). \tag{2.305}$$

Equation (2.304) can be written uniformly as

$$\boxed{\int_D \underline{d}\omega = \oint_{\partial D} \omega} \tag{2.306}$$

where D is a domain of dimension p and ω is a $(p-1)$-form.

We start now with the 'higher' theory of exterior differential calculus. Let U and V be open domains in $\mathbb{R}^n$ and $\mathbb{R}^m$, respectively. Let ϕ and g act as shown in the diagram,

$$
\begin{array}{ccc}
U & \xrightarrow{\ \phi\ } & V \\
\phi^*g \downarrow & & \downarrow g \\
\mathbb{R} & & \mathbb{R}
\end{array}
\tag{2.307}
$$

An induced *pullback* map ϕ^*g from U to $\mathbb{R}$ is defined by

$$
\boxed{\phi^*g = g \circ \phi}
\tag{2.308}
$$

ϕ^* maps functions $g : V \to \mathbb{R}$ to functions $f : U \to \mathbb{R}$. The vector spaces of 0-forms on U and V are, respectively,

$$
F^0(U) = \{f\,|\,f : U \to \mathbb{R}\}, \qquad\qquad F^0(V) = \{g\,|\,g : V \to \mathbb{R}\}.
\tag{2.309}
$$

Therefore,

$$
F^0(U) \xleftarrow{\ \phi^*\ } F^0(V).
\tag{2.310}
$$

Let

$$
(y_1, \ldots, y_m) = \phi(x_1, \ldots, x_n).
\tag{2.311}
$$

The function ϕ^* is generalized to act on general p-forms, $1 \le p \le n$, i.e.,

$$
F^p(U) \xleftarrow{\ \phi^*\ } F^p(V)
\tag{2.312}
$$

by defining, for $i = 1, \ldots, n$,

$$
\phi^*(\underline{d}y_i) = \sum_{j=1}^{n} \frac{\partial y_i}{\partial x_j} \, \underline{d}x_j,
\tag{2.313}
$$

and, furthermore,

$$
\begin{aligned}
\phi^*(\omega + \tau) &= \phi^*(\omega) + \phi^*(\tau), \\
\phi^*(\omega\tau) &= \phi^*(\omega)\,\phi^*(\tau).
\end{aligned}
\tag{2.314}
$$

Note that for the one-monomial $g\,\underline{d}x$, the pullback ϕ^* acts on g via (2.308) and on $\underline{d}x$ via (2.313). Thus, for 1-forms,

$$\phi^*\left(\sum_{i=1}^{m} g_i(y_1, \ldots, y_m)\,\underline{d}y_i\right)$$

$$= \sum_{i=1}^{m}\sum_{j=1}^{n} g_i(y_1(x_1, \ldots, x_n), \ldots, y_m(x_1, \ldots, x_n))\,\frac{\partial y_i}{\partial x_j}\,\underline{d}x_j. \qquad (2.315)$$

For the two-monomial $\underline{d}y_i\,\underline{d}y_j$,

$$\phi^*(\underline{d}y_i\,\underline{d}y_j) = \sum_{k=1}^{n}\sum_{l=1}^{n}\frac{\partial y_i}{\partial x_k}\frac{\partial y_j}{\partial x_l}\,\underline{d}x_k\,\underline{d}x_l$$

$$= \frac{1}{2}\sum_{k=1}^{n}\sum_{l=1}^{n}\left(\frac{\partial y_i}{\partial x_k}\frac{\partial y_j}{\partial x_l} - \frac{\partial y_i}{\partial x_l}\frac{\partial y_j}{\partial x_k}\right)\,\underline{d}x_k\,\underline{d}x_l. \qquad (2.316)$$

Similarly, for $m = n$, and for a Jacobian J of the transformation $x \to y$,

$$\underline{d}y_1 \ldots \underline{d}y_m = J\,\underline{d}x_1 \ldots \underline{d}x_n. \qquad (2.317)$$

Consider the diagram

$$\begin{array}{ccccc}
U & \xrightarrow{\ \phi\ } & V & \xrightarrow{\ \psi\ } & W \\
{\scriptstyle \phi^*\psi^*h}\Big\downarrow & & \Big\downarrow{\scriptstyle \psi^*h} & & \Big\downarrow{\scriptstyle h} \\
\mathbb{R} & & \mathbb{R} & & \mathbb{R}
\end{array} \qquad (2.318)$$

For $\vec{x} \in U$,

$$[(\psi \circ \phi)^*h](\vec{x}) = h[(\psi \circ \phi)(\vec{x})] = h[\psi(\phi(\vec{x}))]$$
$$= [\psi^*h](\phi(\vec{x})) = [\phi^*(\psi^*h)](\vec{x}) = [(\phi^* \circ \psi^*)h](\vec{x}) \qquad (2.319)$$

or

$$(\psi \circ \phi)^* = \phi^* \circ \psi^*. \qquad (2.320)$$

Somewhat faster, this is obtained as follows:

$$\phi^* \circ \psi^*h = \psi^*h \circ \phi = h \circ \psi \circ \phi = (\psi \circ \phi)^*h. \qquad (2.321)$$

This holds for 0-forms. For general p-forms, it is easily shown by induction that $(\psi \circ \phi)^* = \phi^* \circ \psi^*$ maps $F^p(W)$ to $F^p(U)$,

$$F^p(U) \xleftarrow{\ \phi^*\ } F^p(V) \xleftarrow{\ \psi^*\ } F^p(W). \qquad (2.322)$$

The first serious theorem is then as follows.

Theorem

$$\boxed{\underline{d}(\phi^*\omega) = \phi^*(\underline{d}\omega)}$$ (2.323)

This is an important consistency theorem: Let ϕ be a coordinate change. Then from (2.323), differentiation and coordinate change are interchangeable.

Proof by induction. Induction starts at zero-monomials, $g : V \to \mathbb{R}$. Here

$$
\begin{aligned}
\phi^*(\underline{d}\,g) &= \phi^*\left(\sum_{i=1}^{m} \frac{\partial g(y_1, \ldots, y_m)}{\partial y_i}\, \underline{d}y_i\right) \\
&= \sum_{i=1}^{m}\sum_{j=1}^{n} \frac{\partial g(y_1(x_1, \ldots, x_n), \ldots, y_m(x_1, \ldots, x_n))}{\partial y_i}\, \frac{\partial y_i}{\partial x_j}\, \underline{d}x_j \\
&= \sum_{i=1}^{m}\sum_{j=1}^{n} \frac{\partial(g \circ \phi(x_1, \ldots, x_n))}{\partial x_j}\, \underline{d}x_j \\
&= \sum_{j=1}^{n} \frac{\partial}{\partial x_j}\left(\sum_i (\phi^*g)(x_1, \ldots, x_n)\right) \underline{d}x_j \\
&= \underline{d}(\phi^*g).
\end{aligned}
$$ (2.324)

Note that ϕ^* acts not on $\partial/\partial y_i$, but only on differential forms. The induction step is as follows: Let ω be a p-monomial,

$$
\begin{aligned}
\omega &= g\, \underline{d}y_{i_1}\, \underline{d}y_{i_2} \ldots \underline{d}y_{i_p} \\
&= g\, \underline{d}(y_{i_1}\, \underline{d}y_{i_2} \ldots \underline{d}y_{i_p}) \\
&= g\, \underline{d}\eta.
\end{aligned}
$$ (2.325)

Here η is a $(p-1)$-form, for which it is already proved (induction assumption) that

$$\underline{d}(\phi^*\eta) = \phi^*(\underline{d}\eta).$$ (2.326)

Then,

$$
\begin{aligned}
\underline{d}(\phi^*\omega) &= \underline{d}((\phi^*g)(\phi^*\underline{d}\eta)) \\
&= \underline{d}((\phi^*g)(\underline{d}(\phi^*\eta))) \\
&= (\underline{d}(\phi^*g))(\underline{d}(\phi^*\eta)) \\
&= (\phi^*\underline{d}g)(\phi^*\underline{d}\eta) \\
&= \phi^*(\underline{d}g\, \underline{d}\eta) \\
&= \phi^*(\underline{d}(g\, \underline{d}\eta)) \\
&= \phi^*(\underline{d}\omega).
\end{aligned}
$$ (2.327)

Bott and Tu (1982), p. 19 give a condensed proof of this theorem. In our above calculation, the use of (2.313) is somewhat cumbersome. Let $\vec{y} = \phi(\vec{x})$, and

$$y_i = \phi_i(\vec{x}), \tag{2.328}$$

where ϕ_i is the i-th component of ϕ. Then (2.313) can be abbreviated,

$$\phi^*(\underline{d}y_i) = \sum_{j=1}^{n} \frac{\partial y_i}{\partial x_j} \, \underline{d}x_j = \underline{d}\phi_i. \tag{2.329}$$

Bott and Tu then have

$$
\begin{aligned}
\underline{d}\phi^*(g \, \underline{d}y_{i_1} \dots \underline{d}y_{i_p}) &= \underline{d}[\phi^*(g) \, \phi^*(\underline{d}y_{i_1}) \dots \phi^*(\underline{d}y_{i_p})] \\
&= \underline{d}[(g \circ \phi) \, \underline{d}\phi_{i_1} \dots \underline{d}\phi_{i_p}] \\
&= \underline{d}(g \circ \phi) \, \underline{d}\phi_{i_1} \dots \underline{d}\phi_{i_p}.
\end{aligned} \tag{2.330}
$$

On the other hand,

$$
\begin{aligned}
\phi^*\underline{d}(g \, \underline{d}y_{i_1} \dots \underline{d}y_{i_p}) &= \phi^*\left(\sum_{j=1}^{n} \frac{\partial g}{\partial y_j} \, \underline{d}y_j \, \underline{d}y_{i_1} \dots \underline{d}y_{i_p} \right) \\
&\stackrel{(a)}{=} \sum_{j=1}^{n} \left(\frac{\partial g}{\partial y_j} \circ \phi \right) \underline{d}\phi_j \, \underline{d}\phi_{i_1} \dots \underline{d}\phi_{i_p} \\
&\stackrel{(b)}{=} \sum_{j=1}^{n} \sum_{i=1}^{m} \left(\frac{\partial g}{\partial y_j} \circ \phi \right) \frac{\partial y_j}{\partial x_i} \, \underline{d}x_i \, \underline{d}\phi_{i_1} \dots \underline{d}\phi_{i_p} \\
&= \sum_{i=1}^{m} \sum_{j=1}^{n} \left(\frac{\partial y_j}{\partial x_i} \frac{\partial g}{\partial y_j} \circ \phi \right) \underline{d}x_i \, \underline{d}\phi_{i_1} \dots \underline{d}\phi_{i_p} \\
&\stackrel{(c)}{=} \sum_{i=1}^{m} \frac{\partial(g \circ \phi)}{\partial x_i} \, \underline{d}x_i \, \underline{d}\phi_{i_1} \dots \underline{d}\phi_{i_p} \\
&= \underline{d}(g \circ \phi) \, \underline{d}\phi_{i_1} \dots \underline{d}\phi_{i_p}.
\end{aligned} \tag{2.331}
$$

(a) ϕ^* acts on the derivative function $\partial g / \partial y_i$.
(b) Follows from (2.329).
(c) Chain rule. ϕ in $f \circ \phi$ changes coordinates, $\vec{x} \to \vec{y}$.

We have now all tools at hand to prove the converse Poincaré lemma.

Proof (The converse Poincaré lemma) Consider a domain $U \subset \mathbb{R}^n$, with $\vec{x} = (x_1, \dots, x_n) \in U$. Let $I = [0, 1] \subset \mathbb{R}$. For $t \in I$, let moreover

$$s_t : U \to I \times U,$$
$$\vec{x} \to (t, \vec{x}). \tag{2.332}$$

The function s_t is a *section* of constant t in $I \times U$. This specific s takes the role of the general ϕ considered so far. The induced map s_t^* on p-forms with arbitrary $p \geq 0$,

$$s_t^* : F^p(I \times U) \to F^p(U), \tag{2.333}$$

is defined by

$$s_t^*\big(f(t, \vec{x}) \ \underline{d}x_{i_1} \ldots \underline{d}x_{i_p}\big) = f(t; \vec{x}) \ \underline{d}x_{i_1} \ldots \underline{d}x_{i_p},$$
$$s_t^*\big(f(t, \vec{x}) \ \underline{d}t \ \underline{d}x_{i_1} \ldots \underline{d}x_{i_{p-1}}\big) = 0. \tag{2.334}$$

Here $f(t; \vec{x})$ means that t is treated as constant parameter, not as a variable. Define an operation K for all $p \geq 1$,

$$K : F^p(I \times U) \to F^{p-1}(U), \tag{2.335}$$

by

$$K(f(t, x_1, \ldots, x_n) \ \underline{d}x_{i_1} \ldots \underline{d}x_{i_p}) = 0, \tag{2.336}$$
$$K(f(t, x_1, \ldots, x_n) \ \underline{d}t \ \underline{d}x_{i_1} \ldots \underline{d}x_{i_{p-1}}) = \left(\int_0^1 f(t, x_1, \ldots, x_n) \ \underline{d}t \right) \underline{d}x_{i_1} \ldots \underline{d}x_{i_{p-1}}.$$

K shall be linear, $K(\alpha \omega + \beta \tau) = \alpha \, K \omega + \beta \, K \tau$, with $\alpha, \beta \in \mathbb{R}$ and p-forms ω, τ. We show that K and s_t obey for any p-form ω,

$$\boxed{K(\underline{d}\omega) + \underline{d}(K\omega) = s_1^* \, \omega - s_0^* \, \omega} \tag{2.337}$$

It suffices to show this for p-monomials ω.

(i) Let ω not contain $\underline{d}t$,

$$\omega = f(t, \vec{x}) \ \underline{d}x_{i_1} \ldots \underline{d}x_{i_p}. \tag{2.338}$$

Then $K\omega = 0$, and therefore $\underline{d}K\omega = 0$. On the other hand,

$$\underline{d}\omega = \frac{\partial f}{\partial t} \ \underline{d}t \ \underline{d}x_{i_1} \ldots \underline{d}x_{i_p} + \Omega, \tag{2.339}$$

where Ω does not contain $\underline{d}t$, but $p + 1$ differentials $\underline{d}x_i$. Then

$$K(\underline{d}\omega) = \left(\int_0^1 \frac{\partial f}{\partial t} \, \underline{d}t \right) \underline{d}x_{i_1} \dots \underline{d}x_{i_p}$$
$$= [f(1, \vec{x}) - f(0, \vec{x})] \, \underline{d}x_{i_1} \dots \underline{d}x_{i_p}. \tag{2.340}$$

Furthermore,

$$s_0^* \, \omega = f(0; \vec{x}) \, \underline{d}x_{i_1} \dots \underline{d}x_{i_p},$$
$$s_1^* \, \omega = f(1; \vec{x}) \, \underline{d}x_{i_1} \dots \underline{d}x_{i_p}, \tag{2.341}$$

so (2.337) is true for this ω (';' instead of ',' is only syntax).

(ii) Let ω contain $\underline{d}t$,

$$\omega = f(t, \vec{x}) \, \underline{d}t \, \underline{d}x_{i_1} \dots \underline{d}x_{i_{p-1}}. \tag{2.342}$$

Then by (2.334),

$$s_0^*(\omega) = 0,$$
$$s_1^*(\omega) = 0. \tag{2.343}$$

And

$$K(\underline{d}\omega) = K \sum_{j \neq i_l} \frac{\partial f}{\partial x_j} \, \underline{d}x_j \, \underline{d}t \, \underline{d}x_{i_1} \dots \underline{d}x_{i_{p-1}}$$
$$= -K \sum_{j \neq i_l} \frac{\partial f}{\partial x_j} \, \underline{d}t \, \underline{d}x_j \, \underline{d}x_{i_1} \dots \underline{d}x_{i_{p-1}}$$
$$= -\sum_{j \neq i_l} \int_0^1 \frac{\partial f}{\partial x_j} \, \underline{d}t \, \underline{d}x_j \, \underline{d}x_{i_1} \dots \underline{d}x_{i_{p-1}}, \tag{2.344}$$

where l in i_l runs from 1 to $p - 1$. On the other hand,

$$\underline{d}(K\omega) = \underline{d} \left(\int_0^1 f(t, \vec{x}) \, \underline{d}t \right) \underline{d}x_{i_1} \dots \underline{d}x_{i_{p-1}}$$
$$= \sum_{j \neq i_l} \frac{\partial}{\partial x_j} \left(\int_0^1 f(t, \vec{x}) \, \underline{d}t \right) \underline{d}x_j \, \underline{d}x_{i_1} \dots \underline{d}x_{i_{p-1}}$$
$$= \sum_{j \neq i_l} \int_0^1 \frac{\partial f}{\partial x_j} \, \underline{d}t \, \underline{d}x_j \, \underline{d}x_{i_1} \dots \underline{d}x_{i_{p-1}}, \tag{2.345}$$

so (2.337) is again true. □

$\underline{d}K + K\underline{d}$ maps closed forms to exact forms and therefore induces zero in cohomology. Such a K is called a *homotopy operator*. (Bott and Tu 1982, p. 34)

Consider now the mappings

$$U \xrightarrow{\; s_t \;} I \times U \xrightarrow{\; \psi \;} U \tag{2.346}$$

with a homotopy ψ. U shall be contractible,

$$\psi(1, \vec{x}) = \vec{x},$$
$$\psi(0, \vec{x}) = 0. \tag{2.347}$$

In the last equation, the origin is put in the contraction point $\vec{a}$. Let

$$\mathbb{1}(\vec{x}) = \vec{x},$$
$$\mathbb{O}(\vec{x}) = 0. \tag{2.348}$$

With $s_1(\vec{x}) = (1, \vec{x})$ and $s_0(\vec{x}) = (0, \vec{x})$,

$$\psi \circ s_1 = \mathbb{1},$$
$$\psi \circ s_0 = \mathbb{O}. \tag{2.349}$$

For 0-forms then,

$$(s_1^* \circ \psi^*) \, f(\vec{x}) \overset{(2.320)}{=} (\psi \circ s_1)^* \, f(\vec{x})$$
$$\overset{(2.308)}{=} f(\psi \circ s_1(\vec{x}))$$
$$\overset{(2.332)}{=} f(\psi(1, \vec{x}))$$
$$\overset{(2.347)}{=} f(\vec{x}). \tag{2.350}$$

Since $\psi \circ s_1 = \mathbb{1}$, it is also clear from (2.313) that

$$(s_1^* \circ \psi^*) \, \underline{d} x_i = \underline{d} x_i. \tag{2.351}$$

Thus, for p-forms $\eta \in F^p(U)$, for all $p \geq 0$,

$$(s_1^* \circ \psi^*) \, \eta(\vec{x}) = \eta(\vec{x}). \tag{2.352}$$

Correspondingly, for 0-forms,

$$(s_0^* \circ \psi^*) \, f(\vec{x}) = f(0). \tag{2.353}$$

Write $\psi_0(\vec{x}) = \psi(0, \vec{x})$, and let

$$p_i : U \to \mathbb{R}, \qquad \vec{x} \to x_i \tag{2.354}$$

be a coordinate projection. Then

$$\psi_0^* \, \underline{d}x_i \;=\; \psi_0^* \, \underline{d}p_i(\vec{x})$$

$$\overset{(2.323)}{=} \; \underline{d}\psi_0^* \, p_i(\vec{x})$$

$$\overset{(2.308)}{=} \; \underline{d}p_i(\psi_0(\vec{x}))$$

$$\overset{(2.347)}{=} \; \underline{d}p_i(0)$$

$$= \; \underline{d}0$$

$$\overset{(2.295)}{=} \; 0. \tag{2.355}$$

For this calculation, see also p. 34 in Bott and Tu (1982), fifth line from bottom. Therefore, for p-forms η with $p \geq 1$,

$$(s_0^* \circ \psi^*) \, \eta(\vec{x}) = 0. \tag{2.356}$$

Inserting now in (2.337) for ω, the form

$$\omega = \psi^* \eta \tag{2.357}$$

gives

$$K(\underline{d}(\psi^*\eta)) + \underline{d}(K(\psi^*\eta)) = s_1^*(\psi^*\eta) - s_0^*(\psi^*\eta). \tag{2.358}$$

From (2.323), (2.352), and (2.355), and abbreviating $\alpha = K(\psi^*\eta)$,

$$K(\psi^*\underline{d}\eta) + \underline{d}\alpha = \eta, \tag{2.359}$$

for p-forms η with $p \geq 1$. If η is closed, $\underline{d}\eta = 0$, then

$$\eta = \underline{d}\alpha. \tag{2.360}$$

Thus, closed forms η with degree $p \geq 1$ are exact.   □

This proof is taken from Flanders (1963), pp. 27–29. A similar proof is given in Bott and Tu (1982), p. 33. They use the projection $\pi(t, \vec{x}) = \vec{x}$ in place of the homotopy ψ. In fluid dynamics, one often is dealing with contractible spaces (but see Sect. 3.8). In algebraic topology, this is considered as the trivial case:

> A measure of the size of the space of "interesting" solutions is the definition of the de Rham cohomology. The q-th de Rham cohomology of $\mathbb{R}^n$ is the vector space $H^q(\mathbb{R}^n) = \{$closed q-forms$\}/\{$exact q-forms$\}$. (Bott and Tu 1982, p. 15)

The converse Poincaré lemma states that $H^q(\mathbb{R}^n) = \varnothing$. The above proof uses contractibility of U only at a late stage. Instead, contractibility can be used from the start to show $\underline{d}\eta = 0 \rightarrow \eta = \underline{d}\alpha$. By this, the introduction of starred function ϕ^* and ψ^* can be avoided. Idea is to build up the contractible $\mathbb{R}^n$ using $\mathbb{R}^n = \mathbb{R}^{n-1} \times \mathbb{R}$ and induction. A corresponding proof can be found in Flanders (1967).

References

Batchelor, G.K. 1952. The effect of homogeneous turbulence on material lines and surfaces. *Proceedings of the Royal Society of London A* 213: 349.

Batchelor, G.K. 1967. *An introduction to fluid dynamics*. Cambridge: Cambridge University Press.

Bateman, H. 1944. *Partial differential equations of mathematical physics*. New York: Dover.

Benedetto, D., and M. Pulvirenti. 1992. From vortex layers to vortex sheets. *SIAM Journal on Applied Mathematics* 52: 1041.

Bott, R., and L.W. Tu. 1982. *Differential forms in algebraic topology*. New York: Springer.

Broer, L.J.F. 1974. On the Hamiltonian theory of surface waves. *Applied Scientific Research* 29: 430.

Case, K.M. 1961. Hydrodynamic stability and the inviscid limit. *Journal of Fluid Mechanics* 10: 420.

Chorin, A.J., and J.E. Marsden. 1990. *A mathematical introduction to fluid mechanics*. New York: Springer.

Clebsch, A. 1859. Ueber die Integration der hydrodynamischen Gleichungen. *Journal für die reine und angewandte Mathematik* 56: 1.

Cocke, W.J. 1969. Turbulent hydrodynamic line stretching: Consequences of isotropy. *Physics of Fluids* 12: 2488.

Dussan, V.E.B. 1976. On the difference between a bounding surface and a material surface. *Journal of Fluid Mechanics* 75: 609.

Flanders, H. 1963. *Differential forms with applications to the physical sciences*. New York: Academic Press; New York: Dover (1989).

Flanders, H. 1967. Differential forms. In *Studies in global geometry and analysis*, ed. S.-S. Chern, 27. Washington: Mathematical Association of America.

Friedrichs, K.O. 1934. Über ein Minimumproblem für Potentialströmungen mit freiem Rande. *Mathematische Annalen* 109: 60.

Hargreaves, R. 1908. A pressure integral as kinetic potential. *Philosophical Magazine Series 6*, 16: 436.

Herglotz, G. 1985. *Vorlesungen über die Mechanik der Kontinua*. Leipzig: B.G. Teubner.

Huang, K. 1963. *Introduction to statistical physics*. New York: Wiley.

Hundhausen, A.J. 1972. *Coronal expansion and solar wind*. Berlin: Springer.

Jackiw, R. 2002. *Lectures on fluid dynamics*. New York: Springer.

Jänich, K. 2001. *Vector analysis*. Berlin: Springer.

Kotschin, N.J., I.A. Kibel, and N.W. Rose. 1954. *Theoretische Hydromechanik*, vol. I. Berlin: Akademie-Verlag.

Kraichnan, R.H. 1974. On Kolmogorov's inertial-range theories. *Journal of Fluid Mechanics* 62: 305.

Lamb, H. 1932. *Hydrodynamics*. Cambridge: Cambridge University Press; New York: Dover (1945).

Lichtenstein, L. 1929. *Grundlagen der Hydromechanik*. Berlin: Springer.

Lighthill, J. 1986. *An informal introduction to theoretical fluid mechanics*. Oxford: Clarendon Press.

Love, A.E. 1901. Hydrodynamik. In *Encyklopädie der mathematischen Wissenschaften*, vol. 4–3, 48. Leipzig: Teubner (digital at GDZ Göttingen).

Luke, J.C. 1967. A variational principle for a fluid with a free surface. *Journal of Fluid Mechanics* 27: 395.

Markushevich, A.I. 1965. *Theory of functions of a complex variable*, vol. I. Englewood Cliffs: Prentice-Hall.

Meyer, R.E. 1971. *Introduction to mathematical fluid dynamics*. New York: Wiley; Mineola: Dover (2007).

Miles, J.W. 1977. On Hamilton's principle for surface waves. *Journal of Fluid Mechanics* 83: 153.

Milne-Thomson, L.M. 1968. *Theoretical hydrodynamics*, 5th ed. London: Macmillan; New York: Dover (1996).

Norman, M.L., and K.-H. Winkler. 1986. 2-D Eulerian hydrodynamics with fluid interfaces, self-gravity and rotation. In *NATO advanced research workshop on astrophysical radiation hydrodynamics*, ed. K.-H. Winkler and M.L. Norman, 187. Dordrecht: D. Reidel Publishing Co.

Orszag, S.A. 1970. Comments on 'Turbulent hydrodynamic line stretching: Consequences of isotropy'. *Physics of Fluids* 13: 2203.

Prager, W. 1961. *Einführung in die Kontinuumsmechanik*. Basel: Birkäuser Verlag.

Prosperetti, A. 2011. *Advanced mathematics for applications*. Cambridge: Cambridge University Press.

Proudman, J. 1953. *Dynamical oceanography*. London: Methuen & Co.

Rosenhead, L. 1954. Introduction—The second coefficient of viscosity: A brief review of fundamentals. *Proceedings of the Royal Society of London A* 226: 1.

Saffman, P.G. 1995. *Vortex dynamics*. Cambridge: Cambridge University Press.

Saffman, P.G., and G.R. Baker. 1979. Vortex interactions. *Annual Reviews in Fluid Mechanics* 11: 95.

Schaefer, H. 1967. Das Cosserat-Kontinuum. *Zeitschrift für angewandte Mathematik und Mechanik* 47: 485.

Sierpinski, W. 1958. *Cardinal and ordinal numbers*. Polska Akademia Nauk. Monografie matematyczne, vol. 34. Warszawa.

Stokes, G.G. 1845. On the theories of the internal friction of fluids in motion, and of the equilibrium and motion of elastic solids. *Transactions of the Cambridge Philosophical Society* 8: 287 (1849); in *Mathematical and physical papers by G.G. Stokes*, vol. 1, 75. Cambridge: Cambridge University Press (1880).

Stone, J.M., and M.L. Norman. 1992. ZEUS-2D: A radiation magnetohydrodynamics code for astrophysical flows in two space dimensions. I. The hydrodynamic algorithms and tests. *Astrophysical Journal Supplement Series* 80: 753.

Tanaka, M., and S. Kida. 1993. Characterization of vortex tubes and sheets. *Physics of Fluids A* 5: 2079.

Truesdell, C. 1952. The mechanical foundations of elasticity and fluid dynamics. *Journal of Rational Mechanics and Analysis* 1: 125.

Truesdell, C. 1966. *The elements of continuum mechanics*. Berlin: Springer.

Truesdell, C., and R.G. Muncaster. 1980. *Fundamentals of Maxwell's kinetic theory of a simple monatomic gas*. New York: Academic Press.

Truesdell, C., and W. Noll. 1965. The non-linear field theories of mechanics. In *Handbuch der Physik = Encyclopedia of Physics*, vol. 2/3–3, 1, ed. S. Flügge and C. Truesdell. Berlin: Springer.

Weatherburn, C.E. 1924. *Advanced vector analysis*. London: G. Bell & Sons.

Whitham, G.B. 1965. A general approach to linear and non-linear dispersive waves using a Lagrangian. *Journal of Fluid Mechanics* 22: 273.

Whitham, G.B. 1967. Non-linear dispersion of water waves. *Journal of Fluid Mechanics* 27: 399.

Winkler, K.-H., M.L. Norman, and D. Mihalas. 1984. Adaptive-mesh radiation hydrodynamics. I. The radiation transport equation in a completely adaptive coordinate system. *Journal of Quantitative Spectroscopy & Radiative Transfer* 31: 473.

Wright, M.C.M. 2006. Green function or Green's function? *Nature Physics* 2: 646.

Chapter 3
Flows in the Complex Plane

One of the primary goals in this book is to obtain analytical solutions for the equations of motion of fluids. In order that such solutions are found, fluid motion is often restricted to be stationary and two dimensional (2-D) in a plane. All fluid fields depend then on the x- and y-coordinates only, and the fluid velocity has no z-component, $\vec{u} = u_x(x, y)\,\hat{x} + u_y(x, y)\,\hat{y}$. The motion of a fluid parcel is restricted to a planar layer, and the whole body of fluid is made up of layers stacked one upon the other. With the usual identification of the vector space $\mathbb{R}^2$ and the field $\mathbb{C}$ of complex numbers, complex function theory becomes available for 2-D fluid dynamics. In this chapter, stationary fluid dynamics in 2-D is framed as a problem of complex function theory, by introducing a complex potential Φ for the complex speed field w. We discuss some rather elementary topics from potential theory (Laplace equation, boundary conditions, and maximum property). An Appendix gives an overview of theorems from complex function theory on analytic and meromorphic functions without proofs. We discuss the Schwarz–Christoffel theorem on conformal mappings, which is applied in Chap. 5 to flows with jets, wakes, and cavities. Finally, a brief introduction to the Riemann surfaces is given, which will find applications in later chapters.

3.1 The Laplace Equation

The flow velocity field $\vec{u}$ shall be divergence-free and irrotational,

$$\nabla \cdot \vec{u} = 0, \qquad\qquad \nabla \times \vec{u} = 0. \tag{3.1}$$

The fluid shall be inviscid, $\nu = 0$, and the flow stationary, $\partial/\partial t \equiv 0$. Let

$$\rho = \rho(x, y), \qquad\qquad \vec{u} = u(x, y)\,\hat{x} + v(x, y)\,\hat{y}. \tag{3.2}$$

© Springer Nature Switzerland AG 2019
A. Feldmeier, *Theoretical Fluid Dynamics*, Theoretical and Mathematical Physics,
https://doi.org/10.1007/978-3-030-31022-6_3

The flow consists then of identical planar xy-layers stacked in the z-direction. One of these layers is selected and one says that the flow is 'in the xy-plane.' Irrotational flow with $\nabla \times \vec{u} = 0$ implies, from the converse Poincaré lemma for contractible $\mathbb{R}^2$,

$$\vec{u} = \nabla\phi, \tag{3.3}$$

with a *velocity* potential $\phi(x, y)$. A constant offset $\vec{u} + \vec{c}$ is irrelevant here because of Galilean invariance. For a solenoidal velocity field $\vec{u}$, the potential ϕ obeys then the Laplace equation,

$$0 = \nabla \cdot \vec{u} = \nabla \cdot \nabla\phi = \Delta\phi. \tag{3.4}$$

In the xy-plane, this becomes

$$\frac{\partial^2\phi}{\partial x^2} + \frac{\partial^2\phi}{\partial y^2} = 0. \tag{3.5}$$

Alternatively, one can start with $\nabla \cdot \vec{u} = 0$. Then, again from the converse Poincaré lemma,

$$\vec{u} = \nabla \times \vec{\psi}. \tag{3.6}$$

For irrotational flows,

$$0 = \nabla \times \vec{u} = \nabla \times (\nabla \times \vec{\psi}) = \nabla(\nabla \cdot \vec{\psi}) - \Delta\vec{\psi}. \tag{3.7}$$

The vector fields $\vec{u} = \nabla \times \vec{\psi}$ and $\vec{\psi}$ are orthogonal to each other. Since $\vec{u}$ has x- and y-components only, it follows that

$$\vec{\psi} = \psi(x, y)\,\hat{z}, \tag{3.8}$$

which gives

$$\nabla \cdot \vec{\psi} = (\hat{x}\,\partial_x + \hat{y}\,\partial_y) \cdot \psi(x, y)\,\hat{z} = 0. \tag{3.9}$$

Thus, (3.7) becomes $\Delta\vec{\psi} = \hat{z}\,\Delta\psi = 0$, or in the xy-plane,

$$\frac{\partial^2\psi}{\partial x^2} + \frac{\partial^2\psi}{\partial y^2} = 0. \tag{3.10}$$

The scalar field $\psi(x, y)$ is called the *streamfunction*, and plays a central role in the following. Although both ϕ and ψ obey the Laplace equation, they are not identical, $\phi \neq \psi$, as will become clear below. Finally,

$$\vec{u} = \nabla \times \psi\hat{z} = (\hat{x}\partial_x + \hat{y}\partial_y) \times \psi\hat{z} = \frac{\partial\psi}{\partial y}\,\hat{x} - \frac{\partial\psi}{\partial x}\,\hat{y}. \tag{3.11}$$

Comparing this with $\vec{u} = u\hat{x} + v\hat{y} = \nabla\phi$ from (3.3) gives

$$u = \frac{\partial\psi}{\partial y} = \frac{\partial\phi}{\partial x},$$

$$v = -\frac{\partial\psi}{\partial x} = \frac{\partial\phi}{\partial y}. \tag{3.12}$$

These equations will be identified later as the Cauchy–Riemann equations for the complex speed potential $\Phi = \phi + i\psi$.

3.2 Green's Theorems

The Laplace equation is a central topic of *potential theory*. A standard if somewhat outdated reference to potential theory is the book by Kellogg (1929). A real-valued scalar field f is *harmonic* in a domain V of $\mathbb{R}^3$ if all second derivatives of f exist in V and are continuous, and if f obeys the Laplace equation $\Delta f = 0$ in V.

For arbitrary, differentiable scalar fields f and g, Gauss' theorem gives

$$\int_V dV \, \nabla \cdot (g\,\nabla f) = \oint_{\partial V} d\vec{a} \cdot (g\,\nabla f) = \oint_{\partial V} da \, g \, \frac{\partial f}{\partial n}, \tag{3.13}$$

with area element $d\vec{a} = da\,\hat{n}$ and directional derivative $\partial/\partial n = \hat{n}\cdot\nabla$ pointing away from V. Let $g = 1$ and assume that f is harmonic, $\Delta f = 0$. Then the left side of (3.13) is $\int dV \, \nabla \cdot \nabla f = \int dV \, \Delta f = 0$, and (3.13) becomes

$$\boxed{\oint_{\partial V} da \, \frac{\partial f}{\partial n} = 0} \tag{3.14}$$

This is Green's first theorem for an arbitrary harmonic function f. Returning to general f and g, by the product rule for derivatives,

$$\int_V dV \, \nabla \cdot (g\,\nabla f) = \int_V dV \, (\nabla g \cdot \nabla f + g\Delta f). \tag{3.15}$$

From (3.13) to (3.15), it follows that

$$\int_V dV \, (\nabla g \cdot \nabla f + g\Delta f) = \oint_{\partial V} da \, g \, \frac{\partial f}{\partial n}. \tag{3.16}$$

Interchanging f and g,

$$\int_V dV \, (\nabla f \cdot \nabla g + f\Delta g) = \oint_{\partial V} da \, f \, \frac{\partial g}{\partial n}. \tag{3.17}$$

Subtracting the last two equations gives

$$\boxed{\int_V dV \, (f \Delta g - g \Delta f) = \oint_{\partial V} da \, \left(f \frac{\partial g}{\partial n} - g \frac{\partial f}{\partial n} \right)}$$

(3.18)

This is Green's integral theorem, with $\partial/\partial n$ the derivative in the direction normal to the surface element of ∂V, and pointing away from V. For harmonic functions f, g, it becomes

$$\oint_{\partial V} da \, \left(f \frac{\partial g}{\partial n} - g \frac{\partial f}{\partial n} \right) = 0.$$

(3.19)

3.3   The Dirichlet and Neumann Boundary Conditions

This section follows Jackson (1998), p. 37. Assume $\Delta f = 0$ and $g = f$ in (3.16). Then

$$\int_V dV \, (\nabla f)^2 = \oint_{\partial V} da \, f \frac{\partial f}{\partial n}.$$

(3.20)

If $f = 0$ or $\partial f/\partial n = 0$ for each point of ∂V, then

$$\int_V dV \, (\nabla f)^2 = 0.$$

(3.21)

Since $(\nabla f)^2 \geq 0$ in each point, this can only hold if $\nabla f = 0$ in all of V. Therefore, f is constant in V. If $f = 0$ on ∂V, then $f = 0$ in V by continuity of f. On the other hand, if $\partial f/\partial n = 0$ on ∂V, then $f = $ const in V. Only the derivatives of harmonic functions have physical relevance. Thus, the constant in the latter case is irrelevant. One arrives thus at the following.

Theorem *Let f be a harmonic function in a domain $V \subset \mathbb{R}^3$. For each point of ∂V, let $f = 0$ or $\partial f/\partial n = 0$. Then $f = $ const in V.*   □

Theorem (The Dirichlet and Neumann boundary conditions) *Let f and g be harmonic in a domain $V \subset \mathbb{R}^3$. For each point of ∂V, let $f = g$ or $\partial f/\partial n = \partial g/\partial n$. Then $f = g + C$ in V, with some constant C.*

Proof The function $f - g$ is harmonic, since $\Delta(f - g) = \Delta f - \Delta g = 0$. Furthermore, $f - g = 0$ or $\partial(f - g)/\partial n = 0$ on ∂V. Thus, $f - g = 0$ or $f - g = C$ in V, according to the last theorem.   □

If the harmonic function f is specified on the boundary ∂V, one has the *Dirichlet* boundary conditions; if, on the other hand, the derivative $\partial f/\partial n$ normal to the boundary ∂V is specified, one has the *Neumann* boundary conditions. If the constant C in the last theorem is irrelevant, it follows that *a harmonic function is determined by* either the *Dirichlet* or *Neumann boundary conditions* on a domain boundary. This is a remarkable result, given that the Laplace equation is of second order, and seems to require boundary conditions for both f and $\partial f/\partial n$. Actually, posing the Dirichlet *and* Neumann boundary conditions for harmonic functions will, in general, lead to inconsistencies. There can also be the mixed Dirichlet and Neumann boundary conditions on ∂V. For example, the Neumann boundary conditions can hold on an ocean bed, and the Dirichlet boundary conditions on the surface of the ocean.

3.4 Mean Value and Maximum Property

Harmonic functions are characterized by two important properties, the *mean value* and the *maximum* property. The following holds for $\mathbb{R}^3$, the case $\mathbb{R}^2$ is discussed below. We follow the derivation given in Jänich (2001), p. 178. Let $\vec{r}$ be the position vector from the origin. For any differentiable function f, define a new function f_s, the *dilation* of f, by

$$f_s(\vec{r}) = f(s\vec{r}), \qquad s \in \mathbb{R}. \tag{3.22}$$

A short calculation shows that

$$\Delta f_s = s^2 (\Delta f)_s, \tag{3.23}$$

where on the right, $(\Delta f)_s$ is the dilation of Δf. Thus, if f is harmonic, then f_s is also harmonic. Define now

$$I_s = \oint_{S_r} da \; f_s, \tag{3.24}$$

where S_r is the sphere with radius r centered at the origin. Then

$$\begin{aligned}
\frac{dI_s}{ds} &= \oint_{S_r} da \; \frac{d}{ds} f(s\vec{r}) \\
&= \oint_{S_r} da \; \nabla_{s\vec{r}} f(s\vec{r}) \cdot \frac{d(s\vec{r})}{ds} \\
&= \oint_{S_r} da \; \frac{1}{s} \nabla_{\vec{r}} f(s\vec{r}) \cdot \vec{r} \\
&= \frac{r}{s} \oint_{S_r} d\vec{a} \cdot \nabla f_s(\vec{r}) \\
&= \frac{r}{s} \int_{B_r} dV \; \Delta f_s(\vec{r}) = 0,
\end{aligned} \tag{3.25}$$

for any harmonic function f, and B_r is the open ball with radius r. Here $\nabla_{\vec{r}} f$ is the gradient of f with respect to the variable $\vec{r}$ (if other variables are present). Therefore, $I_1 = I_\epsilon$, and for $\epsilon \to 0$,

$$\oint_{S_r} da\, f(\vec{r}) = \oint_{S_r} da\, f_\epsilon(\vec{r}) = \oint_{S_r} da\, f(\epsilon \vec{r}) = f(0) \oint_{S_r} da. \tag{3.26}$$

Thus, for any harmonic function f,

$$\boxed{f(0) = \frac{1}{4\pi r^2} \oint_{S_r} da\, f(\vec{r})} \tag{3.27}$$

We have found so far that $f(0)$ is the surface mean value of f over any sphere. This is called the *mean value property* of harmonic functions. Due to the importance of this result, an alternative derivation is given, following Axler et al. (2001), p. 5. These authors start from Green's theorem (3.18); the integration domain shall be

$$V = B_r(0) \setminus B_\epsilon(0), \tag{3.28}$$

which is the ball with radius r, having a hole with radius ϵ at its center. Then $\partial V = S_\epsilon \cup S_r$. Note that on S_ϵ, the outward normal points toward the origin. Let $g = 1/r$, thus $\Delta g = 0$ in V. Furthermore, f shall be harmonic. Then (3.19) gives

$$\oint_{S_r} da\, \left[\frac{f}{r^2} + \frac{1}{r} \frac{\partial f}{\partial n} \right] = \oint_{S_\epsilon} da\, \left[\frac{f}{\epsilon^2} + \frac{1}{\epsilon} \frac{\partial f}{\partial n} \right], \tag{3.29}$$

or, bringing the second term on the left to the right side, and the first term on the right to the left side,

$$\frac{1}{r^2} \oint_{S_r} da\, f(\vec{r}) - \frac{4\pi \epsilon^2}{\epsilon^2} f(0) = -\frac{1}{r} \oint_{S_r} da\, \frac{\partial f}{\partial n} + \frac{1}{\epsilon} \oint_{S_\epsilon} da\, \frac{\partial f}{\partial n}. \tag{3.30}$$

The right side vanishes according to Green's first theorem (3.14). Thus, one arrives again at (3.27),

$$f(0) = \frac{1}{4\pi r^2} \oint_{S_r} da\, f(\vec{r}). \tag{3.31}$$

Since the origin is arbitrary, this equation holds for any point. For $\mathbb{R}^2$ instead of $\mathbb{R}^3$, the corresponding result is $f(0)$ which is the (line) average of f over any circle with center 0.

We come to the second property of harmonic functions.

Theorem (Maximum property) *Let $V \subset \mathbb{R}^3$ be an open and connected domain, and $f : V \to \mathbb{R}$. If f is harmonic in V and has an extremum in V, then f is constant. Thus, extrema of nonconstant f must lie on the boundary ∂V.*

Proof (Jänich 2001, p. 179) Let F be a maximum of f in V, i.e., $f(\vec{r}) \leq F$ for all $\vec{r} \in V$. For a minimum of f, consider instead $-f$ in the following. Let $U \subset V$ be the set of points $\vec{r}_0$ with $f(\vec{r}_0) = F$. The set $U = f^{-1}(F)$ is closed in V since $\{F\}$ is closed in $\mathbb{R}$ and f is continuous. This employs the so-called *relative* topology of the subspace V: a set U is open in $V \subset \mathbb{R}^3$ if $U = U' \cap V$, with a set U' that is open in $\mathbb{R}^3$. (Note that since $V = \mathbb{R}^3 \cap V$, where $\mathbb{R}^3$ is open in $\mathbb{R}^3$, V is open in V. Furthermore, the empty set $\varnothing$ is open in V, thus its complement in V, which is V, is closed in V. Thus, the set V is both open and closed in the space V, as must be.) Let now $\vec{r}_0 \in U$ be a point in this pre-image U of F. Choose an $\epsilon > 0$ such that the sphere $S_\epsilon(\vec{r}_0)$ is a subset of V; since V is open in $\mathbb{R}^3$, such an ϵ exists. From the mean value property (3.31), it holds for this $\vec{r}_0 \in U$ that

$$F = f(\vec{r}_0) = \frac{1}{4\pi\epsilon^2} \oint_{S_\epsilon(\vec{r}_0)} da \, f(\vec{r}) \leq F. \tag{3.32}$$

The latter inequality holds since $f(\vec{r}) \leq F$ in V. Equality must hold in (3.32), but it can only hold if $f(\vec{r}) = F$ on the *whole sphere* $S_\epsilon(\vec{r}_0)$. The same argument applies for all spheres with radius less than ϵ. Thus, $f(\vec{r}) = F$ in the whole open ball $B_\epsilon(\vec{r}_0)$. By definition of U, $B_\epsilon(\vec{r}_0)$ therefore belongs to U. Since such an ϵ exists for all $\vec{r}_0 \in U$, U is open in V. Above, however, U was also found to be closed in V. Since V is connected and $U \subset V$, it follows that $U = V$. □

This result allows us to give an alternative proof of the theorem that the Dirichlet boundary conditions are sufficient to obtain a unique solution (up to a constant) for the Laplace equation $\Delta f = 0$ in a domain V.

Theorem (The Dirichlet boundary conditions, again) *Let V be a compact, connected domain of $\mathbb{R}^3$. Let $f, g : V \to \mathbb{R}$ be continuous and f, g be harmonic in $V \backslash \partial V$. If $f = g$ on ∂V, then $f = g$ everywhere in V.*

Proof (Jänich 2001, p. 180) The function $f - g$ is continuous in a compact set V. Therefore, $f - g$ attains a minimum $\vec{r}_0$ and a maximum $\vec{r}_1$ in V,

$$f(\vec{r}_0) - g(\vec{r}_0) \leq f(\vec{r}) - g(\vec{r}) \leq f(\vec{r}_1) - g(\vec{r}_1), \tag{3.33}$$

for all $\vec{r} \in V$. Now $f - g$ is harmonic in $V \backslash \partial V$. According to the maximum property, two alternatives exist: (a) $\vec{r}_0 \in \mathring{V}$ or $\vec{r}_1 \in \mathring{V}$ (the open interior of V), thus $f - g = $ const in $\mathring{V}$. Since $f = g$ on ∂V and f, g are continuous, it follows that $f = g$ in V; (b) alternatively, both $\vec{r}_0, \vec{r}_1 \subset \partial V$. Then $f(\vec{r}_0) = g(\vec{r}_0)$ and $f(\vec{r}_1) = g(\vec{r}_1)$. From (3.33), $0 \leq f(\vec{r}) - g(\vec{r}) \leq 0$ for all $\vec{r} \in V$. Thus, again $f = g$ in V, as in case (a). □

3.5 Logarithmic Potential

In this section, the Green's function for the Laplace equation in the two-dimensional plane is derived. This has important applications in later chapters, especially in Sect. 8.12 on waves behind a rotating barrier. Let $f, g : \mathbb{R}^2 \to \mathbb{R}$ be real-valued functions in the real plane. For $\vec{x}$ a (bound) position vector in $\mathbb{R}^2$, let

$$g(\vec{x}) = \ln \frac{1}{|\vec{x}|}. \tag{3.34}$$

We show that $\Delta g = 0$ for $|\vec{x}| \neq 0$ (here Δ is the Laplace operator with respect to $\vec{x}$). To see this, introduce polar coordinates (we do not change the function name g),

$$g(r, \varphi) = \ln \frac{1}{r}, \tag{3.35}$$

with $|\vec{x}| = r$. Then, with $\Delta g = \nabla \cdot \nabla g$,

$$\nabla g = \hat{r} \, \partial g / \partial r = -\hat{r}/r,$$
$$-\nabla \cdot (\hat{r}/r) = -r^{-1} \partial / \partial r \, (r\hat{r}/r) = 0. \tag{3.36}$$

For $r > 0$, the function g can thus be considered as a velocity potential, giving a flow velocity field $\vec{u} = -\hat{r}/r$. This flow is directed radially inward, and there is a mass sink at $r = 0$. Let $A \subset \mathbb{R}^2$ be a simply connected area with boundary curve $\partial A = C_1$. Let $\vec{x}$ be the position vector to an arbitrary point in A, $\vec{y}$ the position vector to a boundary point on C_1, and let f be harmonic in A. A disk $B_\epsilon(\vec{x})$ with radius ϵ and centered at $\vec{x}$ is excluded from A. Let C_2 be the boundary of this disk. Then $g = \ln |\vec{x} - \vec{y}|^{-1}$ is harmonic in $A \backslash B_\epsilon(\vec{x})$. Green's theorem (3.19) was derived for harmonic functions in $\mathbb{R}^3$. It is easily adapted to harmonic functions in $\mathbb{R}^2$ by replacing the closed area integral $\oint da$ in $\mathbb{R}^3$ by a closed line integral $\oint dl$ in $\mathbb{R}^2$ (for this, see also Sect. 3.7), yielding

$$0 = \oint_{C_1 \cup C_2} dl \left(f(\vec{y}) \frac{\partial}{\partial n} \ln \frac{1}{|\vec{x} - \vec{y}|} - \frac{\partial f(\vec{y})}{\partial n} \ln \frac{1}{|\vec{x} - \vec{y}|} \right), \tag{3.37}$$

with infinitesimal arclength dl along C_1 and C_2. Using polar coordinates in the integral along C_2 gives

$$0 = \oint_{C_1} dl \left(f(\vec{y}) \frac{\partial}{\partial n} \ln \frac{1}{|\vec{x} - \vec{y}|} - \frac{\partial f(\vec{y})}{\partial n} \ln \frac{1}{|\vec{x} - \vec{y}|} \right) + \int_0^{2\pi} \epsilon \, d\varphi \left(\frac{f}{\epsilon} - \frac{\partial f}{\partial r} \bigg|_\epsilon \ln \epsilon \right). \tag{3.38}$$

For $\epsilon \to 0$, the function value $f(\vec{x})$ at the disk center can be taken outside this integral. We also assume that $\partial f / \partial r$ remains finite on the circle C_2. Then the second term in the integral along C_2 vanishes, since $\lim_{\epsilon \to 0} \epsilon \ln \epsilon = 0$, and thus

$$f(\vec{x}) = -\frac{1}{2\pi} \oint_{C_1} dl \left(f(\vec{y}) \frac{\partial}{\partial n} \ln \frac{1}{|\vec{x} - \vec{y}|} - \frac{\partial f(\vec{y})}{\partial n} \ln \frac{1}{|\vec{x} - \vec{y}|} \right). \tag{3.39}$$

This gives $f(\vec{x})$ for any $\vec{x} \in A$ from $f(\vec{y})$ and $\partial f(\vec{y})/\partial n$ on the boundary line ∂A. We know already that the Dirichlet boundary conditions for $f(\vec{y})$ on ∂A are sufficient to calculate $f(\vec{x})$ in A.

The logarithm appearing in these equations has a highly significant meaning, and is the Green's function of the Laplace equation in the two-dimensional plane. This Green's function G is defined by

$$\Delta G(\vec{x}) = -\delta(\vec{x}). \tag{3.40}$$

For the Laplace equation in $\mathbb{R}^3$, it was found in Eq. (2.197) that

$$G(\vec{x}) = \frac{1}{4\pi |\vec{x}|}. \tag{3.41}$$

Our present claim is that in the plane,

$$G(\vec{x}) = \frac{1}{2\pi} \ln \frac{1}{|\vec{x}|}. \tag{3.42}$$

Indeed, $\Delta G = 0$ for $\vec{x} \neq 0$ in (3.42) was already shown above. We demonstrate now that (3.40) also holds with G from (3.42) when the point $\vec{x} = 0$ is included, which is a singularity of (3.42). To this end, one integrates over an infinitesimal disk $B_\epsilon(0)$. Gauss' theorem in the plane gives then

$$\int_{B_\epsilon(0)} da \, \Delta G(\vec{x}) = \oint_{\partial B_\epsilon(0)} dl \, \hat{n} \cdot \nabla G(\vec{x})$$

$$= -\frac{1}{2\pi} \oint dl \, \hat{x} \cdot \nabla \ln |\vec{x}|$$

$$= -\frac{1}{2\pi} \oint dl \, \frac{\hat{x} \cdot \hat{x}}{|\vec{x}|}$$

$$= -\frac{1}{2\pi \epsilon} \oint_{\partial B_\epsilon(0)} dl$$

$$= -1 = -\int_{B_\epsilon(0)} da \, \delta(\vec{x}). \tag{3.43}$$

This implies indeed (3.40), since ϵ is arbitrary. Green's theorem (3.39) can now be written (where the derivative $\partial/\partial n$ is in the direction normal to the tangent vector $d\vec{l}$ to C, and points away from the area enclosed by C),

$$\boxed{f(\vec{x}) = -\oint_C dl \left(f(\vec{y}) \frac{\partial G(\vec{x} - \vec{y})}{\partial n} - \frac{\partial f(\vec{y})}{\partial n} G(\vec{x} - \vec{y}) \right)} \tag{3.44}$$

where

$$\Delta_{\vec{x}} \, G(\vec{x} - \vec{y}) = -\delta(\vec{x} - \vec{y}). \tag{3.45}$$

(Here $\Delta_{\vec{x}}$ is the Laplace operator with respect to $\vec{x}$.) Equation (3.44) is also obtained directly by inserting (3.45) into Green's integral theorem (3.18) for a harmonic function f.

3.6 Dirichlet's Principle

The Laplace equation, $\Delta\phi = 0$, describes a potential problem, and potentials in physics obey a minimum potential energy principle. One may thus ask whether the Laplace equation itself can be derived from an extremal principle. Indeed, Dirichlet found in the nineteenth century that the Laplace equation can be obtained from a variational problem, later termed *Dirichlet's principle* (see especially Garding 1977). We consider here first the 2-D and then the 3-D Laplace equation. Let $\phi(x, y)$ be a twice continuously differentiable scalar field. Then Dirichlet's integral is defined as

$$D[\phi] = \iint_G dx\ dy\ (\phi_x^2 + \phi_y^2). \tag{3.46}$$

The integration domain G is specified further below. Dirichlet postulates that $\delta D = 0$ under variations $\delta\phi(x, y)$ of ϕ. The variational principle leads to an Euler–Lagrange differential equation, as follows:

$$\begin{aligned}
0 &= \delta \iint dx\ dy\ (\phi_x^2 + \phi_y^2) \\
&= 2 \iint dx\ dy\ (\phi_x\ \delta(\phi_x) + \phi_y\ \delta(\phi_y)) \\
&= 2 \iint dx\ dy\ (\phi_x\ (\delta\phi)_x + \phi_y\ (\delta\phi)_y) \\
&= -2 \iint dx\ dy\ (\phi_{xx} + \phi_{yy})\ \delta\phi. \tag{3.47}
\end{aligned}$$

In the last line, integration by parts was used. All variations $\delta\phi$ are assumed to vanish on the boundary ∂G of G. The fundamental lemma of the calculus of variations gives then

$$\phi_{xx} + \phi_{yy} = \Delta\phi = 0. \tag{3.48}$$

The classical form of Dirichlet's principle is thus that $\delta D[\phi] = 0$ implies $\Delta\phi = 0$, i.e., that $\delta D[\phi] = 0$ *in G is sufficient for* $\Delta\phi = 0$ *in G*. This, however, does not allow in full generality to replace a boundary value problem by a variational problem, which would also require that $\Delta\phi = 0$ implies $\delta D = 0$, i.e., that $\delta D = 0$ is necessary for $\Delta\phi = 0$. The latter statement is not true in general, and a counterexample of a solvable boundary value problem for the Laplace equation for which the Dirichlet variational problem is not applicable was given by Hadamard (see Courant 1950, p. 10): consider the series

$$g(\varphi) = \sum_{n=1}^{\infty}(a_n \cos(n\varphi) + b_n \sin(n\varphi)), \tag{3.49}$$

and, for $0 \leq r < 1$, the series

$$\phi(r, \varphi) = \sum_{n=1}^{\infty} r^n (a_n \cos(n\varphi) + b_n \sin(n\varphi)). \tag{3.50}$$

For any given coefficients a_n and b_n, the latter series for ϕ converges. The series for g in (3.49) may be divergent, however, since the factor r^n, which for $r < 1$ and $n \to \infty$ converges to 0, is missing. It is easily shown that

$$\Delta\phi = 0. \tag{3.51}$$

As boundary condition on ϕ choose

$$\phi(1, \varphi) = g(\varphi). \tag{3.52}$$

For a circle with radius $r < 1$, Eq. (3.50) gives after a short calculation,

$$D_r[\phi] = \pi \sum_{n=1}^{\infty} n r^{2n} (a_n^2 + b_n^2) \tag{3.53}$$

(the subscript r indicates the parameter r on which D depends), which is again convergent for all a_n and b_n. For Dirichlet's integral $D[\phi]$, with the open interior of the unit disk as domain G,

$$D[\phi] = \lim_{r \to 1} D_r[\phi] = \pi \sum_{n=1}^{\infty} n(a_n^2 + b_n^2), \tag{3.54}$$

which may again be divergent. In order that $D[\phi]$ exists (as a bound functional) and thus Dirichlet's principle is applicable,

$$\sum_{n=1}^{\infty} n(a_n^2 + b_n^2) < \infty \tag{3.55}$$

must hold. Due to the factor of n under the sum, this is more restrictive than what is required for (3.49) to converge, i.e., for the function g of boundary values to be bound. Thus, there are continuous functions g for which (3.55) is divergent. For a specific example of such a function g, see Courant (1950), p. 10. In the case considered, the boundary value problem is solvable, but $\delta G = 0$ is not applicable.

This example suggests that a more restrictive use of Dirichlet's principle is required, if boundary value problems are to be replaced by variational problems.

Let the domain G be the interior of the unit circle. Let $g(\varphi)$ be a function with the following properties:

(a) g is continuous on the closure of G,
(b) g is piecewise continuously differentiable in G, and
(c) $D[g]$ is finite.

Then Dirichlet's principle holds in the following form: Functions $\phi(r, \varphi)$ shall share the properties (a) and (b) with $g(\varphi)$, and $\phi(1, \varphi) = g(\varphi)$. Then the function ϕ for which $\delta D[\phi] = 0$ is exactly that for which $\Delta\phi = 0$. The elementary proof can be found in Courant (1950), pp. 11–13. The property (c) ensures that (3.55) holds for the coefficients a_n and b_n of g.

A special case of Dirichlet's principle that applies to fluid dynamics is *Kelvin's principle*.

Theorem (Kelvin's principle) *Let V be an open, contractible domain of $\mathbb{R}^3$. Let $\vec{u}$ be a divergence-free velocity field in V, $\nabla \cdot \vec{u} = 0$, of a fluid with constant density. The field $\vec{u}$ shall allow a decomposition $\vec{u} = \vec{w} + \vec{v}$, where $\nabla \times \vec{w} = 0$ and $\nabla \cdot \vec{v} = 0$ (The Helmholtz decomposition for vector fields that are twice continuously differentiable). The Neumann boundary conditions are applied on ∂V of V,*

$$u_n = \hat{n} \cdot \vec{u} = h, \tag{3.56}$$

where $\hat{n}$ is the normal vector to the boundary ∂V and h is an arbitrary continuous function on ∂V. Then an irrotational flow $\vec{u} = \nabla\phi$ minimizes the kinetic energy of the flow in the volume V.

Proof Since V is contractible, $\nabla \times \vec{w} = 0$ implies $\vec{w} = \nabla\phi$ by the converse Poincaré lemma from Sect. 2.21, for some scalar field ϕ. We want to show that an irrotational velocity field $\nabla\phi$ minimizes the kinetic energy of the flow. In order to do so, demand as boundary conditions for ϕ and $\vec{v}$ on ∂V,

$$\hat{n} \cdot \nabla\phi = h,$$
$$\hat{n} \cdot \vec{v} = 0. \tag{3.57}$$

In calculating the flow kinetic energy for arbitrary (test) fields ϕ and $\vec{v}$ in the decomposition $\vec{u} = \nabla\phi + \vec{v}$, validity of Eq. (3.57) can always be ensured as follows: perform a transformation $\phi \to \phi' = \phi + \psi$ and $\vec{v} \to \vec{v}' = \vec{v} - \nabla\psi$, with some scalar field ψ that obeys the Laplace equation, $\Delta\psi = 0$. This transformation (i) maintains the Helmholtz decomposition of $\vec{u}$ in V (i.e., $\nabla \times \nabla\phi' = 0$ and $\nabla \cdot \vec{v}' = 0$) and (ii) it allows to realize the boundary conditions (3.57) for the transformed fields,

$$\hat{n} \cdot (\nabla\phi + \nabla\psi) = h,$$
$$\hat{n} \cdot (\vec{v} - \nabla\psi) = 0. \tag{3.58}$$

Indeed, adding these equations gives $\hat{n} \cdot (\nabla\phi + \vec{v}) = \hat{n} \cdot \vec{u} = h$, according to the assumption (3.56) of the theorem. Subtracting the equations gives

$$\hat{n} \cdot \nabla \psi = \frac{1}{2} \left[h - \hat{n} \cdot (\nabla \phi - \vec{v}) \right].$$ (3.59)

The right side is a new, continuous function, h', and (3.59) is a Neumann boundary condition for the solution ψ of the Laplace equation $\Delta \psi = 0$ in V, which has a unique solution. The total kinetic energy of a flow velocity field $\vec{u}$ is given by the functional

$$T[\vec{u}] = \frac{1}{2} \int_V dV \, \vec{u} \cdot \vec{u}.$$ (3.60)

Inserting $\vec{u} = \nabla \phi + \vec{v}$ gives

$$
\begin{aligned}
T[\vec{u}] &- T[\nabla \phi] - T[\vec{v}] \\
&= \int_V dV \, \vec{v} \cdot \nabla \phi \\
&= \int_V dV \left[\nabla \cdot (\phi \vec{v}) - \phi \, \nabla \cdot \vec{v} \right] \\
&= \int_V dV \, \nabla \cdot (\phi \vec{v}) \\
&= \oint_{\partial V} d\vec{a} \cdot (\phi \vec{v}) \\
&= \oint_{\partial V} da \, \phi \, \hat{n} \cdot \vec{v} = 0,
\end{aligned}
$$ (3.61)

using (3.57). Therefore,

$$T[\vec{u}] = T[\nabla \phi] + T[\vec{v}].$$ (3.62)

Since the functional T is positive definite, $T \geq 0$, it follows that

$$T[\vec{u}] \geq T[\nabla \phi].$$ (3.63)

Hence, the irrotational velocity field $\nabla \phi$ minimizes kinetic energy. □

For this theorem, see also Lamb (1932), Sect. 45. So far, a solenoidal vector field was assumed, $\nabla \cdot \vec{u} = 0$, and it was found that T is minimized if $\vec{u}$ is irrotational, $\nabla \times \vec{u} = 0$. The complementary implication also holds.

Theorem *Let density be constant and $\vec{u}$ be an irrotational vector field, $\nabla \times \vec{u} = 0$. Let $V \subset \mathbb{R}^3$ be an open, contractible domain, thus $\vec{u} = \nabla \phi$ in V. Let*

$$\partial \phi / \partial n = u_n = h \qquad \text{on } \partial V,$$ (3.64)

with arbitrary continuous function h. Then a divergence-free flow maximizes the functional

$$S[\vec{u}] = -T[\vec{u}] + \oint_{\partial V} da \, \phi h.$$ (3.65)

Proof Consider another irrotational flow, $\vec{v} = \nabla\psi$, which also obeys

$$\partial\psi/\partial n = v_n = h \qquad \text{on } \partial V, \tag{3.66}$$

and, furthermore, is solenoidal,

$$\nabla \cdot \vec{v} = \Delta\psi = 0. \tag{3.67}$$

Then

$$S[\vec{v}] = -T[\vec{v}] + \oint_{\partial V} da \,\, \psi h, \tag{3.68}$$

and

$$
\begin{aligned}
S[\vec{v}] - S[\vec{u}] &= -T[\vec{v}] + T[\vec{u}] + \oint_{\partial V} da \,\, (\psi - \phi) \, h \\
&= \quad \cdots \quad + \oint_{\partial V} da \,\, (\psi - \phi) \, \frac{\partial\psi}{\partial n} \\
&= \quad \cdots \quad + \oint_{\partial V} d\vec{a} \cdot (\psi - \phi) \nabla\psi \\
&= \quad \cdots \quad + \int_V dV \,\, \nabla \cdot \left[(\psi - \phi) \nabla\psi \right] \\
&= \quad \cdots \quad + \int_V dV \,\, \left[\nabla\psi \cdot \nabla\psi - \nabla\phi \cdot \nabla\psi + (\psi - \phi) \, \Delta\psi \right] \\
&= -T[\vec{v}] + T[\vec{u}] + 2T[\vec{v}] - \int_V dV \,\, \vec{u} \cdot \vec{v} \\
&= T[\vec{u} - \vec{v}] \geq 0.
\end{aligned}
\tag{3.69}
$$

Therefore, the divergence-free field $\vec{v}$ indeed maximizes S. For this theorem, see Serrin (1959), p. 161. The existence of complementary variational principles was already proved by Euler. For the history of Dirichlet's principle, see Kellogg (1929), p. 280; Burkhardt and Meyer (1900), p. 494; and Monna (1975). A further interesting application of this principle is in the water wave problem: Eqs. (2.265)–(2.269) can be put in the form of Dirichlet's principle. The latter gets then a 'dynamical extension,' see Miles (1977), p. 154.

3.7 Streamfunction

We return to flows in the xy-plane, and look into the physical meaning of the stream-function ψ. From (3.12), it follows that $u = \psi_y$ and $v = -\psi_x$, thus

$$\nabla\psi \cdot \vec{u} = (\hat{x}\,\psi_x + \hat{y}\,\psi_y) \cdot (\hat{x}\,\psi_y - \hat{y}\,\psi_x) = \psi_x\psi_y - \psi_y\psi_x, \tag{3.70}$$

and therefore

$$\nabla\psi \cdot \vec{u} = 0. \tag{3.71}$$

In the xy-plane, the velocity vector field $\vec{u}$ is thus tangent to the isocontours $\psi =$ const of the streamfunction. The isocontours of the streamfunction ψ agree with the streamlines of the flow (this is the reason for the name streamfunction), and since a stationary flow is assumed in this chapter, the isocontours of ψ also agree with the pathlines of the flow. Thus, for an inviscid, solenoidal, irrotational, stationary flow in a plane, *pathlines are isocontours of the streamfunction ψ*.

The following is taken from Lamb (1932, p. 62) and Zermelo (1902). The flow shall be stationary and two dimensional in the plane, the velocity field $\vec{u}$ shall be divergence-free, and the surface mass density σ shall be constant. Let C be a fixed, closed curve with length differential $\mathrm{d}s$ in the plane of the flow. Its curve normal $\hat{n}$ shall point away from the area enclosed by C, and shall lie in the flow plane. The infinitesimal area $\mathrm{d}A$ covered in time $\mathrm{d}t$ by fluid parcels inflowing or outflowing through C is given by

$$\mathrm{d}A = \oint_C \mathrm{d}s \, \hat{n} \cdot \vec{u} \, \mathrm{d}t. \tag{3.72}$$

Then $\sigma \, \mathrm{d}A$ is the mass leaving or entering the area enclosed by C in $\mathrm{d}t$. Since C is fixed and the flow is stationary, $\sigma \, \mathrm{d}A = 0$. This can also be derived in a more formal manner. To this end, the divergence operator in two-dimensional space is introduced,

$$\mathrm{d}a \, \nabla \cdot \vec{u} = \oint_{\partial \mathrm{d}a} \mathrm{d}s \, \hat{n} \cdot \vec{u}. \tag{3.73}$$

Gauss' theorem in the plane is then

$$\int_A \mathrm{d}a \, \nabla \cdot \vec{u} = \oint_{\partial A} \mathrm{d}s \, \hat{n} \cdot \vec{u}. \tag{3.74}$$

The proof proceeds by covering A with a net of infinitesimal meshes, cf. Sect. 2.8. Then with $C = \partial A$, (3.72) implies that

$$\mathrm{d}A = \int_A \mathrm{d}a \, \nabla \cdot \vec{u} \, \mathrm{d}t = 0, \tag{3.75}$$

since the vector field $\vec{u}$ is solenoidal, $\nabla \cdot \vec{u} = 0$ in the plane (there is no dependence on z), thus indeed $\mathrm{d}A = 0$. Let now $\vec{x}_0$ and $\vec{x}$ be (vectors to) any two points, where $\vec{x}_0$ is fixed and $\vec{x}$ is variable. Furthermore, let C_1 and C_2 be two arbitrary curves from $\vec{x}_0$ to $\vec{x}$. With $\mathrm{d}A = 0$, Eq. (3.72) implies

$$\int_{C_1} \mathrm{d}s \, \hat{n} \cdot \vec{u} = \int_{C_2} \mathrm{d}s \, \hat{n} \cdot \vec{u}. \tag{3.76}$$

Thus, this integral does not depend on the curve chosen, and a single-valued function ψ can be defined, for arbitrary curves C from $\vec{x}_0$ to $\vec{x}$, by

$$\psi(x, y) = \int_C \mathrm{d}s \, \hat{n} \cdot \vec{u}. \tag{3.77}$$

Rankine (1864) showed that the ψ so defined is indeed the streamfunction. To see this, let $d\vec{s} = (dx, dy)$ be the line element along C (in the counterclockwise, positive sense). Then $ds\,\hat{n} = (dy, -dx)$ is the outward-pointing curve normal, and

$$\psi(x, y) = \int_C (u\,dy - v\,dx). \tag{3.78}$$

The partial differentials of the function ψ are then

$$\frac{\partial\psi}{\partial x} = -v, \qquad\qquad \frac{\partial\psi}{\partial y} = u. \tag{3.79}$$

This is Eq. (3.12), and thus ψ from (3.77) and (3.10) are identical. Finally, consider a curve C that is an isocontour, $\psi = $ const, of ψ. Along C, $\hat{n} \cdot \vec{u} = 0$ from (3.77), and hence the normal $\hat{n}$ of C is normal to $\vec{u}$. In the plane, this means $\vec{u}$ is tangent to C, i.e., to the isocontour $\psi = $ const, which is therefore a streamline.

3.8   Vorticity on a Sphere

In this short section, a famous theorem by Zermelo (1902) is proved, which states that any divergence-free 2-D flow on a sphere must have vorticity, or, in a well-known trope, that one cannot 'comb' a vector field on a sphere without creating a whorl.

Theorem (Zermelo) *Let $\vec{u}$ be a two-dimensional, solenoidal, and smooth (such that Gauss' theorem applies) velocity field of an inviscid fluid on a sphere. Then somewhere $\nabla \times \vec{u} \neq 0$.*

Proof Let r be the radius of the sphere, and ϑ and φ be co-latitude and azimuth. The length differentials on the sphere are

$$r\,d\vartheta \qquad \text{and} \qquad r\sin\vartheta\,d\varphi = s\,d\varphi. \tag{3.80}$$

The area element on the sphere is $da = r\,d\vartheta\,s\,d\varphi$. The velocity vector is $\vec{u} = (u, v)$. Gauss' theorem on the sphere gives, cf. (3.74),

$$\oint_{\partial A}(us\,d\varphi - vr\,d\vartheta) = \iint_A r\,d\vartheta\,s\,d\varphi\,\frac{1}{rs}\left[\frac{\partial(su)}{\partial\vartheta} + \frac{\partial(rv)}{\partial\varphi}\right] = 0, \tag{3.81}$$

since $\nabla \cdot \vec{u} = 0$ for divergence-free flows. A streamfunction ψ as in (3.78) can thus be defined,

$$\psi(\vartheta, \varphi) = \int_C (us\,d\varphi - vr\,d\vartheta), \tag{3.82}$$

which is Eq. (1a) in Zermelo (1902). Its partial derivatives are

$$\frac{1}{r}\frac{\partial\psi}{\partial\vartheta}=-v, \qquad\qquad \frac{1}{s}\frac{\partial\psi}{\partial\varphi}=u. \tag{3.83}$$

Furthermore, for the vorticity,

$$\nabla\times\vec{u}=\frac{1}{rs}\left[\frac{\partial(sv)}{\partial\vartheta}-\frac{\partial(ru)}{\partial\varphi}\right]\hat{r}, \tag{3.84}$$

where $\hat{r}$ points away from the sphere. Inserting (3.83) gives

$$\nabla\times\vec{u}=-\frac{1}{rs}\left[\frac{\partial}{\partial\vartheta}\left(\frac{s}{r}\frac{\partial\psi}{\partial\vartheta}\right)+\frac{r}{s}\frac{\partial^2\psi}{\partial\varphi^2}\right]\hat{r}. \tag{3.85}$$

The following calculation corresponds to Eq. (5) on p. 210 in Zermelo (1902). First, inserting (3.85),

$$\oint_S d\vec{a}\cdot(\psi\,\nabla\times\vec{u})=-\int_0^\pi d\vartheta\,\frac{r}{s}\int_0^{2\pi}d\varphi\,\psi\,\frac{\partial^2\psi}{\partial\varphi^2}-\int_0^{2\pi}d\varphi\int_0^\pi d\vartheta\,\psi\,\frac{\partial}{\partial\vartheta}\left(\frac{s}{r}\frac{\partial\psi}{\partial\vartheta}\right). \tag{3.86}$$

Integration by parts for both integrals gives

$$\oint_S d\vec{a}\cdot(\psi\,\nabla\times\vec{u})=\int_0^\pi d\vartheta\,\frac{r}{s}\int_0^{2\pi}d\varphi\,\left(\frac{\partial\psi}{\partial\varphi}\right)^2$$
$$+\int_0^{2\pi}d\varphi\int_0^\pi d\vartheta\,\frac{s}{r}\left(\frac{\partial\psi}{\partial\vartheta}\right)^2-\int_0^{2\pi}d\varphi\left[\sin\vartheta\,\psi\,\frac{\partial\psi}{\partial\vartheta}\right]_{\vartheta=0}^\pi. \tag{3.87}$$

There are no boundary terms with respect to the φ-integration, since points with $\varphi=0$ and $\varphi=2\pi$ are identical. For the ϑ-integration, however, two boundary terms at $\vartheta=0$ and $\vartheta=\pi$ are included in the last term of (3.87). If ψ and $\partial\psi/\partial\vartheta$ have no singularities at the North and South Poles $\vartheta=0$ and $\vartheta=\pi$, respectively, these boundary terms give no contribution. Finally, from (3.83),

$$\oint_S d\vec{a}\cdot(\psi\,\nabla\times\vec{u})=\iint d\vartheta\,d\varphi\,\,rs\,(u^2+v^2)=\oint_S da\,(u^2+v^2). \tag{3.88}$$

The latter integral $\oint da\,(u^2+v^2)$ is larger than zero for nonvanishing $\vec{u}$. Thus, the vorticity $\nabla\times\vec{u}$ on the left side of (3.88) cannot vanish everywhere on the sphere.

$\square$

3.9   Complex Speed and Potential

Let $(x, y) \to z = x + iy$ be the usual isomorphism between the vector space $\mathbb{R}^2$ and the complex field $\mathbb{C}$. The velocity vector (u, v) in the two-dimensional plane $\mathbb{R}^2$ is mapped to an *analytic* function w on $\mathbb{C}$,

$$w(z) = u - iv. \tag{3.89}$$

The reason for the minus sign becomes clear in (3.97). Let ϕ and ψ be the velocity potential and streamfunction, respectively, and introduce a complex potential Φ by

$$\Phi = \phi + i\psi. \tag{3.90}$$

Especially, Φ shall be *analytic*, i.e., $\mathrm{d}\Phi/\mathrm{d}z$ is unique at each point z. Writing $\Phi = \Phi(z(x, y))$ and using the chain rule gives

$$\frac{\partial}{\partial y}(\phi + i\psi) = \frac{\partial \Phi}{\partial y} = \frac{\mathrm{d}\Phi}{\mathrm{d}z}\frac{\partial z}{\partial y} = i\frac{\mathrm{d}\Phi}{\mathrm{d}z}\frac{\partial z}{\partial x} = i\frac{\partial \Phi}{\partial x} = i\frac{\partial}{\partial x}(\phi + i\psi), \tag{3.91}$$

where x, y, ϕ, and ψ are real. Equating real and imaginary parts in the first and last terms gives

$$\begin{aligned}\frac{\partial \phi}{\partial x} &= \frac{\partial \psi}{\partial y}, \\[6pt] \frac{\partial \phi}{\partial y} &= -\frac{\partial \psi}{\partial x}.\end{aligned} \tag{3.92}$$

These are the Cauchy–Riemann equations. They appeared first in 1752 in a paper by d'Alembert on hydrodynamics. In 1776, they were used by Euler, see Grattan-Guinness (1994), p. 419. The Cauchy–Riemann equations are *necessary* for Φ to be analytic. They are also *sufficient*, if all the derivatives $\phi_x, \phi_y, \psi_x, \psi_y$ are continuous. For the simple proof, see Copson (1935, p. 41). An alternative derivation of the Cauchy–Riemann equations is as follows. With $\Phi' = \mathrm{d}\Phi/\mathrm{d}z = a + ib$,

$$\mathrm{d}\Phi = \Phi'\,\mathrm{d}z = (a + ib)(\mathrm{d}x + i\mathrm{d}y) = a\,\mathrm{d}x - b\,\mathrm{d}y + i(b\,\mathrm{d}x + a\,\mathrm{d}y), \tag{3.93}$$

while also

$$\mathrm{d}\Phi = \mathrm{d}(\phi + i\psi) = \phi_x\mathrm{d}x + \phi_y\mathrm{d}y + i(\psi_x\mathrm{d}x + \psi_y\mathrm{d}y). \tag{3.94}$$

Comparing coefficients in these two equations gives

$$\begin{aligned}a &= \phi_x = \psi_y = u, \\ b &= -\phi_y = \psi_x = -v,\end{aligned} \tag{3.95}$$

which is again (3.92). Furthermore, there is now $\Phi' = a + ib = u - iv$, or

$$\boxed{w = \frac{d\Phi}{dz}} \tag{3.96}$$

This can be derived directly, without introducing a and b:

$$
\begin{aligned}
\frac{d\Phi}{dz} &= \frac{d\phi + id\psi}{dx + idy} \\
&= \frac{\phi_x\,dx + \phi_y\,dy + i\psi_x\,dx + i\psi_y\,dy}{dx + idy} \\
&\overset{(2.92)}{=} \frac{\phi_x\,dx - i\phi_y\,idy - i\phi_y\,dx + i\phi_x\,dy}{dx + idy} \\
&= \frac{(\phi_x - i\phi_y)dx + i(\phi_x - i\phi_y)dy}{dx + idy} \\
&= \phi_x - i\phi_y = u - iv = w.
\end{aligned}
\tag{3.97}
$$

The Cauchy–Riemann equations give the Laplace equations,

$$
\begin{aligned}
\Delta\phi &= \phi_{xx} + \phi_{yy} = \psi_{yx} - \psi_{xy} = 0, \\
\Delta\psi &= \psi_{xx} + \psi_{yy} = -\phi_{yx} + \phi_{xy} = 0,
\end{aligned}
\tag{3.98}
$$

which are (3.4) and (3.10). Furthermore,

$$\phi_x\psi_x + \phi_y\psi_y = -\phi_x\phi_y + \phi_y\phi_x = 0, \tag{3.99}$$

or

$$\nabla\phi \cdot \nabla\psi = 0. \tag{3.100}$$

Therefore,

$$\phi + i\psi \text{ is analytic} \qquad \leftrightarrow \qquad \nabla\phi \perp \nabla\psi. \tag{3.101}$$

The direction '$\leftarrow$' holds, if all partial derivatives $\phi_x, \phi_y, \psi_x, \psi_y$ are continuous. Thus, the contours of ϕ and ψ are normal to each other. Let us assume now that $w = u - iv$ is analytic. The Cauchy–Riemann equations $u_x = -v_y$ and $u_y = v_x$ give

$$u_x + v_y = \nabla \cdot \vec{u} = 0, \tag{3.102}$$

and

$$(v_x - u_y)\,\hat{z} = \nabla \times \vec{u} = 0. \tag{3.103}$$

To summarize, if $w = u - iv$ is an arbitrary analytic function, then (u, v) is the velocity field of a stationary, solenoidal, and irrotational flow in a plane. In polar coordinates (φ grows in the counterclockwise direction),

$$w = qe^{-i\varphi}. \tag{3.104}$$

Then $\vec{u} = (u, v)$ is given by, with $w = u - iv$,

$$u = q \cos \varphi, \qquad\qquad v = q \sin \varphi. \tag{3.105}$$

We close with a remark. The complex conjugate of $z = x + iy$ is

$$\bar{z} = x - iy. \tag{3.106}$$

This is not an independent variable, since $(x, y) \to z$ is already a one-to-one mapping. But $\bar{z}(z)$ can be understood as a complex function. However, this function is not analytic, since the Cauchy–Riemann equations do not hold, $\partial x / \partial x = 1 \neq -1 = -\partial y / \partial y$.

3.10   Analytic Functions. Conformal Transformation

A complex function $f : D \to \mathbb{C}$ on an open domain D (see below) of the complex plane is *holomorphic* if it is *complex differentiable* on D. Holomorphic functions are found to be identical to complex *analytic* functions, i.e., for each $x \in D$, a neighborhood exists (actually, the largest disk centered at x lying still fully in D) in which f can be written as convergent infinite power series, i.e., equals its own Taylor series, and is thus infinitely often complex differentiable.

A *domain* D is a connected and open subset of $\mathbb{C}$ (we use the standard definitions of *open set* and *neighborhood*). The domain D is *connected* if any two points of D can be joined by a curve in D. Note that D can still have holes. If this is not the case, then D is *simply connected*. More precisely, in a simply connected domain, every closed, simple curve (without self-intersection) is homotopic to a point, i.e., can be contracted to a point within the domain, see Sect. 2.21. For the complex plane, a bounded domain is simply connected if and only if its boundary is connected, i.e., consists either of a single continuum or a single point, see Golusin (1957), p. 3 and Basye (1935), p. 352, corollary 1. Let $\bar{\mathbb{C}} = \mathbb{C} \cup \{\infty\}$ be the *extended* complex plane with the point at infinity included. The extended complex plane has no boundary (point) and is simply connected. Then a domain D is *simply connected* if and only if $\bar{\mathbb{C}} \backslash D$ is connected, see Ahlfors (1966), p. 139 or the above corollary in Basye (1935). The exterior of a disk is not simply connected in $\mathbb{C}$, since its $\bar{\mathbb{C}}$-complement is the disk plus the point at infinity (which are not connected).

The identification of the whole 'circle' at infinity in the Euclidean plane with a single point ∞ at infinity in the extended complex plane makes the mapping

$w = 1/z$ one-to-one everywhere, including $z = 0$, and allows to replace $\lim_{z \to \infty} f(z)$ by $\lim_{z \to 0} f(\frac{1}{z})$ and vice versa. For any $R > 0$ (actually $R \gg 1$), the points z with $|z| > R$ form a *neighborhood* of ∞. The point ∞ is then a limit point of any sequence z_i with $i \in \mathbb{N}$ that grows without bound. Remarkably, the extended complex plane with a circular hole D around, say, $z = 0$ is a simply connected domain; any closed curve C around D can be contracted to a single point without crossing D, by expanding all of C toward the single point at infinity (see Lu 2002, p. 16).

A complex function $f : D \to \mathbb{C}$ is *analytic* in a domain D if f is complex differentiable at each point of D. A function $f : D \to \mathbb{C}$ is analytic at a point c if there is an open neighborhood U of c in which $f|U$ is analytic. For this definition, see, e.g., Remmert (1992), p. 45. The open neighborhood is required to form the difference quotient.

Integrals in the complex plane are performed along continuous curves $C : [a, b] \subset \mathbb{R} \to \mathbb{C}$ and $t \to z(t)$. All curves are assumed to be *piecewise differentiable*, i.e., there is only a finite number of points along C at which dz/dt does not exist (or is not unique). For a *closed* curve, $z(a) = z(b)$. A *cycle* or *circuit* is a continuous, piecewise differentiable, closed curve. (When referring to closed curves, we will often mean cycles, avoiding the latter, somewhat technical term.) Theorems for integrals taken along cycles also hold for integrals taken along *rectifiable* curves, which are curves that have a well-defined length.

The Appendix lists some standard theorems from complex function theory for analytic and so-called meromorphic functions without proofs.

A unique and important property of analytic functions is that they are *conformal* maps, i.e., preserve the angle between two crossing curves. However, the curves themselves are usually rotated under an analytic map, and distances along the curves are stretched or contracted. For the precise formulation, let $\zeta = f(z)$ be an analytic map, and C_1 and C_2 two curves that cross at z_0 with an inner angle α. The curves $f(C_1)$ and $f(C_2)$ cross at $f(z_0)$ under an angle α'. If now $f'(z_0) \neq 0$, then $\alpha = \alpha'$: analytic maps are *angle-preserving*.

To see this, write in an infinitesimal neighborhood of z_0,

$$d\zeta = f'(z_0)\, dz. \tag{3.107}$$

Let $f'(z_0) = c + is$ with $c, s \in \mathbb{R}$, and $d\zeta = d\xi + id\eta = f'\, dz = (c + is)(dx + idy)$. Identifying $\mathbb{C}$ with $\mathbb{R}^2$, this can be written as

$$\begin{pmatrix} d\xi \\ d\eta \end{pmatrix} = \begin{pmatrix} c & -s \\ s & c \end{pmatrix} \cdot \begin{pmatrix} dx \\ dy \end{pmatrix}. \tag{3.108}$$

Let $c = a \cos \alpha$ and $s = a \sin \alpha$, with $a = \sqrt{c^2 + s^2}$ and $\tan \alpha = s/c$. Then (3.108) describes a rotation of (dx, dy) by an angle α and stretching (resp. contraction) by a factor a. Thus, the multiplication of an infinitesimally short segment dz (along a curve) with the complex number $f'(z_0)$ is equivalent to a rotation (by an angle α) and stretching (by a factor a) of the segment dz. Since α and a are the same for all dz along all curves in an infinitesimal neighborhood of z_0, angles between crossing curves (i.e., angles between their tangents at the crossing point or between straight lines through infinitesimally close points on the curves) are preserved under the map f. Note that for $f' = 0$, also $a = 0$, and α is then undetermined. For a full proof of this theorem on conformal mapping, see Knopp (1996) and Cartan (1963).

The graph $y(x)$ in the Cartesian xy-plane helps to visualize a real-valued function $f : x \mapsto y$. Similarly, to gain a picture of a complex function $g : z \mapsto w$, one examines how g maps certain curves (e.g., boundaries of domains) from the complex z-plane to the complex w-plane. This technique will be utilized in Chap. 5 where conformal mappings between four different complex planes are considered.

3.11 The Schwarz–Christoffel Theorem

In Chap. 5, jets, wakes, and cavities in fluids will be treated. The theorem from complex function theory that allows to find analytic solutions in these three cases is the Schwarz–Christoffel theorem, which is treated here. Let A be a simply connected, non-empty domain of $\mathbb{C}$ which is not all of $\mathbb{C}$. Let B be the open unit disk in $\mathbb{C}$ with the origin as center. Then the *Riemann mapping theorem* states that there exists an isomorphism $f : A \to B$, where f is analytic on A and f^{-1} is analytic on B. The isomorphism f can be made unique by demanding $f(z_0) = 0$ and $f'(z_0) = |f'(z_0)|\, e^{i\varphi}$, for an arbitrary point $z_0 \in A$ and angle φ with $0 \leq \varphi < 2\pi$. The theorem is nonconstructive, since it does not give a method to determine the function f. It thus becomes interesting to find examples of complex functions that map certain simple geometrical shapes into the unit disk (or vice versa). Schwarz (1869) found that the function $z(t)$ with the derivative

$$\frac{\mathrm{d}z}{\mathrm{d}t} = \frac{1}{\sqrt{1 - t^4}} \tag{3.109}$$

maps a circle in the complex t-plane to a square in the complex z-plane, and the interior of the circle to the interior of the square. The Jacobi *elliptic function* sn with variable z and parameter k is defined by

$$z = \int_0^{\mathrm{sn}(z,k)} \frac{\mathrm{d}t}{\sqrt{1 - t^2}\,\sqrt{1 - k^2 t^2}}. \tag{3.110}$$

Writing (3.109) as

$$z = \int_0^t \frac{\mathrm{d}t'}{\sqrt{1 - t'^2}\,\sqrt{1 - i^2 t'^2}}, \tag{3.111}$$

and thus for Schwarz's map,

$$t = \mathrm{sn}(z, i). \tag{3.112}$$

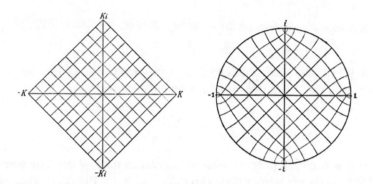

Fig. 3.1 Conformal map (3.112) from a square in the z-plane to a circle in the t-plane. Reproduced from Schwarz (1869)

This is an example of the Riemann mapping theorem, since (3.112) maps a square in the complex z-plane conformally to the unit circle in the complex t-plane. The corresponding figures from Schwarz's original paper from 1869 are shown in Fig. 3.1.

The *Schwarz–Christoffel theorem* gives an explicit formula for an analytic mapping $f : A \to B$ from the upper half-plane A in the complex t-plane to the interior B of an arbitrary closed *polygon* in the complex z-plane. Furthermore, f maps the real t-axis to the boundary of the z-polygon. (Note that neither A nor B is the unit disk from the Riemann mapping theorem.) The formula for this mapping is derived in Schwarz's paper from 1869; the relevant papers by Christoffel appeared between 1868 and 1870. An introductory exposition of the Schwarz–Christoffel theorem with many examples is given by Walker (1964).

As a very simple example toward the theorem, consider the map $z = \sqrt{t}$ from the complex t- to the z-plane. In the following, always the positive branch of the square root is understood. Any $t > 0$ is then mapped to $\sqrt{t} > 0$ and $t < 0$ to $i\sqrt{-t}$. Thus, the real t-axis is mapped to a corner at C with opening angle $\pi/2$. For complex t in the upper half-plane, $t = ae^{i\varphi}$, with $a > 0$ and $0 < \varphi < \pi$. Then $z = \sqrt{t} = \sqrt{a}\, e^{i\varphi/2}$ lies in the first quadrant of the z-plane. Thus, $z = \sqrt{t}$ maps the upper t-half-plane to points z lying within the corner at C. Generally, $z = t^{1/n}$ maps the real t-axis to a corner with opening angle π/n, and the upper t-half-plane is mapped to points z lying within this corner.

We come to a central argument of Schwarz (1869). Let $f : t \to z$ map the upper t-half-plane to a polygon in the z-plane. Polygons that differ only in size, position, and rotation angle are (geometrically) *similar*. Schwarz demands that the function f is determined by a differential equation. The size, position, and rotation angle of the polygon shall *not* enter this equation. Instead, they shall appear as integration constants only. Now if the function $z(t)$ with $t \in \mathbb{R}$ describes a polygon, then $Z = bz + a$ with $a, b \in \mathbb{C}$ is a similar polygon. Namely, a gives a translation and b gives a rotation and stretching/contraction of the geometrical figure. Thus, $f : t \to z$ and $g : t \to Z$ shall obey the same differential equation. The above expression for z and Z gives $g(t) = bf(t) + a$, and therefore

$$\frac{g''}{g'} = \frac{f''}{f'} = \frac{z''}{z'},$$

(3.113)

where a prime indicates differentiation with respect to t. For the function $z = b(t - t_0)^\alpha$, which has a corner at t_0 with interior angle $\alpha\pi$,

$$\frac{z''}{z'} = -\frac{1 - \alpha}{t - t_0}.$$

(3.114)

Consider now a closed polygon with m corners in the z-plane. The corners are located at z_k, and have interior angles $\alpha_k \pi$ $(0 < \alpha_k < 2)$. Then the correct generalization of (3.114) is

$$\frac{z''}{z'} = -\sum_{k=1}^{m} \frac{1 - \alpha_k}{t - t_k},$$

(3.115)

where the t_k, with $z_k = f(t_k)$, are appropriate values on the real t-axis. For the proof, see Nehari (1952), p. 189, Markushevich (1967), p. 322, and Kellogg (1929), p. 372. Note that $t_k < t_l$ for $k < l$ and $z_1, z_2, \ldots, z_m$ are the polygon corners when one moves along the polygon in the mathematically positive sense. Equation (3.115) can be integrated once. Introducing an auxiliary function $h = z'$ gives

$$\frac{dh}{h} = -\sum_{k=1}^{m} \frac{1 - \alpha_k}{t - t_k} \, dt,$$

(3.116)

and thus

$$\begin{aligned}
\ln h &= -\sum_{k} (1 - \alpha_k) \int \frac{dt}{t - t_k} \\
&= \sum_{k} [-(1 - \alpha_k) \ln(t - t_k)] + \ln b \\
&= \sum_{k} \ln \left[(t - t_k)^{-(1 - \alpha_k)} \right] + \ln b,
\end{aligned}$$

(3.117)

with an integration constant b. This gives the Schwarz–Christoffel formula

$$\boxed{\frac{dz}{dt} = b \prod_{k=1}^{m} \frac{1}{(t - t_k)^{1 - \alpha_k}}}$$

(3.118)

The factor $b \in \mathbb{C}$ rotates and stretches the polygon as required. We briefly discuss the geometrical significance of the integral

$$z(t) = b \int_{-\infty}^{t} \frac{dt'}{(t' - t_1)^{1-\alpha_1} \dots (t' - t_m)^{1-\alpha_m}}, \qquad (3.119)$$

taken along the real t-axis. Assume for simplicity that $b = 1$. For any given t', the factors $t' - t_1$ to $t' - t_m$ are either positive or negative (or zero). The integral can acquire a phase $\varphi \neq 0$, i.e., $z(t) = |z| e^{i\varphi}$, only if at least one of these factors is negative, through the exponentiation with $1 - \alpha_k$ (e.g., if $\alpha_k = 1/2$). Assume that the integral was calculated up to some point $t > t_{l-1}$, and has phase φ_{l-1} there. Integrating further up to the point t_l, the phase φ does not change but remains φ_{l-1}, since none of the factors $t' - t_1$ to $t' - t_m$ changes sign. Thus, the point $z(t)$ in the z-plane during integration in the interval from t_{l-1} to t_l moves along a straight line at an angle φ_{l-1} with the real x-axis. When t' passes through t_l, the factor $t' - t_l$ changes sign from negative to positive. Since $(-1)^{\alpha_l - 1} = e^{i\pi(\alpha_l - 1)}$, the phase of the integral changes from φ_{l-1} to

$$\varphi_l = \varphi_{l-1} + (1 - \alpha_l)\pi \qquad (3.120)$$

when t' passes through t_l. For all $t < t_{l+1}$, then, the point $z(t)$ moves along a straight line at an angle φ_l with the real x-axis in the z-plane. In Fig. 3.2, the angle between the two straight z-lines of the polygon from t_{l-1} to t_l and from t_l to t_{l+1} is $\alpha_l \pi$. Namely, from the figure, this angle is $\pi - (\varphi_l - \varphi_{l-1})$, which according to (3.120) is $\pi - (1 - \alpha_l)\pi$, i.e., $\alpha_l \pi$. This holds for all possible $0 < \alpha_l < 2$.

Thus, when t' runs from $-\infty$ to $+\infty$ along the real t-axis, passing through all $t_k < \infty$, the image-point $z(t)$ moves along a polygon with m interior angles α_k, as was intended. For more details on this geometrical interpretation of the integral (3.119), see Markushevich (1965), pp. 128–130.

Fig. 3.2 Interior angle $q = \pi - (\varphi_l - \varphi_{l-1})$ of the polygon (bold lines) described by $z(t)$ in (3.119), at the corner $z(t_l)$

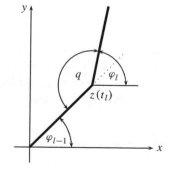

In order that a polygon with interior corner angles $\alpha_k\pi$ is indeed closed,

$$\sum_{k=1}^{m}(1 - \alpha_k)\pi = 2\pi \tag{3.121}$$

must hold. Namely, $(1 - \alpha_k)\pi$ (not $\alpha_k\pi$) is the turning angle at a corner (think of a car turning at a crossroad), and a full turn corresponds to an angle of 2π. For example, for the three turning angles of a triangle, $(\pi - \alpha) + (\pi - \beta) + (\pi - \gamma) = 2\pi$, or $\alpha + \beta + \gamma = \pi$ for the interior angles.

To (3.119), an integration constant a can be added, and this corresponds to a translation of the polygon in the z-plane. In summary, $z(t)$ maps the straight line Re t to the boundary of the polygon. The half-plane Im $t > 0$ is mapped to the *interior* of the polygon. Two of the $t_k \in \mathbb{R}$ can be chosen arbitrarily, by adapting $b \in \mathbb{C}$. The remaining $m - 2$ of the t_k are then fixed and have to be calculated from the known z_k using (3.118). None of the t_k in (3.118) may be infinite; a restraint that will be lifted in the next section.

How is $t \to \infty$, the point at infinity, mapped to the polygon, if none of the t_k is infinite? In (3.118), all the t_k can be neglected if $t \to \infty$, giving

$$\frac{dz}{dt} = b \prod_{k=1}^{m} \frac{1}{t^{1-\alpha_k}}. \tag{3.122}$$

Then from (3.121),

$$\frac{dz}{dt} = \frac{b}{t^2}, \tag{3.123}$$

and therefore, if $t \to \infty$,

$$z = -\frac{b}{t} + a = a. \tag{3.124}$$

Thus, $t \to \infty$ is mapped to the translation point $z = a$. This is illustrated for a triangle in Fig. 3.3. Note that the points A_-, A_+, and the line A in the Euclidean plane correspond to one and the same point at infinity in the complex plane. If $A, B, C, \ldots$ are points in the t-plane, we name their image in the z-plane under the Schwarz–Christoffel mapping again $A, B, C, \ldots$ Furthermore, points at infinity in the complex plane will *not* be marked with special symbols like A_∞ in the following, or by interrupting lines that lead to them with dots, since their special significance will be clear from the given context.

Fig. 3.3 Mapping of A at $t \to \infty$

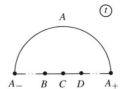
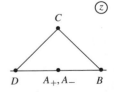

3.12 Mapping of Semi-infinite and Infinite Strips

In problems from fluid dynamics, it is often necessary to map the half-plane $\text{Im } t > 0$ to an infinite or semi-infinite strip in the z-plane (the reason for this becomes clear in Chap. 5). In order to do so, one of the t_k, say the last, t_m, is moved to infinity. This leads to a problem in the Schwarz–Christoffel formula, with the right side becoming zero. To remove this problem, we follow Nehari (1952), p. 192 and perform a variable transformation, $t \to s$, with

$$t = t_m - \frac{1}{s}. \tag{3.125}$$

For complex variables t and s, this is a *linear* transformation, see Bieberbach (1921). The arbitrary point t_m corresponds to $s_m \to \infty$. Equation (3.118) then becomes, with $dt = s^{-2}\, ds$,

$$
\begin{aligned}
dz &= \frac{b\,ds}{s^2}\, \frac{1}{(-s)^{-(1-\alpha_m)}}\, \prod_{k=1}^{m-1} \frac{1}{(t_m - t_k - s^{-1})^{1-\alpha_k}} \\
&= \frac{b\,ds}{s^2}\, \frac{1}{(-s)^{(1-\alpha_1)+\cdots+(1-\alpha_{m-1})-2}}\, \prod_{k=1}^{m-1} \frac{1}{(t_m - t_k)^{1-\alpha_k}\left(1 - \dfrac{1}{s(t_m - t_k)}\right)^{1-\alpha_k}} \\
&= b\,ds\, \prod_{k=1}^{m-1} \frac{1}{(t_k - t_m)^{1-\alpha_k}}\, \prod_{l=1}^{m-1} \frac{1}{\left(s - \dfrac{1}{t_m - t_l}\right)^{1-\alpha_l}} \\
&= c\,ds\, \prod_{k=1}^{m-1} \frac{1}{(s - s_k)^{1-\alpha_k}}, \tag{3.126}
\end{aligned}
$$

using (3.121) in the second line. The constant $c = b\prod_{k=1}^{m-1}(t_k - t_m)^{-(1-\alpha_k)}$ is finite since none of the $t_m - t_1$ to $t_m - t_{m-1}$ is zero. Furthermore, $s_k = (t_m - t_k)^{-1}$ for $k = 1$ to $m - 1$. Comparison with (3.118) shows that the term with $t_m \to \infty$ dropped out of the Schwarz–Christoffel formula, i.e., the product from 1 to m is replaced by a product from 1 to $m - 1$. Apart from this, the transformation $t \to s$ leaves the formula unchanged. In the following, a constant b and variable t will be used throughout, even if a transformation $t \to s$ was applied.

The semi-infinite strip is considered first, see Fig. 3.4. The points B and C shall correspond to $t = -1$ and 1, respectively (the t-values of two z-points can be chosen freely, see above). A and D are the points at infinity both in the t- and z-planes.

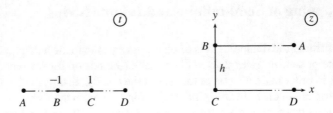

Fig. 3.4 Mapping to a semi-infinite strip

The semi-infinite strip in the z-plane corresponds to a triangle (polygon) with two sides of infinite length. The interior angles are $\pi/2$ at both B and C, and zero at the point $A = D$ where the parallels meet at infinity. Note especially that there are *not* two right angles at A and D. Equation (3.118) gives, with the point $A = D$ not considered (transformation $t \to s$),

$$\frac{dz}{dt} = \frac{b}{\sqrt{t+1}\,\sqrt{t-1}} = \frac{b}{\sqrt{t^2-1}}. \tag{3.127}$$

This can be integrated, giving

$$z = b \, \cosh^{-1} t + a. \tag{3.128}$$

Below it is shown that for a strip of width h,

$$z = \frac{h}{\pi} \, \cosh^{-1} t \quad \text{or} \quad t = \cosh \frac{\pi z}{h}. \tag{3.129}$$

Then for a point B at $z = ih$, $t = \cosh(i\pi) = \cos \pi = -1$, and $t = 1$ for C, as desired.

Next, we map the half-plane $\text{Im} \, t > 0$ to an infinite strip in the z-plane, as shown in Fig. 3.5. To map the points A, C, D, F to the point at infinity in the z-plane, the transformation $t \to s$ from (3.125) is performed and s is renamed to t. The points A

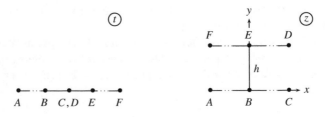

Fig. 3.5 Mapping to an infinite strip

and F at $t \to \mp\infty$ do then no longer appear in the Schwarz–Christoffel formula. The point $C = D$ shall lie at $t = 0$ and $z \to \infty$. The normalization is fixed by assuming $t = -1$ and $t = 1$ for B and E, respectively. The polygon in the z-plane has now two parallel sides only, both of infinite length, which meet at $A = F$ and $C = D$ with interior angles both zero, see Nehari (1952), p. 196. Nehari observes that the use of the Schwarz–Christoffel formula for this degenerate geometric figure needs further justification. The latter is brought about in Sect. 47 of Markushevich (1967). Note that the sum of turning angles is again $\pi + \pi = 2\pi$, as in (3.121). Equation (3.118) reads, for one single corner with $\alpha = 0$ at $C = D$,

$$\frac{dz}{dt} = \frac{b}{t}. \tag{3.130}$$

This result can also be derived directly, without the transformation $t \to s$ that makes $t \to \infty$ drop out of the Schwarz–Christoffel formula. Indeed, let $t = a$ be the distance from $t = 0$ to the points A and F. Then (3.118) gives, for the *two* corners at $A = F$ and $C = D$,

$$\frac{dz}{dt} = \frac{\tilde{b}}{t(t-a)} = \frac{\tilde{b}}{at}\frac{a}{t-a}. \tag{3.131}$$

One demands (for the limits, see the website by Richard Fitzpatrick, http://farside.ph.utexas.edu/teaching/336L/Fluidhtml/node88.html)

$$\frac{a}{t-a} \to -1 \qquad \text{and} \qquad \frac{\tilde{b}}{a} \to -b, \tag{3.132}$$

where b shall be finite. This gives again, as in (3.130),

$$\frac{dz}{dt} = \frac{b}{t}, \tag{3.133}$$

or after integration,

$$z = b \ln t + a, \tag{3.134}$$

where a and b are finite. Fixing a and b in accordance with Fig. 3.5 leads to

$$z = \frac{h}{\pi} \ln t + ih \qquad \text{or} \qquad t = -e^{\pi z/h}. \tag{3.135}$$

After these few examples, we refer to Betz (1948), pp. 345–353 for a table of the Schwarz–Christoffel maps of fundamental geometric figures.

3.13 The Riemann Surfaces

Analytic functions are power series with only positive exponents, whereas meromorphic functions are Laurent's series that have also negative exponents. A next step of generalization is to consider series of the form

$$f(z) = a_0 + a_1 \, z^{1/n} + a_2 \, z^{2/n} + a_3 \, z^{3/n} + \cdots \tag{3.136}$$

and

$$g(z) = a_{-m} \, z^{-m/n} + a_{-m+1} \, z^{(-m+1)/n} + \cdots + a_{-1} \, z^{-1/n} + a_0 + a_1 \, z^{1/n} + \cdots , \tag{3.137}$$

where $n, m \in \mathbb{N}$ and $z, a_k \in \mathbb{C}$. Then f and g are multivalued functions (with n values). To make f and g single-valued, n 'copies' of the z-plane are introduced. These so-called Riemann's *sheets* are laid one atop of the other. The n sheets of z are put in one-to-one relation with the n branches of f and g.

As a standard example, consider the function $w = \sqrt{z}$, where again the positive value of the root is always understood. In order that the full w-plane is covered, $z = re^{i\varphi}$ with $0 \le \varphi < 4\pi$ is required. The full w-plane shall be covered in order to obtain all possible values of the root. The corresponding z-values are put on two different *sheets* z_+ and z_- (Osgood 1907, p. 311): for $0 \le \varphi < 2\pi$, put $z_+ = z$, and for $2\pi \le \varphi < 4\pi$, put $z_- = z$. With this trick of introducing two z-planes, the map $f : z \to w$ is one-to-one. Especially, for $2\pi \le \varphi < 4\pi$ and $\tilde{\varphi} = \varphi - 2\pi$,

$$\sqrt{z_-} = \sqrt{re^{i\varphi}} = \sqrt{re^{2\pi i} e^{i\tilde{\varphi}}} = -\sqrt{re^{i\tilde{\varphi}}}, \tag{3.138}$$

thus the z_--sheet corresponds to the negative branch of the root. Continuity of f can be achieved by connecting the sheets properly. A *cut*—a new class of curves—is introduced on each sheet, starting at $z = 0$ and extending along the positive x-axis (this is an irrelevant detail) to infinity, see Fig. 3.6. Each cut has two *banks* (similar to the banks of a river). For example, for the positive real axis, the bank $\dot{x}$ has $x > 0$ and $y = 0_+$, and the bank $\ddot{x}$ has $x > 0$ and $y = 0_-$. The concept of a bank of a cut or slit generalizes to curves in the plane of the two one-sided limits $\lim_{\epsilon \to 0} f(x - \epsilon)$ and $\lim_{\epsilon \to 0} f(x + \epsilon)$ toward a point x from calculus on the real line.

How to achieve then continuity of $f = w$ on the two Riemann z-sheets? Start on the $\dot{x}_+$-bank of the z_+-sheet at $w = \sqrt{r}$ (see Fig. 3.6). Proceed on this sheet along a circle toward $\varphi \to 2\pi$ and $w \to -\sqrt{r}$ at the opposite bank $\ddot{x}_+$. Obviously, there is a discontinuity of w (sign change!) between the banks $\ddot{x}_+$ and $\dot{x}_+$. Therefore, *crossover* from the bank $\ddot{x}_+$ on the z_+-sheet to the bank $\dot{x}_-$ on the z_--sheet, where $w = -\sqrt{r}$. This transition is continuous! Proceed on the z_--sheet along a circle toward $\varphi \to 4\pi$ and $w \to \sqrt{r}$ at the bank $\ddot{x}_-$. Crossover again with continuous $w = \sqrt{z}$ to the $\dot{x}_+$-bank on the z_+-sheet, which was the starting point.

Fig. 3.6 The Riemann surface for $\sqrt{z}$, with branch points at 0 and ∞

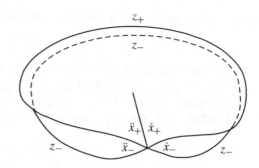

In this way, all possible values $w \in \mathbb{C}$ of the square root are encountered on the two Riemann z-sheets that cross along the positive x-axis. No double-valuedness occurs along this cut: every z-argument lies on one and only one of the two sheets. This configuration is termed a *Riemann surface*. As is easily seen, the function $w = \sqrt{z}$ is even differentiable on the Riemann surface for all z except $z = 0$. Therefore, $\sqrt{z}$ is analytic on the Riemann surface, except at the so-called *branch point* (cf. below) $z = 0$. For more details on this example, see Gamelin (2001).

Note that in some books, the argument runs somewhat differently: for the function $w = z^2$ (hence $z = \sqrt{w}$), it is observed then that two w-planes are required to map in a one-to-one fashion all w-values obtained from $z \in \mathbb{C}$. After exchanging $z \leftrightarrow w$ this gives: for $w = \sqrt{z}$, two z-planes are required to fill one complete w-plane.

It is obviously not possible to map a Riemann surface with two sheets one-to-one to $\mathbb{C}$. Bieberbach (1921), p. 230 gives a proof for ≥ 4 branch points.

The somewhat arbitrary surface in Fig. 3.6 can be constructed in a well-defined procedure as follows (Neumann 1884, p. 66). Consider two circular cones with a common footpoint. The mantles of the cones shall touch along a straight line from the origin. The opening angles of both cones are $\frac{1}{2}(\pi - \epsilon)$. Then the angle between the cone axes is $\pi - \epsilon$. For $\epsilon \to 0$, the mantles of the cones resemble Fig. 3.6.

The origin and the point at infinity in Fig. 3.6 are *branch points*. At a branch point, a multivalued function has a discontinuity on every infinitesimal cycle around this point. The *order* of a branch point is the number of sheets it connects minus one. This is not necessarily the total number of sheets in a Riemann surface. On a cut connecting two branch points, the sheets are crosswise connected. Except for their two end points, the cuts where sheets cross can be chosen largely arbitrary, yet the cut is not allowed to intersect with itself. Since the cut is rather unsubstantial, it is often not drawn in figures. Moving on a Riemann surface and crossing sheets at cuts, the function f is everywhere continuous.

Next, *uniformizing* parameters are introduced. The simplest way to obtain two Riemann's sheets $z_{\pm}$ is by putting $z = t^2$, $t \in \mathbb{C}$: when t covers the $\mathbb{C}$-plane once, z covers the $\mathbb{C}$-plane twice. In general, for an n-valued function $f(z)$, let

$$z = a + t^n, \tag{3.139}$$

$$f(z) = F(t) \tag{3.140}$$

hold in a neighborhood of $t = 0$, where $F(t)$ is single-valued for $t \in \mathbb{C}$. As before, the function $f(z)$ is multivalued for $z \in \mathbb{C}$, but is single-valued if there are n copies of the z-plane (i.e., n Riemann's sheets) as introduced by (3.139). Compare this with (3.136) for $a = 0$. For the point at infinity, use instead

$$z = t^{-n},$$

$$f(z) = F(t), \tag{3.141}$$

in a neighborhood of $t = 0$. One defines (Weyl 1913; Bieberbach 1921): a multivalued function $f(z)$ is *locally uniformized* by t, if, for some $m, n \in \mathbb{N}$,

$$f(z) = F(t) = a_{-m} t^{-m} + \cdots + a_0 + a_1 t + a_2 t^2 + \cdots, \tag{3.142}$$

where

$$z = a + t^n \qquad \text{or} \qquad z = t^{-n} \tag{3.143}$$

in a neighborhood of $t = 0$. Continuity and analyticity were defined for single-valued functions; by (3.143) they are also defined for multivalued f. The latter equation remains valid at branch points.

As next example after $\sqrt{z}$, consider the function (see Neumann 1884, p. 74)

$$w(z) = \sqrt{(z - z_1)(z - z_2)(z - z_3)}, \tag{3.144}$$

with arbitrary but fixed numbers $z_1, z_2, z_3 \in \mathbb{C}$. For $k = 1, 2, 3$, let

$$z - z_k = r_k\, e^{i\varphi_k}. \tag{3.145}$$

Then

$$w = \sqrt{r_1 r_2 r_3}\, e^{i(\varphi_1 + \varphi_2 + \varphi_3)/2}, \tag{3.146}$$

where again the positive value is taken for the root. As before, two different sheets z_+ and z_- are introduced. These sheets are in contact at the branch points z_1, z_2, and z_3 where $w = 0$. Let C be a closed curve about one and only one of the three points z_k. Then $\Delta\varphi_k = \pm 2\pi n$ (with $n \in \mathbb{N}$, and Δ is now the difference symbol, not the Laplace operator) when moving along C, starting and ending at the same point. On the other hand, $\Delta\varphi_l = 0$ for $l \neq k$. This is clear from elementary geometry; for details on this 'variation of the argument' see Neumann (1884), p. 76 and Nevanlinna (1953), p. 12. Starting at an arbitrary $z \in C$ at a value a of the function w, the value $-a$ of w is reached after moving along C by $\Delta\varphi = 2\pi$, i.e., w is discontinuous. This can be avoided by passing from one z-sheet to the other somewhere along C. If C encloses two of the z_k, say z_1 and z_2 but not z_3, then w will not change sign

Fig. 3.7 The Riemann surface for (3.144)

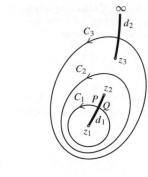

Fig. 3.8 Crossing of sheets along a line d to make w continuous

when going around C according to (3.146), since $\Delta\varphi_1 = \Delta\varphi_2 = 2\pi$ and $\Delta\varphi_3 = 0$. Finally, if C encloses z_1, z_2, and z_3, the function w will again change sign along C. Now let d_1 (see Fig. 3.7) be any line connecting the branch points z_1 and z_2, and let d_2 connect the branch points z_3 and ∞. Any closed curve C_1 about z_1 crosses d_1. Let P and Q be the points on the two banks of d_1 at this crossing point. By (3.146), the function w has opposite signs in P and Q.

Next, let P and Q be opposite points on the banks of d_2 at a crossing with a closed curve C_3. Again w has opposite signs in P and Q. Finally, on any curve C_2 about z_1 and z_2 but not z_3 in Fig. 3.7, w is continuous. Therefore, w has opposite signs on the banks of d_1 and d_2. One connects the banks of d_1 and d_2 on the two Riemann sheets crosswise, as shown in Fig. 3.8. Demanding that f is continuous along both the segments P_+Q_- and P_-Q_+ (of a curve C) determines how the two values of w with opposite sign are put in correspondence with the z-values of the two sheets (once a single argument–value pair (z_+, w_+) was chosen). In this way, f becomes continuous everywhere on the Riemann surface.

As a final example, consider the function (Neumann 1884, p. 85)

$$w(z) = \left(\frac{z - z_1}{z - z_2}\right)^{1/3}, \tag{3.147}$$

with arbitrary $z_1, z_2 \in \mathbb{C}$. For $k = 1, 2$ let

$$z - z_k = r_k \, e^{i\varphi_k}, \tag{3.148}$$

so that

$$w = \left(\frac{r_1}{r_2}\right)^{1/3} e^{i(\varphi_1 - \varphi_2)/3}, \tag{3.149}$$

Fig. 3.9 Connecting three sheets across the curve d to make w continuous

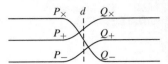

Fig. 3.10 Connecting three sheets across curves d and d'

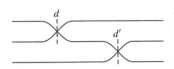

where in $(r_1/r_2)^{1/3}$, the positive, real-valued root is meant. The function w is zero in $z = z_1$ and has a pole at $z = z_2$. For all other values of z, w has three different values,

$$w, \quad \eta w, \quad \eta^2 w, \qquad \text{with} \qquad \eta = e^{2\pi i/3}.$$

These w-values are put in one-to-one relation with three z-sheets: w, ηw, and $\eta^2 w$ with the upper $(z_\times)$, middle (z_+), and lower (z_-) sheets, respectively. Let C be a closed curve about z_1 but not z_2. Going around C in positive direction once, $\Delta\varphi_1 = 2\pi$ and $\Delta\varphi_2 = 0$. Next, let C enclose z_2 but not z_1. Then $\Delta\varphi_1 = 0$ and $\Delta\varphi_2 = -2\pi$. If C encloses both z_1 and z_2, the total change is $\Delta\varphi_1 + \Delta\varphi_2 = 0$. Moving along *any* closed curve C once in the positive direction, the change in w from (3.149) is

$$w_e = w_s\, \eta^{m_1 - m_2}, \tag{3.150}$$

with start and end values w_s and w_e, and $m_k = 1$ or 0 if z_k is enclosed or not. On any C not crossing a cut d from z_1 to z_2, the function w is continuous. Along the cut d, the three sheets are connected crosswise. Let the points P and Q lie opposite to each other on the banks of d. Along a closed curve C about z_1 but not z_2, w changes by a factor of η. With function values w in the upper sheet, the appropriate correspondence of the z-values from the three sheets with the three w-values is

$$
\begin{aligned}
P_\times &: w, & Q_\times &: \eta w, \\
P_+ &: \eta w, & Q_+ &: \eta^2 w, \\
P_- &: \eta^2 w, & Q_- &: \eta^3 w = w.
\end{aligned}
\tag{3.151}
$$

The three sheets are connected across d as shown in Fig. 3.9. Then w is everywhere continuous on this Riemann surface.

It is not necessary that the curve d is the same in all three sheets. Instead, sheets can be connected along some d in the upper and middle sheets, and the middle and lower sheets are connected along another curve d'. The connection between sheets appears then as shown in Fig. 3.10.

To summarize, the function $((z - z_1)/(z - z_2))^{1/3}$ is one-to-one and continuous on a Riemann surface with three sheets connected along two cuts (possibly identical) between branch points z_1 and z_2.

A function $f(z)$ is *rational* if, with $a_k, b_k \in \mathbb{C}$ and $m, n \in \mathbb{N}$ and for all $z \in \mathbb{C}$,

$$f(z) = \frac{a_0 + a_1 z + \cdots + a_m z^m}{b_0 + b_1 z + \cdots + b_n z^n}. \tag{3.152}$$

A function $f(z)$ is *algebraic* of degree n if, with rational functions g_0 to g_n and for all $z \in \mathbb{C}$,

$$g_0(z) + g_1(z) f(z) + \cdots + g_n(z) f^n(z) = 0. \tag{3.153}$$

The elementary considerations of the present section lead, after further developments, to the following fundamental theorem.

Theorem (cf. Appendix) *Let f be a meromorphic function on a Riemann surface with n sheets. Then f is an algebraic function of degree n.*

Proof Neumann (1884), Chap. 5, with a classical argument on p. 121. This argument is also given in Hurwitz and Courant (1964), p. 393.

In the following, the Riemann surfaces will be encountered again in three different cases: (i) for the re-entrant jet cavity, Sect. 5.9; (ii) for a flow against a tilted wedge, Sect. 5.10; and (iii) for the Burgers equation, Sect. 12.8.

References

Ahlfors, L.V. 1966. *Complex analysis*, 2nd ed. New York: McGraw-Hill.

Axler, S., P. Bourdon, and W. Ramey. 2001. *Harmonic function theory*, 2nd ed. New York: Springer.

Basye, R.E. 1935. Simply connected sets. *Transactions of the American Mathematical Society* 38: 341.

Betz, A. 1948. *Konforme Abbildung*. Berlin: Springer.

Bieberbach, L. 1921. *Lehrbuch der Funktionentheorie, Band I, Elemente der Funktionentheorie*. Leipzig: B.G. Teubner; Wiesbaden: Springer Fachmedien.

Burkhardt, H., and W.F. Meyer. 1900. Potentialtheorie. In *Encyklopädie der mathematischen Wissenschaften*, vol. 2-1-1, 464. Leipzig: Teubner (digital at GDZ Göttingen).

Cartan, H.P. 1963. *Elementary theory of analytic functions of one or several complex variables*. Reading: Addison-Wesley; New York: Dover (1995).

Copson, E.T. 1935. *Theory of functions of a complex variable*. Oxford: Oxford University Press.

Courant, R. 1950. *Dirichlet's principle, conformal mapping, and minimal surfaces*. New York: Interscience Publishers.

Gamelin, T.W. 2001. *Complex analysis*. Berlin: Springer.

Garding, L. 1977. *Encounter with mathematics*. New York: Springer.

Golusin, G.M. 1957. *Geometrische Funktionentheorie*. Berlin: Deutscher Verlag der Wissenschaften.

Grattan-Guinness, I. (ed.). 1994. *Companion encyclopedia of the history and philosophy of the mathematical sciences*, vol. 1. London: Routledge.

Hurwitz, A., and R. Courant. 1964. *Funktionentheorie*. Berlin: Springer.

Jackson, J.D. 1998. *Classical electrodynamics*, 3rd ed. New York: Wiley.

Jänich, K. 2001. *Vector analysis*. Berlin: Springer.

Kellogg, O.D. 1929. *Foundations of potential theory*. Berlin: Springer.

Knopp, K. 1996. *Theory of functions*. Mineola: Dover.

Lamb, H. 1932. *Hydrodynamics*. Cambridge: Cambridge University Press; New York: Dover (1945).

Lu, J.-K., S.-G. Zhong, and S.-Q. Liu. 2002. *Introduction to the theory of complex functions*. Singapore: World Scientific.

Markushevich, A.I. 1965. *Theory of functions of a complex variable*, vol. II. Englewood Cliffs: Prentice-Hall.

Markushevich, A.I. 1967. *Theory of functions of a complex variable*, vol. III. Englewood Cliffs: Prentice-Hall.

Miles, J.W. 1977. On Hamilton's principle for surface waves. *Journal of Fluid Mechanics* 83: 153.

Monna, A.F. 1975. *Dirichlet's principle: A mathematical comedy of errors and its influence on the development of analysis*. Utrecht: Oosthoek, Scheltema & Holkema.

Nehari, Z. 1952. *Conformal mapping*. New York: McGraw-Hill; New York: Dover (1975).

Neumann, C. 1884. *Vorlesungen über Riemann's Theorie der Abel'schen Integrale*, 2nd ed. Leipzig: Teubner.

Nevanlinna, R. 1953. *Uniformisierung*. Berlin: Springer.

Osgood, W.F. 1907. *Lehrbuch der Funktionentheorie*, Erster Band. Leipzig: Teubner.

Rankine, W.J. 1864. On plane water-lines in two dimensions. *Philosophical Transactions of the Royal Society of London A* 154: 369.

Remmert, R. 1992. *Funktionentheorie 1*, 3rd ed. Berlin: Springer.

Schwarz, H.A. 1869. Ueber einige Abbildungsaufgaben. *Journal für die reine und angewandte Mathematik* 70: 105; in *Gesammelte Mathematische Abhandlungen*, vol. 2, 65. Berlin: Springer (1890).

Serrin, J. 1959. Mathematical principles of classical fluid mechanics. In *Handbuch der Physik = Encyclopedia of Physics*, vol. 3/8-1, Strömungsmechanik 1, ed. S. Flügge, and C.A. Truesdell, 125. Berlin: Springer.

Walker, M. 1964. *The Schwarz-Christoffel transformation and its applications – a simple exposition*. New York: Dover.

Weyl, H. 1913. *Die Idee der Riemannschen Fläche*. Leipzig: Teubner.

Zermelo, E. 1902. Hydrodynamische Untersuchungen über die Wirbelbewegungen in einer Kugelfläche. Erste Mitteilung. *Zeitschrift für Mathematik und Physik* 47: 201.

Chapter 4
Vortices, Corner Flow, and Flow Past Plates

Vortices are among the most fascinating features of fluid dynamics: the flow bends around 'on itself,' with the possibility of becoming self-regulatory. This is indeed the case in the turbulent cascade of ever smaller eddies whirling in larger eddies, which has universal properties on all lengthscales. We do not deal with turbulence in this book, but consider in the present chapter certain cases of single and multiple vortices, and certain related structures. After some preliminaries on vortices, the first few sections deal with corner flows of inviscid and viscous fluids, with a strong bending of streamlines similar to that found in vortices; a later section treats corner eddies. Flows with nonvanishing vorticity past flat plates and cylinders are treated, leading to the Kutta–Joukowski formula for the vertical lift resulting from circulation. The stability properties (beyond first order) of infinite double rows of vortices, named the Kármán vortex streets, are analyzed. The last section deals with a completely different 'vortex,' the accretion disk of viscous gas spiraling on near-Keplerian orbits onto a central star.

4.1 Straight Vortex

Consider a two-dimensional, solenoidal flow of an inviscid fluid in the complex plane. The logarithmic potential (2.34) is generalized to complex z,

$$\Phi = \ln \frac{1}{z}. \tag{4.1}$$

Then $z = r\,e^{i\varphi}$ gives $\Phi = \ln r - i\varphi = \phi + i\psi$, thus $\phi = \ln r$ and $\psi = -\varphi$, and

$$u_r = \frac{\partial \phi}{\partial r} = \frac{1}{r}, \qquad\qquad u_\varphi = \frac{\partial \phi}{r\,\partial \varphi} = 0. \tag{4.2}$$

This velocity field corresponds to radial outflow from a source.

© Springer Nature Switzerland AG 2019
A. Feldmeier, *Theoretical Fluid Dynamics*, Theoretical and Mathematical Physics,
https://doi.org/10.1007/978-3-030-31022-6_4

Fig. 4.1 A whirlpool with diameter $\approx$2.5 m in Lake Texoma on the border of Oklahoma and Texas. The swirl is the result of conduits below the water surface built to accomplish a release of water. Photograph: U. S. Army Corps of Engineers, Tulsa District, YouTube video at https://www.youtube.com/watch?v=5hRSvVfAha0

If a complex function $w(z)$ is analytic in a domain, so is $iw(z)$. Namely, let $w = f + ig$ and therefore $iw = -g + if$. The Cauchy–Riemann equations $f_x = g_y$ and $g_x = -f_y$ hold for w, or $-g_x = f_y$ and $f_x = g_y$, which are the Cauchy–Riemann equations for iw, thus iw is analytic. With ① from above, also $i①$ is a complex potential, with speeds

$$u_r = -\frac{\partial \psi}{\partial r} = 0, \qquad\qquad u_\varphi = -\frac{\partial \psi}{r\, \partial \varphi} = \frac{1}{r}. \qquad (4.3)$$

This corresponds to a circular flow in the positive φ-direction, which is called a *straight vortex*. 'Straight' refers here to the notion that the flow pattern in the complex plane is repeated identically in the normal direction to the plane, giving a cylindrical flow pattern with no flow in the direction of the cylinder axis—quite in contrast to whirlpools, see Figs. 4.1 and 4.2, which have nonvanishing radial and vertical flow velocities.

The curl of the flow is, excluding the singularity at $r = 0$,

$$\nabla \times \vec{u} = \left(\frac{\partial (r u_\varphi)}{r\, \partial r} - \frac{\partial u_r}{r\, \partial \varphi} \right) \hat{z} = 0. \qquad (4.4)$$

Using Stokes' theorem, the circulation Γ is then zero along any closed curve that does *not* have the origin $r = 0$ as an interior point. On the other hand, for a circle that does enclose the origin,

$$\Gamma = \int_0^{2\pi} (r\, d\varphi\, \hat{\varphi}) \cdot (\hat{\varphi}/r) = 2\pi. \qquad (4.5)$$

This holds for *any* closed curve that has the origin as interior point, as seen from a calculation similar to that in the proof of Kelvin's theorem (2.125). Let K_1 and K_2 be

Fig. 4.2 A ca. 12-mile-high dust devil on Mars, captured with NASA's Mars Reconnaissance Orbiter.   Photograph: NASA Goddard Space Flight Center, orbiter images courtesy of NASA/JPL/MSSS. URL: http://svs.gsfc.nasa.gov/11265. An animation is available at https://svs.gsfc.nasa.gov/11265

two arbitrary curves around $r = 0$ and enclosing areas A_1 and A_2 such that $A_2 \subset A_1$ (especially, K_1 can be a circle). Then

$$\oint_{K_1} d\vec{l} \cdot \vec{u} - \oint_{K_2} d\vec{l} \cdot \vec{u} = \int_{A_1} d\vec{a} \cdot (\nabla \times \vec{u}) - \int_{A_2} d\vec{a} \cdot (\nabla \times \vec{u})$$

$$= \int_{A_1 \setminus A_2} d\vec{a} \cdot (\nabla \times \vec{u}) = 0, \qquad (4.6)$$

since the ring-shaped area $A_1 \setminus A_2$ does not include the origin $r = 0$, thus $\nabla \times \vec{u} = 0$ in this ring from (4.4).

The complex potential for a straight vortex with circulation Γ is thus

$$\Phi = \frac{i\Gamma}{2\pi} \ln \frac{1}{z} \qquad (4.7)$$

Taking the maximum diameter of the curve enclosing the origin to be infinitesimally small implies that the vorticity is concentrated in the origin. Let da be the infinitesimal area in the complex plane enclosed by this infinitesimal curve. Then, with $\omega = |\vec{\omega}|$,

$$da\, \omega = da\, |\nabla \times \vec{u}| = \oint_{\partial da} d\vec{l} \cdot \vec{u} = \Gamma, \qquad (4.8)$$

or

$$\omega\, da = \Gamma. \qquad (4.9)$$

For finite Γ and since $da \to 0$, vorticity at the origin tends to infinity, and is zero everywhere else. The transformation from polar coordinates r, φ to Cartesian coordinates gives for the vector field of the straight vortex

$$(u, v) = \frac{\Gamma}{2\pi} \frac{1}{x^2 + y^2} (-y, x). \tag{4.10}$$

Introducing for the flow in the xy-plane a streamfunction ψ via

$$u = \frac{\partial \psi}{\partial y}, \qquad\qquad v = -\frac{\partial \psi}{\partial x}, \tag{4.11}$$

one reads off from (4.10),

$$\psi = \frac{\Gamma}{2\pi} \ln \frac{1}{\sqrt{x^2 + y^2}}, \tag{4.12}$$

which corresponds to (4.7).

4.2   Corner Flow

Consider the flow in a corner formed by two solid, planar walls that meet at an opening angle α. The complex plane cuts normally through the walls, which appear then as two straight lines with an inner angle α; they shall meet at $z = 0$. On the walls, the normal component of the fluid velocity must vanish. For inviscid fluids, the tangential fluid speed at the wall is arbitrary, and for viscous fluids, it is zero by definition. We first follow Lamb (1932, p. 68). The simplest analytic functions are the powers

$$\Phi(z) = az^n, \tag{4.13}$$

with $a \in \mathbb{R}$ and $n \in \mathbb{N}$. Using polar coordinates, this becomes

$$az^n = a(re^{i\varphi})^n = ar^n e^{in\varphi} = ar^n (\cos(n\varphi) + i \sin(n\varphi)) = \phi + i\psi, \tag{4.14}$$

giving

$$\phi = ar^n \cos(n\varphi),$$
$$\psi = ar^n \sin(n\varphi). \tag{4.15}$$

The streamlines are given by $\psi = $ const. From (4.15), $\psi = 0$ if either $\varphi = 0$ or $\varphi = \pi/n$. But the points in the complex plane with $\varphi = 0$ and $\varphi = \pi/n = \alpha$ form two straight lines through the origin—as do the solid walls! The wall angle α is now

Fig. 4.3 Streamlines for a
corner flow

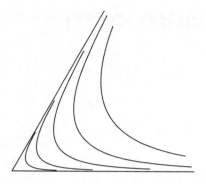

restricted to fractions of π. Inserting $n = \pi/\alpha$ in (4.15) gives

$$\phi = ar^{\frac{\pi}{\alpha}} \cos \frac{\pi\varphi}{\alpha},$$
$$\psi = ar^{\frac{\pi}{\alpha}} \sin \frac{\pi\varphi}{\alpha}. \tag{4.16}$$

This ψ obeys the boundary conditions appropriate for a flow in a corner with opening angle $\alpha = \pi/n$. Furthermore, ψ is differentiable and obeys the Laplace equation. It is thus the *unique* solution of the boundary value problem. Figure 4.3 shows the streamlines $\psi = $ const for this flow, as calculated from $\psi = $ const using (4.15).

The components of the flow velocity in polar coordinates are

$$u_r = \frac{\partial\phi}{\partial r} = \frac{\pi a}{\alpha} r^{\frac{\pi}{\alpha}-1} \cos \frac{\pi\varphi}{\alpha},$$
$$u_\varphi = \frac{1}{r}\frac{\partial\phi}{\partial\varphi} = -\frac{\pi a}{\alpha} r^{\frac{\pi}{\alpha}-1} \sin \frac{\pi\varphi}{\alpha}. \tag{4.17}$$

Therefore, the total flow speed is

$$u = \sqrt{u_r^2 + u_\varphi^2} = \frac{\pi a}{\alpha} r^{\frac{\pi}{\alpha}-1}. \tag{4.18}$$

This means that for $\alpha < \pi$, $u = 0$ at $r = 0$. In the opposite case of $\alpha > \pi$, the fluid speed reaches infinity at the origin. This is an example of a theorem by Helmholtz: *In a potential flow, the velocity at an obtuse corner approaches infinity.*

Due to the importance of corner flows, we give a second derivation of these results, now without using complex functions and for the case of a flow past a rectangular corner (Fig. 4.4), i.e., with $\alpha = 3\pi/2$. The calculation follows Landau and Lifshitz (1987, § 10, ex. 6). The flow region is given by $r > 0$ and $0 \leq \varphi \leq 3\pi/2$. The Laplace equation for ϕ reads in polar coordinates,

$$\Delta\phi = \phi_{rr} + \phi_r/r + \phi_{\varphi\varphi}/r^2 = 0. \tag{4.19}$$

Fig. 4.4 Flow past a corner

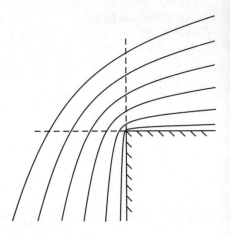

At the walls, the normal speed component is zero, or

$$\partial\phi/\partial\varphi = 0 \quad \text{for} \quad \varphi = 0 \quad \text{and} \quad \varphi = 3\pi/2. \tag{4.20}$$

These boundary conditions are fulfilled if

$$\phi(r, \varphi) = A(r)\,\cos(2\varphi/3). \tag{4.21}$$

Inserting this in (4.19) gives the ordinary differential equation

$$r^2 A'' + r A' - \frac{4}{9} A = 0. \tag{4.22}$$

Assuming

$$A(r) = ar^n, \tag{4.23}$$

Equation (4.22) implies

$$n(n-1) + n - 4/9 = 0, \tag{4.24}$$

or

$$n = \pm 2/3. \tag{4.25}$$

The solution $n = -2/3$ is omitted, and thus

$$u_r = \frac{2a}{3r^{1/3}}\,\cos(2\varphi/3),$$

$$u_\varphi = -\frac{2a}{3r^{1/3}}\,\sin(2\varphi/3), \tag{4.26}$$

which is again (4.17), for $\alpha = 3\pi/2$. The total fluid speed is

$$u(r) = \frac{2a}{3r^{1/3}},$$ (4.27)

and approaches infinity for $r \to 0$.

4.3 Corner Flow with Viscosity

The last section showed that the flows of inviscid fluids in corners are rather featureless, e.g., streamlines have no inflexion points. It may be expected that viscous fluids cannot show more intricate flow features than inviscid fluids, since friction decelerates fluid parcels and presumably reduces flow complexity. This argument is however wrong, and it was early suggested that viscosity may cause flow instability and 'excite' complex flow (Reynolds 1883; Heisenberg 1924; see discussion in Drazin and Reid 1981, p. 124ff). We show in this section that a viscosity-dominated flow near an acute corner shows more complex dynamics than the corresponding flow of an inviscid fluid, in that an infinite sequence of ever smaller eddies in geometrical progression forms near the corner. This result, though, is somewhat academic, and already the observation of two such eddies poses a challenge.

If the fluid density ρ and the viscosity coefficient ν are constant, the Navier–Stokes Equations (2.100) become, for a two-dimensional, planar flow,

$$uu_x + vu_y = -p_x/\rho + \nu\Delta u,$$
$$uv_x + vv_y = -p_y/\rho + \nu\Delta v.$$ (4.28)

Introducing the streamfunction ψ from (4.11),

$$u = \psi_y, \qquad\qquad v = -\psi_x,$$ (4.29)

this becomes

$$\psi_y\psi_{xy} - \psi_x\psi_{yy} = -p_x/\rho + \nu\Delta\psi_y,$$
$$-\psi_y\psi_{xx} + \psi_x\psi_{xy} = -p_y/\rho - \nu\Delta\psi_x.$$ (4.30)

Differentiating the first equation with respect to y, the second with respect to x, and subtracting the equations,

$$\psi_y\psi_{xyy} - \psi_x\psi_{yyy} + \psi_y\psi_{xxx} - \psi_x\psi_{xxy} = \nu\Delta\psi_{yy} + \nu\Delta\psi_{xx},$$ (4.31)

where terms $\psi_{xy}\psi_{xx}$ cancel, as do terms $\psi_{xy}\psi_{yy}$. Thus

$$\Delta\Delta\psi = \nu^{-1}\big[\psi_y\partial_x(\psi_{xx} + \psi_{yy}) - \psi_x\partial_y(\psi_{xx} + \psi_{yy})\big],$$ (4.32)

or

$$\Delta\Delta\psi = \nu^{-1}(\psi_y\partial_x - \psi_x\partial_y)\Delta\psi. \tag{4.33}$$

If viscosity is *very large* and dominates the flow dynamics, then

$$\boxed{\Delta\Delta\psi = \Delta^2\psi = 0} \tag{4.34}$$

This is the so-called Stokes equation for flows dominated by viscosity. An alternative derivation of this equation is as follows. For slow flows dominated by viscosity, inertia terms can be neglected. The equation of motion for stationary flows is then, for $\rho = $ const (see Davis and O'Neill 1977),

$$\rho^{-1}\nabla p = \nu\Delta\vec{u}. \tag{4.35}$$

Taking the curl on both sides, and using $\vec{u} = \nabla \times \vec{\psi}$ (see (3.6)) gives

$$0 = \Delta\nabla \times (\nabla \times \vec{\psi})$$
$$= \Delta\nabla(\nabla \cdot \vec{\psi}) - \Delta\Delta\vec{\psi}. \tag{4.36}$$

For 2-D flows, $\nabla \cdot \vec{\psi} = 0$, see (3.9), and therefore again

$$\Delta\Delta\psi = 0, \tag{4.37}$$

for motion in a plane. In polar coordinates,

$$\Delta\psi = \frac{1}{r}\frac{\partial}{\partial r}\left(r\frac{\partial\psi}{\partial r}\right) + \frac{1}{r^2}\frac{\partial^2\psi}{\partial\varphi^2}. \tag{4.38}$$

Inserting here $\Delta\psi$ for ψ gives

$$\Delta\Delta\psi = \frac{1}{r}\frac{\partial}{\partial r}\left(r\frac{\partial}{\partial r}\left[\frac{1}{r}\frac{\partial}{\partial r}\left(r\frac{\partial\psi}{\partial r}\right) + \frac{1}{r^2}\frac{\partial^2\psi}{\partial\varphi^2}\right]\right)$$
$$+ \frac{1}{r^2}\frac{\partial^2}{\partial\varphi^2}\left[\frac{1}{r}\frac{\partial}{\partial r}\left(r\frac{\partial\psi}{\partial r}\right) + \frac{1}{r^2}\frac{\partial^2\psi}{\partial\varphi^2}\right]. \tag{4.39}$$

Performing the r-differentiation except in the very first term gives

$$\Delta^2\psi = \frac{1}{r}\frac{\partial}{\partial r}\left(r\frac{\partial}{\partial r}\left(\frac{1}{r}\frac{\partial}{\partial r}\left(r\frac{\partial\psi}{\partial r}\right)\right)\right) + \frac{1}{r^4}\frac{\partial^4\psi}{\partial\varphi^4}$$
$$+ \frac{2}{r^2}\frac{\partial^4\psi}{\partial r^2\partial\varphi^2} - \frac{2}{r^3}\frac{\partial^3\psi}{\partial r\partial\varphi^2} + \frac{4}{r^4}\frac{\partial^2\psi}{\partial\varphi^2}. \tag{4.40}$$

This can be solved (partially) by separation, $\psi(r, \varphi) = g(r)f(\varphi)$. More specifically, one looks for solutions of the form

$$\psi(r, \varphi) = r^m f(\varphi). \tag{4.41}$$

We do not insert this directly in (4.40), which would result in lengthy expressions. Instead (see Rayleigh 1911), inserting (4.41) in (4.38) gives

$$\Delta(r^m f(\varphi)) = \left(m^2 + \frac{d^2}{d\varphi^2}\right) (r^{m-2} f(\varphi)). \tag{4.42}$$

Since m is a constant,

$$\Delta(\Delta(r^m f(\varphi))) = \left(m^2 + \frac{d^2}{d\varphi^2}\right) \Delta(r^{m-2} f(\varphi)). \tag{4.43}$$

Using again (4.42), now with $r^{m-2} f(\varphi)$ instead of $r^m f(\varphi)$, gives

$$\Delta(r^{m-2} f(\varphi)) = \left((m-2)^2 + \frac{d^2}{d\varphi^2}\right) (r^{m-4} f(\varphi)). \tag{4.44}$$

Therefore,

$$\Delta^2(r^m f(\varphi)) = r^{m-4} \left(m^2 + \frac{d^2}{d\varphi^2}\right) \left((m-2)^2 + \frac{d^2}{d\varphi^2}\right) f(\varphi). \tag{4.45}$$

Then $\Delta^2 \psi = 0$ implies that for $\psi(r, \varphi) = r^m f(\varphi)$,

$$\left(m^2 + \frac{d^2}{d\varphi^2}\right) \left((m-2)^2 + \frac{d^2}{d\varphi^2}\right) f(\varphi) = 0. \tag{4.46}$$

The solution of this is (see Rayleigh 1911, p. 191),

$$\psi(r, \varphi) = r^m \left[A \cos(m\varphi) + B \sin(m\varphi) + C \cos((m-2)\varphi) + D \sin((m-2)\varphi)\right]. \tag{4.47}$$

We consider again flows in a corner made up of two straight walls. The walls are now located at polar angles $\varphi = \pm\alpha$, and $0 \leq r < \infty$. The boundary conditions are as follows:

(i) The normal speed $u_\phi = -\partial\psi/\partial r$ vanishes on both walls,

$$\frac{\partial\psi(r, \pm\alpha)}{\partial r} = 0, \tag{4.48}$$

or, with $\psi = r^m f(\varphi)$,

$$\psi(r, \pm\alpha) = 0. \tag{4.49}$$

(ii) For viscous fluids, the tangential wall speed $u_r = r^{-1} \, \partial \psi / \partial \varphi$ vanishes,

$$\frac{\partial \psi (r, \varphi)}{r \, \partial \varphi}\bigg|_{\varphi = \pm \alpha} = 0. \tag{4.50}$$

Thus, for $\Delta \Delta \psi = 0$, on each wall there is one Dirichlet's *and* one Neumann's boundary condition. This matches the result from Sect. 3.3 that the Laplace equation, $\Delta \psi = 0$, requires either a Dirichlet or a Neumann boundary condition on (every part of) the boundary of a domain. Furthermore, we assume that in the upper wedge, $0 < \varphi < \alpha$, the flow is directed toward the corner, and that it is directed outward in the lower wedge, $0 < \varphi < \alpha$. This suggests to assume

$$\begin{aligned} u_r(r, \varphi) &= -u_r(r, -\varphi), \\ u_\varphi(r, \varphi) &= u_\varphi(r, -\varphi), \end{aligned} \tag{4.51}$$

or

$$\begin{aligned} \frac{\partial \psi}{\partial \varphi}(r, \varphi) &= -\frac{\partial \psi}{\partial \varphi}(r, -\varphi), \\ \frac{\partial \psi}{\partial r}(r, \varphi) &= \frac{\partial \psi}{\partial r}(r, -\varphi). \end{aligned} \tag{4.52}$$

Using this in (4.47) gives

$$\psi(r, \varphi) = r^m \left[A \cos(m\varphi) + C \cos((m - 2)\varphi) \right]. \tag{4.53}$$

Inserting this result in the boundary conditions (4.49) and (4.50) gives, in matrix notation,

$$\begin{pmatrix} \cos(m\alpha) & \cos((m - 2)\alpha) \\ m \sin(m\alpha) & (m - 2) \sin((m - 2)\alpha) \end{pmatrix} \cdot \begin{pmatrix} A \\ C \end{pmatrix} = 0. \tag{4.54}$$

Putting the determinant of the matrix to zero,

$$(m - 2) \cos(m\alpha) \, \sin((m - 2)\alpha) = m \sin(m\alpha) \, \cos((m - 2)\alpha). \tag{4.55}$$

This expression is made more symmetric by replacing m by $n = m - 1$, giving

$$(n - 1) \cos((n + 1)\alpha) \, \sin((n - 1)\alpha) = (n + 1) \sin((n + 1)\alpha) \, \cos((n - 1)\alpha), \tag{4.56}$$

or

$$\begin{aligned} &\sin((n + 1)\alpha) \, \cos((n - 1)\alpha) + \cos((n + 1)\alpha) \, \sin((n - 1)\alpha) \\ &= n \cos((n + 1)\alpha) \, \sin((n - 1)\alpha) - n \sin((n + 1)\alpha) \, \cos((n - 1)\alpha). \end{aligned} \tag{4.57}$$

Using

$$\sin(x \pm y) = \sin x \, \cos y \pm \cos y \, \sin x, \tag{4.58}$$

this becomes

$$\boxed{\sin(2n\alpha) = -n \sin(2\alpha)} \tag{4.59}$$

This equation was first derived by Dean and Montagnon (1949), see their equation (4). These authors also give an alternative derivation using complex variables z and $\bar{z}$. For a given opening angle 2α at the corner, Eq. (4.59) may have different solutions for the real number $n > 0$. One is mainly interested in the solution with the smallest n for the following reason. The azimuthal flow speed is $u_\varphi = -\partial \psi / \partial r$. Its limiting value as $r \to 0$ is determined by the smallest power n in r^{n+1}; and only close to the corner at small r, viscosity can be expected to dominate the flow. Dean and Montagnon observe that (4.59) has no real solution $n \in \mathbb{R}$ for angles α below critical value, $\alpha < \alpha_c$, thus n and ψ are complex for $\alpha < \alpha_c$. The critical opening angle is found to be quite large, $2\alpha_c \approx 146°$, and $\alpha < \alpha_c$ can be assumed. Now $\Delta^2 \psi = 0$ is a linear equation, thus $\Delta^2 \operatorname{Re} \psi = \Delta^2 \operatorname{Im} \psi = 0$, and we demand that the physical, i.e., real speed v, is determined by $\operatorname{Re} \psi$. Writing

$$n = p + iq, \tag{4.60}$$

with $p, q \in \mathbb{R}$, gives after substitution into (4.59),

$$\sin(2\alpha p) \, \cosh(2\alpha q) = -p \sin(2\alpha),$$
$$\cos(2\alpha p) \, \sinh(2\alpha q) = -q \sin(2\alpha). \tag{4.61}$$

From the first line in (4.54), with $n = m - 1$,

$$C = -A \, \frac{\cos((n+1)\alpha)}{\cos((n-1)\alpha)}. \tag{4.62}$$

Inserting this in (4.53),

$$\psi(r, \varphi) = B r^{n+1} \left[\cos((n+1)\varphi) \, \cos((n-1)\alpha) - \cos((n-1)\varphi) \, \cos((n+1)\alpha) \right], \tag{4.63}$$

with $B = A / \cos((n-1)\alpha)$. Abbreviating then

$$\xi = \cos((n-1)\alpha) - \cos((n+1)\alpha), \tag{4.64}$$

it holds on the symmetry axis $\varphi = 0$ that

$$\psi(r, 0) = B \xi r^{n+1}. \tag{4.65}$$

The normal speed on this symmetry axis is

$$
\begin{aligned}
u_\varphi(r, 0) &= -\frac{\partial \psi(r, 0)}{\partial r} \\
&= -B(n+1)\xi r^n \\
&= -Br^p(n+1)\xi r^{iq} \\
&= -Br^p(n+1)\xi e^{iq \ln r} \\
&= -Cr^p e^{i(q \ln r + \zeta)},
\end{aligned}
\tag{4.66}
$$

with abbreviations

$$
\begin{aligned}
C &= B\,|(n+1)\xi|, \\
\zeta &= \arg\,[(n+1)\xi].
\end{aligned}
\tag{4.67}
$$

Only the real part is of interest,

$$
u_\varphi(r, 0) = -Cr^p \cos(q \ln r + \zeta).
\tag{4.68}
$$

where C, p, q, and ζ are constants. For $r \to 0$, $\ln r \to -\infty$. Thus, for $r \to 0$, $\cos(q \ln r + \zeta)$ changes sign infinitely often, as does $u_\varphi(r, 0)$. *Therefore, the flow crosses the x-axis infinitely often as $r \to 0$.* From

$$
q \ln r_k + \zeta = -\left(k + \frac{1}{2}\right)\pi, \qquad k = 0, 1, 2, \ldots
\tag{4.69}
$$

follows

$$
\frac{r_k}{r_{k-1}} = e^{-\pi/q} = \text{const.}
\tag{4.70}
$$

The locations $r_1, r_2, \ldots$ of midplane crossings form therefore a geometric sequence. The fluid fills the whole region between the walls. This suggests that in each interval $[r_{k-1}, r_k]$, the flow is essentially an eddy. The turnabout direction of fluid is opposite in neighboring eddies. From (4.68), $|u_\varphi(r, 0)|$ has a maximum at $q \ln r_{k-1/2} + \zeta = -k\pi$. Then

$$
\left|\frac{u_\varphi(r_{k+1/2}, 0)}{u_\varphi(r_{k-1/2}, 0)}\right| = \left(\frac{r_{k+1/2}}{r_{k-1/2}}\right)^p = e^{-\pi p/q} = \text{const,}
\tag{4.71}
$$

using again (4.68). The circulation Γ of an eddy scales with ru_φ. Thus, the circulation also drops in geometric progression with $r \to 0$. Finally, the streamlines $\psi = \text{const}$ of the flow can be determined. Instead of the constant ξ from (4.64), introduce a function

$$
\xi(\varphi) = \cos((n+1)\varphi)\,\cos((n-1)\alpha) - \cos((n-1)\varphi)\,\cos((n+1)\alpha).
\tag{4.72}
$$

From (4.63), then,

$$
\begin{aligned}
\psi(r, \varphi) &= \mathrm{Re}\left[Br^{n+1}\,\xi(\varphi)\right] \\
&= \mathrm{Re}\left[Br^{p+1}e^{iq\ln r}\,\xi(\varphi)\right. \\
&= Br^{p+1}\left[\cos(q\ln r)\,\mathrm{Re}\,\xi(\varphi) - \sin(q\ln r)\,\mathrm{Im}\,\xi(\varphi)\right].
\end{aligned}
\tag{4.73}
$$

Especially for the so-called *dividing* streamline with $\psi = 0$, it follows that

$$
q\ln r = \mathrm{atan}\left(\frac{\mathrm{Re}\,\xi(\varphi)}{\mathrm{Im}\,\xi(\varphi)}\right).
\tag{4.74}
$$

From (4.73), the streamline $r(\varphi)$ and thus $y(x)$ can be calculated numerically. Figure 4.5 shows the flow in a corner with $2\alpha = 30°$. Here the values $p = 8.06$ and $q = 4.20$ from Table 1 in Moffatt (1964) were chosen.

Moffatt (1964) also considers the limit $\alpha \to 0$, for which the corner turns into parallel walls. An eddy sequence still forms in this limit. The eddies have then equal diameter, but decreasing strength (see Fig. 7 in Moffatt). Davis and O'Neill (1977) consider a cylinder 'lying' on a plane. The 2-D flow region is then limited by a circular arc and a straight line; the arc and the straight line meet in a cusp. If the cylinder rotates, an infinite eddy sequence is set up toward the cusp. This configuration allows to control the eddy sequence by changing the rotation speed of the cylinder. The same problem was already solved in parts by Schubert (1967). Finally, a physical explanation for the formation of an eddy sequence in terms of flow *separation* can be found in Davis and O'Neill (1977), p. 552.

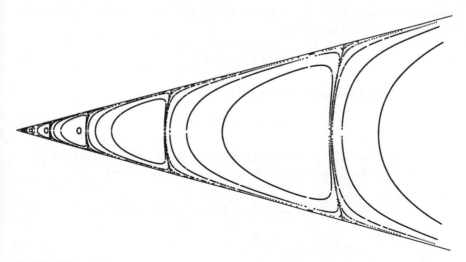

Fig. 4.5 Eddies in the corner flow of a viscous fluid. The red line is for $\psi = 0$

4.4   Flow Past a Flat Plate

All analytic functions $\Phi(z)$ are complex flow potentials. Thus, if $z = z(\zeta)$ is an analytic function and $\Phi(z)$ is the flow potential in the variable z, then $\Phi \circ z$ (function composition) is a flow potential in the variable ζ. For example, let $\zeta = a + ib$ with $a \geq 0$ and $b \in [0, 2\pi]$ be elliptic coordinates, defined by

$$x = \cosh a \, \cos b,$$
$$y = \sinh a \, \sin b. \tag{4.75}$$

This gives

$$\frac{x^2}{\cosh^2 a} + \frac{y^2}{\sinh^2 a} = \cos^2 b + \sin^2 b = 1, \tag{4.76}$$

and

$$\frac{x^2}{\cos^2 b} - \frac{y^2}{\sin^2 b} = \cosh^2 a - \sinh^2 a = 1. \tag{4.77}$$

Thus, the coordinate lines $a = $ const and $b = $ const are ellipses and hyperbolas, respectively,

$$\frac{x^2}{k^2} \pm \frac{y^2}{l^2} = 1, \tag{4.78}$$

with abbreviations $k = \cosh a$, $l = \sinh a$ and $k = \cos b$, $l = \sin b$, respectively. The value $a = 0$ gives the degenerate ellipse $x = \cos b$, $y = 0$, which is the straight line $x \in [-1, 1]$, $y = 0$. The idea is now that this straight line can resemble a flat plate, immersed as an obstacle into the fluid flow. The plate is a streamline, and thus for the streamfunction on the plate let

$$\psi(a = 0, b) = 0. \tag{4.79}$$

To fulfill this equation, a factorization in the form

$$\psi(a, b) = f(b) \sinh a \tag{4.80}$$

is applied. At spatial infinity, the velocity vector (u, v) shall approach the value $(1, 1)$. This means that the plate is tilted at $45°$ against the far flow. Then

$$u = \frac{\partial \psi(a \to \infty, b)}{\partial y} = 1 = \frac{\partial}{\partial y}(y + g(x)),$$
$$v = -\frac{\partial \psi(a \to \infty, b)}{\partial x} = 1 = -\frac{\partial}{\partial x}(-x + h(y)), \tag{4.81}$$

where new functions $g(x)$ and $h(y)$ of one variable were introduced, $\partial g/\partial y = 0 = \partial h/\partial x$. From $y + g(x) = -x + h(y)$ (both sides are identical to $\psi(a \to \infty, b)$), it follows immediately that, apart from constants,

$$g(x) = -x, \qquad\qquad h(y) = y. \qquad\qquad (4.82)$$

On substituting (4.75), this gives

$$\psi(a \to \infty, b) = -\cosh a \,\cos b + \sinh a \,\sin b. \qquad (4.83)$$

This, however, is in conflict with (4.80), which has no term $\cosh a$. The solution to this problem is as follows: Eq. (4.83) holds for $a \to \infty$ only. Yet, in this limit, one cannot distinguish between $\cosh a$ and $\sinh a$, both approaching $\frac{1}{2} e^a$ then. Equation (4.83) is therefore identical to

$$\psi(a \to \infty, b) = \sinh a(\sin b - \cos b). \qquad (4.84)$$

This streamfunction has the right asymptotics, and it obeys the factorization (4.80). This means, it obeys the boundary conditions at $a \to \infty$ and $a = 0$. Being an analytic function, it is thus the unique solution for the streamfunction of a flow past a tilted plate at *all* locations,

$$\psi(a, b) = \sinh a(\sin b - \cos b). \qquad (4.85)$$

For more details on this solution, see Lamb (1932, p. 86) and Milne-Thomson (1968, p. 172). Still more details can be found in Kotschin et al. (1954, p. 249). The streamlines can be given in explicit form: with $\eta = y - \psi$ follows, directly from (4.75), (4.76), (4.77), and (4.85),

$$x = \eta \sqrt{1 + \frac{1}{y^2 + \eta^2}}. \qquad (4.86)$$

The curves $x(y)$ for different constant values of ψ are shown in Fig. 4.6.

One streamline has *stagnation* points on the plate, where $\psi = 0$. Thus, $\psi = 0$ on this streamline, too. Away from stagnation points, (4.85) implies $\cos b = \sin b = 1/\sqrt{2}$

Fig. 4.6 Parallel flow against a plate tilted at 45°. Far away from the plate, the incoming and outgoing flows have $\vec{u} = (1, 1)$. Two stagnation points occur on the plate

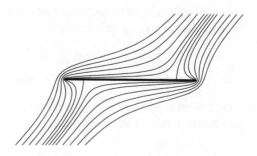

in order to have $\psi = 0$. Inserting this in (4.77), the streamline with stagnation points is given by the hyperbolas $x^2 - y^2 = \frac{1}{2}$. Finally, calculating the fluid speed from the streamfunction (4.85) gives that this speed is infinite at both the terminating corners of the plate.

4.5   The Blasius and Kutta–Joukowski Theorems

In this short section, we derive the *lift* force on a body in a frictionless fluid, which is the reason why airplanes fly. Since the lift force is upward and thus normal to the horizontal flow direction, one may surmise that vorticity ('turnaround' flow) plays a role. This is indeed the case: the lift force is proportional to the flow circulation around the body (e.g., a wing).

The flow of an incompressible fluid shall be stationary and two dimensional in the plane, with negligible friction. A solid body is immersed as an obstacle in the flow, appearing as a forbidden flow area A in the flow plane. All cross sections of the body in the direction normal to the plane shall be identical. The flow past the body is treated again in the complex z-plane. At the surface of the body, the velocity $\vec{u} = (u, v)$ has to be tangent to this surface. Thus, if dx and dy are the differentials along the boundary curve $C = \partial A$, then

$$\frac{dy}{dx} = \frac{v}{u} \quad \text{or} \quad u \, dy = v \, dx. \tag{4.87}$$

Let $w = u - iv$ be the complex speed. Then (see Chorin and Marsden 1990, p. 53 for the following)

$$\begin{aligned} w^2 \, dz &= (u - iv)^2 \, (dx + i dy) \\ &= (u^2 + v^2)(dx - i dy). \end{aligned} \tag{4.88}$$

Since $u^2 + v^2 \in \mathbb{R}$, it follows that

$$\overline{w^2 \, dz} = (u^2 + v^2) \, dz. \tag{4.89}$$

This will be significant in the calculations below. In vector notation, the total pressure force on the body in the $\mathbb{R}^2$-plane is

$$\vec{F} = -L \oint_{C=\partial A} ds \, \hat{n} \, p. \tag{4.90}$$

Here L is the length of the body in the direction normal to the flow plane, $L \, ds$ is the area element on the body surface, and $\hat{n}$ is the normal to C. Evidently,

$$ds \, \hat{n} = (dy, -dx), \tag{4.91}$$

Let $f = (F_x + i F_y)/L$ be the complex force per unit length (e.g., of a wing). Then (4.90) can be transformed to a complex integral, using the rules

$$(F_x, F_y) \rightarrow F_x + i F_y = Lf,$$
$$(dy, -dx) \rightarrow dy - i dx. \tag{4.92}$$

This gives

$$f = -\oint_C (dy - i dx)\, p$$
$$= i \oint_C (dx + i dy)\, p. \tag{4.93}$$

The boundary contour C of the body is a streamline, and thus the Bernoulli equation $p + \frac{1}{2}\rho(u^2 + v^2) = c$ (no gravity) with a constant c along the streamline C can be applied, and gives

$$f = -\frac{1}{2} i\rho \oint_C (u^2 + v^2)\, dz. \tag{4.94}$$

The constant c gives no contribution to f, since $c \oint_C dz = 0$. Using (4.89), it follows finally that

$$\boxed{f = -\frac{1}{2} i\rho \oint_C \overline{w^2}\, dz} \tag{4.95}$$

This formula is called the *Blasius theorem*.

If f is a *lift* (force), i.e., acts normal to the flow direction, then f can be related to the *circulation* Γ around the immersed body. The circulation $\Gamma = \oint_C d\vec{l} \cdot \vec{u}$ along the boundary C of the obstacle, on which $u\, dy = v\, dx$ holds, is

$$\Gamma = \oint_C (u\, dx + v\, dy) = \oint_C (u - iv)\,(dx + i dy), \tag{4.96}$$

or in complex variables,

$$\Gamma = \oint_C w\, dz. \tag{4.97}$$

Assume that the flow has constant horizontal speed $(U, 0)$ far from the body. This implies that the complex speed can be expanded in a Laurent series without any positive powers of z,

$$w = U + \frac{a}{z} + \frac{b}{z^2} + \cdots \tag{4.98}$$

Using Cauchy's integral formula (A.5),

$$\Gamma = \oint w\, dz = \oint \left(U + \frac{a}{z} + \frac{b}{z^2} + \cdots \right) dz = a \oint \frac{dz}{z} = 2\pi i a. \tag{4.99}$$

Inserting this in (4.98) gives for the velocity-square,

$$w^2 = U^2 + \frac{\Gamma U}{\pi i z} + \cdots \qquad (4.100)$$

For the complex force function f in (4.95), then,

$$\begin{aligned}
f &= -\frac{i\rho}{2} \oint_C \overline{w^2}\, dz \\
&= \frac{\rho \Gamma U}{2\pi} \oint_C \overline{dz/z} \\
&= \frac{\rho \Gamma U}{2\pi} \oint_C dz/z \\
&= -i\rho \Gamma U.
\end{aligned} \qquad (4.101)$$

The imaginary unit $i = (0, 1)$ can be identified with the unit *upward* normal $\hat{n}$. Thus, in vector notation,

$$\boxed{\vec{F} = -\rho L \Gamma U\, \hat{n}} \qquad (4.102)$$

This force on the body is *normal* to the far-flow velocity $(U, 0)$, and is proportional to the latter and to the circulation around the body. For $U > 0$ and $\Gamma < 0$, Eq. (4.102) gives a vertical *lift* of the body. In an inertial frame where $u = 0$ at the lower edge y_- of the body, $u > 0$ at the upper edge y_+. From the Bernoulli equation for irrotational flow, i.e., with the global Bernoulli constant, the pressure at y_+ is smaller than that at y_-, i.e., the body is indeed lifted upward.

It should be emphasized that (4.102) holds only if nowhere on the surface of the body cavity formation takes place. This assumption is inherent in the above derivation by assuming that the contour line C along which the circulation is calculated is the body surface. Cavity formation is treated in Chap. 5, among other subjects.

In contrast to the lift, a *drag* force acts *in* the flow direction, trying to carry the immersed object with the flow. We have thus found that in stationary flow of an inviscid fluid, no drag force acts on a solid body. This is the famous d'Alembert paradox. Its origin lies in the assumption of inviscid, slipping boundary conditions. The flow is tangent at the body surface, but the flow speed is not reduced by the presence of the body.

4.6 Plane Flow Past a Cylinder

Consider the stationary, divergence-free, irrotational flow of an inviscid fluid in the complex plane. At large distances from the origin, the flow shall be uniform. The corresponding complex speed potential is, with $U \in \mathbb{R}$,

$$\Phi(z) = Uz, \tag{4.103}$$

giving $w = d\Phi/dz = U$ as desired. A cylinder with radius a is put into the flow at the origin $z = 0$. The cylinder axis is normal to the complex plane. Then the complex potential is

$$\Phi(z) = Uz + \frac{Ua^2}{z}. \tag{4.104}$$

This is seen as follows:

(i) The cross section of the cylinder with the z-plane is a circle C. On this circle, $a^2 = x^2 + y^2 = (x + iy)(x - iy) = z\bar{z}$, or $\bar{z} = a^2/z$. Then (4.104) gives $\Phi = Uz + U\bar{z} = 2Ux$ on the circle C, which is real-valued. With $\Phi = \phi + i\psi$ follows $\psi = 0$ on C. Since isocontours $\psi = \text{const}$ are streamlines, C is a streamline, as intended.

(ii) $\Phi(z) \to Uz$ for $|z| \to \infty$, as in (4.103).

Thus, Φ from (4.104) obeys the correct boundary conditions at $|z| = a$ and $|z| \to \infty$. The potential Φ is analytic for $z \neq 0$, thus $\Delta\phi = \Delta\psi = 0$ (from the Cauchy–Riemann equations): Eq. (4.104) solves the Laplace equation for Neumann boundary conditions on the circle and at infinity. Equation (4.104) is an example of the *circle theorem* (Milne-Thomson 1940), which is stated and proved in Milne-Thomson (1968). The streamlines $\psi = \text{const}$ are symmetric with respect to the transformation $x \to -x$ and $y \to -y$. To (4.104), a straight vortex is added at $z = 0$, see (4.7), giving

$$\Phi(z) = \left(z + \frac{a^2}{z}\right) U + \frac{i\Gamma}{2\pi} \ln \frac{1}{z}. \tag{4.105}$$

On the circle C, $z = ae^{i\varphi}$, and the last term gives a streamfunction $\psi = (2\pi i)^{-1}\Gamma \ln a$, which is still constant on C. For $|z| \to \infty$, however, the flow has now circulation $\Gamma \neq 0$, whereas $\Gamma = 0$ in (4.104). Three streamline patterns for the potential (4.105) for different values of Γ are shown in Figs. 4.7, 4.8, and 4.9. One streamline can cross the circle at two (or one) stagnation point(s). The figures show how the vortex at $z = 0$ dominates over the potential flow U at increasing Γ. In the last figure, part of the fluid is trapped near the cylinder and orbits it along quasi-elliptic streamlines.

Fig. 4.7 Flow with circulation past a cylinder, with two stagnation points. Circulation $\Gamma/2\pi = aU$

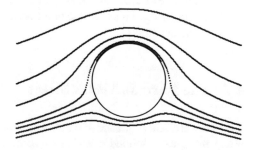

Fig. 4.8 Flow past a
cylinder, with one stagnation
point. Circulation
$\Gamma/2\pi = 2aU$

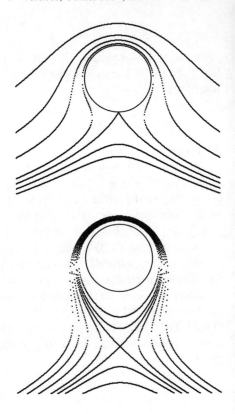

Fig. 4.9 Flow past a
cylinder. Circulation
$\Gamma/2\pi = 3aU$

Consider then the conformal map

$$f(z) = z + \frac{1}{z}. \tag{4.106}$$

Its singular points are, by definition, at $\mathrm{d}f/\mathrm{d}z = 0$ or $z_\pm = \pm 1$. Let C be a circle with center $x > 0$, $y > 0$ that passes through the point $z_- = -1$. Then the curve $f(C)$ resembles the profile of a thin airfoil (Kutta 1902). The streamlines and lift past a cylinder are known from the above formulas. By (4.106) then, they can be obtained for the Kutta airfoil. This estimate of lift forces on airplanes played an important role in early aerodynamics.

4.7　The Kármán Vortex Street

Given is a straight sequence of parallel and identical straight line vortices positioned along the x-axis, termed a *vortex chain* or *vortex row*. The motion is stationary, inviscid, solenoidal, and irrotational, and is treated in the complex plane, with all vortex axes normal to the plane. The chain has infinite extent and the vortex centers $(x_k, 0)$ are at constant distance d,

$$x_k = kd, \qquad k \in \mathbb{Z}. \qquad (4.107)$$

Summing over the complex velocities ('vector addition') $w(z) = d\Phi/dz$ of individual vortices with Φ from (4.7) gives the total velocity law of the vortex row,

$$w(z) = \frac{\Gamma}{2\pi i} \sum_{k=-\infty}^{\infty} \frac{1}{z + kd}. \qquad (4.108)$$

Using the well-known series expansion of the cotangent,

$$\pi \cot(\pi z) = \sum_{k=-\infty}^{\infty} \frac{1}{z + k}, \qquad (4.109)$$

this becomes

$$\boxed{w(z) = \frac{\Gamma}{2id} \cot \frac{\pi z}{d}} \qquad (4.110)$$

This holds everywhere except at the locations $z = kd, k \in \mathbb{Z}$ of the vortex singularities. It will be shown in Sect. 6 that a line vortex moves along with the local fluid velocity $\vec{u}$ that it encounters at its core singularity. This means also that any vortex will *induce* motion of another vortex that resides somewhere in its velocity field, and vice versa. For the vortex chain, it is clear form elementary kinematics that at vortex cores $z = kd$, the speeds induced by vortices at $(k - m)d$ and $(k + m)d$ compensate each other. Thus,

$$w(kd) = 0, \qquad k \in \mathbb{Z} \qquad (4.111)$$

may be assumed. Note, however, that the pairwise cancellation of terms in an infinite sum is a delicate matter.

Next, consider *two* parallel vortex chains in the x-direction, each with a vortex distance d; the lateral separation (y-direction) of the two chains is ih. This configuration is termed a *Kármán vortex street*. The basic flow pattern is shown in Fig. 4.10. It will be found below that the vortices in the two streets must spin in opposite directions, and the two rows must be shifted against each other by $\Delta x = 0$ or $\Delta x = \pm d/2$ in the x-direction, in order that no relative motion of the vortex chains occurs. Double vortex streets with $\Delta x = \pm d/2$ typically form behind cylindrical obstacles in a flow, i.e., circular obstacles in 2-D flow.

> The process by which a vortex street is formed is often called "eddy shedding", but this name may not always be appropriate. In the case of flow past a circular cylinder, it is applicable when the Reynolds number exceeds about 100. Then the attached eddies are periodically shed from the cylinder to form the vortices of the street. Whilst the eddy on one side is being shed, that on the other side is re-forming. (Tritton 1988, p. 27)

Many examples of the Kármán vortex streets can be found on the Internet. A particularly striking case is shown in the photographs, Figs. 4.11 and 4.12, taken

Fig. 4.10 The Kármán vortex street behind an obstacle. Photograph by Jürgen Wagner, at https://de.wikipedia.org/wiki/Datei:Karmansche_Wirbelstr_große_Re.JPG

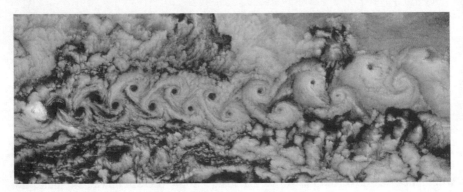

Fig. 4.11 The Kármán vortex street in a wind flow across Jan Mayen Island. Size covered is 365×158 km. Photograph: NASA. Satellite image taken June 2001

with a NASA satellite in the years 2001 and 2009, respectively. The images show Jan Mayen's island in the North Atlantic. The Beerenberg volcano on the island interrupts the air flow and gives rise to a double vortex row. Clouds that are dragged into the vortices spin ever faster when approaching the vortex core, until they are eventually torn apart and give free sight on the ocean surface below.

Tubular structures like chimneys, UHF antenna masts, supporting pillars, or tethered cables may be induced to oscillations in strong winds by the formation of double vortex streets behind these elongated structures. The oscillation is normal to the wind direction, as is intuitively clear from the pattern of the double vortex street. If the oscillation is close to the resonance frequency of the structure, complete destruction may result. In order to reduce vortex-induced vibration, turbulence generators are put on the structure. They have the shape of a helix and are termed the Scruton strakes, see Fig. 4.13.

The Kármán street is modeled as two parallel vortex chains. Let z_+ and z_- be the positions of the two vortices in the upper and lower chains, respectively, that lie closest to $z = 0$. To obtain the velocity law of this configuration, z on the right side

Fig. 4.12 Jan Mayen Island, with Greenland to the northwest. Photograph: NASA. Satellite image taken February 2009

of (4.110) is replaced first by $z - z_+$, then by $z - z_-$, then the resulting equations are added. The complex speed of the two vortex chains is then

$$w(z) = \frac{\Gamma_+}{2id} \cot \frac{\pi(z - z_+)}{d} + \frac{\Gamma_-}{2id} \cot \frac{\pi(z - z_-)}{d}. \qquad (4.112)$$

Each chain induces the other chain to move. Consider the vortex with center at z_+. There is no speed induced on it from the other vortices in its own chain, thus

$$w(z_+) = \frac{\Gamma_-}{2id} \cot \frac{\pi(z_+ - z_-)}{d}. \qquad (4.113)$$

Similarly, for the lower vortex with center at z_-, the velocity induced by the upper chain is

$$w(z_-) = \frac{\Gamma_+}{2id} \cot \frac{\pi(z_- - z_+)}{d}. \qquad (4.114)$$

There is no relative motion of the two chains if $w(z_-) = w(z_+)$, or

$$\Gamma_+ = -\Gamma_-, \qquad (4.115)$$

thus the vortices of the two chains must spin in opposite direction. Furthermore, the chains shall only move in the x-direction, not at some angle with this direction. This

Fig. 4.13 The Helical Scruton strakes on chimneys, to prevent vortex-induced vibrations in the normal direction to an incoming wind. Photograph: Author

means that

$$\text{Im } w(z_+) = \text{Im } w(z_-) = 0. \tag{4.116}$$

Separating real and imaginary parts in the complex function $\cot z$, these conditions give (for details of the simple calculation see Kotschin et al. 1954, p. 189)

$$\sin \frac{2\pi(z_+ - z_-)}{d} = 0. \tag{4.117}$$

Thus, either $z_+ - z_- = 0$ or $z_+ - z_- = \pm d/2$ must hold. Since it is often observed in nature, only the latter case is treated in the following. Note that so far there is no constraint on the separation ih of the two vortex streets. This will follow from a stability analysis of the Kármán vortex street.

In the rest of this section, the *stability* of vortex chains under small perturbations of the positions of the line vortices is analyzed. The perturbations are assumed to be spatially periodic, i.e., to repeat themselves along the vortex street. Furthermore, the perturbations shall be of small scale, i.e., affect neighboring vortices differently: for large-scale perturbations affecting whole groups of neighboring vortices similarly, one expects that the vortices of a group can be replaced by one new, 'effective' vortex that is perturbed relative to its adjacent effective vortices. We start with the single vortex chain, and derive the evolution of perturbations δz in vortex positions. If these $|\delta z|$ decrease with time, the configuration is called stable, otherwise unstable.

We closely follow Kotschin et al. (1954), pp. 191–197, and assume that all vortices with an odd index k remain at their position, whereas all vortices with an even index are shifted by the same complex, infinitesimal δz. This results in *two* straight and parallel vortex chains. In each of these chains, the vortex distance is $2d$. Each of the two chains sets the other chain into motion, according to the equations

$$\frac{d\bar{z}_\pm}{dt} = w(z_\pm) = \frac{\Gamma}{4id} \cot \frac{\pi(z_\pm - z_\mp)}{2d}, \qquad (4.118)$$

where $\Gamma = \Gamma_+ = \Gamma_-$. Note that $\bar{z}$ appears on the left side, since $w = u - iv = \dot{x} - i\dot{y} = \dot{\bar{z}}$. Abbreviate

$$\zeta = \frac{\pi(z_\pm - z_\mp)}{2d}. \qquad (4.119)$$

Subtracting then the two Eqs. (4.118) gives

$$\frac{d\bar{\zeta}}{dt} = \frac{\pi\Gamma}{4id^2} \cot \zeta. \qquad (4.120)$$

Taking the complex conjugate of this equation gives

$$\frac{d\zeta}{dt} = -\frac{\pi\Gamma}{4id^2} \cot \bar{\zeta}, \qquad (4.121)$$

and dividing the two equations,

$$\frac{d\bar{\zeta}}{d\zeta} = -\frac{\cot \zeta}{\cot \bar{\zeta}}, \qquad (4.122)$$

or

$$\cot \zeta \, d\zeta + \cot \bar{\zeta} \, d\bar{\zeta} = 0. \qquad (4.123)$$

This can be written with total differentials,

$$d \ln \sin \zeta + d \ln \sin \bar{\zeta} = 0, \qquad (4.124)$$

and thus can be integrated to give

$$\ln \sin \zeta + \ln \sin \bar{\zeta} = \ln c, \qquad (4.125)$$

with constant c, or

$$\sin \zeta \, \sin \bar{\zeta} = c. \qquad (4.126)$$

With $\zeta = \xi + i\eta$, this becomes after minor algebra,

$$\sin^2 \xi + \sinh^2 \eta = c. \qquad (4.127)$$

The curves (ξ, η) in the z-plane resulting from (4.127) give the vortex trajectories from mutual induction of motion. A discussion of these curves shows (see Kotschin

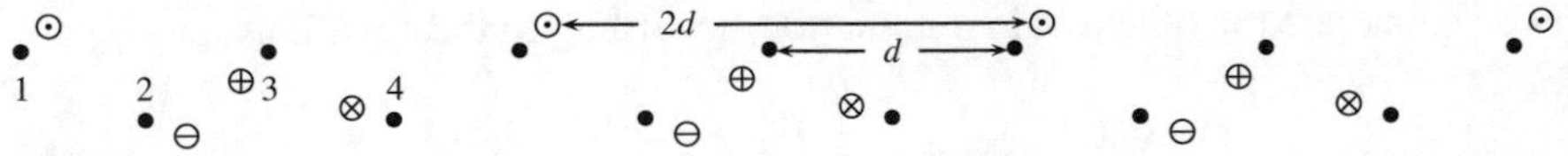

Fig. 4.14 Periodic perturbation of a Kármán vortex street. Unperturbed vortex cores are marked with bullets • at horizontal distance d. The perturbations δz_k of four subsequent vortices are independent, and repeat periodically to the left and right. This results in four parallel vortex streets, with vortex cores marked $\odot$, $\ominus$, $\oplus$, $\otimes$, each street with horizontal vortex distance $2d$. The δz_k in the figure are chosen to obey (4.137)

et al. 1954, p. 191) that all δz are unstable in this configuration. The single straight vortex chain is therefore unconditionally unstable.

The next major issue is to analyze the stability of the Kármán vortex street. The vortices in the upper chain shall carry odd indices $\ldots, z_{-3}, z_{-1}, z_1, z_3, \ldots$, and the vortices in the lower chain shall carry even indices $\ldots, z_{-2}, z_0, z_2, \ldots$ The vortices in the upper and lower chains shall have circulation Γ and $-\Gamma$, respectively. Furthermore, let $z_1 < z_2 < z_3 < z_4$ (see Fig. 4.14), and only the vortex street with a distance $z_2 - z_1 = d/2 + ih$ between vortices in the unperturbed upper and lower chains is treated. These four vortices experience independent displacements $\delta z_1, \delta z_2, \delta z_3, \delta z_4$. The perturbations of this vortex quartet repeat themselves periodically to the left and right along the street. The vortex distance in each of the four chains is $2d$. We follow now the analysis given by Schmieden (1936), who considered this perturbation configuration first. The complex speed of the unperturbed vortex street is

$$w(z) = \frac{\Gamma}{4id} \left(\cot \frac{\pi(z - z_1)}{2d} - \cot \frac{\pi(z - z_2)}{2d} + \cot \frac{\pi(z - z_3)}{2d} - \cot \frac{\pi(z - z_4)}{2d} \right).$$
$$(4.128)$$

The relations between z_1 and z_4 in the unperturbed street are simple, and will be employed below. The vortex speeds in the four chains are

$$\frac{d\bar{z}_1}{dt} = \frac{\Gamma}{4id} \left(-\cot \frac{\pi(z_1 - z_2)}{2d} + \cot \frac{\pi(z_1 - z_3)}{2d} - \cot \frac{\pi(z_1 - z_4)}{2d} \right),$$

$$\frac{d\bar{z}_2}{dt} = \frac{\Gamma}{4id} \left(+\cot \frac{\pi(z_2 - z_1)}{2d} + \cot \frac{\pi(z_2 - z_3)}{2d} - \cot \frac{\pi(z_2 - z_4)}{2d} \right),$$

$$\frac{d\bar{z}_3}{dt} = \frac{\Gamma}{4id} \left(+\cot \frac{\pi(z_3 - z_1)}{2d} - \cot \frac{\pi(z_3 - z_2)}{2d} - \cot \frac{\pi(z_3 - z_4)}{2d} \right),$$

$$\frac{d\bar{z}_4}{dt} = \frac{\Gamma}{4id} \left(+\cot \frac{\pi(z_4 - z_1)}{2d} - \cot \frac{\pi(z_4 - z_2)}{2d} + \cot \frac{\pi(z_4 - z_3)}{2d} \right), \quad (4.129)$$

where infinite terms cot 0 were dropped, i.e., there shall be no self-induced speed of the street. According to the analysis above, this can be achieved by a Galilei transformation in the x-direction. Equation (4.129) is Eq. (4) in Domm (1956), after renaming $z_2 \leftrightarrow z_3$. Each vortex position z_k is perturbed by δz_k, with $k = 1, 2, 3, 4$, and on the left side of (4.129), $\delta(d\bar{z}_k/dt) = d(\delta\bar{z}_k)/dt$ is used. This gives, to first order in the perturbations,

$$\frac{d\,\delta\bar{z}_1}{dt} = \frac{\pi\Gamma}{8id^2}\left(\frac{\delta z_1 - \delta z_2}{\sin^2(\pi(z_1 - z_2)/2d)} - \frac{\delta z_1 - \delta z_3}{\sin^2(\pi(z_1 - z_3)/2d)} + \frac{\delta z_1 - \delta z_4}{\sin^2(\pi(z_1 - z_4)/2d)}\right),$$
$$(4.130)$$

plus three more equations of the same type. Introduce a nondimensional number

$$\alpha = \frac{\pi}{4} + \frac{i\pi}{2}\frac{h}{d}. \tag{4.131}$$

Then for the unperturbed configuration with vortex distance d in each chain and separation ih of the two chains,

$$z_1 - z_2 = -\frac{d}{2} + ih, \qquad \frac{\pi(z_1 - z_2)}{2d} = \alpha - \frac{\pi}{2},$$

$$z_1 - z_3 = -d, \qquad \frac{\pi(z_1 - z_3)}{2d} = -\frac{\pi}{2},$$

$$z_1 - z_4 = -\frac{3d}{2} + ih, \qquad \frac{\pi(z_1 - z_4)}{2d} = \alpha - \pi,$$

$$z_2 - z_3 = -\frac{d}{2} - ih, \qquad \frac{\pi(z_2 - z_3)}{2d} = -\alpha,$$

$$z_2 - z_4 = -d, \qquad \frac{\pi(z_2 - z_4)}{2d} = -\frac{\pi}{2},$$

$$z_3 - z_4 = -\frac{d}{2} + ih, \qquad \frac{\pi(z_3 - z_4)}{2d} = \alpha - \frac{\pi}{2}. \tag{4.132}$$

Inserting this in (4.130) and the corresponding other three equations gives

$$\frac{d\,\delta\bar{z}_1}{dt} = \frac{\gamma}{i}\left(\frac{\delta z_1 - \delta z_2}{\cos^2\alpha} - (\delta z_1 - \delta z_3) + \frac{\delta z_1 - \delta z_4}{\sin^2\alpha}\right),$$

$$\frac{d\,\delta\bar{z}_2}{dt} = \frac{\gamma}{i}\left(-\frac{\delta z_2 - \delta z_1}{\cos^2\alpha} - \frac{\delta z_2 - \delta z_3}{\sin^2\alpha} + (\delta z_2 - \delta z_4)\right),$$

$$\frac{d\,\delta\bar{z}_3}{dt} = \frac{\gamma}{i}\left(-(\delta z_3 - \delta z_1) + \frac{\delta z_3 - \delta z_2}{\sin^2\alpha} + \frac{\delta z_3 - \delta z_4}{\cos^2\alpha}\right),$$

$$\frac{d\,\delta\bar{z}_4}{dt} = \frac{\gamma}{i}\left(-\frac{\delta z_4 - \delta z_1}{\sin^2\alpha} + (\delta z_4 - \delta z_2) - \frac{\delta z_4 - \delta z_3}{\cos^2\alpha}\right), \tag{4.133}$$

where

$$\gamma = \frac{\pi \Gamma}{8d^2}. \tag{4.134}$$

A remarkable property of this system is that the number

$$\bar{C} = \delta \bar{z}_1 + \delta \bar{z}_3 - \delta \bar{z}_2 - \delta \bar{z}_4 \tag{4.135}$$

is conserved during perturbative motion and growth,

$$\frac{d\bar{C}}{dt} = 0. \tag{4.136}$$

According to Saffman (1995), p. 136, this is a consequence of momentum conservation. In the following, only such perturbation configurations are considered for which

$$\delta z_3 = -\delta z_1,$$
$$\delta z_4 = -\delta z_2. \tag{4.137}$$

Other patterns are possible, and our results will be limited by this restriction. Equation (4.137) means that adjacent vortices in each chain move in opposite directions. For this pattern, thus, the last two equations in (4.133) become identical to the first two. The remaining two equations are

$$\frac{d\,\delta \bar{z}_1}{dt} = -\gamma(iA\,\delta z_1 + B\,\delta z_2),$$
$$\frac{d\,\delta \bar{z}_2}{dt} = \gamma(B\,\delta z_1 + iA\,\delta z_2), \tag{4.138}$$

with real numbers

$$A = \frac{1}{\sin^2 \alpha} + \frac{1}{\cos^2 \alpha} - 2 = \frac{4}{\cosh^2 \pi h/d} - 2,$$
$$B = \frac{i}{\sin^2 \alpha} - \frac{i}{\cos^2 \alpha} = 4\,\frac{\sinh(\pi h/d)}{\cosh^2(\pi h/d)}, \tag{4.139}$$

where α from (4.131) was inserted. Put now

$$\delta z_1 = \xi_1 + i\eta_1,$$
$$\delta z_2 = \xi_2 + i\eta_2, \tag{4.140}$$

which gives

$$\begin{aligned}
\dot{\xi}_1 &= \gamma(A\eta_1 - B\xi_2), \\
\dot{\eta}_1 &= \gamma(A\xi_1 + B\eta_2), \\
\dot{\xi}_2 &= \gamma(B\xi_1 - A\eta_2), \\
\dot{\eta}_2 &= -\gamma(B\eta_1 + A\xi_2).
\end{aligned} \tag{4.141}$$

Let $\dot{\xi}_1 = \omega\xi_1$, $\dot{\eta}_1 = \omega\eta_1$, $\dot{\xi}_2 = \omega\xi_2$, and $\dot{\eta}_2 = \omega\eta_2$, and thus

$$\begin{pmatrix} \omega & -\gamma A & \gamma B & 0 \\ -\gamma A & \omega & 0 & -\gamma B \\ -\gamma B & 0 & \omega & \gamma A \\ 0 & \gamma B & \gamma A & \omega \end{pmatrix} \cdot \begin{pmatrix} \xi_1 \\ \eta_1 \\ \xi_2 \\ \eta_2 \end{pmatrix} = \begin{pmatrix} 0 \\ 0 \\ 0 \\ 0 \end{pmatrix}. \tag{4.142}$$

The determinant of this matrix must vanish, giving

$$\omega^4 - 2\gamma^2(A^2 - B^2)\,\omega^2 + \gamma^4(A^2 + B^2)^2 = 0. \tag{4.143}$$

The four roots of this equation are

$$\omega = \pm\gamma(A \pm iB). \tag{4.144}$$

From $\dot{\xi}_1 = \omega\xi_1$ follows $\xi_1 \sim e^{\omega t}$, etc. Thus, there is exponential growth and instability, unless

$$A = 0, \tag{4.145}$$

or, according to (4.139), $\cosh^2 \pi h/d = 2$, or

$$\boxed{\sinh \frac{\pi h}{d} = 1} \tag{4.146}$$

This is a famous result first derived by Kármán (1911). Roughly,

$$\frac{h}{d} \approx 0.2806 \tag{4.147}$$

for the unique ratio of the separation of the two streets to the vortex separation within a street that is stable to first order. This result is also derived in Lamb (1932), pp. 225–228.

We turn to a major achievement by Dolaptschiew (1937) and derive a stability criterion for the Kármán street from perturbations up to *second order*. Any perturbation analysis that can be carried out beyond first order is quite unique. Second-order perturbations for the Kármán street were first considered by Schmieden (1936). We follow here the calculation given by Domm (1956), and introduce perturbations δz_1

to δz_4 in (4.129) and perform a Taylor expansion up to second order using (4.131). This gives

$$
\cot \frac{\pi(z_1 + \delta z_1 - z_2 - \delta z_2)}{2d} = -\tan\alpha - \frac{\pi}{2d}\frac{\delta z_1 - \delta z_2}{\cos^2\alpha} - \frac{\pi^2}{4d^2}\frac{\sin\alpha}{\cos^3\alpha}(\delta z_1 - \delta z_2)^2,
$$

$$
\cot \frac{\pi(z_1 + \delta z_1 - z_3 - \delta z_3)}{2d} = -\frac{\pi}{2d}(\delta z_1 - \delta z_3),
$$

$$
\cot \frac{\pi(z_1 + \delta z_1 - z_4 - \delta z_4)}{2d} = \cot\alpha - \frac{\pi}{2d}\frac{\delta z_1 - \delta z_4}{\sin^2\alpha} + \frac{\pi^2}{4d^2}\frac{\cos\alpha}{\sin^3\alpha}(\delta z_1 - \delta z_4)^2,
$$

$$
\tag{4.148}
$$

and similarly for the remaining nine cot-terms. Subtracting the unperturbed system (4.129) from this, (4.133) now becomes, up to second order,

$$
\dot{\overline{\delta z_1}} = \frac{\gamma}{i}\left(-\delta z_1 + \delta z_3 + \frac{\delta z_1 - \delta z_4}{s^2} + \frac{\delta z_1 - \delta z_2}{c^2} - \frac{lc}{s^3}(\delta z_1 - \delta z_4)^2 + \frac{ls}{c^3}(\delta z_1 - \delta z_2)^2\right),
$$

$$
\dot{\overline{\delta z_2}} = \frac{\gamma}{i}\left(\delta z_2 - \delta z_4 - \frac{\delta z_2 - \delta z_3}{s^2} - \frac{\delta z_2 - \delta z_1}{c^2} - \frac{lc}{s^3}(\delta z_2 - \delta z_3)^2 + \frac{ls}{c^3}(\delta z_2 - \delta z_1)^2\right),
$$

$$
\dot{\overline{\delta z_3}} = \frac{\gamma}{i}\left(-\delta z_3 + \delta z_1 + \frac{\delta z_3 - \delta z_2}{s^2} + \frac{\delta z_3 - \delta z_4}{c^2} - \frac{lc}{s^3}(\delta z_3 - \delta z_2)^2 + \frac{ls}{c^3}(\delta z_3 - \delta z_4)^2\right),
$$

$$
\dot{\overline{\delta z_4}} = \frac{\gamma}{i}\left(\delta z_4 - \delta z_2 - \frac{\delta z_4 - \delta z_1}{s^2} - \frac{\delta z_4 - \delta z_3}{c^2} - \frac{lc}{s^3}(\delta z_4 - \delta z_1)^2 + \frac{ls}{c^3}(\delta z_4 - \delta z_3)^2\right),
$$

$$
\tag{4.149}
$$

where the following abbreviations were used,

$$
s = \sin\alpha, \qquad\qquad c = \cos\alpha, \qquad\qquad l = \frac{\pi}{2d}. \tag{4.150}
$$

Note that the indices k and l in $\delta z_k - \delta z_l$ agree in the second and fourth columns, and also in the third and fifth columns. Equations (4.149) correspond to Eqs. (6) in Domm (1956). The conservation law (4.136) still holds, $d\bar{C}/dt = 0$. The reason is that in $\bar{C}$ and only in $\bar{C}$, nonlinear terms cancel out. This suggests to consider *combinations* of δz_1 to δz_4 with two sign pairs,

$$
C = \delta z_1 - \delta z_2 + \delta z_3 - \delta z_4,
$$

$$
D = \delta z_1 + \delta z_2 - \delta z_3 - \delta z_4,
$$

$$
E = \delta z_1 - \delta z_2 - \delta z_3 + \delta z_4. \tag{4.151}
$$

Both in D and E, half of the nonlinear terms still cancel out. Domm makes the following interesting observation: for D and E, *linear* evolution equations hold, i.e., including perturbations to first order only. Namely, directly from (4.149),

$$\dot{D} = \frac{2\gamma}{i}\left(\frac{s^2}{c^2} + \frac{ls}{c^3}\,C\right)E,$$

$$\dot{E} = \frac{2\gamma}{i}\left(\frac{c^2}{s^2} - \frac{lc}{s^3}\,C\right)D, \tag{4.152}$$

where C is known from above to be a constant. The numbers C, D, E determine the distances of vortices via the equations

$$\delta z_1 - \delta z_4 = (C + D)/2,$$
$$\delta z_1 - \delta z_2 = (C + E)/2,$$
$$\delta z_3 - \delta z_2 = (C - D)/2,$$
$$\delta z_3 - \delta z_4 = (C - E)/2. \tag{4.153}$$

A necessary condition for stability is that these distances are bounded. A short calculation using (4.131) for α gives

$$sc = \cosh\,\pi h/d = \sqrt{2},$$

$$\frac{s^2}{c^2} = \frac{1 + i\sinh\,\pi h/d}{1 - i\sinh\,\pi h/d} = i. \tag{4.154}$$

In both equations, the Kármán criterion (4.146) was used, since we look here only for new stability criteria from second order, assuming that the first-order stability conditions are obeyed. Then (4.152) becomes

$$\dot{D} = \ \ 2\gamma(1 + \lambda C)E,$$
$$\dot{E} = -2\gamma(1 - \lambda C)D, \tag{4.155}$$

where

$$\lambda = l/\sqrt{2}. \tag{4.156}$$

Writing $C = C_1 + iC_2$, $D = D_1 + iD_2$, and $E = E_1 + iE_2$, Eq. (4.155) becomes

$$\begin{pmatrix}\dot{D}_1 \\ \dot{D}_2 \\ \dot{E}_1 \\ \dot{E}_2\end{pmatrix} = 2\gamma\begin{pmatrix} 0 & 0 & 1 + \lambda C_1 & -\lambda C_2 \\ 0 & 0 & -\lambda C_2 & -1 - \lambda C_1 \\ -1 + \lambda C_1 & -\lambda C_2 & 0 & 0 \\ -\lambda C_2 & 1 - \lambda C_1 & 0 & 0\end{pmatrix}\cdot\begin{pmatrix}D_1 \\ D_2 \\ E_1 \\ E_2\end{pmatrix}. \tag{4.157}$$

Let again $\dot{D}_1 = \omega D_1, \ldots, \dot{E}_2 = \omega E_2$, giving the linear homogeneous system

$$\begin{pmatrix} \Omega & 0 & -1-\lambda C_1 & \lambda C_2 \\ 0 & \Omega & \lambda C_2 & 1+\lambda C_1 \\ 1-\lambda C_1 & \lambda C_2 & \Omega & 0 \\ \lambda C_2 & -1+\lambda C_1 & 0 & \Omega \end{pmatrix} \cdot \begin{pmatrix} D_1 \\ D_2 \\ E_1 \\ E_2 \end{pmatrix} = 0, \qquad (4.158)$$

where

$$\Omega = \frac{\omega}{2\gamma}. \qquad (4.159)$$

The determinant of the matrix in (4.158) must vanish. This leads, after some lengthy expressions, to the simple equation

$$\left[\Omega^2 + 1 - \lambda^2 (C_1^2 + C_2^2) \right]^2 + 4\lambda^2 C_2^2 = 0, \qquad (4.160)$$

or

$$\Omega = \pm \left[-1 + \lambda^2 (C_1^2 + C_2^2) \right) \pm 2i\lambda C_2 \right]^{1/2}. \qquad (4.161)$$

To have stability, ω and thus Ω must be purely imaginary. This implies

$$\boxed{C_2 = 0 \qquad \text{and} \qquad |\lambda C_1| < 1} \qquad (4.162)$$

The latter inequality means

$$\frac{|\xi_1 + \xi_3 - \xi_2 - \xi_4|}{d} < \frac{2\sqrt{2}}{\pi}, \qquad (4.163)$$

which is fulfilled since small perturbations were assumed. The equation $C_2 = \eta_1 + \eta_3 - \eta_2 - \eta_4 = 0$, however, is a new condition for the stability of the Kármán street. It says that changes δh in the width h of the street by perturbations δz_k lead to instability. This is *not* identical to the statement derived from a first-order analysis that the width of the street must have a unique value (4.146): to first order, small perturbations δz_k about this unique value for the unperturbed street are indeed stable, but they are stable to second order only when $C_2 = 0$. The condition $C_2 = 0$ poses a restriction on the allowed perturbations δz_k themselves, not on the parameters of the unperturbed street as does (4.146). Perturbations can usually not be controlled to obey equations like $C_2 = 0$, but arise randomly. Therefore, Domm (1956) concludes that the Kármán vortex street is unconditionally unstable to second order, and that double vortex rows obeying (4.146) are only the least unstable ones. Remarkably though, the quite stable Kármán streets with a ratio $h/d \approx 0.28$ are found in nature and in experiments.

4.8 Corner Eddy

In this section, we consider the flow of an inviscid fluid with finite vorticity, $\nabla \times \vec{u} \neq 0$, following a paper by Yih (1959). The most prominent feature of the flow is the occurrence of two corner eddies. The flow domain is a semi-infinite, straight canal of finite width in the two-dimensional plane with Cartesian coordinates x and y, see Fig. 4.15. This figure shows only the upper half of the configuration, which is mirror-symmetric with respect to the x-axis. For $x \to -\infty$, the streamlines are parallel straight lines. At $x = 0$, the motion is stopped at a solid wall across the canal, and the fluid is forced into a sink at the point $x = y = 0$ on the wall (i.e., in 3-D into a straight slit normal to the flow plane).

The flow is stationary and divergence-free, but is *not* irrotational. The flow region is given by $-\infty < x < 0$ and $-1 < y < 1$, with $y = 0$ as symmetry axis. With the z-direction normal to the plane, let

$$\vec{u} = \nabla \times \psi \hat{z} \tag{4.164}$$

in order to fulfill $\nabla \cdot \vec{u} = 0$. The vorticity is then, with $\nabla \cdot \psi \hat{z} = \partial_z \psi(x, y)\hat{z} = 0$, given by

$$\vec{\omega} = \nabla \times \vec{u} = \nabla \times \nabla \times \psi \hat{z} = -\Delta \psi \hat{z} + \nabla(\nabla \cdot \psi \hat{z}) = -\Delta \psi \hat{z}. \tag{4.165}$$

With $\vec{\omega} = \omega \hat{z}$, the two-dimensional Poisson equation is obtained,

$$\frac{\partial^2 \psi}{\partial x^2} + \frac{\partial^2 \psi}{\partial y^2} = -\omega, \tag{4.166}$$

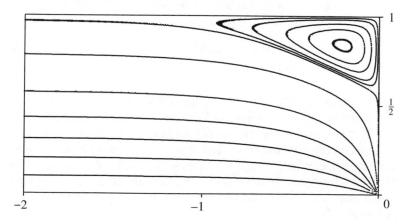

Fig. 4.15 Yih's corner eddy after Eqs. (4.192) and (4.200)

where $\omega = \omega(\psi)$. Yih (1959) equips the flow with vorticity by assuming for the speed $\vec{u} = (u, 0)$ far from the wall

$$u(x \to -\infty, y) = \cos \frac{\pi y}{2}, \tag{4.167}$$

using again normalized units. The horizontal speed u vanishes on the canal walls $y = \pm 1$, and attains a maximum on the symmetry axis $y = 0$, as is the case for viscous fluids. Here, however, (4.167) enters only as boundary condition for an inviscid fluid with vorticity. The streamfunction is then

$$\psi(x \to -\infty, y) = \int_0^y dy' \cos \frac{\pi y'}{2} = \frac{2}{\pi} \sin \frac{\pi y}{2}. \tag{4.168}$$

The vorticity fed into the system at $x \to -\infty$ is thus

$$\omega = -\Delta\psi = -\partial_y^2 \psi = \frac{\pi^2}{4} \psi. \tag{4.169}$$

Therefore, Eq. (4.166) becomes, for $x \to -\infty$,

$$\boxed{\frac{\partial^2 \psi}{\partial x^2} + \frac{\partial^2 \psi}{\partial y^2} = -\frac{\pi^2}{4} \psi} \tag{4.170}$$

This is a linear equation, which will be solved in the rest of this section: Yih makes the strong assumption that (4.170) actually holds for *all* x, since else no analytic solution can be obtained. The boundary conditions for (4.170) are

$$\psi(x \to -\infty, -1 \le y \le 1) = 2\pi^{-1} \sin(\pi y/2), \tag{4.171}$$

$$\psi(x, \pm 1) = \pm 2\pi^{-1}, \tag{4.172}$$

$$\psi(0, 0 < y \le 1) = +2\pi^{-1}, \tag{4.173}$$

$$\psi(0, -1 \le y < 0) = -2\pi^{-1}. \tag{4.174}$$

The first line repeats (4.168). The second line follows from the fact that the canal walls are streamlines with $\psi = \text{const}$, and from continuity at corners. The third line expresses that the normal speed at the solid wall at $x = 0$ vanishes, $u = \partial\psi/\partial y = 0$, and uses again continuity at a corner. The fourth line expresses the same at the lower half of the terminating wall. The point $(0, 0)$ is the point sink, or a straight slit in 3-D. All the fluid escapes through this slit. The flow has to turn from the upper and lower canal walls to this sink. This leads to the formation of corner eddies near the points $(0, 1)$ and $(0, -1)$.

We solve now (4.170). First, note that $\psi(x \to -\infty, y)$ from (4.171) does indeed obey (4.170). To ensure thus the boundary condition (4.171), put

$$\psi(x, y) = \psi_0(y) + \psi_1(x)\psi_2(y), \tag{4.175}$$

where

$$\psi_0(y) = \psi(x \to -\infty, y). \tag{4.176}$$

Inserting this into (4.170) gives

$$\psi_2\psi_{1xx} + \psi_1\psi_{2yy} = -\frac{\pi^2}{4}\,\psi_1\psi_2. \tag{4.177}$$

The corresponding homogeneous equation is, after division by $\chi_1\chi_2$ (a possible value of zero poses no problem here),

$$\frac{\chi_{1xx}}{\chi_1} + \frac{\chi_{2yy}}{\chi_2} = 0, \tag{4.178}$$

The first term on left depends only on x and the second only on y. To have them equal for all x and y, both terms must be constant, viz.,

$$\frac{\chi_{1xx}}{\chi_1} = m^2, \qquad \frac{\chi_{2yy}}{\chi_2} = -m^2, \tag{4.179}$$

with a constant m. From the boundary conditions above, χ_2 is odd in y, thus

$$\chi_1(x) = e^{mx}, \tag{4.180}$$
$$\chi_2(y) = \sin(my). \tag{4.181}$$

The appropriate boundary conditions for χ_1 and χ_2 are

$$\chi_1(-\infty) = 0, \tag{4.182}$$
$$\chi_2(\pm 1) = 0. \tag{4.183}$$

Equation (4.182) is fulfilled if $m > 0$ in (4.180), and (4.183) is identically satisfied if $m = n\pi$ in (4.181), with $n \in \mathbb{Z}$, and takes together, $m = n\pi$ with $n \in \mathbb{N}$. The solution of the homogeneous equation is therefore

$$\chi_1(x)\chi_2(y) = \sum_{n=1}^{\infty} C_n \sin(n\pi y)\, e^{n\pi x}. \tag{4.184}$$

To solve the inhomogeneous equation, *variation of the constant* is used,

$$\psi_1(x)\psi_2(y) = \sum_{n=1}^{\infty} C_n(x) \sin(n\pi y)\, e^{n\pi x}. \tag{4.185}$$

Insertion into (4.170) gives

$$0 = \sum_{n=1}^{\infty} \left(n^2\pi^2 C_n + 2n\pi C_n' + C_n'' - n^2\pi^2 C_n + \frac{\pi^2}{4} C_n \right) \sin(n\pi y)\, e^{n\pi x}, \quad (4.186)$$

where $C_n' = \mathrm{d}C_n/\mathrm{d}x$. Orthogonality of the functions $\sin(n\pi y)$ leads to

$$C_n'' + 2\pi n C_n' + \frac{\pi^2}{4} C_n = 0. \quad (4.187)$$

We try to solve this by

$$C_n(x) = c_n\, e^{N\pi x}, \quad (4.188)$$

where $N = N(n)$ is a function of n to be found. Substitution into (4.187) gives

$$N^2 + 2nN + \frac{1}{4} = 0, \quad (4.189)$$

or

$$N = -n \pm \sqrt{n^2 - 1/4}. \quad (4.190)$$

Inserting (4.188) and (4.190) into (4.185) leads to the expression

$$\psi_1(x)\,\psi_2(y) = \sum_{n=1}^{\infty} c_n \sin(n\pi y)\, e^{\pi x \sqrt{n^2 - 1/4}}, \quad (4.191)$$

where the plus sign of the square root is taken to obey (4.182). The full solution for the streamfunction is therefore

$$\boxed{\psi(x, y) = \frac{2}{\pi} \sin\frac{\pi y}{2} + \sum_{n=1}^{\infty} c_n \sin(n\pi y)\, e^{\pi x \sqrt{n^2 - 1/4}}} \quad (4.192)$$

The constants c_n are determined from the remaining boundary conditions (4.173) and (4.174) on the wall at $x = 0$ across the flow. For $x = 0$ and $y > 0$, (4.192) becomes then

$$\sum_{n=1}^{\infty} c_n \sin(n\pi y) = \frac{2}{\pi}\left(1 - \sin\frac{\pi y}{2} \right). \quad (4.193)$$

This equation can be solved by the Fourier series expansion of the right side. Let $\eta = \pi y$. Then the function $\sin(\eta/2)$ has period 4π. We perform an expansion not in terms of $\sin(n\eta)$, but of $|\sin(n\eta)|$. The variable y is restricted to $y \in [0, 1]$ instead of $[-1, 1]$. Then the standard Fourier transform formula gives

$$\left|\sin\frac{\eta}{2}\right| = \sum_{n=1}^{\infty} b_n |\sin(n\eta)|, \tag{4.194}$$

where

$$
\begin{aligned}
b_n &= \frac{1}{\pi} \int_{-\pi}^{\pi} d\eta \left|\sin(n\eta) \ \sin\frac{\eta}{2}\right| \\
&= \frac{2}{\pi} \int_0^{\pi} d\eta \ \sin(n\eta) \ \sin\frac{\eta}{2} \\
&= \frac{2}{\pi} \left[\frac{\sin((n-1/2)\eta)}{2n-1} - \frac{\sin((n+1/2)\eta)}{2n+1}\right]_0^{\pi} \\
&= \frac{2}{\pi} \left(\frac{(-1)^{n+1}}{2n-1} - \frac{(-1)^n}{2n+1}\right) \\
&= (-1)^{n+1} \frac{2}{\pi} \frac{4n}{4n^2-1}.
\end{aligned} \tag{4.195}
$$

Thus, for $\eta \in [0, \pi]$,

$$-\frac{2}{\pi}\sin\frac{\eta}{2} = \frac{4}{\pi^2}\sum_{n=1}^{\infty}(-1)^n \frac{4n}{4n^2-1} \sin(n\eta). \tag{4.196}$$

Next, we expand the 1 in the bracket on the right side of (4.193) in a Fourier series,

$$1 = \sum_{n=1}^{\infty} b_n |\sin(n\eta)|, \tag{4.197}$$

with the Fourier coefficients

$$b_n = \frac{2}{\pi} \int_0^{\pi} d\eta \ \sin(n\eta) = \frac{2}{\pi n}\left(1 - (-1)^n\right). \tag{4.198}$$

The right side of (4.193) now becomes, for $y \in [0, 1]$,

$$
\begin{aligned}
\frac{2}{\pi}\left(1 - \sin\frac{\pi y}{2}\right) &= \frac{4}{\pi^2}\sum_{n=1}^{\infty}\left(\frac{1}{n} + (-1)^n\left[\frac{4n}{4n^2-1} - \frac{1}{n}\right]\right)\sin(n\pi y) \\
&= \frac{4}{\pi^2}\sum_{n=1}^{\infty}\left(\frac{1}{n} + \frac{(-1)^n}{n(4n^2-1)}\right)\sin(n\pi y).
\end{aligned} \tag{4.199}
$$

Comparing this with the left side of (4.193) gives finally

$$\boxed{c_n = \frac{4}{\pi^2}\left(\frac{1}{n} + \frac{(-1)^n}{n(4n^2-1)}\right)} \tag{4.200}$$

This, together with (4.192), establishes the full solution. The streamlines are again the isocontours $\psi = \text{const}$ of the streamfunction ψ. Figure 4.15 shows a scatter plot of this solution. Here the first 200 terms in the sum (4.192) were added up. The $-x$-axis corresponds to $\psi = 0$. The spacing is $\Delta\psi = 0.1$. The last open streamline has $\psi = 0.636$ and the first closed has $\psi = 0.638$.

Yih (1959) observes that the corner eddy is fully separated from the free flow arriving from $x \to -\infty$. Thus, there is no reason that the vorticity of the eddy and of the flow agree, and details of the eddy flow may look different from Fig. 4.15. Fraenkel (1961) discusses general conditions for these corner eddies.

4.9 Angular Momentum Transport

We consider a rotational flow that is quite different from the examples treated so far. In astrophysics, *accretion disks* are highly flattened flows of dust, gas, and/or charged plasma around a central, massive object, often a star. The planetary system of Sun evolved from such an accretion disk. Figure 4.16 shows an observation of a protoplanetary disk.

The flatness of the disk is a consequence of rotation and contraction. An initially rather isotropic cloud of interstellar gas starts to collapse under its own gravitational pull. The angular momentum of the cloud is conserved during this collapse. Thus, any initial angular speed of the cloud, however small, is enhanced dramatically during a collapse that can shrink the cloud by orders of magnitude in diameter. With increasing angular speed, the cloud gets more and more flattened into an accretion disk. Our

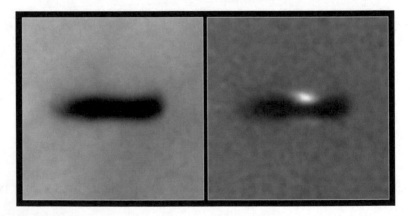

Fig. 4.16 Edge-on view of a protoplanetary disk of dust in the Orion Nebula, observed with the Hubble Space Telescope in 1995. The central star is largely hidden by the dust in the disk, but can be seen in the right picture taken through a special filter. Photograph: Mark McCaughrean (Max-Planck-Institute for Astronomy) and C. Robert O'Dell (Rice University), NASA and ESA. http://hubblesite.org/image/359/news_release/1995-45

main subject here is the angular momentum transport in the disk. As a result of the many parcel–parcel collisions that take place during the contraction process, the gas, dust, or plasma is on nearly circular, Keplerian orbits. This was already suggested by I. Kant. The interaction of individual gas rings in the accretion disk is, therefore, small, and it is reasonable to model the angular momentum transport as being caused by friction from viscosity of the gas. The physical origin of this viscosity is a central topic of current research; for example, the *magnetorotational instability* could make the flow turbulent. By losing angular momentum, gas drifts radially inward until it reaches the central star. We thus consider in this section a flow dominated by viscosity, as was the case for the flow into an acute corner treated in Sect. 4.3. The following assumptions are made:

– All gravitating mass resides in a central object.
– Self-gravity of the gas in the disk is negligible.
– Gas orbits the central object in a single plane.
– The gas is cold, and gas pressure can be neglected.
– The gas moves on almost circular orbits.
– The azimuthal speed obeys a Kepler law, $u_\varphi = C/\sqrt{r}$.
– Radial motion is slow, $u_r \ll u_\varphi$, and is caused by friction alone.

The following quotes capture the essentials of the dynamics resulting from friction between adjacent gas rings:

> The fast-moving interior will tend to drag forward the slower-moving exterior, and thus will increase its energy and make it recede from the Sun. Thus the outer parts will be slowly expelled from the system. The inner parts, on the other hand, will have their motion delayed, and will therefore gradually fall into the Sun. (Jeffreys 1924, p. 55)

> The angular momentum is steadily concentrated onto a small fraction of the mass which orbits at greater and greater radii while the rest is accreted onto the central body. (Lynden-Bell and Pringle 1974)

> Eventually all of the matter initially in a ring ends up at the origin and all of the angular momentum is carried to infinite radius by none of the mass. (Pringle 1981)

First, the continuity equation for stationary accretion disks is derived. Consider a gas ring of radial extent dr and mass dM located at the radius r in a thin disk. To formulate the dynamics of flat, two-dimensional gas rings, a surface density σ is introduced by

$$\sigma(t, r) = \frac{dM}{2\pi r \, dr}. \tag{4.201}$$

The angular momentum of a ring of gas is (we use polar coordinates)

$$dM \, r \, u_\varphi(r) = 2\pi r \, dr \, \sigma \, r \, r\omega(r). \tag{4.202}$$

For the azimuthal speed, $u_\varphi(r) = r\omega(r)$ is adopted throughout, with the Kepler angular speed ω. Division by $2\pi r \, dr$ gives a surface angular momentum density,

$$r^2 \sigma \omega. \tag{4.203}$$

Indeed, σ and $r^2 \sigma \omega$ are the two fundamental dynamical quantities of the disk. Since $\omega(r)$ is given, $r^2 \sigma \omega$ is *not* a fundamental unknown field, but can be calculated from σ. The ring mass dM changes by in- and outflow with radial speed u_r,

$$\frac{\partial}{\partial t}(r \, dr \sigma(t, r)) = r u_r(t, r) \sigma(t, r) - (r + dr) u_r(t, r + dr) \sigma(t, r + dr), \tag{4.204}$$

where both sides were divided by 2π. With $\partial r / \partial t = 0$ (independent coordinates), it follows that

$$\frac{\partial \sigma}{\partial t} + \frac{1}{r} \frac{\partial}{\partial r}(r u_r \sigma) = 0. \tag{4.205}$$

Next, the equation of motion is derived for an accretion disk that is subject to viscous stress. The key idea is indicated in the above quote from Jeffreys, and made more specific by Lynden-Bell and Pringle (1974):

> An element of mass $dM = 2\pi r \, \sigma \, dr$ [our symbols] which lies initially between r and $r + dr$ will normally have its angular momentum increased by the couple from the faster rotating material inside it, and will have its angular momentum decreased by shearing past the material further out. (Lynden-Bell and Pringle 1974, p. 608)

We consider thus the mutual shearing of gas rings around the central object, and derive an equation of motion from the corresponding angular momentum balance due to advection of angular momentum, and to its destruction by friction. Note, incidentally, that the disk gas is still treated as a Boltzmann continuum where individual fluid parcels respond to shear stresses only—which occur here at the outer and inner rims of a gas ring—and not as a Cosserat continuum where *point* parcels with rotational degrees of freedom ('spin') respond to couple stresses, which cannot happen in a gas of monopole particles. We start with the case of pure advection of angular momentum with speed u_r in radial direction. The law of angular momentum conservation is then simply

$$\frac{\partial}{\partial t}(r \, dr r^2 \, \sigma \omega(r)) = r u_r(t, r) r^2 \, \sigma(t, r) \omega(r)$$
$$- (r + dr) u_r(t, r + dr)(r + dr)^2 \sigma(t, r + dr) \omega(r + dr), \tag{4.206}$$

again without 2π. Therefore, for pure advection,

$$\frac{\partial}{\partial t}(r^2 \sigma \omega) + \frac{1}{r} \frac{\partial}{\partial r}(r u_r \, r^2 \sigma \omega) = 0. \tag{4.207}$$

Friction between gas rings reduces the total amount of angular momentum, but it can speed up the rotation of one ring at the expense of the other. We consider the *viscous torque*: Newton's friction force F_v scales with the velocity gradient normal to the flow direction, say

$$F_v = \mu A \frac{du_x}{dy}, \tag{4.208}$$

where A is the area of contact. To translate this to fluids, a kinematic viscosity is introduced, $v = \mu/\sigma$ (notice that the area mass density σ is used here, where in Sect. 2.8 the volume mass density was used). Instead of fluid parcels, gas rings are considered; thus, μA is replaced by $v\sigma\, 2\pi r$, giving

$$F_v = 2\pi r v \sigma \frac{du_\varphi}{dr}. \tag{4.209}$$

Furthermore, one postulates that if a fluid rotates like a solid body (so-called rigid body rotation; imagine that the fluid is frozen), $\omega = $ const, then there is no friction between adjacent fluid rings. Therefore, (4.209) is modified to

$$F_v = 2\pi r v \sigma r \frac{d\omega}{dr}. \tag{4.210}$$

The viscous torque is given by $rF_v(r)$, and a gas ring experiences the difference in torques at its inner and outer rims. Equation (4.206) then becomes, again dropping 2π,

$$\frac{\partial}{\partial t}(r^3 dr\sigma\omega(r)) = r^3 u_r(t, r)\sigma(t, r)\omega(r)$$
$$- (r + dr)^3 u_r(t, r + dr)\sigma(t, r + dr)\omega(r + dr)$$
$$- vr^3\sigma(t, r)\frac{d\omega}{dr} + v\,(r + dr)^3\sigma(t, r + dr)\frac{d\omega}{dr}, \tag{4.211}$$

or

$$r^2\omega\frac{\partial\sigma}{\partial t} + \frac{1}{r}\frac{\partial}{\partial r}(r^3\omega u_r\sigma) = \frac{1}{r}\frac{\partial}{\partial r}\left(vr^3\frac{d\omega}{dr}\sigma\right). \tag{4.212}$$

The sign on the right agrees with that in the Navier–Stokes equation. We turn now to the solution of Eqs. (4.205) and (4.212),

$$\dot{\sigma} + \frac{1}{r}(ru_r\,\sigma)' = 0, \tag{4.213}$$

$$r^2\omega\dot{\sigma} + \frac{1}{r}(r^2\omega\,ru_r\,\sigma)' = \frac{1}{r}(vr^3\omega'\sigma)', \tag{4.214}$$

where $\dot{\sigma} = \partial\sigma/\partial t$ and $\sigma' = \partial\sigma/\partial r$. Equation (4.213) can be written as

$$r^2\omega\dot{\sigma} + \frac{1}{r}(r^2\omega)'\,ru_r\,\sigma + \frac{1}{r}r^2\omega\,(ru_r\,\sigma)' = \frac{1}{r}(vr^3\omega'\sigma)'. \tag{4.215}$$

Inserting here $(ru_r\,\sigma)'/r = -\dot{\sigma}$ from (4.213) gives

$$(r^2\omega)' \, ru_r \, \sigma = (vr^3\omega'\sigma)'.$$ (4.216)

Substituting then $ru_r \, \sigma$ from this equation back into (4.213) gives

$$\frac{\partial \sigma}{\partial t} = -\frac{1}{r} \frac{\partial}{\partial r} \left[\frac{1}{(r^2\omega)'} \frac{\partial}{\partial r} (vr^3\omega'\sigma) \right].$$ (4.217)

Thus, the unknown speed u_r is eliminated, and a diffusion-type equation for σ results. For the Kepler motion about a point mass M, $r\omega^2 = GM/r^2$ or $\omega = \sqrt{GM/r^3}$ with G the gravitational constant. Then (4.217) takes the form

$$\frac{\partial \sigma}{\partial t} = \frac{3}{r} \frac{\partial}{\partial r} \left[\sqrt{r} \frac{\partial}{\partial r} (v\sqrt{r}\sigma) \right].$$ (4.218)

In the following, it is assumed that v is a constant. Furthermore, let

$$\Sigma = r^{1/4} \sigma.$$ (4.219)

Then, after a simple calculation, (4.218) becomes an equation of the *diffusion* type,

$$\boxed{\frac{r^2}{3v} \dot{\Sigma} = r^2\Sigma'' + r\Sigma' - \frac{1}{16} \Sigma}$$ (4.220)

It is well known that diffusion leads to an exponential decay of concentration gradients in time. Assuming thus that a factor e^{-st} appears in Σ, the time derivative $\dot{\Sigma}$ can be replaced by $-s\Sigma$, giving

$$r^2\Sigma'' + r\Sigma' + \left(\frac{r^2 s}{3v} - \frac{1}{16} \right) \Sigma = 0.$$ (4.221)

The *Bessel* differential equation, on the other hand, is

$$x^2 y'' + xy' + \left(x^2 - n^2 \right) y = 0,$$ (4.222)

where $y = y(x)$ and $y' = dy/dx$. For non-integer n (this is important here), the solution of (4.222) is

$$y(x) = a J_n(x) + b J_{-n}(x),$$ (4.223)

where the constants a and b are determined from the boundary conditions. Here J_n is the Bessel function of order n (see Abramowitz and Stegun 1964)

$$J_n(x) = \sum_{k=0}^{\infty} \frac{(-1)^k}{k! \, \Gamma(n+k+1)} \left(\frac{x}{2} \right)^{n+2k}.$$ (4.224)

This equation holds for all $n \in \mathbb{R}$, not just $n \in \mathbb{N}$. The Gamma function is defined by

$$\Gamma(x) = \int_0^\infty e^{-t} t^{x-1} dt, \tag{4.225}$$

where, from integration by parts,

$$\Gamma(x+1) = x\,\Gamma(x). \tag{4.226}$$

Then $\Gamma(1) = 1$ implies $\Gamma(n+1) = n!$ for integer n. Equation (4.221) for the accretion disk is transformed to the Bessel equation by introducing a new variable

$$x = r \sqrt{\frac{s}{3\nu}}. \tag{4.227}$$

This gives

$$x^2 \frac{d^2 \Sigma}{dx^2} + x \frac{d\Sigma}{dx} + \left(x^2 - \frac{1}{16}\right)\Sigma = 0, \tag{4.228}$$

with the solution

$$\Sigma(x) = a J_{1/4}(x) + b J_{-1/4}(x). \tag{4.229}$$

The function $J_{-1/4}(x)$ diverges at the disk center $x \to 0$, thus put $b = 0$. The decay rate s for diffusion is not fixed, and thus an integration is performed over all possible values, giving

$$\sigma(t, r) = \int_0^\infty d\sqrt{\frac{s}{3\nu}} \left\{ -st \, \frac{1}{r^{1/4}} \, a\left(\sqrt{\frac{s}{3\nu}}\right) J_{1/4}\left(r\sqrt{\frac{s}{3\nu}}\right) \right\}. \tag{4.230}$$

The variable $\sqrt{s/3\nu}$ has units of an inverse length. A wavenumber k is therefore introduced,

$$k = \sqrt{s/3\nu}. \tag{4.231}$$

This gives, as was first derived by Lüst (1952),

$$\boxed{\sigma(t, r) = \int_0^\infty dk \, e^{-3\nu k^2 t} \frac{1}{r^{1/4}} a(k) J_{1/4}(kr)} \tag{4.232}$$

The next major step was taken by Lynden-Bell and Pringle (1974), who solved (4.220) for the initial values

$$\sigma(0, r) = \frac{M}{2\pi R} \delta(r - R), \tag{4.233}$$

which corresponds to an infinitesimal gas ring of mass M at radius R. Lynden-Bell and Pringle derive the secular (i.e., slow, quasi-steady) evolution of the ring under the action of its inherent friction. In order to calculate the coefficients $a(k)$ for these initial values, (4.230) must be 'inverted.' This is achieved by the remarkable Fourier–Bessel theorem,

$$\int_0^\infty dx \left[\int_0^\infty dk \, a(k) \, J_n(kx) \, \sqrt{kx} \right] J_n(k'x) \, \sqrt{k'x} = a(k'). \qquad (4.234)$$

Here k, k', and x are independent variables. From (4.232),

$$r^{1/4} \, \sigma(0, r) = \int_0^\infty dk \, \frac{a(k)}{\sqrt{kr}} \, J_{1/4}(kr) \, \sqrt{kr}. \qquad (4.235)$$

Multiplying both sides by $r J_{1/4}(k'r)$ and integrating over r yields

$$\int_0^\infty dr \, r^{5/4} \, \sigma(0, r) \, J_{1/4}(k'r) =$$

$$= \int_0^\infty dr \left[\int_0^\infty dk \, \frac{a(k)}{\sqrt{kk'}} \, J_{1/4}(kr) \, \sqrt{kr} \right] J_{1/4}(k'r) \, \sqrt{k'r} = \frac{a(k')}{k'}, \qquad (4.236)$$

where (4.234) was applied. Inserting the initial values (4.233) gives

$$a(k') = \frac{M}{2\pi R} \, k' \int_0^\infty dr \, r^{5/4} \, J_{1/4}(k'r) \, \delta(r - R) = \frac{M}{2\pi} \, k' R^{1/4} \, J_{1/4}(k'R). \quad (4.237)$$

Substituting this in (4.232) gives finally

$$\sigma(t, r) = \frac{M}{2\pi} \left(\frac{R}{r} \right)^{1/4} \int_0^\infty dk \, k \, e^{-3\nu k^2 t} \, J_{1/4}(kR) \, J_{1/4}(kr). \qquad (4.238)$$

The integral can be found in tables of the Hankel transforms: Eq. (23), p. 51 in Sect. 8.11 of Erdélyi (1954) gives

$$\boxed{\sigma(t, r) = \frac{M}{\pi R^2} \, \frac{1}{\tau x^{1/4}} \, \exp\left(-\frac{1 + x^2}{\tau} \right) I_{1/4}\left(\frac{2x}{\tau} \right)} \qquad (4.239)$$

see Eq. (2.13) in Pringle (1981) or Eq. (26) in Lynden-Bell and Pringle (1974). Here $x = r/R$, which is not to be confused with x from (4.227), and τ is a timescale,

$$\tau = 12\nu t / R^2. \qquad (4.240)$$

Furthermore, I_n is the *modified* Bessel function,

$$I_n(x) = \sum_{k=0}^\infty \frac{1}{k! \, \Gamma(n + k + 1)} \left(\frac{x}{2} \right)^{n+2k}. \qquad (4.241)$$

Fig. 4.17 Viscous evolution of a gas ring orbiting a point mass in arbitrary linear units. Times shown are $\tau = 0.002$ (highest peak), 0.02, 0.2, and 2

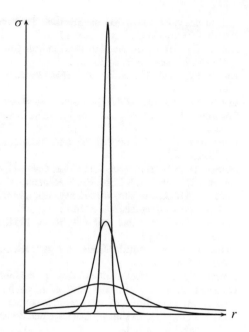

For non-integer n, the function $y(x) = aI_n(x) + bI_{-n}(x)$ is the general solution of the differential equation

$$x^2 y'' + xy' - (x^2 + n^2)y = 0. \tag{4.242}$$

Figure 4.17 shows the evolution of the gas ring. The ring spreads in the course of time, and the maximum of the surface density σ drifts inward. This agrees with the general ideas quoted at the beginning of the section.

We close this chapter with a reference to the bibliography of vortex dynamics by Meleshko and Aref (2007). Besides giving a brief historical overview, this lists 1035 papers and books, including their full titles, that appeared between 1858 and 1956, so that further interesting examples on vortex dynamics can be easily found.

References

Abramowitz, M., and I.A. Stegun. 1964. *Handbook of mathematical functions*. New York: Dover.

Chorin, A.J., and J.E. Marsden. 1990. *A mathematical introduction to fluid mechanics*. New York: Springer.

Davis, A.M.J., and M.E. O'Neill. 1977. Separation in a slow linear shear flow past a cylinder and a plane. *Journal of Fluid Mechanics* 81: 551.

Dean, W.R., and P.E. Montagnon. 1949. On the steady motion of viscous liquid in a corner. *Mathematical Proceedings of the Cambridge Philosophical Society* 45: 389.

Dolaptschiew, Bl. 1937. Über die Stabilität der Kármánschen Wirbelstraße. *Zeitschrift für ange-wandte Mathematik und Mechanik* 17: 313.

Domm, U. 1956. Über die Wirbelstraßen von geringster Intensität. *Zeitschrift für angewandte Mathematik und Mechanik* 36: 367.

Drazin, P.G., and W.H. Reid. 1981. *Hydrodynamic stability*. Cambridge: Cambridge University Press.

Erdélyi, A. (ed.). 1954. *Tables of integral transforms*, vol. II. New York: McGraw-Hill.

Fraenkel, L.E. 1961. On corner eddies in plane inviscid shear flow. *Journal of Fluid Mechanics* 11: 400.

Heisenberg, W. 1924. Über Stabilität und Turbulenz von Flüssigkeitsströmen. *Annalen der Physik IV. Folge* 74: 577.

Jeffreys, H. 1924. *The Earth*. Cambridge: Cambridge University Press.

Kármán, Th., von. 1911. Ueber den Mechanismus des Widerstandes, den ein bewegter Körper in einer Flüssigkeit erfährt. *Nachrichten von der Gesellschaft der Wissenschaften zu Göttingen, Mathematisch-Physikalische Klasse* 1911: 509.

Kotschin, N.J., I.A. Kibel, and N.W. Rose. 1954. *Theoretische Hydromechanik*, vol. I. Berlin: Akademie-Verlag.

Kutta, W.M. 1902. Auftriebskräfte in strömenden Flüssigkeiten. *Illustrierte aeronautische Mit-teilungen* 6: 133.

Lamb, H. 1932. *Hydrodynamics*. Cambridge: Cambridge University; New York: Dover (1945).

Landau, L.D., and E.M. Lifshitz. 1987. *Fluid mechanics, Course of theoretical physics*, vol. 6, 2nd ed. Amsterdam: Elsevier Butterworth-Heinemann.

Lüst, R. 1952. Die Entwicklung einer um einen Zentralkörper rotierenden Gasmasse. *Zeitschrift für Naturforschung A* 7: 87.

Lynden-Bell, D., and J.E. Pringle. 1974. The evolution of viscous discs and the origin of the nebular variables. *Monthly Notices of the Royal Astronomical Society* 168: 603.

Meleshko, V.V., and H. Aref. 2007. A bibliography of vortex dynamics 1858–1956. *Advances in Applied Mechanics* 41: 197.

Milne-Thomson, L.M. 1940. Hydrodynamical images. *Mathematical Proceedings of the Cambridge Philosophical Society* 36: 246.

Milne-Thomson, L.M. 1968. *Theoretical hydrodynamics*, 5th ed. London: Macmillan; New York: Dover (1996).

Moffatt, H.K. 1964. Viscous and resistive eddies near a sharp corner. *Journal of Fluid Mechanics* 18: 1.

Pringle, J.E. 1981. Accretion discs in astrophysics. *Annual Review of Astronomy and Astrophysics* 19: 137.

Rayleigh, Lord. 1911. Hydrodynamical notes. *Philosophical Magazine Series 6*, 21: 177.

Reynolds, O. 1883. An experimental investigation of the circumstances which determine whether the motion of water shall be direct or sinuous, and of the law of resistance in parallel channels. *Philosophical Transactions of the Royal Society of London* 174: 935.

Saffman, P.G. 1995. *Vortex dynamics*. Cambridge: Cambridge University Press.

Schmieden, C. 1936. Zur Theorie der Kármánschen Wirbelstraße. *Ingenieur-Archiv = Archive of Applied Mechanics* 7: 215.

Schubert, G. 1967. Viscous flow near a cusped corner. *Journal of Fluid Mechanics* 27: 647.

Tritton, D.J. 1988. *Physical fluid dynamics*. Oxford: Oxford University Press.

Yih, C.-S. 1959. Two solutions for inviscid rotational flow with corner eddies. *Journal of Fluid Mechanics* 5: 36.

Chapter 5
Jets, Wakes, and Cavities

5.1 Free Streamlines

In the foregoing chapter, we encountered boundaries of solid bodies and obstacles that are fully covered by streamlines of the flow. There is no necessity that one single streamline follows the whole boundary of an obstacle. Instead, the streamline can detach from the solid body; it is then called a *free streamline*. Three standard flow scenarios with free streamlines are jets, wakes, and cavities, shown schematically in Fig. 5.1. The subsequent Figs. 5.2, 5.3, and 5.4 show examples of these flow types. From a theoretical perspective, the interest in free streamlines lies in the fact that they pose boundary value problems where the shape of the boundary is not known aforehand but has to be determined as a part of the solution of the problem. This is also true for free surface waves, which are the subject of Chap. 9.

Jets form when fluids leave vessels through orifices into free space or into fluid at rest. A main question concerns the ratio A/A_0 of the asymptotic jet area to the orifice area. A classical result by J.-C. Borda (1733–1799) is $A/A_0 = 1/2$. A simple derivation of this from conservation laws is given in Birkhoff and Zarantonello (1957, p. 15). An example of an astrophysical jet is shown in Fig. 5.2.

Wakes form in flows past obstacles. The fluid in a wake does not take part in the gross motion of the fluid. A surface discontinuity occurs between the inner fluid at rest in the wake and the outer fluid flow past the solid body. Typically, eddies occur behind the obstacle and at the rim of the wake. Wakes are the main cause of fluid resistance to moving objects. A photograph of a wake behind a tidal plant is shown in Fig. 5.3.

Cavities form when bodies move at high speed through fluids. Cavities are filled with fluid vapor. At acute corners of an obstacle, the flow speed can become very large. By the Bernoulli equation, the pressure would then become negative. This is avoided by streamlines detaching from the obstacle, which leads to the formation of a cavity (or a wake). Cavity formation, or cavitation, can be a severe technical problem due to the damage inflicted on propellers or valves by the implosion of

© Springer Nature Switzerland AG 2019
A. Feldmeier, *Theoretical Fluid Dynamics*, Theoretical and Mathematical Physics,
https://doi.org/10.1007/978-3-030-31022-6_5

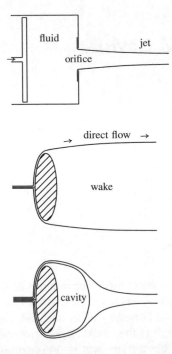

Fig. 5.1   Sketch of a jet, wake, and cavity

Fig. 5.2   Elliptical galaxy M87 emitting a relativistic jet from its active galactic core. The length of the jet is approximately 5,000 light years. Image Source: NASA and STScI. Image data obtained in 1998 with the Hubble Space Telescope by J. A. Biretta, W. B. Sparks, F. D. Macchetto, and E. S. Perlman (STScI). Composite image compiled by the Hubble Heritage team. See Wikipedia entries 'Astrophysical jet' and 'Messier 87'

Fig. 5.3 Wake behind the tidal plant Strangford in Northern Ireland. Photograph: Fundy, in Wikipedia entry 'SeaGen'

Fig. 5.4 Cavity formation behind an axisymmetric headform. Reproduced from Brennen (1970)

vapor bubbles. An example of cavity formation is shown in Fig. 5.4. For an extensive discussion of free streamlines for fluid cavities, see Chap. 8 in Brennen (2013).

The theory of jets and cavities was initiated by Helmholtz (1868). A central issue is here to find the shape of the free streamlines detaching from an obstacle, with a special interest in the precise location of the detachment. Often a complicated analysis is necessary to find this location. Alternatively, bodies with acute corners can be used to enforce detachment. The location of flow separation is then imposed on the flow.

On segments of a streamline that are attached to an obstacle in the flow, the fluid velocity $\vec{u}$ is tangential to the surface of the obstacle. This is a purely kinematic boundary condition. By contrast, along a *free* streamline, i.e., the detached part of the streamline that is in contact with the solid body, we show that the *fluid pressure is constant*, which gives a dynamic boundary condition. The condition $p = $ const is actually used to define free streamlines, see Batchelor (1952), p. 493. To see that p must be constant, note that in jets the free streamline separates moving fluid passing through the orifice from a vacuous region, or from air or fluid at rest. Similarly, in wakes and cavities, the free streamline separates a region of fluid moving about an obstacle from a region of fluid that does not take part in this motion and can be assumed to be at rest, or may even be replaced by vacuum, air, or vapor at rest. A free streamline, therefore, separates regions of moving fluid from regions where $\vec{u} = 0$. If no other forces than fluid pressure are present, then the pressure is constant in the fluid at rest or even zero in a vacuous region. The pressure is therefore constant (or zero) along a free streamline. Furthermore, the Bernoulli equation $p/\rho + \frac{1}{2}u^2 =$ const implies for $p = $ const and $\rho = $ const that *the speed* $|\vec{u}|$ (obviously not the vector $\vec{u}$ in general) *is constant along a free streamline.* This result will play a central role in deriving the shape of free streamlines in the following sections. Note that along the solid body surface, the fluid speed $|\vec{u}|$ is in general not constant, but the fluid will accelerate or decelerate there.

In the following, assume again an inviscid fluid with $\nabla \cdot \vec{u} = 0$ in the complex plane.

Good references for the material in this chapter are the book by Birkhoff and Zarantonello (1957) and the review article by Gilbarg (1960).

5.2 Flow Past a Step

We start with a simple example that shows how the Schwarz–Christoffel formula is used to solve flow problems in the complex plane. In the last chapter, flows past single corners were considered, and one may wonder what happens if *two* corners follow one upon the other, e.g., when the flow encounters a material step as shown in Fig. 5.5. (Note that this is neither a jet nor a wake nor a cavity.)

The step is a polygonal boundary with two corners, having interior angles $\alpha_1 = 3\pi/2$ and $\alpha_2 = \pi/2$. The Schwarz–Christoffel formula (3.118) in Sect. 3.11 gives then, when assuming $t_1 = -1$ and $t_2 = 1$ at the two corners,

$$\frac{dz}{dt} = b \sqrt{\frac{t+1}{t-1}} = \frac{b(t+1)}{\sqrt{t^2-1}}, \tag{5.1}$$

with some constant $b \in \mathbb{C}$. This can be integrated to yield

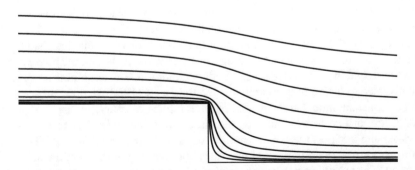

Fig. 5.5 Calculated streamlines past a step in a flow

$$z = b\left(\sqrt{t^2 - 1} + \cosh^{-1} t\right) + a. \tag{5.2}$$

Assume $z = 0$ at the corner with $\alpha_2 = \pi/2$, and $z = ih$ at the corner with $\alpha_1 = 3\pi/2$, giving

$$0 = b \cosh^{-1}(1) + a = a, \tag{5.3}$$

and

$$ih = b \cosh^{-1}(-1) + a = b\pi, \tag{5.4}$$

and therefore

$$z = \frac{ih}{\pi}\left(\sqrt{t^2 - 1} + \cosh^{-1} t\right). \tag{5.5}$$

Let now $t = r + is$. For the complex potential $\Phi(t) = \phi + i\psi$ we make the strong assumption that ϕ depends only on r and ψ only on s,

$$\Phi(t) = \phi(r) + i\psi(s). \tag{5.6}$$

Then $\phi = \text{const}$ and $\psi = \text{const}$ are straight lines parallel to the imaginary s- and the real r-axis, respectively. The Cauchy–Riemann equations become $\partial_r \phi = \partial_s \psi$. Using the analytic mapping (singularities excluded) $t \rightarrow z$ from (5.5), the analytic potential $\Phi(t)$ in the t-plane (with $\Delta_t \Phi(t) = 0$) allows to define an analytic potential $\tilde{\Phi}$ in the z-plane (with $\Delta_z \tilde{\Phi}(z) = 0$) by

$$\tilde{\Phi}(z) = \Phi(t) \quad \text{for} \quad z = z(t). \tag{5.7}$$

Choose $\psi(s = 0) = 0$ along the real r-axis in the complex t-plane. The step in the z-plane is the image under (5.5) of this r-axis. With $\tilde{\Phi} = \tilde{\phi} + i\tilde{\psi}$ it follows that $\tilde{\psi}(z) = 0$ along the step in the z-plane: the z-values along the step correspond to t-values on the real r-axis alone, on which $\psi = 0$. Thus the complex potential $\tilde{\Phi}(z)$ also obeys the correct boundary condition for a flow past a step in the z-plane: the step is a streamline. More generally, images under (5.5) of the lines $\psi = \text{const}$ in the complex t-plane (i.e., horizontal straight lines) are the (curved) streamlines in the complex z-plane. This can be expressed in general terms as follows:

> Any transformation from one complex variable z_1 to another z_2 will transform the solution of one potential problem described by the first variable to the solution of another potential problem described by the second variable. [...] This coordinate transformation $z_1 = g(z_2)$ must be chosen so that it will carry the complex potential geometry $\Phi_1(z_1)$ of the original problem into a simpler complex potential geometry $\Phi_2(z_2) = \Phi_1(g(z_2)) = \Phi_1(z_1)$. (Panofsky and Phillips 1962, p. 61, with minor adaptions)

As is customary in physics (but not usually in mathematics), we will in the following write Φ again instead of $\tilde{\Phi}$ for the transformed function. The streamlines $y(x)$ from

$s = \text{Im } t = \text{const}$ and therefore $\psi(s) = \text{const}$ in (5.5) are shown in Fig. 5.5. Note that horizontal lines $s \to \infty$ are mapped to horizontal streamlines $y \to \infty$, i.e., the step occurs in a flow of infinite width. The case of a step in a canal of finite width is treated by Newman (1965), using a method developed by Havelock (1929). The analysis leads here to an integral equation that can be solved numerically.

5.3 Complex Potential and Speed Plane

In the example of the last section, we assumed that the potential isocontours $\phi = \text{const}$ and $\psi = \text{const}$ are identical to the coordinate isocontours $s = \text{const}$ and $r = \text{const}$ in the plane of the complex variable $t = r + is$, and performed an analytic Schwarz–Christoffel transformation $t \to z$ to the complex z-plane in which a step occurs in the flow. This simple approach can be made applicable to many more examples by the following generalization. Assume that the flow domain is simply connected, and that all maps are one-to-one. Then standard conformal mapping theorems apply.

The *two* independent Schwarz–Christoffel transformations are considered now. The first is from an auxiliary t-plane to the complex potential plane Φ, and the second from the t-plane to the complex speed plane $\ln w$. The boundary of the flow obstacle shall correspond to the real axis of the t-plane. From these equations then, t is eliminated. Since $w = d\Phi/dz$, a first-order differential equation for Φ results. This can often be solved analytically to give the streamlines $\psi = \text{const}$.

The image of a Schwarz–Christoffel transformation must be a polygon. First, for the Φ-plane of the complex potential, this is indeed the case, as was discussed in the previous section where streamlines with $\psi = \text{const}$ were straight lines parallel to the real ϕ-axis. Often, ψ is put to zero on the free streamline. The potential ϕ can then act as a curve parameter along the isocontours $\psi = \text{const}$. However, it will often be necessary for symmetry reasons to map the real axis of the t-plane to a semi-infinite or infinite strip in the Φ-plane, so the mapping $t \to \Phi$ is not wholly trivial.

Second, for the speed logarithm $\ln w$, one argues as follows. Let $w = qe^{-i\varphi}$ and thus

$$\ln w = \ln q - i\varphi. \tag{5.8}$$

The $\ln w$-plane has then a real q-axis and an imaginary $-i\varphi$-axis. On straight lines $\varphi = \text{const}$ parallel to the q-axis, the velocity vector does not change its direction during the motion, as is the case on each straight segment of the polygonal boundary of the obstacle (kinematic boundary condition). On the other hand, $\ln q = \text{const}$ and $q = \text{const}$ on straight lines parallel to the $-i\varphi$-axis, as is the case on a free streamline that has detached from the obstacle (dynamic boundary condition). Combining these results, the streamline $\psi = 0$ that follows initially the boundary of the obstacle and then, after separation, is a free streamline, corresponds to a line in the $\ln w$-plane that has either straight horizontal segments (obstacle boundary) or straight vertical segments (free streamline).

Once the two Schwarz–Christoffel maps $\Phi(t)$ and $\ln w(t)$ are found, t can be eliminated from them in order to obtain $\ln w(\Phi)$. Then, with

$$\frac{d\Phi}{dz} = w = e^{\ln(w(\Phi))}, \tag{5.9}$$

the streamlines are obtained from integration (quadrature) as

$$z = \int d\Phi \, e^{-\ln(w(\Phi))} = \int dt \, \frac{d\Phi}{dt} \, e^{-\ln(\tilde{w}(t))}, \tag{5.10}$$

where $\tilde{w}(t) = w(\Phi)$ for $\Phi = \Phi(t)$.

5.4 Outflow from an Orifice

The above method is applied now to the problem of horizontal outflow from an orifice, first solved by Helmholtz (1868) and Kirchhoff (1869). The flow geometry is shown schematically in Fig. 5.6. Note that for flows in the z-plane, the orifice corresponds to a long slit in 3-D, not to a circular hole.

The jet far from the orifice shall become parallel to the x-axis, with the asymptotic jet speed reaching a limiting value $(U, 0)$ with constant U. Four complex variables and their associated planes are introduced: (i) $z = x + iy$ gives the streamlines $y(x)$, (ii) $\Phi = \phi + i\psi$ is the complex potential, (iii) $\ln(w/U) = \ln(q/U) - i\varphi$ is the logarithm of the complex speed, and (iv) t is the auxiliary variable in the Schwarz–Christoffel transform, to be eliminated from the equations.

Note that $w = qe^{-i\varphi}$, due to the minus sign from (3.104) in Sect. 3.9. We alert the reader to the fact that, in order to have here $+i\varphi$, many texts consider $\ln(U/w)$ instead of $\ln(w/U)$. In the following, $-i\varphi$ is maintained, which is also the choice of Schmieden (1929). The Cauchy–Riemann equations are $u = \phi_x = \psi_y$ and $v =$

Fig. 5.6 Outflow from an orifice with walls AB and $A'B'$. At B and B', the velocity is vertically down and upward, respectively

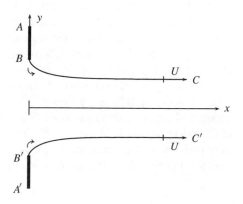

Fig. 5.7 Three complex
planes for Φ, t, and $\ln(w/U)$

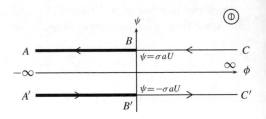

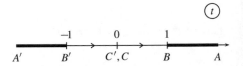

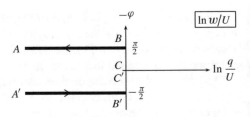

$\phi_y = -\psi_x$. For the complex variable t, the real axis Re t is mapped to the boundary
of the vessel. Figure 5.7 shows the Φ-, t-, and $\ln(w/U)$-planes of the problem. The
walls of the vessel are drawn again as bold lines.

The fluid volume, i.e., the complex domain considered, has boundaries
$A'B'C'CBAA'$ in all the complex planes. The points A and A' lie at infinity in
all planes, and C and C' lie either at infinity (z- and Φ-plane) or at zero (t- and
$\ln w/U$-plane). Note, again, that corresponding points in different complex planes have
identical names. Fluid moves along the walls AB and $A'B'$, and then freely along
BC and $B'C'$. Our task is to determine the curve BC in the z-plane. This z-plane
will only appear at the very end of the calculation, when employing $d\Phi/dz = w$.

First, consider the Schwarz–Christoffel map from t to Φ. Let us set $\phi \rightarrow -\infty$
at A and A', $\phi = 0$ at B and B', and $\phi \rightarrow \infty$ at C and C'. At C, $u = U = \psi_y$ and
$v = 0 = -\psi_x$, thus $\psi = Uy$ at C. Let a and σa, respectively, be the half-width of the
jet at the orifice and far away from it. Then at C, $y = \sigma a$ and $\psi = \sigma aU$. Since ψ is
constant along streamlines, $\psi = \sigma aU$ all along the line ABC. Similarly, $y = -\sigma a$
at C' and $\psi = -\sigma aU$ along $A'B'C'$. The relevant mapping $t \rightarrow \Phi$ for the present
flow problem is thus from the upper half of the t-plane to an infinite strip in the
Φ-plane. For this, from (3.130) in Sect. 3.12,

$$\frac{d\Phi}{dt} = \frac{b_1}{t},\qquad\qquad (5.11)$$

since the interior angle at the corner at $C = C'$ in the Φ-plane is zero. The 'opposite' corner $A = A'$ is left out of consideration, since $t \to \pm\infty$ there ($t \to s$ transformation in Sect. 3.12). The constant b_1 is real since the polygon is not rotated in the Φ-plane. Thus,

$$\Phi = b_1 \ln t + a_1. \tag{5.12}$$

The constants a_1 and b_1 are easily found: at B, $t = 1$ by assumption, and using $\ln 1 = 0$, this yields (see the uppermost panel in Fig. 5.7)

$$\Phi_B = 0 + i\sigma a U = a_1. \tag{5.13}$$

At B' then, $t = -1$. Using $\ln(-1) = i\pi$ (since $e^{i\pi} = -1$) gives

$$\Phi_{B'} = 0 - i\sigma a U = i\pi b_1 + i\sigma a U, \tag{5.14}$$

yielding $b_1 = -2\sigma a U / \pi$. At C and C', $t = 0$, thus $\phi_{C,C'} = +\infty$ with $\ln 0 = -\infty$. Therefore

$$\boxed{\Phi = \sigma a U \left(-\frac{2}{\pi} \ln t + i \right)} \tag{5.15}$$

Next, consider the Schwarz–Christoffel map from t to $\ln(w/U)$. From Fig. 5.7, this is clearly a map to a semi-infinite strip, with two corners with interior angle $\alpha = \pi/2$, corresponding to $t = \pm 1$. Thus from (3.127) in Sect. 3.12,

$$\frac{d \ln(w/U)}{dt} = \frac{b_2}{\sqrt{t+1}\sqrt{t-1}}, \tag{5.16}$$

implying by integration

$$\ln \frac{w}{U} = b_2 \cosh^{-1} t + a_2. \tag{5.17}$$

At B, $t = 1$, which, with $\cosh^{-1} 1 = 0$, determines a_2,

$$\ln \frac{w}{U} = \ln \frac{q}{U} - i\varphi = 0 + \frac{i\pi}{2} = a_2. \tag{5.18}$$

At B' then, $t = -1$, and $\cosh^{-1}(-1) = i\pi$, since $\cosh(i\pi) = \cos\pi = -1$. Thus,

$$\ln \frac{w}{U} = 0 - \frac{i\pi}{2} = i\pi b_2 + \frac{i\pi}{2}, \tag{5.19}$$

or $b_2 = -1$. At C and C', $t = 0$, and $\cosh^{-1} 0 = i\pi/2$, since $\cosh(i\pi/2) = \cos(\pi/2) = 0$. With this, $\ln(w/U)_{C,C'} = -i\pi/2 + i\pi/2 = 0$, as required. Therefore

$$\ln \frac{w}{U} = - \cosh^{-1} t + \frac{i\pi}{2}$$

(5.20)

The results (5.15) and (5.20) allow to derive $\Phi(z)$ and thus $\psi(z)$. This allows calculation of the streamlines z for constant ψ. Writing

$$\ln \frac{w}{U} = \ln \left(\frac{1}{U} \frac{d\Phi}{dz} \right) = - \cosh^{-1} t + \frac{i\pi}{2} = - \ln \left(t + \sqrt{t^2 - 1} \right) + \frac{i\pi}{2},$$

(5.21)

which implies, since $- \ln(a) = \ln(1/a)$ and $\ln(ib) = \ln(b) + i\pi/2$,

$$\frac{1}{U} \frac{d\Phi}{dz} = \frac{i}{t + \sqrt{t^2 - 1}}.$$

(5.22)

From (5.15),

$$\ln t = \frac{i\pi}{2} - \frac{\pi \Phi}{2\sigma a U},$$

(5.23)

or

$$t = i e^{-\pi \Phi / (2\sigma a U)}.$$

(5.24)

Inserting this into (5.22) and rearranging,

$$dz = \frac{d\Phi}{iU} \left(t + \sqrt{t^2 - 1} \right) = \frac{d\Phi}{U} \left(e^{-\pi \Phi / (2\sigma a U)} + \sqrt{e^{-\pi \Phi / (\sigma a U)} + 1} \right)$$

(5.25)

results, where a choice was required for the sign of the square root. It remains to perform the integration corresponding to (5.10); this is done here for the two streamlines only that touch the rim of the orifice. Here $\psi = \pm \sigma a U = \text{const}$, and ϕ shall vary from 0 to $+\infty$. With $d\Phi = d\phi$ follows

$$dz = \frac{d\phi}{U} \left(-i e^{-\pi \phi / (2\sigma a U)} + \sqrt{1 - e^{-\pi \phi / (\sigma a U)}} \right).$$

(5.26)

Introducing the abbreviation

$$X = \frac{\pi \phi}{2\sigma a U},$$

(5.27)

and separating real and imaginary parts, this becomes

$$dx = \frac{2\sigma a}{\pi} \, dX \, \sqrt{1 - e^{-2X}}, \tag{5.28}$$

$$dy = -\frac{2\sigma a}{\pi} \, dX \, e^{-X}. \tag{5.29}$$

Integrating the latter equation subject to the condition that $y = a$ for $X = 0$ ($\phi = 0$) gives

$$a - y = \frac{2\sigma a}{\pi} \left(1 - e^{-X}\right). \tag{5.30}$$

For $X \to \infty$ the jet approaches its asymptotic half-width $y = \sigma a$. Inserting this, the last equation becomes

$$a(1 - \sigma) = \frac{2\sigma a}{\pi}, \tag{5.31}$$

yielding

$$\boxed{\sigma = \frac{\pi}{\pi + 2}} \tag{5.32}$$

This contraction factor $\sigma \approx 0.61$ of the jet from a slit of infinite length compares well with Borda's result $\sigma = 0.5$ for the contraction factor for outflow from a circular hole. A new variable u is introduced by

$$X = -\ln u, \qquad dX = -\frac{du}{u}. \tag{5.33}$$

Integrating then from $X' = 0$ (corresponding to $\phi = 0$ at the orifice outlet B) to X, that is from $u' = 1$ to u, Eq. (5.28) becomes, after inserting (5.32) for σ,

$$x = \frac{a}{1 + \frac{\pi}{2}} \int_u^1 du \, \frac{\sqrt{1 - u^2}}{u}, \tag{5.34}$$

which yields

$$x = \frac{a}{1 + \frac{\pi}{2}} \left[\ln \left(\frac{1 + \sqrt{1 - u^2}}{u} \right) - \sqrt{1 - u^2} \right]. \tag{5.35}$$

This function is plotted in Fig. 5.6.

We close with some historical remarks. Helmholtz (1868) was the first to treat fluid jets. Kirchhoff (1869) essentially proposed the analytical method used above. However, he did not have the Schwarz–Christoffel theorem at hand! His insight was that in the complex speed plane, fluid motion with $q = \text{const}$ is along circular arcs. Kirchhoff maps therefore a strip in the Φ-plane to a *sickle* in the w-plane.

[...] sehen wir auch ϕ und ψ als die rechtwinkligen Coordinaten eines Punktes in einer Ebene, die wir die Φ-Ebene nennen wollen, an, machen über die Gebiete von Φ und w^{-1} geeignete Annahmen, suchen die Relation zwischen Φ und w^{-1}, durch welche diese beiden Gebiete in den kleinsten Teilchen ähnlich auf einander abgebildet werden, und berechnen z aus dieser mit Hülfe von $dz/d\Phi = w^{-1}$. [...] Als Gebiet von Φ wollen wir nun einen Streifen annehmen [...], als Gebiet von w^{-1} eine Sichel. [Wir] können die Gleichung zwischen Φ und w^{-1} aufstellen, durch welche das eine dieser Gebiete auf dem andern abgebildet wird; dieselbe bestimmt Φ und w^{-1} als einwerthige Functionen von einander.[1] (Kirchhoff 1897, p. 291f)

We have replaced here Kirchhoff's symbols with ours. Kirchhoff's map is his Eq. (6), with constants K, C, C', and $|w| \leq 1$,

$$\left(\frac{1-w}{1+w}\right)^2 = K \, \frac{e^\Phi - C}{e^\Phi - C'}. \tag{5.36}$$

5.5   A Simple Wake Model

Wakes are of central importance for fluid resistance to motion. In a stationary flow of an inviscid fluid, there is no drag force on solid bodies, if the flow at infinity is unaffected by the body. Helmholtz found that wakes could resolve this so-called d'Alembert's paradox, if the wake extends from the solid body to infinity. The work done in moving the body is then equal to work done on the wake.

D'Alembert's paradox no longer applies since the effective body is now [with an open wake] infinite. (Brennen 2013, Chap. 8)

Cavities, on the other hand, would not solve the paradox, since they have finite extent and the cavity can be replaced by a solid extension of the body itself.

We consider what is probably the simplest wake model, a parallel fluid stream of speed U that hits a plate at normal incidence, leading to a stagnation point A at the center of the plate. Our calculation follows Gilbarg (1960), p. 328, Birkhoff and Zarantonello (1957), p. 27, Milne-Thomson (1968), p. 318, and Lamb (1932), p. 100. Figure 5.8 shows the z-, Φ-, t-, and $\ln(w/U)$-plane. The wake speed itself is trivial, $w = 0$, and one is interested only in the flow region outside the wake. The boundary of this domain is the curve $ABCC'B'A'A$. Going along this curve in the positive direction, the enclosed flow domain lies to the left, as should be. The sequence of points A, B, C along the real t-axis is reversed compared to that for the jet; there are

---

[1] We also consider ϕ and ψ as rectilinear coordinates of a point in a plane, which we will call the Φ-plane, make appropriate assumptions about the domains of Φ and w^{-1}, look for the relation between Φ and w^{-1}, through which both domains are mapped to each other similar in the smallest parts, and [we] calculate z from this [relation] with the help of $dz/d\Phi = w^{-1}$. As the domain of Φ we will now assume a strip, as domain of w^{-1} a sickle. [We] can write down the equation between Φ and w^{-1}, by which one of the domains is mapped to the other; this determines Φ and w^{-1} as single-valued functions of each other.

now three instead of formerly two singularities: two at B and B' again, with $t_B = 1$ and $t_{B'} = -1$, and a new singularity at $A = A'$ with $t_A = t_{A'} = 0$.

The polygon in the Φ-plane is novel in this example. The wake domain limited by the flat plate and the free streamlines (i.e., by the closed curve $A'B'C'CBAA'$) is a superfluous region in the complex z-plane, since $w = 0$ and thus $\Phi = \text{const}$; choose $\Phi = 0$. The streamline $\psi = 0$ extends from $\phi \to -\infty$ to the point A on the plate, which is also identified with $\phi = 0$. The line $\phi > 0$, $\psi = 0$ cannot be reached by the

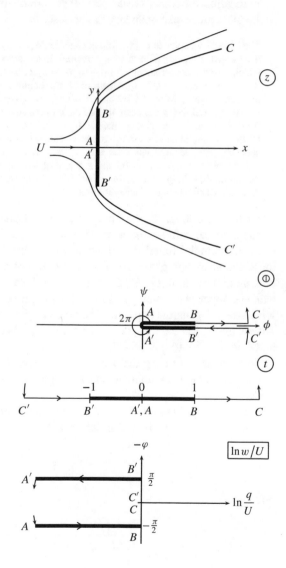

Fig. 5.8 A simple wake model

flow and is taken out of the plane; the complex plane has then a *slit*. The flat plate plus the subsequent free streamlines, i.e., the curves ABC and $A'B'C'$, are put on the two banks of this slit, shown for clarity at a small distance in Fig. 5.8. Thus the two *branches* ABC and $A'B'C'$, into which the incoming streamline $x < 0$, $y = 0$ splits at the stagnation point $A = A'$, are mapped in a one-to-one manner onto the two *banks* of the slit $\phi > 0$, $\psi = 0$. In the following quote from Courant (1950), the k constants $c_1, \ldots, c_k$ for ψ can be replaced by just one single constant $\psi = 0$, in order to adjust the content to the current situation; also 'boundary' in the quote refers to the plate plus wake boundary in our case.

> The boundary curves of G [a k-fold connected domain; see below] are parts of streamlines $\psi = $ const. It is plausible that the streamlines are smooth analytic closed curves, along which ϕ varies monotonically from $-\infty$ to $+\infty$, with the exception of those k streamlines $\psi = c_1, \psi = c_2, \ldots, \psi = c_k$, which reach the boundaries [the flat plate in point A], split there into two branches [ABC and $A'B'C'$], and pass around the boundary in different directions until they meet again to lead back to the source. Each streamline $\psi = c$, except for $c = c_1, c_2, \ldots, c_k$, is then mapped biuniquely onto a full line $\psi = c$ in the Φ-plane. Of these exceptional streamlines, the parts coinciding with the boundary are mapped onto slits, i.e., straight segments $\psi = c_1, \psi = c_2, \ldots, \psi = c_k$ in the Φ-plane, in such a way that the two edges [banks ABC and $A'B'C'$] of the slit correspond to the two branches of the streamline along the corresponding boundary curve of G. (Courant 1950, p. 46; with omissions, and adapted to our symbols.)

What we have touched upon here is the conformal mapping of k-fold connected domains in the complex plane to 'canonical' parallel-slit domains. Most books on advanced complex function theory treat this at some length, as direct generalization of the Riemann mapping theorem. The *slit theorem* establishes that any k-fold connected domain in the complex plane that is not the full plane, i.e., a domain with $k - 1$ holes, can be mapped conformally to a full complex plane with k parallel slits in it, some of which may degenerate to points. The proof is given in Golusin (1957), pp. 178–182. Alternatively, see Nehari (1952), Chap. VII, p. 347, Ahlfors (1966), pp. 243–253, and Courant (1950), pp. 56–57.

Note in the following that the complex speed $w(\Phi)$ is discontinuous across the slit in the Φ-plane, and that the interior angle of the polygon at $A = A'$ is 2π in this plane.

Finally, the polygon in the $\ln(w/U)$-plane is identical to that of the last example, the outflow from an orifice. The Schwarz–Christoffel transformations are therefore

$$\frac{d\Phi}{dt} = b_1 t \tag{5.37}$$

and

$$\frac{d \ln(w/U)}{dt} = \frac{b_2}{\sqrt{t-1} \; t \; \sqrt{t+1}}. \tag{5.38}$$

Note that the stagnation point A at $t_A = 0$ has to be considered also in the Schwarz–Christoffel formula (5.38), unlike the point A in the deep interior of the vessel in Sect. 5.4, where $t_A \to -\infty$. The interior angle is 0 at the corner $A = A'$ in the $\ln(w/U)$-plane, leading to the extra factor of t in the denominator of (5.38) as compared to (5.16). Integration over t gives in (5.37) and (5.38)

$$\Phi = \frac{1}{2} b_1 t^2 + a_1 \tag{5.39}$$

and

$$\ln \frac{w}{U} = -i b_2 \cosh^{-1}\left(-\frac{1}{t}\right) + a_2. \tag{5.40}$$

(One can perform the calculation alternatively with the real-valued function arcsin.) To calculate a_1 and b_1, the values of Φ and t at the points A, A' and C, C' as indicated in Fig. 5.8 are used in (5.39),

$$\Phi_{A,A'} = 0 = \frac{b_1}{2} \cdot 0 + a_1,$$

$$\Phi_{C,C'} = \infty = \frac{b_1}{2} \cdot \infty + a_1, \tag{5.41}$$

thus $a_1 = 0$ and $b_1 \in \mathbb{R}_+$; the calculation of a specific value for b_1 is postponed to (5.50). Next, for a_2 and b_2, using $t_{A'} = -\epsilon$ and $t_A = \epsilon$ with $\epsilon \to 0$ (see Fig. 5.8), Eq. (5.40) gives

$$\ln \frac{w}{U}\bigg|_{A'} = -\infty + \frac{i\pi}{2} = -i b_2 \cosh^{-1}(+\infty) + a_2 = -i b_2 \infty + a_2,$$

$$\ln \frac{w}{U}\bigg|_{A} = -\infty - \frac{i\pi}{2} = -i b_2 \cosh^{-1}(-\infty) + a_2 = -i b_2 (\infty + i\pi) + a_2,$$

$$\tag{5.42}$$

from which

$$a_2 = \frac{i\pi}{2}, \qquad b_2 = -i \tag{5.43}$$

is obtained, and therefore

$$\ln \frac{w}{U} = -\cosh^{-1}\left(-\frac{1}{t}\right) + \frac{i\pi}{2}. \tag{5.44}$$

This gives also correct values at the other points,

$$\ln \left.\frac{w}{U}\right|_{B'} = \frac{i\pi}{2} = 0 + \frac{i\pi}{2} = -\cosh^{-1}(+1) + \frac{i\pi}{2},$$

$$\ln \left.\frac{w}{U}\right|_{B} = -\frac{i\pi}{2} = -i\pi + \frac{i\pi}{2} = -\cosh^{-1}(-1) + \frac{i\pi}{2}, \qquad (5.45)$$

$$\ln \left.\frac{w}{U}\right|_{C,C'} = 0 = -\frac{i\pi}{2} + \frac{i\pi}{2} = -\cosh^{-1}(\pm 0) + \frac{i\pi}{2}.$$

Equation (5.44) can be rearranged as

$$-\frac{1}{t} = \cosh\left(\frac{i\pi}{2} - \ln\frac{w}{U}\right). \qquad (5.46)$$

Using $\cosh(x) = \frac{1}{2}(e^x + e^{-x})$, this is readily seen to give

$$-\frac{1}{t} = \frac{iU}{w} - \frac{iw}{U}. \qquad (5.47)$$

This is a quadratic equation in w/U,

$$\left(\frac{w}{U}\right)^2 + \frac{2i}{t}\frac{w}{U} - 1 = 0, \qquad (5.48)$$

with solutions

$$\frac{w}{U} = -\frac{i}{t}\left(1 \mp \sqrt{1 - t^2}\right). \qquad (5.49)$$

At the stagnation point A, $t_A = 0$ and $w_A = 0$. Thus, the minus sign must be chosen in (5.49). Now b_1 can be determined. With $w = d\Phi/dz$, and using (5.37) for $d\Phi/dz$ and (5.49) for w,

$$\frac{dz}{dt} = \frac{dz}{d\Phi}\frac{d\Phi}{dt} = \frac{1}{w}\frac{d\Phi}{dt} = \frac{ib_1 t^2}{U(1 - \sqrt{1 - t^2})}. \qquad (5.50)$$

Multiplying numerator and denominator with $1 + \sqrt{1 - t^2}$ and using $(1 - \sqrt{1 - t^2})(1 + \sqrt{1 - t^2}) = t^2$ gives

$$\frac{dz}{dt} = \frac{ib_1}{U}\left(1 + \sqrt{1 - t^2}\right). \qquad (5.51)$$

Integrating this from B to B' and assuming that the plate (with end points $t = -1$ and $t = 1$ in the t-plane) has length L in the z-plane gives

$$iL = \int_{-iL/2}^{iL/2} dz = \frac{ib_1}{U}\int_{-1}^{1} dt\,(1 + \sqrt{1 - t^2}) = \frac{ib_1}{U}\left(2 + \frac{\pi}{2}\right). \qquad (5.52)$$

Therefore,

$$b_1 = \frac{LU}{2 + \frac{\pi}{2}}, \tag{5.53}$$

and (5.39) becomes, with $a_1 = 0$,

$$\Phi = \frac{LU}{4 + \pi} t^2. \tag{5.54}$$

This fixes the value of Φ at B and B', where $t = \pm 1$. From (5.49) and (5.54) follows

$$-\frac{i}{t}\left(1 - \sqrt{1 - t^2}\right) = \frac{w}{U} = \frac{1}{U}\frac{d\Phi}{dz} = \frac{1}{U}\frac{d\Phi}{dt^2}\frac{dt^2}{dz} = \frac{L}{4 + \pi}\frac{dt^2}{dz}. \tag{5.55}$$

Next, let

$$\tau = t^2 - 1. \tag{5.56}$$

Then (5.55) takes the form

$$d\tau \frac{\sqrt{1 + \tau}}{1 + i\sqrt{\tau}} = \frac{4 + \pi}{iL} dz, \tag{5.57}$$

where the branch $\sqrt{-1} = -i$ of the root was taken. Multiplying numerator and denominator on the left side with $1 - i\sqrt{\tau}$ gives

$$d\tau \frac{1 - i\sqrt{\tau}}{\sqrt{1 + \tau}} = \frac{4 + \pi}{iL} dz. \tag{5.58}$$

Consider then the free streamline that encloses the wake, for which $t \in \mathbb{R}$. On BC and $B'C'$, $t^2 \geq 1$ (see Fig. 5.8) and thus $\tau \geq 0$, and (5.58) gives

$$\int_0^\tau d\tau \frac{1 - i\sqrt{\tau}}{\sqrt{1 + \tau}} = \frac{4 + \pi}{iL}\left[x + i\left(y - \frac{L}{2}\right)\right]. \tag{5.59}$$

Separating real and imaginary parts, an elementary integration yields

$$\frac{4 + \pi}{L}x = \sqrt{\tau(1 + \tau)} - \ln(\sqrt{\tau} + \sqrt{1 + \tau}), \tag{5.60}$$

$$\frac{4 + \pi}{L}\left(y - \frac{L}{2}\right) = 2\sqrt{1 + \tau} - 2. \tag{5.61}$$

Solving (5.61) for τ, and inserting the result into (5.60),

$$\bar{x} = \left(1 + \frac{\bar{y}}{2}\right)\sqrt{\bar{y} + \frac{\bar{y}^2}{4}} - \ln\left(1 + \frac{\bar{y}}{2} + \sqrt{\bar{y} + \frac{\bar{y}^2}{4}}\right) \tag{5.62}$$

is obtained, where

$$\bar{x} = \frac{4 + \pi}{L}\, x, \qquad \bar{y} = \frac{4 + \pi}{L}\left(y - \frac{L}{2}\right). \tag{5.63}$$

This is the curve BC shown in the uppermost panel of Fig. 5.8. Asymptotically, for large x, this becomes $\bar{y} = 2\sqrt{\bar{x}}$.

5.6 The Riabouchinsky Cavity

Consider again a flow past a plate, but now a cavity shall form behind the plate instead of the wake from the last section. The basic, ingenious idea is due to Riabouchinsky (1921): the wake is *forced* to close behind the plate by introducing a second plate parallel to the first plate. To make the problem tractable, the flow is assumed to be symmetric with respect to the x- *and* y-axes. A central demand for the Riabouchinsky cavity is that the vertical flow speed v shall vanish at the point $C = C'$.

Figure 5.9 shows the z-, Φ-, and $\ln(w/V)$-planes, where V is the constant speed on the free streamline. The domain enclosed by the curve $PABCC'B'A'P'P$, which lies outside the cavity, is mapped to the upper half of the Φ-plane. Note the difference to the foregoing section, where the full z-plane exclusive of the wake was mapped to the whole Φ-plane exclusive of the positive ϕ-axis. Choose again $\psi = 0$ for the limiting, free streamline of the cavity. Then an auxiliary t-plane is not necessary, since the real line $\phi \in \mathbb{R}$ of the Φ-plane can be used directly in the mapping $\phi \to \ln(w/V)$, just as for the flow past a step in Sect. 5.2. Let $\phi_A = -1$ and $\phi_{A'} = 1$, and $\phi_B = -k$ and $\phi_{B'} = k$, where $k < 1$ is a free parameter of the model. For the $\ln(w/V)$-plane, note that at large distances from the plates, the flow has horizontal speed U. On the streamline BC from plate to plate, the flow has speed $V > U$. The reason is that vanishing pressure in the cavity leads to an increasing speed according to the Bernoulli equation. The polygon in the $\ln(w/V)$-plane has therefore a new corner at $P = P'$, which does not appear in the Schwarz–Christoffel formula since ϕ_P and $\phi_{P'}$ correspond to the point at infinity (transformation $t \to s$ in Sect. 3.12). However, the *two* corners BAP and $B'A'P'$ at A and A' with finite $\phi_A = -1$ and $\phi_{A'} = 1$ and each with interior angle 0, as well as the two corners with interior angle $\pi/2$ at B and B', have to be included. Therefore,

$$\frac{\mathrm{d}\ln(w/V)}{\mathrm{d}\phi} = \frac{b}{(\phi - 1)\,\sqrt{\phi - k}\,\sqrt{\phi + k}\,(\phi + 1)} = \frac{b}{(\phi^2 - 1)\,\sqrt{\phi^2 - k^2}}. \tag{5.64}$$

Fig. 5.9 The Riabouchinsky cavity

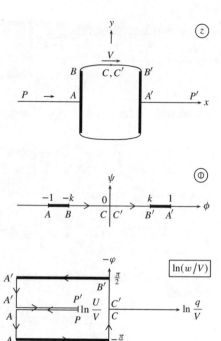

The free streamline BC corresponds to the interval

$$\phi \in [-k, 0].\qquad(5.65)$$

The potential ϕ shall be the curve variable along BC. Integrate over $d\phi$, starting at the plate corner B where $\phi = -k$ (see Fig. 5.9) and $\ln(w/V) = -i\pi/2$ (since $w = qe^{-i\varphi}$ with $q = V$ and $\varphi = \pi/2$), up to an arbitrary point with coordinate $\phi \le 0$ and speed logarithm $\ln(w/V) \le 0$ along the free streamline. Introducing a new variable,

$$y = \phi^2 - 1,\qquad(5.66)$$

Equation (5.64) becomes then

$$\int_{-i\pi/2}^{\ln(w/V)} d\ln\frac{w}{V} = \frac{b}{2}\int_{k^2-1}^{\phi^2-1}\frac{dy}{y\,\sqrt{y^2+(2-k^2)y+1-k^2}}.\qquad(5.67)$$

The integral is standard, and

$$\ln \frac{w}{V} + \frac{i\pi}{2} = \frac{b}{2\sqrt{1-k^2}} \left[\cosh^{-1} \left(\frac{2 - 2k^2 + (2-k^2)y}{k^2 y} \right) \right]_{y=k^2-1}^{\phi^2 - 1}$$

$$= \frac{b}{2\sqrt{1-k^2}} \left[\cosh^{-1} \left(\frac{(2-k^2)\phi^2 - k^2}{k^2(\phi^2 - 1)} \right) - i\pi \right] \qquad (5.68)$$

is obtained. At the point C, $\phi = 0$ and $\ln(w/V) = 0$. This yields

$$\frac{i\pi}{2} = \frac{b}{2\sqrt{1-k^2}} (\cosh^{-1}(1) - i\pi) = -\frac{i\pi b}{2\sqrt{1-k^2}}, \qquad (5.69)$$

thus $b = -\sqrt{1 - k^2}$, and

$$\ln \frac{w}{V} = -\frac{1}{2} \cosh^{-1} \left(\frac{k^2 - (2-k^2)\phi^2}{k^2(1 - \phi^2)} \right). \qquad (5.70)$$

Applying on both sides of this equation $\cosh(x) = \frac{1}{2}(e^x + e^{-x})$ gives the quadratic equation

$$\frac{w^4}{V^4} - 2\frac{k^2 - (2-k^2)\phi^2}{k^2(1 - \phi^2)} \frac{w^2}{V^2} + 1 = 0, \qquad (5.71)$$

with solution

$$\frac{w^2}{V^2} = \frac{k^2 - (2-k^2)\phi^2 \pm 2i\phi\sqrt{(1-k^2)(k^2 - \phi^2)}}{k^2(1 - \phi^2)}. \qquad (5.72)$$

Separate real and imaginary parts using $w = u - iv$,

$$\frac{u^2 - v^2}{V^2} = \frac{k^2 - (2-k^2)\phi^2}{k^2(1 - \phi^2)}, \qquad (5.73)$$

$$\frac{uv}{V^2} = -\frac{\phi\sqrt{(1-k^2)(k^2 - \phi^2)}}{k^2(1 - \phi^2)}. \qquad (5.74)$$

The sign in (5.74) was chosen so that $uv > 0$ for $\phi \le 0$, as is indicated in Fig. 5.9. Furthermore, (5.73) gives $u = 0$ and $v = V$ at $\phi = -k$, and $u = V$ and $v = 0$ at $\phi = 0$, also in agreement with Fig. 5.9. Multiplying (5.74) by V/v gives an expression for u/V. Inserting this in (5.73) yields

$$\left(ukV^{-1}\sqrt{1-\phi^2} \right)^4 - (k^2 - (2-k^2)\phi^2)\left(ukV^{-1}\sqrt{1-\phi^2} \right)^2 - \phi^2(1-k^2)(k^2 - \phi^2) = 0, \qquad (5.75)$$

which is a quadratic equation in $\left(ukV^{-1}\sqrt{1-\phi^2} \right)^2$, with solution

$$\left(ukV^{-1}\sqrt{1-\phi^2} \right)^2 = \frac{1}{2}\left[k^2 - (2-k^2)\phi^2 \pm \sqrt{(k^2 - k^2\phi^2)^2} \right]. \qquad (5.76)$$

Choosing the positive root (which is seen below to give the correct solution) yields

$$u = \frac{\partial \phi}{\partial x} = \sqrt{\frac{k^2 - \phi^2}{1 - \phi^2}} \frac{V}{k}, \tag{5.77}$$

and from (5.74) directly,

$$v = \frac{\partial \phi}{\partial y} = \sqrt{\frac{1 - k^2}{1 - \phi^2}} \frac{|\phi|V}{k}. \tag{5.78}$$

This is Eq. (8.2) in Gilbarg (1960, p. 335), where however dz/df must be replaced by df/dz due to a misprint. Equations (5.77) and (5.78) allow to find the length of the cavity as function of k. This leads to elliptic functions, for which we refer to Gilbarg (1960).

From (5.77) and (5.78) follows

$$\frac{dy}{dx} = \frac{v}{u} = \frac{|\phi|\sqrt{1 - k^2}}{\sqrt{k^2 - \phi^2}}, \tag{5.79}$$

which is easily integrated numerically. The results for three different k-values are shown in Fig. 5.10. Finally, a relation between k and the plate length L is obtained as follows. The interval $-1 \le \phi \le -k$ corresponds to the segment from A to B. Assuming therefore $\phi^2 > k^2$ in (5.72) gives

$$\frac{w^2}{V^2} = \frac{k^2 - (2 - k^2)\phi^2 \pm 2\phi\sqrt{(1 - k^2)(\phi^2 - k^2)}}{k^2(1 - \phi^2)}. \tag{5.80}$$

The right side is a real number, thus $\text{Im } w = -2uv = 0$. This is due to the horizontal speed $u = 0$ along the vertical plate. From $v^2 = -\text{Re } w^2$, and integrating from $\phi = 0$ to k, $L(k)$ is found. This gives expressions in elliptic functions, which are not further specified here. In Fig. 5.10, formulas (8.6) and (8.7) from Gilbarg (1960) were used to find L.

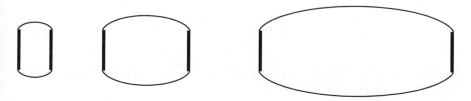

Fig. 5.10 Three examples of the Riabouchinsky cavities, for $k = 0.8, 0.9$, and 0.95

5.7   Levi-Civita's Method

So far, the upper half-plane of the complex t-plane was mostly mapped to strips in the Φ- and $\ln w$-plane. The calculations were quite involved, and it may thus be of advantage to have maps to (or between) other geometrical figures. We consider in this section mappings from half-planes to semicircles and from semicircles to semicircles. This is a key idea of a method proposed by Levi-Civita (1907).

An advantage of using semicircles is that they allow to work directly in the complex w-plane instead of the complex $\ln w$-plane considered so far. Free streamlines have $|w| = $ const and thus correspond to circular arcs in the w-plane. The latter statement can be reversed: any function $w(\tau)$ that is analytic on the unit τ-circle gives a free streamline. The ratio u/v of the velocity components in the x- and y-directions is constant along any straight segment of the polygonal boundary curve of an obstacle immersed in the flow. Thus, the streamline on the obstacle boundary corresponds to straight line segments in the w-plane, since $w = u - iv$. For the simplest flow examples (cavities and wakes), straight line segments and circular arcs can be assembled to semicircles.

A new variable τ is introduced, replacing the variable t used so far, and semicircles in τ are considered instead of half-planes in t. The real axis in the t-plane corresponds to a circle with infinite radius. Every circle is uniquely determined by three points that lie on it. Thus, the map $t \leftrightarrow \tau$ is uniquely determined by mapping three points. Let $t_0 = -1$, $t_1 = 0$, and $t_2 = 1$ correspond to $\tau_0 = 1$, $\tau_1 = i$, and $\tau_2 = -1$, respectively. This is achieved by (the inverse map)

$$t = -\frac{1}{2}\left(\tau + \frac{1}{\tau}\right), \tag{5.81}$$

see Eq. (12) in Levi-Civita (1907). Furthermore, (5.81) maps the *upper* half-plane of the variable t to the *interior* of the semicircle with radius 1 in the upper half-plane of the variable τ: indeed, for $\tau = ae^{i\varphi}$ with $a < 1$ and $0 < \varphi < \pi$, it follows that $-\mathrm{Im}\,(\tau + \tau^{-1}) = -(a - a^{-1})\sin\varphi > 0$.

Equipped with this new complex variable τ, we consider again the simple wake behind a flat plate from Sect. 5.5. The solution is fully determined by the complex wake potential Φ and the complex wake speed w. Writing (5.54) for Φ as $4bt^2$, and using (5.81) for $t(\tau)$ gives

$$\Phi = b\left(\tau + \frac{1}{\tau}\right)^2. \tag{5.82}$$

Inserting (5.81) in (5.49), i.e., $w/U = -i\left(1 - \sqrt{1 - t^2}\right)/t$, gives, after an elementary calculation, the two solutions

$$\frac{w}{U} = \frac{i - \tau}{i + \tau}, \qquad \frac{w}{U} = -\frac{i + \tau}{i - \tau}. \tag{5.83}$$

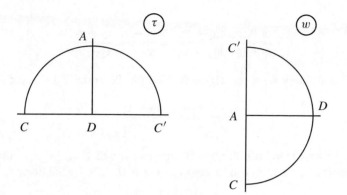

Fig. 5.11 Linear map $w = \dfrac{i - \tau}{i + \tau}$ between two semicircles

Choose the first one and put $U = 1$,

$$w = \frac{i - \tau}{i + \tau}. \tag{5.84}$$

For $\tau \in \mathbb{R}$, this implies

$$w\bar{w} = \frac{i - \tau}{i + \tau} \frac{i + \tau}{i - \tau} = 1, \tag{5.85}$$

which means that the interval $\tau \in [-1, 1]$ is mapped to the circular w-arc CDC' in Fig. 5.11. The inverse map to (5.84) is

$$\tau = i \frac{1 - w}{1 + w}. \tag{5.86}$$

For $w \in i\mathbb{R}$, $\tau\bar{\tau} = 1$, thus $w \in [-i, i]$ is mapped to the circular τ-arc CAC', see again Fig. 5.11. Therefore, (5.84) maps a semicircle to a semicircle. It is also easily seen that (5.84) maps the interior of one semicircle to that of the other.

We close this section with a short summary on *linear* complex maps, for which $w(\tau)$ in (5.84) and $\tau(w)$ in (5.86) are examples (the inverse of a linear map is a linear map). For a general introduction to linear complex maps, see Nehari (1952) and Bieberbach (1921). A complex linear map $\tau \to w$ is given by

$$w = \frac{a\tau + b}{c\tau + d}, \qquad ad - bc \neq 0, \tag{5.87}$$

where $a, b, c, d \in \mathbb{C}$ are constants. If $ad - bc = 0$ then $w = $ const. Linear maps $\tau \to w$ map circles in the τ-plane to circles in the w-plane. Again, circles with infinite radius correspond to straight lines. A linear map between any three pairs of points is obtained as follows. Linear complex maps leave the *double ratio* invariant, which means

$$\frac{w - w_1}{w - w_3} \frac{w_2 - w_3}{w_2 - w_1} = \frac{\tau - \tau_1}{\tau - \tau_3} \frac{\tau_2 - \tau_3}{\tau_2 - \tau_1} \tag{5.88}$$

for pairwise different τ_1, τ_2, τ_3. To see this, let (cf. Bieberbach 1921, p. 57)

$$z = \frac{\tau - \tau_1}{\tau - \tau_3} \frac{\tau_2 - \tau_3}{\tau_2 - \tau_1}, \tag{5.89}$$

which is a linear map with $z(\tau_1) = 0$, $z(\tau_2) = 1$, and $z(\tau_3) \to \infty$. Furthermore, arbitrary points w_1, w_2, w_3 can be mapped to $z = 0, 1, \infty$ by the linear map

$$z = \frac{w - w_1}{w - w_3} \frac{w_2 - w_3}{w_2 - w_1}. \tag{5.90}$$

Equations (5.89) and (5.90) give indeed (5.88). For example, to obtain the map (5.84), choose from Fig. 5.11,

$$\begin{array}{ccc}
\tau_1 = 1 & & w_1 = i \\
\tau_2 = i & \to & w_2 = 0 \\
\tau_3 = -1 & & w_3 = -i.
\end{array} \tag{5.91}$$

Then (5.88) gives

$$-\frac{w - i}{w + i} = \frac{\tau - 1}{\tau + 1} \frac{i + 1}{i - 1}. \tag{5.92}$$

After some simplifications, (5.84) is indeed obtained.

5.8 Kolscher's Cusped Cavity

In applying complex function theory to find solutions for free streamlines, we turn now to *singularities* and address the question of whether *cusps* can form at the trailing edge of a cavity. This issue is of some importance since there were claims in the early twentieth century that such cusps cannot occur. The solution discussed in the following was first obtained by M. Kolscher (1940) in his dissertation. There is little information available about Kolscher, and it seems that he perished in WW II. A short discussion of Kolscher's solution is also given in Gilbarg (1960), p. 338.

Figure 5.12 shows the basic flow geometry in the z-, Φ-, and τ-planes. A parallel flow from infinity hits a flat plate at normal incidence. Kolscher (1940) allows for detachment from the back of the plate in points C and C'. A terminating cusp of the cavity behind the plate is postulated to occur in D. A slit, now of finite length, is excluded from the real ϕ-axis of the Φ-plane. This slit is taken to be symmetric with respect to the ψ-axis; the figure shows the remaining banks of the slit. In the τ-plane,

Fig. 5.12 Cusped cavity behind a flat plate (Kolscher 1940). The free streamline detaches in points C and C' on the backside of the plate, behind the corners at B and B'. Plus and minus signs in the τ-plane indicate whether $\varphi = +\pi/2$ or $-\pi/2$, see (5.107)

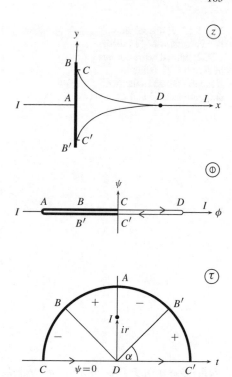

a semicircle is introduced, as was discussed in the last section. The plate end points B' and B are mapped to $e^{i\alpha}$ and $-e^{-i\alpha}$ in the τ-plane, respectively.

This geometry builds upon a similar but simpler one shown in Fig. 5.13, adapted from a paper by Schmieden (1932). The flow detaches here from the polygonal obstacle in C and C'. The acute corners at B and B' were introduced by Schmieden to force the flow toward the x-axis. This then leads to inflexion points of the free streamline at D and D'. We do not go into the detailed calculations for this example, but only note that Schmieden (1932) maps the point I where $z \to \infty$ to zero in the τ-plane, whereas Kolscher's crucial idea is to map instead D to zero in the τ-plane. Kolscher then puts the point I on the imaginary axis, at a distance $ir < i$ from zero. This comparison shows the great flexibility in setting up the flow geometry in the complex τ-plane.

Below, the Riemann–Schwarz reflection principle is needed in three slightly different versions (actually, version II implies versions I and III).

Fig. 5.13 Enforced flow
detachment after Schmieden
(1932) behind acute corners
at B and B', leading to
inflexion points in the free
streamlines at D and D'

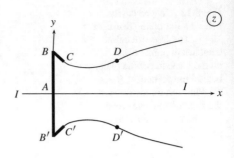

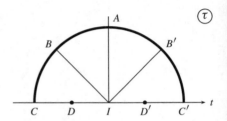

Reflection principle version I, for straight lines

Let D be a domain in the upper half of the complex z-plane, $y > 0$. The (topological)
closure of D shall contain a nonempty interval γ of the real x-axis. Let the complex
function $f : D \rightarrow \mathbb{C}$ have the following three properties: f is analytic on D, continuous on $D \cup \gamma$, and real-valued on $\gamma \subset \mathbb{R}$. It is especially important to check for
the latter property in applications. Define a domain D^* by

$$z \in D^* \quad \leftrightarrow \quad \bar{z} \in D. \tag{5.93}$$

This is the mirror image of D with respect to the real x-axis. The function f is
extended to $z \in D^*$ by

$$f(z) = \overline{f(\bar{z})}. \tag{5.94}$$

The reflection principle states that f is analytic on $D \cup \gamma \cup D^*$. For the proof of this
statement, see Ahlfors (1966), p. 171. The extension of the domain of f from D to
$D \cup \gamma \cup D^*$ is referred to as an *analytic continuation* of f.

Reflection principle version II, for circular arcs

Instead of intervals of the real axis, let γ and $f(\gamma)$ be circular arcs with radii R and
R', respectively. The center of both the corresponding circles shall lie at $z = 0$. The
domain D shall lie on one side of the arc γ and have γ as part of its boundary. Define
D^* by

Fig. 5.14 Geometry for
version II of the reflection
principle. The domain D is
the interior of the semicircle,
the domain D^* its exterior in
the upper half-plane

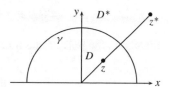

$$z \in D^* \qquad \leftrightarrow \qquad z^* = \frac{R^2}{z} \in D. \tag{5.95}$$

Obviously, $z^{**} = z$, and $z = re^{i\varphi}$ and $z^* = R^2 r^{-1} e^{i\varphi}$ lie on a straight line through the origin, see Fig. 5.14. Let f be analytic on D and continuous on $D \cup \gamma$. For $z \in D^*$ define

$$f(z) = [f(z^*)]^*, \tag{5.96}$$

where f^* is calculated using the arc radius R' in the f-plane. Then f is analytic on $D \cup \gamma \cup D^*$. (The restriction that the arc γ is centered at $z = 0$ can be removed by a translation $z \rightarrow z + a$.)

Reflection principle version III, mixed case

If γ is a circular arc and $f(\gamma)$ a segment of the real axis, then with z^* from (5.95), analytic continuation from D to $z \in D^*$ is performed by

$$f(z) = \overline{f(z^*)}. \tag{5.97}$$

In Fig. 5.12, the semicircle $ABCDC'B'A$ (there is no danger of confusing the domain D with the point D) in the τ-plane is a streamline. Let $\psi = 0$ on this semicircle, thus the complex potential becomes real-valued on it,

$$\Phi(\tau) = \phi(\tau) \in \mathbb{R}, \tag{5.98}$$

Let $\phi(x) \rightarrow \infty$ for $x \rightarrow \infty$ (point I); then $\Phi(\tau)$ has a pole at $\tau = ir$. The straight line segment CD is part of the real axis both in the τ- and the Φ-planes. Thus $\Phi(-ir) = \overline{\Phi(ir)}$ from (5.94) for version I of the reflection principle for straight lines, and therefore Φ has a second pole at $\tau = -ir$. Next, version III of the reflection principle is used: the segment ABC is a quarter circle in the τ-plane, and a real-valued, straight line segment in the Φ-plane. Let the radius of the quarter circle in the τ-plane be $R = 1$. Then

$$(ir)^* = \frac{1}{-ir}, \qquad (-ir)^* = \frac{1}{ir}. \tag{5.99}$$

Since $r < 1$, both points lie outside the τ-circle with radius 1. According to (5.97), the function Φ has poles at these locations too,

$$\Phi(i/r) = \overline{\Phi(ir)}, \qquad \Phi(-i/r) = \overline{\Phi(-ir)}. \tag{5.100}$$

Finally, at the point A the fluid speed must vanish,

$$w = \frac{d\Phi}{dz} = \frac{d\Phi}{d\tau}\frac{d\tau}{dz} = 0. \tag{5.101}$$

This can hold in general only if

$$\left.\frac{d\Phi}{d\tau}\right|_A = \frac{d\Phi}{d\tau}(\pm i) = 0. \tag{5.102}$$

Besides the point A at $\tau = i$, the reflection point $\tau = -i$ was added. A complex potential $\Phi(\tau)$ that obeys (5.102) and has simple poles (i.e., poles of first order; Kolscher considers the simplest possible solution for a cusped cavity) at the four locations $\tau = \pm ir$ and $\pm i/r$ is given by

$$\Phi(\tau) = \frac{b(1 + \tau^2)^2}{(\tau^2 + r^2)(r^2\tau^2 + 1)}, \tag{5.103}$$

where $b \in \mathbb{R}^+$. The derivative is

$$\frac{d\Phi}{d\tau} = -2b(1 - r^2)^2 \frac{(1 - \tau^2)\,\tau\,(1 + \tau^2)}{(\tau^2 + r^2)^2(r^2\tau^2 + 1)^2}. \tag{5.104}$$

Thus the complex potential Φ is known. We turn to the complex fluid speed $w = qe^{-i\varphi}$ and normalize the speed of the incoming stream to $U = 1$, or

$$w(\tau = ir) = 1. \tag{5.105}$$

However, the speed on the free streamline CD behind the plate will again be $V \neq 1$. At the turning points B and B' of the plate,

$$w_B \to \infty, \qquad w_{B'} \to \infty. \tag{5.106}$$

Thus, the function $w(\tau)$ has poles at $\tau = e^{i\alpha}$ and $-e^{-i\alpha}$, where α has been introduced in Fig. 5.12. Furthermore, on the arc $CBAB'C'$, the azimuth φ of the flow direction is as follows:

$$\begin{aligned}
\text{from } C' \text{ to } B' : & \quad \varphi = +\pi/2, \\
B' \text{ to } A : & \quad = -\pi/2, \\
A \text{ to } B : & \quad = +\pi/2, \\
B \text{ to } C : & \quad = -\pi/2.
\end{aligned} \tag{5.107}$$

These values of φ are indicated by plus and minus signs in Fig. 5.12. From Fig. 5.11, the map (5.84) gives purely imaginary values for w along $CBAB'C'$. This corresponds to $\arg(w) = \pm\pi/2$. Thus a complex function $w(\tau)$ is searched for that

(i)   is purely imaginary along $CBAB'C'$,
(ii) obeys the sign conventions (5.107), and
(iii) has poles at $\tau = e^{i\alpha}$ and $-e^{-i\alpha}$.

Kolscher gives the following expression for the speed w that has these three properties:

$$w(\tau) = V \, \frac{i-\tau}{i+\tau} \, \frac{\tau+e^{i\alpha}}{\tau-e^{i\alpha}} \, \frac{\tau-e^{-i\alpha}}{\tau+e^{-i\alpha}} \tag{5.108}$$

The speed $V > 0$ can be expressed in terms of r and $\sin\alpha$ by the relation (5.105). Equation (5.108) fulfills condition (iii) on w. As for (i), observe that for two points $e^{i\alpha}$ and $e^{i\beta}$ on the unit circle,

$$\arg(e^{i\alpha} + e^{i\beta}) = \frac{\alpha+\beta}{2}, \tag{5.109}$$

where arg is the argument of a complex number in polar representation. Therefore, substituting $\tau = e^{i\beta}$ in (5.108),

$$\arg w = \frac{\pi/2+\beta+\pi}{2} - \frac{\pi/2+\beta}{2} + \frac{\beta+\alpha}{2} - \frac{\beta+\alpha+\pi}{2} + \frac{\beta-\alpha+\pi}{2} - \frac{\beta-\alpha}{2} = \frac{\pi}{2}$$

follows. Finally, (5.108) also gives the correct signs according to condition (ii). This can be seen from Fig. 5.15, which shows $w(\tau)$ from (5.108) for τ on the semicircle from Fig. 5.12. The poles at B' and B correspond to $\tau = e^{i\alpha}$ and $-e^{-i\alpha}$, respectively. As a minor detail, note that C and C' in Fig. 5.15 have reversed their position as compared to Fig. 5.11.

This leads now to an interesting mathematical issue. The point I at $z \to \infty$ is put at $\tau = ir$ and is an interior point of the closed contour $\gamma = ABCDC'B'A$ in the τ-plane. At I, corresponding to $\tau = ir$, the derivative $d\Phi/d\tau$ from (5.104) has a pole of second order from division by $(\tau^2 + r^2)^2 = (\tau + ir)^2(\tau - ir)^2$. This is the only singularity in the interior of γ, hence according to the residue theorem (A.29) from Appendix A,

$$0 = \oint dz = \oint_\gamma d\tau \, \frac{d\Phi}{d\tau} \, \frac{1}{w(\tau)}$$

$$= 2\pi i \, \mathrm{Res} \left[\frac{d\Phi}{d\tau} \, \frac{1}{w(\tau)} \right]_{\tau=ir} \tag{5.110}$$

follows. This is the 'Schliessungsbedingung' or closure condition (6) of Kolscher (see also Koppenfels and Stallmann 1959, p. 174). Furthermore, for irrotational flow, the circulation Γ is zero along any closed curve, thus with Eq. (4.97) in Sect. 4.5,

$$0 = \oint dz \, w = \oint d\Phi = \oint_\gamma d\tau \, \frac{d\Phi}{d\tau}$$

$$= 2\pi i \, \text{Res} \left[\frac{d\Phi}{d\tau} \right]_{\tau=ir}. \tag{5.111}$$

For a pole of order $m \geq 1$ at $\tau = ir$, a quite trivial but somewhat awkward calculation gives

$$0 = \text{Res} \left[\frac{d\Phi}{d\tau} \frac{1}{w(\tau)} \right]_{\tau=ir}$$

$$= \frac{1}{m!} \lim_{\tau \to ir} \frac{d^{m-1}}{d\tau^{m-1}} \left[(\tau - ir)^m \frac{d\Phi}{d\tau} \frac{1}{w(\tau)} \right]$$

$$= \frac{1}{w(ir)} \frac{1}{m!} \lim_{\tau \to ir} \frac{d^{m-1}}{d\tau^{m-1}} \left[(\tau - ir)^m \frac{d\Phi}{d\tau} \right]$$

$$+ \frac{1}{m!} \sum_{k=1}^{m-1} \binom{m-1}{k} \lim_{\tau \to ir} \left[\frac{d^k}{d\tau^k} \left(\frac{1}{w} \right) \frac{d^{m-k-1}}{d\tau^{m-k-1}} \left\{ (\tau - ir)^m \frac{d\Phi}{d\tau} \right\} \right]$$

$$= \frac{1}{w(ir)} \, \text{Res} \left[\frac{d\Phi}{d\tau} \right]_{\tau=ir}$$

$$+ \frac{1}{m!} \sum_{k=1}^{m-1} \binom{m-1}{k} \lim_{\tau \to ir} \left[\frac{d^{k-1}}{d\tau^{k-1}} \left(\frac{d}{d\tau} \frac{1}{w} \right) \frac{d^{m-k-1}}{d\tau^{m-k-1}} \left\{ (\tau - ir)^m \frac{d\Phi}{d\tau} \right\} \right]. \tag{5.112}$$

Here, $\text{Res} \, [d\Phi/d\tau]_{ir} = 0$ according to (5.111), and $1/w(ir)$ from (5.108) is finite. Furthermore, for $m = 2$ and thus $k = 1$, the term $[(\tau - ir)^m \, d\Phi/d\tau]_{ir}$ does not vanish, and therefore

Fig. 5.15 $w(\tau)$ from (5.108), for τ on the semicircle from Fig. 5.12. A small offset separates lines that would else coincide

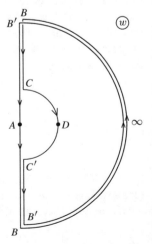

$$\boxed{\left.\frac{d}{d\tau}\frac{1}{w}\right|_{\tau=ir} = 0}$$ (5.113)

This is Kolscher's *regularity* condition (7). For a pole of order $m > 2$, all derivatives $[d^k/d\tau^k(w^{-1})]_{ir}$ for $k = 1$ to $m - 1$ must vanish.

Inserting (5.108) in (5.113) yields a quadratic equation in r^2. Its coefficients depend on the angle α introduced in Fig. 5.12. A single, real-valued solution $0 \leq r \leq 1$ exists only if

$$30° \leq \alpha \leq 90°,$$ (5.114)

see Eq. (10) in Kolscher (1940). For α in this interval, the point I is located at $0 \leq \tau = ir \leq i$, as should be. The value of α controls the diameter of the cavity, as is shown in Fig. 5.16. The integral from (5.110),

$$z = \int^\tau d\tau \, \frac{d\Phi}{d\tau} \frac{1}{w(\tau)},$$ (5.115)

can be performed analytically for $\Phi(\tau)$ and $w(\tau)$ from (5.103) and (5.108), and Kolscher finds in his Eq. (11) for the streamline $z(\tau)$,

$$\frac{z}{c} = \frac{(1 - \tau^4)(2 - \sin\alpha) + 2i(\tau + \tau^3)(2\sin\alpha - 1)}{(1 + \tau^4)(2\sin\alpha - 1) + 2\tau^2(1 + 2\sin\alpha - 2\sin^2\alpha)}$$

$$+ \frac{1}{\cos\alpha}\left(i \ln \frac{\tau - e^{-i\alpha}}{\tau + e^{i\alpha}} + \frac{\pi}{2} - \alpha\right),$$ (5.116)

where, for a plate that extends from $z = -i$ to $z = i$,

$$c = \frac{\sin 2\alpha}{\cos^2\alpha - 2\sin\alpha \,\, \ln\sin\alpha}.$$ (5.117)

Note that τ is real-valued along the free streamline. Thus (5.116) is a parameterization of the streamline $\psi = 0$ in terms of $\tau \in \mathbb{R}$. Separating the logarithm in (5.116) into real and imaginary parts gives

$$i \ln \frac{\tau - e^{-i\alpha}}{\tau + e^{i\alpha}} = -\text{atan}\frac{b}{a} + i \ln\sqrt{a^2 + b^2},$$ (5.118)

where

$$a = \frac{\sin^2\alpha - \cos^2\alpha + \tau^2}{1 + 2\tau\cos\alpha + \tau^2}, \qquad b = \frac{2\sin\alpha \,\cos\alpha}{1 + 2\tau\cos\alpha + \tau^2}.$$ (5.119)

The streamline $(x(\tau), y(\tau))$ from (5.116) is plotted for three values of α in Fig. 5.16.

Kolscher thus disproved earlier claims by Brillouin (1911) and Villat (1913) that cusped cavities cannot exist. Their argument runs as follows. In Cartesian coordi-

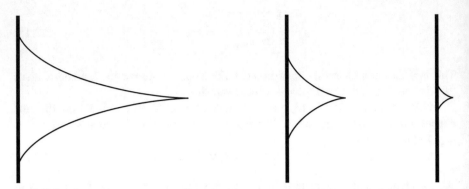

Fig. 5.16 Kolscher's cusped cavity, from left to right for $\sin\alpha = \frac{1}{\sqrt{2}}$, 0.9, and 0.99

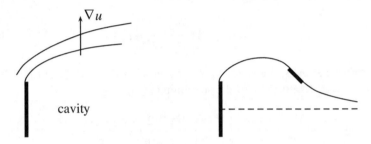

Fig. 5.17 Left: convex cavity. Right: transition convex–concave by tilted plate

nates, the fluid acceleration for a stationary flow is given by

$$a_i = u_k \frac{\partial u_i}{\partial x_k} \overset{(*)}{=} u_k \frac{\partial u_k}{\partial x_i} = \frac{1}{2}\frac{\partial}{\partial x_i}(u_k u_k) = u \frac{\partial u}{\partial x_i}, \tag{5.120}$$

where $u^2 = u_k u_k$ and the summation convention is used. In $(*)$ an irrotational flow is assumed, $\partial u_i/\partial x_k = \partial u_k/\partial x_i$. From (5.120), the vector of acceleration $\vec{a}$ points in the direction of the velocity gradient ∇u. This is Brillouin's principle for irrotational flows: streamlines curve in the direction of increasing speed u, see Fig. 5.17. The cavity boundary should therefore be convex with respect to the cavity, and streamlines as in cusped cavities are excluded. Both Schmieden (see Fig. 5.13) and Kolscher (see Fig. 5.12) point a way out of this conclusion, by enforcing concavity through the choice of boundary conditions. A free streamline can also be forced to undergo a transition from convexity to concavity by tilted plates in the flow domain behind the flat plate, see the right panel in Fig. 5.17.

It seems that the first solution for a cusped cavity was found by Cisotti (1932). An interesting cusped cavity is considered by Fabula (1962), who analyzes the flow past a thin airfoil at small angle α of attack. For $\alpha > \alpha^+$, a cavity forms behind the airfoil, which is cusped for $\alpha \to \alpha^+$.

5.9   Re-Entrant Jet Cavity

Figure 5.18 shows a cavity with a stagnation point S in the flow behind a plate, besides the obvious stagnation point A on the plate. Two cavities occur behind the plate, and between them the flow is re-directed toward the plate. Since the velocity is constant on the free streamline, this cannot meet the boundary condition of vanishing normal speed on the plate. To resolve this problem, the jet that is directed toward the plate is put on a second Riemann sheet. This is termed a *re-entrant jet*, first proposed by different authors in 1946.

The solution for this cavity problem is summarized in Gilbarg (1960), p. 332 and Birkhoff and Zarantonello (1957), p. 57. The points A, I, S, J are mapped to $\tau = 0, ia, ib, i$, respectively, with real a and b such that $0 < a < b < 1$. The flat plate is mapped to the interval $[-1, 1]$ of the real axis of the τ-plane, and the free streamline to the upper unit semicircle, see Fig. 5.19. Then Eqs. (20′) and (21) in Birkhoff and Zarantonello are, with $C \in \mathbb{R}^+$,

$$\frac{\mathrm{d}\textcircled{1}}{\mathrm{d}\tau} = -C\tau \, \frac{(\tau^2 + b^2)(\tau^2 + b^{-2})(\tau^2 - 1)}{(\tau^2 + a^2)^2(\tau^2 + a^{-2})^2(\tau^2 + 1)}, \tag{5.121}$$

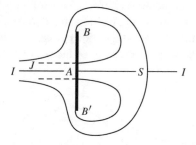

Fig. 5.18 Re-entrant jet cavity

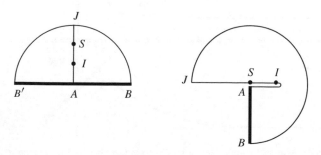

Fig. 5.19 Parameter plane τ (left) and speed plane w (right) for the re-entrant jet cavity. The left panel shows the correspondence to the *whole* z-plane of the flow, and the right panel shows the correspondence to the upper half of the z-plane only. The points B', etc., from the lower half-plane in z are obtained by mirroring the figure at the real axis. An overlap is avoided by distributing corresponding points on the two different Riemann sheets

$$w(\tau) = \frac{\tau(\tau^2 + b^2)}{b^2\tau^2 + 1}. \tag{5.122}$$

These equations again fully specify the flow. The right panel in Fig. 5.19 shows parts of the w-plane.

5.10 Tilted Wedge

We come to a final example of a flow about an obstacle, solved with the conformal mapping technique in a remarkable paper by Cox and Clayden (1958). The flow harbors both a wake and a cavity, see Fig. 5.20. A parallel stream from infinity (e.g., the points I_1 and I_2 in the figure) hits the wedge B_1CB_2. The wedge has an interior angle α and is tilted in such a way that one of its sides is parallel to the incoming flow. At this side, a detached cavity forms, leading to a re-entrant jet that is put on a second Riemann sheet *before* reaching the line CD. Behind the wedge, a wake forms, which is not subject of the following calculations. The incoming streamlines I_1A_1 and I_2A_2 split at the stagnation points A_1 and A_2 on the wedge. The flow can reach infinity I outside the lines $I_1A_1B_1$ and $I_2A_2B_2$.

Figure 5.21 shows the Φ-plane. As before, the splitting of a streamline at a stagnation point is realized by introducing a slit from the finite ϕ-value at the stagnation point to infinity. Since there are two stagnation points, two slits are needed. Trivially, they are parallel, since $\psi = \text{const}$ on each of them. Note that the split streamlines

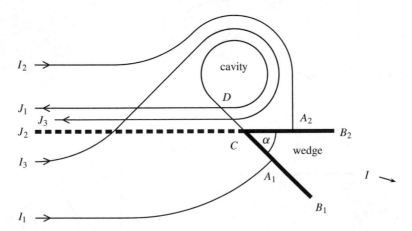

Fig. 5.20 Tilted wedge in a parallel flow after Cox and Clayden (1958), leading to the formation of a wake and a cavity. Full bold line: wedge on the Riemann sheet 1; dashed bold line: boundary on the Riemann sheet 2. All remaining lines are streamlines. I and J are the points at infinity on the first and second Riemann sheets, respectively. The indices at I and J serve only for easier identification of matching points

Fig. 5.21 Φ-plane with two parallel slits

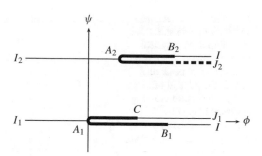

A_1B_1I and A_1CDJ_1 enclose a domain that contains both the cavity and the wake behind the wedge, whereas the split streamlines A_2B_2I and A_2J_2 (the flow is put to the second Riemann sheet before reaching the line CD) enclose a domain that contains the wake and the domain beyond the boundary CJ_2 on the second Riemann sheet. These extended domains (which are not subject of the calculation) are 'lost in' the two excluded slits. There is no need for a second Riemann sheet in the Φ-plane.

Next, we consider the plane $\ln w$ for the complex speed in Fig. 5.22. At the stagnation points A_1 and A_2 the speed is zero; hence, they correspond to the point at infinity. Along the free streamline, the speed is constant and normalized to 1. Therefore, the crucial free streamline CJ corresponds to a finite interval of the imaginary $-i\varphi$-axis (only the direction angle of the streamline changes). The numbers on this $-i\varphi$-axis are clear from the geometry of Fig. 5.20. Figure 5.22 shows a single closed loop on the Riemann surface. The branch points lie at Z' and $Z \to \infty$. One crosses the Riemann sheets, for example, along the dotted line.

As before, we are looking for an algebraic relation $w(\Phi)$ between the complex speed and the complex potential. To this end, both the $\ln w$- and the Φ-planes are mapped to a parameter τ-plane. The conformal mappings $\tau \to \Phi$ and $\tau \to \ln w$ are found by construction/inspection: the loop $B_2 \dots B_2$ in the $\ln w$-plane of Fig. 5.22 is mapped to a *rectangular* boundary in the τ-plane, see Fig. 5.23. For this parameter plane, no second Riemann's sheet is needed. The rectangle corners A_1B_1I and A_2B_2I correspond to streamlines that follow first a side of the wedge and then become free streamlines enclosing a wake and reaching infinity. The curves IA_1 and IA_2 correspond to the incoming streamlines that reach the stagnation points on the wedge. All streamlines that go round the cavity lie in the domain enclosed by IA_1 and IA_2. The curve IJ shows one of them, the shortest. It is rather clear that to fill this whole domain, curves with inflexion points are needed, see also figure (6a) in Cox and Clayden (1958). The present τ-rectangle replaces the upper half-plane of the variable t in our first examples for jets and wakes, and the semicircle of the Levi-Civita method.

All complex planes are defined now, and we turn to the analytic mappings between them. First, for the pair of variables τ and $\ln w$, a finite rectangle in the τ-plane is mapped to an infinite strip in the $\ln w$-plane, or, with a further conformal mapping, to the upper half-plane. Such a mapping is, by definition, achieved by the Jacobi *elliptic functions*. Actually, for the latter, a periodic *tiling* of the τ-plane with rectangles (or, more general, with parallelograms) is assumed, but these rectangle copies do not enter our present considerations.

Fig. 5.22 The Riemann
surface with two sheets for
ln w. Full lines: physical
sheet 1; dashed lines: sheet
2. Branch points are at Z'
and Z (the point at infinity).
Dotted line: sheet crossing.
A closed loop without
self-intersection is given by
$B_2 I B_1 A_1 Z A_1 C J A_2 Z A_2 B_2$

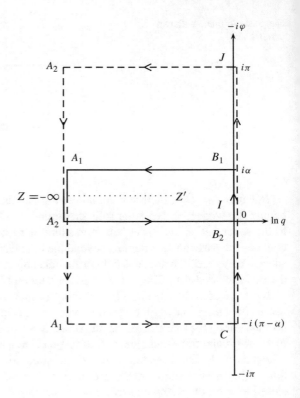

Fig. 5.23 The parameter
τ-plane

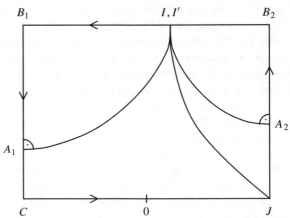

Cox and Clayden (1958) observe (i) that the differential quotient $d \ln w / d\tau$ is purely imaginary on the boundary of the rectangle in the τ-plane. Namely, each boundary line in Fig. 5.22 for the $\ln w$-plane is rotated by $90°$ in Fig. 5.23 for the τ-plane, and rotation by $90°$ corresponds to complex multiplication with i. Furthermore, (ii) the function $d \ln w / d\tau$ has poles at τ_{A_1} and τ_{A_2}, both of which are easily shown to be simple poles. Finally, (iii) $d \ln w / d\tau$ is finite and not equal to zero at τ_I. A function that has the properties (i) to (iii) is given by

$$\frac{d \ln w}{d\tau} = \frac{i T_1 \, \text{sn} \, \tau}{\text{sn} \, \tau - \text{sn} \, \tau_{A_1}} + \frac{i T_2 \, \text{sn} \, \tau}{\text{sn} \, \tau - \text{sn} \, \tau_{A_2}} + iN, \qquad (5.123)$$

with real-valued constants T_1, T_2, N, and the Jacobi elliptic function sn introduced in (3.110) in Sect. 3.11. The point where $d \ln w / d\tau = 0$ is the branch point Z' in Fig. 5.22.

We summarize some elementary facts about elliptic functions without giving the proofs. Elliptic functions are by definition double-periodic in the complex plane,

$$\text{sn}(z + 4K) = \text{sn} \, z,$$
$$\text{sn}(z + 2iK') = \text{sn} \, z, \qquad (5.124)$$

where the complex numbers K and K' are determined by (see below)

$$K = \int_0^1 \frac{dt}{\sqrt{(1 - t^2)(1 - k^2 t^2)}},$$
$$K' = \int_0^1 \frac{dt}{\sqrt{(1 - t^2)(1 - k'^2 t^2)}}, \qquad (5.125)$$

with

$$0 \le k \le 1, \qquad k' = \sqrt{k^2 - 1}. \qquad (5.126)$$

Using a variable substitution $t' = (1 - k'^2 t^2)^{-1/2}$, it follows that

$$K' = \int_1^{1/k} \frac{dt}{\sqrt{(1 - t^2)(1 - k^2 t^2)}}, \qquad (5.127)$$

where t was used again instead of t'. Therefore

$$K + iK' = \int_0^{1/k} \frac{dt}{\sqrt{(1 - t^2)(1 - k^2 t^2)}}, \qquad (5.128)$$

and thus,

$$\text{sn}(K + iK') = \frac{1}{k}. \qquad (5.129)$$

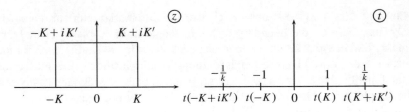

Fig. 5.24 The Jacobi elliptic function $t = \mathrm{sn}(z, k)$

Fig. 5.25 Period rectangle of sn

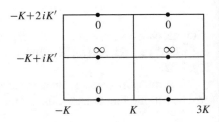

Elliptic functions give the conformal mapping shown in Fig. 5.24.

The function $t = \mathrm{sn}\, z$ maps a rectangle in the z-plane to the upper t-plane, where the rectangle has side lengths $2K$ and K'. The point zero is mapped to zero, and the point iK' is mapped to infinity. The corners $\pm K$ are mapped to ± 1, and the corners $K \pm iK'$ to $\pm 1/k$. Indeed, by the Schwarz–Christoffel formula, the mapping is given by

$$\frac{\mathrm{d}z}{\mathrm{d}t} = \frac{b}{\sqrt{(1 - t^2)(1 - k^2 t^2)}}, \tag{5.130}$$

which, for the indicated position of the rectangle, gives the definition

$$z = \int_0^{\mathrm{sn}(z,k)} \frac{\mathrm{d}t}{\sqrt{(1 - t^2)(1 - k^2 t^2)}}. \tag{5.131}$$

This integral is termed an *elliptic integral* of the first kind. The parameter k in $\mathrm{sn}(z, k)$ is often suppressed. The function sn is odd, $\mathrm{sn}(-z, k) = -\mathrm{sn}(z, k)$. Figure 5.25 shows a so-called period rectangle of sn in the complex plane. This is obtained by mirroring the rectangle in Fig. 5.24 at its upper and right boundaries. Elliptic functions must have poles, else they would be constant. This follows directly from Liouville's theorem in Appendix A.1: analytic functions without any poles in the complex plane are constant. Since all values of sn occur already in the rectangle in Fig. 5.24, so must its poles. (We refer to Copson (1935), p. 350 for the precise statement.) The function sn has two poles of first order, with residues $+A$ and $-A$. The latter fact is easily found from the residue theorem: the line integral along the boundary C of a period rectangle is

$$\frac{1}{2\pi i} \oint_C \mathrm{d}z\, f(z) = \sum_i \mathrm{Res}_i. \tag{5.132}$$

Fig. 5.26 Upper half-plane
$t = \mathrm{sn}\,\tau$

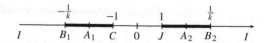

This contour integral is zero, since contributions from opposite sides cancel. More specifically, sn has simple poles with residue $1/k$ at $iK' + 4mK + 2inK'$ ($m, n \in \mathbb{Z}$), and simple poles with residue $-1/k$ at $2K + iK' + 4mK + 2inK'$.

After this brief summary on the function sn, we return to the flow problem of the tilted wedge and detached cavity, and determine $\Phi(\tau)$ by introducing a new variable t defined by $t = \mathrm{sn}\,\tau$. This maps the rectangle in the τ-plane from Fig. 5.23 to the upper t-half-plane in Fig. 5.26.

Cox and Clayden list the following properties of the function $d\Phi/dt$: it has simple zeros at t_{A_1} and t_{A_2}, a simple pole at t_J, and a triple pole at t_I. This leads to the function

$$\frac{d\Phi}{dt} = a\, \frac{(t - t_{A_1})\,(t - t_{A_2})}{(t - t_J)\,(t - t_I)^3}, \tag{5.133}$$

which is easily transformed to the variable τ using $t = \mathrm{sn}\,\tau$. Finally then, using the expressions for $\ln w(\tau)$ and $d\Phi/d\tau$ in Eq. (5.10) in Sect. 5.3, the streamlines of the flow are obtained.

One may wonder how realistic a description of a stationary flow on two different Riemann sheets is, given that there is no way to translate this to a meaningful three-dimensional flow geometry. However, flows resembling the solution of Cox and Clayden are indeed observed, and the cavity is created and destroyed here *periodically* in course of time. It grows, breaks away from the wedge, and forms again. The above stationary model with the two Riemann sheets is a rough approximation for this time-periodic flow.

5.11 Weinstein's Theorem

In the last sections, different solutions for jets, wakes, and cavities were discussed. We return now to the first example, the jet from an orifice, and prove the *uniqueness* of the solution. Even though the relevant differential equation, $\Delta\Phi = 0$, is linear, the boundary conditions are not; thus uniqueness is not trivial since multiple solutions may occur. In general, proving uniqueness (and/or existence of a solution) is a complicated issue and the subject of active current research. The scope of the task shall therefore be reduced to the question whether, for a given differential equation and given boundary conditions, alternative solutions exist in an *infinitesimal* neighborhood of a known solution. This kind of analysis was pioneered by Cauchy in the early nineteenth century.

For the case of a jet from an orifice, Weinstein (1924) proved the uniqueness of the solution in such an infinitesimal neighborhood; his proof is presented in the following. To start off, three complex planes are introduced,

$$z = x + iy,$$
$$\Phi = \phi + i\psi,$$
$$\zeta = \xi + i\eta = \rho e^{i\varphi}. \tag{5.134}$$

Let $-\pi/2 \le \psi \le \pi/2$ and $0 \le \rho \le 1$, and thus $|\xi| \le 1$ and $|\eta| \le 1$. The infinite strip $ABCDEF$ in the Φ-plane is conformally mapped to a quarter circle of radius one in the ζ-plane (see Fig. 5.27) by

$$e^{\Phi} = \frac{i}{2}\left(\zeta + \frac{1}{\zeta}\right). \tag{5.135}$$

(There is a minor sign error in Eq. 1 of Weinstein 1924.) The complex fluid speed is again

$$w = \frac{d\Phi}{dz} = u - iv. \tag{5.136}$$

A new complex function is introduced for later use,

$$\omega = \theta + i\tau, \tag{5.137}$$

and we define

$$w = e^{-i\omega}. \tag{5.138}$$

Then $\bar{w} = u + iv = e^{\tau + i\theta}$, and θ is the flow angle with the x-axis. Along the free streamline of the jet, the normalization $|w| = 1$, or $\tau = 0$, is chosen. The new potential ω may be considered a function of z, Φ or ζ. Following Weinstein, the same symbol ω is used in all three cases, without further distinctions like a tilde, circumflex, etc., and the independent variables indicate then which specific function ω is

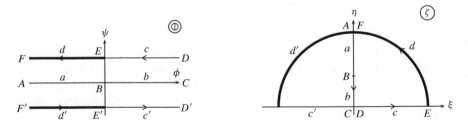

Fig. 5.27 Complex Φ- and ζ-planes, related by the conformal mapping (5.135). The integration domain is the infinite strip and the quarter circle $ABCDEF$, respectively

meant (thus ω without variables is only the name of a certain physical quantity, not a specific function used to calculate it). The same holds for other functions and their real and imaginary parts. For a given vessel shape and orifice size, let Φ_1 be a (hypothetical) second solution in an infinitesimal neighborhood of the known jet solution Φ. This means that at each z, the distance $\delta\Phi(z) = \Phi_1(z) - \Phi(z)$ shall be infinitesimally small. The streamlines for the solution Φ_1 have $\psi_1 = $ const. Thus, the curves $ABCDEF$ in Fig. 5.27 are the same for Φ and Φ_1, however, at an infinitesimal distance.

Theorem *(Weinstein 1924) Let Φ be a complex harmonic function, $\Delta\Phi = 0$, which solves the orifice problem, i.e., obeys the appropriate boundary conditions and is symmetric with respect to the flow axis x,*

$$\overline{\Phi(\overline{z})} = \Phi(z). \tag{5.139}$$

Along the whole free streamline of the jet from the orifice to infinity, let

$$\left| \frac{\partial\theta}{\partial\xi} \right| < \frac{2}{\pi}, \tag{5.140}$$

with spatial variable ξ from (5.134) and jet inclination θ. Furthermore, let $\Phi_1 = \Phi + \delta\Phi$ be another symmetric solution of the orifice problem, where $\delta\Phi(z)$ is for all z infinitesimally close to Φ. Then $\delta\Phi = 0$.

Proof Let Φ and $\Phi_1 = \Phi + \delta\Phi$ be solutions of the orifice problem, where $\delta\Phi$ is an infinitesimal quantity. We have to show that $\delta\Phi = 0$. Since both Φ and Φ_1 are symmetric with respect to the flow axis, let

$$\delta\psi = 0 \qquad \text{on } a, b. \tag{5.141}$$

A point z of the vessel boundary d is mapped by both Φ and Φ_1 to $\psi = \psi_1 = \pi/2$ (but $\phi \neq \phi_1$). Similarly, $\psi = \psi_1 = -\pi/2$ for points along the vessel boundary d', thus

$$\delta\psi = 0 \qquad \text{on } d, d'. \tag{5.142}$$

But this is *not* true along the free streamlines c and c': if the point z lies on the free streamline c for which $\psi = \pi/2$, it will generally *not* lie on the neighboring free streamline with $\psi_1 = \pi/2$. This means that $\psi_1(z) \neq \pi/2$ for this point, and therefore $\delta\psi(z) = \psi_1(z) - \psi(z) \neq 0$ (the isocontours $\psi = $ const 'move through' the point under variation); and similarly $\delta\phi(z) \neq 0$. Weinstein proceeds as follows: for any point z on the free streamline corresponding to the Φ-jet, i.e., on c or c', find the point $z_1 = z + \delta z$ on the alternative solution, i.e., the Φ_1-jet, which has the same value of the complex potential,

$$\Phi(z) = \Phi_1(z_1) = \Phi_1(z + \delta z). \tag{5.143}$$

Condition (5.143) has always a solution for z_1 since $\phi_1(z_1)$ is continuous. Then to first order in δz,

$$\Phi(z) = \Phi_1(z) + \frac{d\Phi_1}{dz}(z)\,\delta z = \Phi_1(z) + \frac{d\Phi}{dz}(z)\,\delta z. \qquad (5.144)$$

Using $\delta\Phi = \Phi_1 - \Phi$, this yields

$$\frac{\delta\Phi}{\delta z} = -\frac{d\Phi}{dz}. \qquad (5.145)$$

We abbreviate $\Phi' = d\Phi/dz$ and $\Phi'_1 = d\Phi_1/dz$, and demand that $|w_1| = |w| = 1$ everywhere on the two different free streamlines. Thus, the speeds $w(z)$ and $w_1(z + \delta z)$, since z and $z + \delta z$ lie indeed on the respective free streamline, differ only by a complex phase with magnitude 1,

$$\Phi'_1(z + \delta z) = e^{i\epsilon}\,\Phi'(z) = (1 + i\epsilon)\,\Phi'(z) \qquad (5.146)$$

to first order in ϵ, since Φ and Φ_1 are infinitesimally close. All equations until (5.153) will hold only on the free streamlines corresponding to the z- and z_1-solutions, without explicit mention. The Taylor series expansion of the first order on the left of (5.146) generates

$$\Phi'_1 + \Phi''\,\delta z = \Phi' + i\epsilon\Phi', \qquad (5.147)$$

where all functions are taken at z, and Φ''_1 was replaced by Φ'' since the second order term $(\delta\Phi)''\,\delta z$ can be neglected. Inserting $\delta z = -\delta\Phi/\Phi'$ from (5.145) and $\Phi'_1 = \Phi' + (\delta\Phi)'$, Eq. (5.147) implies

$$(\delta\Phi)' - (\Phi''/\Phi')\,\delta\Phi = i\epsilon\Phi', \qquad (5.148)$$

and therefore, since ϵ is real,

$$\text{Re}\left[\frac{(\delta\Phi)' - (\Phi''/\Phi')\,\delta\Phi}{\Phi'}\right] = 0. \qquad (5.149)$$

This can be written, with $\ln w = -i\theta$ on the free streamlines c and c', as

$$\text{Re}\left[\frac{d\delta\Phi}{d\Phi} - \frac{d\ln w}{d\Phi}\,\delta\Phi\right] = \text{Re}\left[\frac{d\delta\Phi}{d\Phi} + i\,\frac{d\theta}{d\Phi}\,\delta\Phi\right] = 0. \qquad (5.150)$$

Complex differentiation $d/d\Phi$ can be performed in any direction in the Φ-plane; choosing the ϕ-direction, and thus replacing $d/d\Phi$ by $\partial/\partial\phi$, gives

$$\text{Re}\left[\frac{\partial(\delta\phi)}{\partial\phi} + i\,\frac{\partial(\delta\psi)}{\partial\phi} + i\,\frac{\partial\theta}{\partial\phi}\,(\delta\phi + i\delta\psi)\right] = 0. \qquad (5.151)$$

Taking the real part gives

$$\frac{\partial(\delta\phi)}{\partial\phi} = \frac{\partial\theta}{\partial\phi}\,\delta\psi. \tag{5.152}$$

Finally, using the Cauchy–Riemann equations for analytic $\delta\Theta$,

$$\boxed{\frac{\partial(\delta\psi)}{\partial\psi} = \frac{\partial\theta}{\partial\phi}\,\delta\psi \qquad \text{on } c, c'} \tag{5.153}$$

follows. This is the boundary condition for variations on the free streamline and is to be supplemented by the boundary condition (5.142) on the vessel wall. We assume now $\delta\psi(\xi, \eta = 0) \neq 0$ for $\xi \in [0, 1]$ and derive a contradiction. The following Dirichlet integral for $\delta\psi$ (compare with 3.46 in Sect. 3.6) is performed on the infinite strip $ABCDEF$.

$$
\begin{aligned}
D &\overset{(a)}{=} \int_0^{\pi/2} d\psi \int_{-\infty}^{\infty} d\phi \left[\left(\frac{\partial(\delta\psi)}{\partial\phi} \right)^2 + \left(\frac{\partial(\delta\psi)}{\partial\psi} \right)^2 \right] \\
&\overset{(b)}{=} \int_A da \left[(\partial_\phi, \partial_\psi)\delta\psi \cdot (\partial_\phi, \partial_\psi)\delta\psi \right] \\
&\overset{(c)}{=} \oint_{\partial A} dl\, \frac{\partial(\delta\psi)}{\partial n}\, \delta\psi \\
&\overset{(d)}{=} -\int_{-\infty}^{\infty} d\phi \left[\frac{\partial(\delta\psi)}{\partial\psi}\,\delta\psi \right]_{\psi=0} - \int_{-\infty}^{\infty} d\phi \left[\frac{\partial(\delta\psi)}{\partial\psi}\,\delta\psi \right]_{\psi=\pi/2} \\
&\overset{(e)}{=} -\int_0^{\infty} d\phi \left[\frac{\partial\theta}{\partial\phi}\,(\delta\psi)^2 \right]_{\psi=\pi/2} \\
&\overset{(f)}{=} \int_0^1 d\xi \left[\frac{\partial\theta}{\partial\xi}\,(\delta\psi)^2 \right]_{\eta=0} \\
&\overset{(g)}{<} \frac{2}{\pi} \int_0^1 d\xi\, [\delta\psi(\xi, \eta = 0)]^2 \\
&\overset{(h)}{=} \frac{2}{\pi} \int_0^1 d\rho \left[-\int_0^{\pi/2} d\varphi\, \frac{\partial(\delta\psi)}{\partial\varphi}(\rho, \varphi) \right]^2 \\
&\overset{(i)}{\leq} \int_0^1 d\rho \int_0^{\pi/2} d\varphi \left[\frac{\partial(\delta\psi)}{\partial\varphi}(\rho, \varphi) \right]^2 \\
&\overset{(j)}{\leq} \int_0^1 d\rho \int_0^{\pi/2} d\varphi \left[\left(\frac{\partial(\delta\psi)}{\partial\varphi} \right)^2 \frac{1}{\rho} + \left(\frac{\partial(\delta\psi)}{\partial\rho} \right)^2 \rho \right] \\
&\overset{(k)}{=} D.
\end{aligned} \tag{5.154}
$$

Thus contradiction $D < D$. The above steps (a) to (k) are justified as follows:

(a) That this integral is finite becomes clear in step (g) below.
(b) da is an area element in the $\phi\psi$-plane, and $(\partial_\phi, \partial_\psi)$ is the gradient.
(c) dl is the line element along the boundary ∂A.
    $\partial/\partial n$ points in the normal direction to ∂A.
    Green's first identity is, in the plane and for arbitrary f, g,

$$\int_A da \, (f \Delta g + \nabla f \cdot \nabla g) = \oint_{\partial A} dl \, \left(f \, \frac{\partial g}{\partial n} \right). \tag{5.155}$$

Use $f = g = \delta\psi$ and $\Delta\delta\psi = 0$, since $\Delta\Phi = \Delta\Phi_1 = 0$.
(d) The first minus occurs because $\partial/\partial n = -\partial/\partial\psi$ on a and b.
    For $\phi \to \pm\infty$, there is no contribution from CD and FA.
(e) For $\psi = 0$, $\delta\psi = 0$ according to (5.141).
    For $\psi = \pi/2$ and $\phi < 0$, $\delta\psi = 0$ according to (5.142).
    For $\psi = \pi/2$ and $\phi \geq 0$, the boundary condition (5.153) holds.
(f) The sign is correct according to Fig. 5.27.
    The Jacobi factor $d\phi/d\xi$ from variable substitution is absorbed in

$$\frac{d\phi}{d\xi} \frac{\partial\theta(\phi, \psi)}{\partial\phi} = \frac{\partial\tilde{\theta}(\xi, \eta)}{\partial\xi}, \tag{5.156}$$

where $\tilde{\theta}(\xi, \eta) = \theta(\phi(\xi, \eta), \psi(\xi, \eta))$, etc. As noted, the tilde is dropped.
(g) This crucial '<' holds only if $\delta\psi \neq 0$. Equation (5.140) is used.
(h) From (5.141), one has $\delta\psi(\rho, \pi/2) = 0$ on a and b.
    Write $\delta\psi(\rho, \pi/2) - \delta\psi(\rho, 0)$ as φ-integral of $\partial_\varphi \delta\psi$ on a quarter circle.
(i) For real-valued functions f and g, the Schwarz inequality is

$$\left(\int_0^a dx \, f(x)g(x) \right)^2 \leq \left(\int_0^a dx \, f^2(x) \right) \left(\int_0^a dx \, g^2(x) \right). \tag{5.157}$$

For $g = 1$ then,

$$\left(\int_0^a dx \, f(x) \right)^2 \leq a \int_0^a dx \, f^2(x). \tag{5.158}$$

(j) This holds since the radius $\rho \leq 1$.
(k) D from (a) is written in polar coordinates in (j).

The contradiction that occurs in step (g) can be only avoided by $\delta\psi = 0$ for $0 \leq \xi \leq 1$ and $\eta = 0$. Then $D = 0$ according to step (f). From (a) follows $\delta\psi = 0$ in the whole strip $ABCDEF$. The Cauchy–Riemann equations imply then $\delta\phi = 0$, and thus $\delta\Phi = 0$ in this strip. This concludes the proof.   □

Weinstein (1924) derives then conditions on the geometry of the vessel from which outflow occurs, in order that the condition (5.140) is fulfilled.

The present chapter assumed two central limitations on the flows considered, viz., that they are two dimensional and stationary. It is rather hard to lift the restriction on the dimensionality in obtaining interesting analytical solutions of the Euler equation, and thus we turn now to a more systematic study of time-dependent phenomena in fluids, mostly in the form of propagation (waves, bores, shocks) and growth (instabilities) of flow perturbations.

References

Ahlfors, L.V. 1966. *Complex analysis*, 2nd ed. New York: McGraw-Hill.

Batchelor, G.K. 1952. The effect of homogeneous turbulence on material lines and surfaces. *Proceedings of the Royal Society of London A* 213: 349.

Bieberbach, L. 1921. *Lehrbuch der Funktionentheorie, Band I, Elemente der Funktionentheorie*. Leipzig: B.G. Teubner; Wiesbaden: Springer Fachmedien.

Birkhoff, G.B., and J. Fisher. 1959. Do vortex sheets roll up? *Rendiconti del Circolo Matematico di Palermo, Series 2*, 8: 77.

Birkhoff, G.B., and E.H. Zarantonello. 1957. *Jets, waves and cavities*. New York: Academic Press.

Brennen, C.E. 1970. Cavity surface wave patterns and general appearance. *Journal of Fluid Mechanics* 44: 33.

Brennen, C.E. 2013. *Cavitation and bubble dynamics*. New York: Cambridge University Press.

Brillouin, M. 1911. Les surfaces de glissement de Helmholtz et la résistance des fluides. *Annales de chimie et de physique* 23: 145.

Cisotti, U. 1932. Scie limitate. *Annali della Scuola Normale Superiore di Pisa, Classe di Scienca, Series 2*, 1: 101.

Copson, E.T. 1935. *Theory of functions of a complex variable*. Oxford: Oxford University Press.

Courant, R. 1950. *Dirichlet's principle, conformal mapping, and minimal surfaces*. New York: Interscience Publishers.

Cox, A.D., and W.A. Clayden. 1958. Cavitating flow about a wedge at incidence. *Journal of Fluid Mechanics* 3: 615.

Fabula, A.G. 1962. Thin-airfoil theory applied to hydrofoils with a single finite cavity and arbitrary free-streamline detachment. *Journal of Fluid Mechanics* 12: 227.

Gilbarg, D. 1960. Jets and cavities. In *Handbuch der Physik = Encyclopedia of physics*, vol. 3/9, Strömungsmechanik 3, ed. S. Flügge and C. Truesdell, 311. Berlin: Springer.

Golusin, G.M. 1957. *Geometrische Funktionentheorie*. Berlin: Deutscher Verlag der Wissenschaften.

Havelock, T. 1929. Forced surface-waves on water. *Philosophical Magazine Series 7*, 8: 569.

Helmholtz, H. 1868. Über diskontinuierliche Flüssigkeits-Bewegungen. *Monatsberichte der Königlich preussischen Akademie der Wissenschaften zu Berlin* 23: 215. On discontinuous movements of fluids. *Philosophical Magazine Series 4*, 36: 337.

Kirchhoff, G. 1869. Zur Theorie freier Flüssigkeitsstrahlen. *Journal für die reine und angewandte Mathematik* 70: 289.

Kirchhoff, G. 1897. *Vorlesungen über Mechanik*, Lectures 21 and 22. Leipzig: Teubner.

Kolscher, M. 1940. Unstetige Strömungen mit endlichem Totwasser. *Luftfahrtforschung* (Berlin-Adlershof) vol. 17, Lfg. 5, München: Oldenbourg.

von Koppenfels, W., and F. Stallmann. 1959. *Praxis der konformen Abbildung*. Berlin: Springer.

Lamb, H. 1932. *Hydrodynamics*. Cambridge: Cambridge University Press; New York: Dover (1945).

Levi-Civita, T. 1907. Scie e leggi di resistenza. *Rendiconti del Circolo Matematico di Palermo* 23: 1

Milne-Thomson, L.M. 1968. *Theoretical hydrodynamics*, 5th ed. London: Macmillan; New York: Dover (1996).

Nehari, Z. 1952. *Conformal mapping*. New York: McGraw-Hill; New York: Dover (1975).

Newman, J.N. 1965. Propagation of water waves over an infinite step. *Journal of Fluid Mechanics* 23: 399.

Panofsky, W.K.H., and M. Phillips. 1962. *Classical electricity and magnetism*, 2nd ed. Reading: Addison-Wesley; Mineola: Dover (2005).

Riabouchinsky, D. 1921. On steady fluid motions with free surfaces. *Proceedings of the London Mathematical Society, Series 2*, 19: 206.

Schmieden, C. 1929. Die unstetige Strömung um einen Kreiszylinder. *Ingenieur-Archiv = Archive of Applied Mechanics* 1: 104.

Schmieden, C. 1932. Über die Eindeutigkeit der Lösungen in der Theorie der unstetigen Strömungen. *Ingenieur-Archiv = Archive of Applied Mechanics* 3: 356.

Villat, H.R.P. 1913. Sur le changement d'orientation d'un obstacle dans un courait fluide et sur quelques questions connexes. *Annales de la faculté des sciences de Toulouse série 3*, 5: 375.

Weinstein, A. 1924. Ein hydrodynamischer Unitätssatz. *Mathematische Zeitschrift* 19: 265.

Chapter 6
The Kelvin–Helmholtz Instability

The flows considered so far were mostly stationary, that is, did not depend on time. As long as individual fluid parcels are not followed, the fluid could equally well be frozen. We turn now to flows that show mild modes of excitation in the form of wave propagation or unstable growth. Often, both phenomena are intimately linked: time-periodic, harmonic waves with angular speed ω and amplitude $\sim e^{i\omega t}$ can start to grow or decay exponentially, $\sim e^{\pm|\omega|t}$, once ω turns to be a complex number. Since the dispersion relation that describes waves as well as instabilities is often a nonlinear algebraic equation in ω and the wavenumber k, such switching from real to imaginary angular speeds (or wavenumbers) is quite common. The occurrence of a wave or instability at one ω and one k can well be considered as an excitation of a fluid with one degree of freedom. In this sense, waves and instabilities are the most fundamental nonstationary phenomena in fluids. This chapter treats the most significant fluid instability, the Kelvin–Helmholtz (K–H) instability. It occurs in shear flows and leads to the formation of macroscopic vortices at a mutual distance comparable to the size of the vortices. Figures 6.1, 6.2, and 6.3 show the Kelvin–Helmholtz instability in clouds in Alabama, in Hessen, and on the planet Saturn, respectively. In its fully developed form, the instability leads to fluid mixing and turbulence, which are both of great importance for fluids ranging from blood to gasoline. Furthermore, this instability is universal and occurs on all lengthscales from microscopic to cosmic.

The original version of this chapter was revised: Belated correction from author has been incorporated. The correction to this chapter is available at https://doi.org/10.1007/978-3-030-31022-6_13.

© Springer Nature Switzerland AG, corrected publication 2020
A. Feldmeier, *Theoretical Fluid Dynamics*, Theoretical and Mathematical Physics, https://doi.org/10.1007/978-3-030-31022-6_6

Fig. 6.1 The Kelvin–Helmholtz instability in clouds near Birmingham, Alabama in December 2011. Image source: YouTube video at https://www.youtube.com/watch?v=dcwOn4VeJOE

Fig. 6.2 The Kelvin–Helmholtz clouds in the Rhön, state of Hessen, Germany. The hill toward the right corner is the Milseburg. Note especially the lateral spread of the clouds. Photograph: Sandra Mihm

Fig. 6.3 The Kelvin–Helmholtz instability in the atmosphere of Saturn. Photograph: NASA photojournal at http://photojournal.jpl.nasa.gov/catalog/PIA06502en.wiki

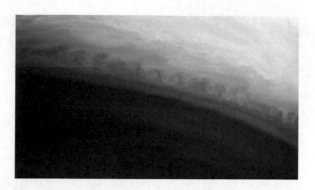

6.1 The Kelvin–Helmholtz Circulation Theorem

In an inviscid fluid of constant density, the most fundamental property of vortex tubes and line vortices is that they are made up of the same fluid parcels at all times, i.e., they move along with the fluid. This is in marked contrast to waves, for which fluid parcels do *not* (with minor exceptions, to be dealt with) partake in the gross wave motion, but oscillate about a fixed position, with the wave (phase) propagating past them. We derive this property of vortex tubes in two theorems.

Theorem (The Kelvin–Helmholtz circulation theorem) *Let $C(t)$ be a closed, time-dependent loop in an inviscid fluid, and Γ the circulation along C,*

$$\Gamma(t) = \oint_{C(t)} d\vec{l} \cdot \vec{u}(t, \vec{r}). \tag{6.1}$$

External forces shall be derivable from a scalar potential V,

$$\vec{g} = -\nabla V, \tag{6.2}$$

and the density is either constant or

$$\rho = \rho(p) \tag{6.3}$$

holds. If the curve $C(t)$ moves along with the fluid parcels that lie on it, then

$$\boxed{\frac{d\Gamma}{dt} = 0} \tag{6.4}$$

The circulation along a closed curve moving with the flow is conserved.

Comment Equation (6.3) is a severe restriction on the thermodynamics of the flow, and holds in isothermal and isentropic flows of an ideal gas: for constant temperature, let $p = p(\rho, T)$ and drop the constant T; for an isentropic flow, the specific entropy S is constant throughout the flow, thus let $p = p(\rho, S)$ and drop S.

Proof For an infinitesimal line element $\mathrm{d}\vec{l}$ along the moving curve $C(t)$, the Lagrange derivative is

$$\frac{\mathrm{d}(\mathrm{d}\vec{l})}{\mathrm{d}t} = (\mathrm{d}\vec{l} \cdot \nabla)\vec{u}. \tag{6.5}$$

Therefore, using the product rule,

$$\frac{\mathrm{d}\Gamma}{\mathrm{d}t} = \frac{\mathrm{d}}{\mathrm{d}t} \oint_C \mathrm{d}\vec{l} \cdot \vec{u} = \oint_C \mathrm{d}\vec{l} \cdot \frac{\mathrm{d}\vec{u}}{\mathrm{d}t} + \oint_C \vec{u} \cdot (\mathrm{d}\vec{l} \cdot \nabla)\,\vec{u}. \tag{6.6}$$

The last term can be simplified,

$$\vec{u} \cdot (\mathrm{d}\vec{l} \cdot \nabla)\,\vec{u} = \vec{u} \cdot \mathrm{d}_l\,\vec{u} = \frac{1}{2}\,\mathrm{d}_l\,(\vec{u} \cdot \vec{u}), \tag{6.7}$$

with differential d_l along C, when $\hat{l}$ is the tangent vector to C and l the curve parameter along C. Equation (6.7) becomes trivial in a local Cartesian frame with $\hat{l} = \hat{x}$ and $\mathrm{d}\vec{l} \cdot \nabla = \mathrm{d}x\,(\mathrm{d}/\mathrm{d}x)$. Thus

$$\oint_C \vec{u} \cdot (\mathrm{d}\vec{l} \cdot \nabla)\,\vec{u} = \frac{1}{2} \oint_C \mathrm{d}_l\,(\vec{u} \cdot \vec{u}) = 0, \tag{6.8}$$

since C is a closed contour, and (6.6) becomes

$$\frac{\mathrm{d}\Gamma}{\mathrm{d}t} = \oint_C \mathrm{d}\vec{l} \cdot \frac{\mathrm{d}\vec{u}}{\mathrm{d}t}. \tag{6.9}$$

Using the Euler equation for an inviscid fluid gives

$$\frac{\mathrm{d}\Gamma}{\mathrm{d}t} = -\oint_C \mathrm{d}\vec{l} \cdot \left[\frac{\nabla p}{\rho} - \vec{g}\right]. \tag{6.10}$$

For $\rho = \rho(p)$, the lemma in Sect. 2.15 yields (with enthalpy H)

$$\frac{\nabla p}{\rho(p)} = \nabla H, \tag{6.11}$$

which is also trivially true for constant ρ. Thus (6.10) becomes, using (6.2),

$$\frac{\mathrm{d}\Gamma}{\mathrm{d}t} = -\oint_C \mathrm{d}\vec{l} \cdot \nabla(H + V) = -\oint_C \mathrm{d}_l\,(H + V), \tag{6.12}$$

and again since C is a closed contour,

$$\frac{d\Gamma}{dt} = 0, \tag{6.13}$$

and the proof is completed. $\qquad\qquad\qquad\qquad\qquad\qquad\qquad\qquad\square$

For this proof of the circulation theorem, see Batchelor (1967), pp. 269 and 275. The proof becomes somewhat less abstract in Cartesian coordinates; for the following, see Lamb (1932), p. 36. Let $\delta\vec{l} = (\delta x, \delta y, \delta z)$ be a line element along the closed curve C, and let $\delta\vec{l}$ be a material line made up of fluid parcels. The rate of change of δx is given by the speed difference δu between the two end points of δx (compare this with the argument made after Eq. (2.32)),

$$d(\delta x) = \delta u\, dt. \tag{6.14}$$

Therefore,

$$\frac{d}{dt}(u\delta x) = u\,\frac{d(\delta x)}{dt} + \frac{du}{dt}\,\delta x = u\,\delta u + \frac{du}{dt}\,\delta x. \tag{6.15}$$

Adding the corresponding expressions for the y- and z-directions and using the Euler equations for du/dt, dv/dt, and dw/dt, gives

$$\frac{d}{dt}(u\,\delta x + v\,\delta y + w\,\delta z) = u\delta u + v\delta v + w\delta w$$
$$- \left(\frac{1}{\rho}\frac{\partial p}{\partial x} + \frac{\partial V}{\partial x}\right)\delta x - \left(\frac{1}{\rho}\frac{\partial p}{\partial y} + \frac{\partial V}{\partial y}\right)\delta y - \left(\frac{1}{\rho}\frac{\partial p}{\partial z} + \frac{\partial V}{\partial z}\right)\delta z. \tag{6.16}$$

The terms in brackets are the total derivatives δp and δV, and therefore

$$\frac{d}{dt}(u\,\delta x + v\,\delta y + w\,\delta z) = u\delta u + v\delta v + w\delta w - \frac{\delta p}{\rho} - \delta V. \tag{6.17}$$

Assuming $\rho = \rho(p)$ and $\delta p/\rho = \delta H$ gives

$$\frac{d}{dt}(u\,\delta x + v\,\delta y + w\,\delta z) = \delta\left[\frac{1}{2}(u^2 + v^2 + w^2) - V - H\right]. \tag{6.18}$$

Here, $u\,\delta x + v\,\delta y + w\,\delta z$ is the Cartesian expression for $\vec{u} \cdot d\vec{l}$. Taking a closed-loop integral $\oint$ on both sides of (6.18) gives again

$$\frac{d\Gamma}{dt} = 0. \tag{6.19}$$

Consider next a vortex tube in the fluid, i.e., the set of all vortex lines through a closed curve to which the vorticity vector is nowhere tangent. The following theorem expresses a far-reaching consequence of the Kelvin–Helmholtz circulation theorem.

Theorem (The Helmholtz theorem for vortex tubes) *Vortex tubes move along with the fluid.*

Proof idea The surface of the vortex tube is foliated with curves of genus one, going once round the tube. All these curves have the same Γ by definition of a vortex tube and by the Kelvin circulation theorem from Sect. 2.11; and all these curves C maintain this Γ in course of time, when C moves with the fluid. Furthermore, the vortex tube can be covered with a net of loops of genus zero, not going round the tube, where each loop encloses an infinitesimal area element $d\vec{a}$ of the tube's surface. Thus, a topological (open) cover of this surface is obtained. Then circulation $\Gamma = 0$ along each of these loops is maintained in time.

> That is to say, the vortex strength of each of the surface elements bounded by these small closed curves [i.e. $\kappa = d\vec{a} \cdot \vec{\omega} = \oint d\vec{l} \cdot \vec{u}$] remains zero, and this is possible only if the material closed curves continue to lie on the surface of a vortex tube without passing round it. (Batchelor 1967, p. 274)

Thus the surface of the vortex tube covered by the circulation-free curves remains circulation-free, i.e., is the surface of *some* vortex tube. Since the circulation round this tube is also constant in time, one can identify this as *one* vortex tube at different times, moving with the fluid. Therefore, in Fig. 6.4, circulation Γ is the same on all the shown vortex tubes at all times.

Fig. 6.4 Evolution of a tornado. The circulation Γ is ideally constant on each of the vortex tubes (the Kelvin circulation theorem, Sect. 2.11) and, according to the Kelvin–Helmholtz circulation theorem, also constant during the evolution of the tornado. This image is created from eight images shot in sequence as a tornado formed north of Minneola, Kansas on May 24, 2016. Photograph: J. Weingart, in Wikipedia entry 'Tornado' at https://en.wikipedia.org/wiki/Tornado

These rather coarse and elementary topological arguments cannot, however, count as a mathematical proof, which would require more refined arguments from differential geometry. Thus Chorin and Marsden (1990), p. 28 consider arguments like the one above as being 'hopelessly incomplete' (see also Sect. 2.11). □

Alternatively to the proof given above, the Kelvin–Helmholtz circulation theorem can also be derived from an action principle, $\delta W = 0$. For fluids, the action W is a functional of density ρ, velocity $\vec{u}$, and specific entropy S. According to Noether's theorem, any continuous one-parameter family of transformations that leaves W invariant corresponds to a constant of motion (a conserved quantity; see Fließbach 1996, Chap. 15 and Hill 1951). For example, the invariance of W under temporal shifts (homogeneity of time), spatial translations (homogeneity of space), and changes of directions (isotropy of space) implies the conservation of energy, momentum, and angular momentum, respectively (see Landau and Lifshitz 1976, Chap. II). Bretherton (1970) shows that the circulation Γ is the conserved quantity associated with the invariance of W under the interchange of two fluid parcels with equal density, velocity, and entropy ('reshuffling of particles which leaves the fields $\vec{u}(\vec{x}, t)$, $\rho(\vec{x}, t)$, and $s(\vec{x}, t)$ unaltered'; Bretherton, p. 24). The central kinematic argument in the proof of this statement is given in Fig. 3 of Bretherton (1970).

6.2 The Bjerknes Circulation Theorem

We drop now the assumption $p = p(\rho)$ and allow for a fluid pressure that depends on arbitrarily many independent thermodynamic variables, $p = p(\rho, T, \ldots)$. The circulation is then in general no longer conserved, and an expression is found for $d\Gamma/dt$ that was first derived by Bjerknes (1898). The above derivation up to Eq. (6.10) is still valid and gives for $\vec{g} = -\nabla V$,

$$\frac{d\Gamma}{dt} = -\oint_C d\vec{l} \cdot \frac{\nabla p}{\rho}. \tag{6.20}$$

Next, an area A can be chosen that has C as boundary $\partial A = C$, giving

$$\frac{d\Gamma}{dt} = -\int_A d\vec{a} \cdot \left(\nabla \times \frac{\nabla p}{\rho} \right). \tag{6.21}$$

For scalar fields a and vector fields $\vec{c}$, the identity

$$\nabla \times (a\vec{c}) = a \, \nabla \times \vec{c} + (\nabla a) \times \vec{c} \tag{6.22}$$

holds. Furthermore, $\nabla \times \nabla \equiv 0$, and Eq. (6.21) becomes

$$\frac{d\Gamma}{dt} = -\int_A d\vec{a} \cdot (\nabla q \times \nabla p), \tag{6.23}$$

Fig. 6.5 Lateral view of a
Bjerknes tube

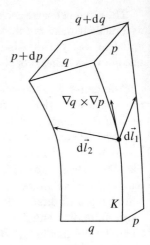

where for easier readability the specific volume (volume per mass), $q = 1/\rho$, is
introduced.

For the barotropic case, the level surfaces for density ρ and pressure $p = p(\rho)$
have identical shape. This is no longer true for a pressure of the form $p(\rho, T, \ldots)$,
with different level surfaces for ρ and p. Let two level surfaces $q = $ const and
$p = $ const cross at the line marked K in Fig. 6.5. Since the cross product $\nabla q \times \nabla p$
points normal to both ∇q and ∇p, it is a tangent vector to the line K. Let c and
f be constant numbers, and dc, df be constant infinitesimals. The four surfaces
$p = c, p = c + dc, q = f$, and $q = f + df$ form then a *Bjerknes tube* of infinitesimal cross section. From now on, these surfaces are denoted by $p, p + dp, q$, and
$q + dq$ simply. The area element $d\vec{a}$ from (6.23) crosses the tube at some arbitrary
angle. Specifically, let $d\vec{a}$ cross the tube walls $p = $ const and $q = $ const along the
vectors $d\vec{l}_1$ and $d\vec{l}_2$, respectively, as shown in Fig. 6.5. Then

$$d\vec{a} = d\vec{l}_1 \times d\vec{l}_2. \tag{6.24}$$

The vector $d\vec{l}_1$ points from the surface q to the surface $q + dq$, and the vector $d\vec{l}_2$
from surface p to surface $p + dp$. The change in circulation is thus

$$\frac{d\Gamma}{dt} = -\int_A (d\vec{l}_1 \times d\vec{l}_2) \cdot (\nabla q \times \nabla p). \tag{6.25}$$

The well-known identity

$$(\vec{a} \times \vec{b}) \cdot (\vec{c} \times \vec{d}) = (\vec{a} \cdot \vec{c})(\vec{b} \cdot \vec{d}) - (\vec{b} \cdot \vec{c})(\vec{a} \cdot \vec{d}) \tag{6.26}$$

leads to

$$\frac{d\Gamma}{dt} = -\int_A (d\vec{l}_1 \cdot \nabla q)(d\vec{l}_2 \cdot \nabla p). \tag{6.27}$$

Here

$$d\vec{l}_1 \cdot \nabla p = d\vec{l}_2 \cdot \nabla q = 0 \tag{6.28}$$

was used since ∇p is orthogonal to the surface $p = \mathrm{const}$ to which $d\vec{l}_1$ is tangential, and similarly for the pair ∇q and $d\vec{l}_2$. With $dq = d\vec{l}_1 \cdot \nabla q$ and $dp = d\vec{l}_2 \cdot \nabla p$ follows

$$\boxed{\frac{d\Gamma}{dt} = -\int_A dq \, dp} \tag{6.29}$$

This is the Bjerknes circulation theorem. The product $dq \, dp$ is the cross-sectional 'area' of the tube. All the Bjerknes tubes can be assumed to have the same dq and dp. Then $d\Gamma/dt$ along the closed curve C is proportional to the number N of the Bjerknes tubes that are enclosed by C.

With Bjerknes' theorem, the basic mechanism of the *passats* can be understood. Figure 6.6 shows a quadrant of a central cross section through the Earth. The isobars $p = \mathrm{const}$ close to the Earth surface are roughly (quarter) circles. An ideal gas law is assumed, $pq \sim T$: for given pressure p, the specific volume q is proportional to the temperature T. Thus, q at the poles is small (cold, dense air), but is large at the equator (warm, expanded air). The air density falls and q increases with the height above ground. Thus, the isochores $q = \mathrm{const}$ are roughly (quarter) ellipses as shown in Fig. 6.6. The figure shows a cross section through a resulting Bjerknes tube, extending in azimuthal direction around the Earth. The circulation that forms along the tube's cross section carries polar air close to the Earth surface toward the equator. At larger heights, back-currents carry equatorial air toward the poles.

Similarly, the heat capacities of sea water and dry coastal ground are different. At day the land heats up more quickly than the sea water. This causes the air density above land to be smaller than above water and leads to tilted isochores. The isobars

Fig. 6.6 Origin of the passats (with atmospheric heights exaggerated). The four bold lines define the cross section of a (non-infinitesimal) Bjerknes tube, which extends in azimuthal direction around the Earth

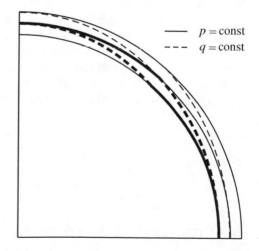

—— $p = \mathrm{const}$

- - - $q = \mathrm{const}$

are again horizontal, as in the example above. A cool wind will then blow from the sea toward land. At night, the land cools faster than the sea water, and a warm wind blows from the land toward the sea. For more details, see Hess (1959), p. 244.

From Bjerknes' theorem, one can directly infer the conservation law $dP/dt=0$ for the so-called *potential vorticity* P in the *adiabatic* and frictionless motion of fluid parcels. The scalar field P was introduced by Rossby in the late 1930s for stratified and rotating fluids and is a quantity with widespread applications in atmosphere-ocean dynamics and meteorology. As was shown by Rossby (1938), P in a continuous stratification is linearly proportional to the absolute circulation C_Γ ('absolute' means that the velocity vector $\vec{u}$ is given in an inertial frame, not in Earth's rotating system) around an *infinitesimally small* (i.e., infinitesimal diameter in every direction) loop Γ that lies within an iso-surface of the stratification, which for the atmosphere are isentropic surfaces (and in the simplest case are horizontal planes). Also in a continuous stratification, P is linearly proportional to the component of absolute vorticity in the direction normal to the iso-surfaces of the stratification (Ertel 1942). For a derivation of these results and a detailed discussion of potential vorticity, we refer to the review article by McIntyre (2015), especially his Eqs. (6)–(10).

6.3   The Kelvin–Helmholtz Instability

We turn now to the physical origin of the Kelvin–Helmholtz instability. This section is intentionally kept short, to demonstrate that the occurrence of such an instability is readily seen. The following explanation follows Batchelor (1967), p. 516. It is also briefly indicated in Drazin and Reid (1981), p. 15.

The straight, horizontal lines in the left and right panels of Fig. 6.7, adapted from Fig. 7.1.3 in Batchelor (1967), show a vortex sheet in side view. Each point of these lines corresponds to the axis of a straight line vortex pointing normally to the paper plane. All vortices spin counterclockwise, in the mathematically positive sense. The sinusoidal line represents the same vortex sheet, deformed, however, by a perturbation of given finite ('macroscopic') wavelength. The basic idea of linear stability analysis is to decompose any given perturbation (perturbations persistently occur in nature due to influences from the environment) into its Fourier components, and then study the individual growth properties of the latter. We assume the equations of motion are linearized; then there is no coupling between the different Fourier modes (mode-coupling in waves and instabilities is rather demanding in mathematical terms). The

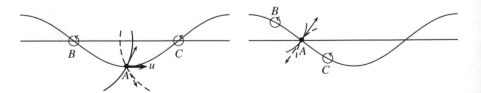

Fig. 6.7 Origin of the Kelvin–Helmholtz instability

unstable growth resulting from the given perturbation is obtained by superposition of the growth of its individual Fourier components. Often it is even sufficient to consider only the components of maximum growth rate, outgrowing all other components.

Consider the line vortex lying at point A in the left panel of the figure. This vortex moves about with an induced velocity that results from adding up the velocity fields of all the other line vortices. The vortex at B contributes a velocity pointing to the upper right (the arc-plus-vector in the figure shows part of a circular streamline of the vortex at B plus the velocity vector at A); the vortex at C contributes a velocity pointing to the lower right (dashed arc-plus-vector). Vector addition gives a total speed u from the vortices at B and C that points horizontally to the right. From the Helmholtz vortex theorem, the line vortex at A will thus start to move to the right. The same is true for the motion induced by all other vortex pairs similar to B and C that lie above A (i.e., in the direction toward the unperturbed vortex sheet): they all induce a motion to the right. Since there are no vortices below A at all, the line vortex at A must thus indeed move to the right.

The same argument holds for any line vortex A' that lies below the zeros of the cosine in the figure, i.e., below the vortices at B and C: such a point A' will have more vortices above it than below it. The former induce motion to the right, the latter to the left; the former dominate, and motion to the right results. By the same argument, if a vortex lies in the upper half of the perturbed sheet (above the zeros), then the motion induced by all other line vortices is to the left.

Finally, consider the vortex lying at point A in the right panel of Fig. 6.7. Each vortex pair B, C induces zero net motion of A by trivial vector addition. Yet, in a sinusoidal perturbation of infinite length, *all* vortices can be grouped in similar pairs with respect to A, which, by symmetry, induce no motion. Thus the vortex lying at A in the right panel remains at rest, as do all the zeros of the perturbed vortex sheet.

The overall motion of vortices in the perturbed sheet is indicated in Fig. 6.8, where the horizontal arrows show the induced vortex velocities. The amazing conclusion is that vortices are *drained* from points marked o, and *accumulate* at points marked •.

The accumulation of microscopic line vortices corresponds to the integration over infinitesimal amounts of vorticity, giving a *finite* vorticity at points marked •. Thus the accumulated line vortices 'become visible' as macroscopic, individual, well-separated vortices.

Now when some accumulation of (positive) vorticity does exist near points like •, there will be a corresponding induced velocity distribution which tends to carry fluid round • in an anti-clockwise sense and thereby to increase the amplitude of the sinusoidal displacement of the sheet vortex. A larger amplitude gives a more rapid accumulation of vorticity near points like •, and so the whole cycle accelerates. (Batchelor 1967, p. 516, with our symbols)

Fig. 6.8   Vortex-induced velocities along a sinusoidal perturbation. Vortices are drained from points marked o, and accumulate at •

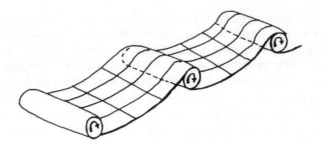

Fig. 6.9 Sketch of a vortex sheet rolling up. Reproduced from Lo et al. (2000)

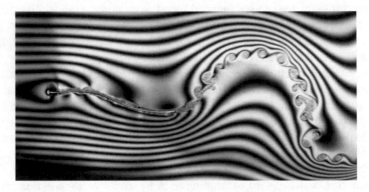

Fig. 6.10 A silk thread flapping in a free-falling, flat soap film. Reproduced from Zhang et al. (2000)

This process of vortex accumulation and induced furling of the vortex sheet is called *vortex sheet roll-up*, see Fig. 6.9, and will be addressed in some detail in the following.

The Kelvin–Helmholtz instability is, somewhat unexpectedly, the origin of a well-known phenomenon, the flapping of flags. The latter is not (primarily) caused by small changes in the wind direction. Instead, a wind floating along a flag gives a shear flow (in case that the highly elastic flag can be treated as remotely resembling a fluid-like medium), with a discrete jump from wind speed u to flag speed 0. This corresponds to a vortex sheet (see Sect. 2.12), and the configuration is prone to the Kelvin–Helmholtz instability. The growth of macroscopic eddies along a silk thread in a parallel flow of soapy water is shown in image 6.10 from an experiment.

6.4   Vortex Chain Perturbation

After these general considerations, we show in a simple kinematic model that perturbations of a vortex chain are unstable for an inviscid fluid. For the following calculations, see Lamb (1932), p. 225. For a straight-line vortex, the radial speed is $u_r = 0$ and the azimuthal speed is $u_\varphi \sim 1/r$. In Cartesian coordinates, a vortex centered at the origin has thus (see Eq. (4.10)) a velocity field given by

$$\begin{pmatrix} u \\ v \end{pmatrix} = \frac{\Gamma}{2\pi r^2} \begin{pmatrix} -y \\ x \end{pmatrix}. \tag{6.30}$$

Introduce now a single vortex chain with vortices at positions, for $a \in \mathbb{R}_+$ and $m \in \mathbb{Z}$,

$$x = ma, \quad y = 0. \tag{6.31}$$

Quite obviously, no mutually induced motions occur in the chain. All the vortices shall now experience a perturbation to new positions

$$x = ma + x_m, \quad y = y_m, \tag{6.32}$$

where x_m, y_m are considered small. The speed induced then in vortex $m = 0$ by all vortices $m \neq 0$ is

$$\begin{aligned} \frac{dx_0}{dt} &= -\frac{\Gamma}{2\pi} \sum_{m \in \mathbb{Z} \setminus \{0\}} \frac{y_0 - y_m}{r_m^2}, \\ \frac{dy_0}{dt} &= \frac{\Gamma}{2\pi} \sum_{m \in \mathbb{Z} \setminus \{0\}} \frac{x_0 - ma - x_m}{r_m^2}, \end{aligned} \tag{6.33}$$

where

$$r_m^2 = (x_0 - ma - x_m)^2 + (y_0 - y_m)^2. \tag{6.34}$$

Keeping only terms linear in the perturbations x_m and y_m in (6.33) gives

$$\begin{aligned} \frac{dx_0}{dt} &= -\frac{\Gamma}{2\pi} \sum_{m \neq 0} \frac{y_0 - y_m}{(ma - (x_0 - x_m))^2 + (y_0 - y_m)^2} \\ &= -\frac{\Gamma}{2\pi} \sum_{m \neq 0} \frac{y_0 - y_m}{m^2 a^2 (1 - 2(x_0 - x_m)/(ma))} \\ &= -\frac{\Gamma}{2\pi} \sum_{m \neq 0} \frac{y_0 - y_m}{m^2 a^2} \left(1 + 2 \frac{x_0 - x_m}{ma} \right) \\ &= -\frac{\Gamma}{2\pi a^2} \sum_{m \neq 0} \frac{y_0 - y_m}{m^2}, \end{aligned} \tag{6.35}$$

and

$$\begin{aligned} \frac{dy_0}{dt} &= -\frac{\Gamma}{2\pi} \sum_{m \neq 0} \frac{ma - (x_0 - x_m)}{(ma - (x_0 - x_m))^2 + (y_0 - y_m)^2} \\ &= -\frac{\Gamma}{2\pi} \sum_{m \neq 0} \frac{ma - (x_0 - x_m)}{m^2 a^2 (1 - 2(x_0 - x_m)/(ma))} \end{aligned}$$

$$= -\frac{\Gamma}{2\pi} \sum_{m \neq 0} \frac{ma - (x_0 - x_m)}{m^2 a^2} \left(1 + 2\,\frac{x_0 - x_m}{ma}\right)$$

$$= -\frac{\Gamma}{2\pi a} \sum_{m \neq 0} \frac{1}{m} - \frac{\Gamma}{2\pi a^2} \sum_{m \neq 0} \frac{x_0 - x_m}{m^2}. \tag{6.36}$$

The first term on the right is indeterminate, but it is clear on physical grounds how to handle it: 'the vortices are to be taken in pairs equidistant from the origin' (Lamb 1932, p. 226),

$$\sum_{m \neq 0} \frac{1}{m} \equiv \sum_{m > 0} \left(\frac{1}{m} - \frac{1}{m}\right) = 0. \tag{6.37}$$

The (remaining) system (6.35) and (6.36) of coupled ordinary linear differential equations is solved by complex exponentials. Assuming thus perturbations of the form

$$x_m = \alpha(t)\,e^{2\pi i m a / \lambda},$$
$$y_m = \beta(t)\,e^{2\pi i m a / \lambda}, \tag{6.38}$$

where λ is a lengthscale of the perturbations, one obtains

$$\frac{d\alpha(t)}{dt} = -\frac{\Gamma}{2\pi a^2}\,\beta(t) \sum_{m \neq 0} \frac{1}{m^2}\left(1 - e^{2\pi i m a / \lambda}\right),$$

$$\frac{d\beta(t)}{dt} = -\frac{\Gamma}{2\pi a^2}\,\alpha(t) \sum_{m \neq 0} \frac{1}{m^2}\left(1 - e^{2\pi i m a / \lambda}\right), \tag{6.39}$$

or

$$\frac{d\alpha(t)}{dt} = -\gamma\beta(t),$$
$$\frac{d\beta(t)}{dt} = -\gamma\alpha(t), \tag{6.40}$$

where, since the sum of the sine terms is zero due to $\sin(x) = -\sin(-x)$,

$$\gamma = \frac{\Gamma}{\pi a^2} \sum_{m > 0} \frac{1 - \cos(2\pi m a / \lambda)}{m^2}. \tag{6.41}$$

Here

$$\sum_{m > 0} \frac{1}{m^2} = \zeta(2) = \frac{\pi^2}{6}, \tag{6.42}$$

where ζ is the Riemann zeta function, and

$$\sum_{m>0} \frac{\cos mx}{m^2} = \frac{\pi^2}{6} - \frac{\pi x}{2} + \frac{x^2}{4} \qquad (0 \le x \le 2\pi) \tag{6.43}$$

is a well-known trigonometric (Fourier) series, see Gradshteyn and Ryzhik (1980), form. 3, Sect. 1.443, p. 39. With $x = 2\pi a/\lambda$ the restriction $0 \le x \le 2\pi$ in this formula becomes

$$\lambda \ge a, \tag{6.44}$$

which is physically meaningful: sinusoidal perturbations with wavelength $\lambda < a$ are not 'resolved' by a vortex row with vortex centers at a distance a. One obtains thus (cf. Lamb 1932, p. 226, Eq. 12),

$$\gamma = \frac{\Gamma \pi}{a^2} \frac{a}{\lambda} \left(1 - \frac{a}{\lambda}\right) \ge 0. \tag{6.45}$$

From (6.40) follows directly $\ddot{\alpha} = \gamma^2 \alpha$ and $\ddot{\beta} = \gamma^2 \beta$, with solutions $\alpha, \beta \sim e^{\pm \gamma t}$. The solutions proportional to $e^{|\gamma| t}$ cannot be excluded, and thus the vortex chain is unstable. All wavelengths $\lambda > a$ are unstable, but the growth rate γ has a maximum for $\lambda = 2a$. For a vortex distance $a \to 0$, sinusoidal perturbations with ever smaller wavelength λ become unstable. And these shortest perturbations grow fastest, and thus dominate the flow. This is indeed confirmed by linear stability analysis, see Sect. 6.6. The infinitesimal limit $a \to 0$ therefore corresponds to the Kelvin–Helmholtz instability.

For $\lambda = 2a$, Eq. (6.38) yields $x_{m+1}/x_m = y_{m+1}/y_m = -1$. Thus for $\lambda = 2a$, neighboring vortices move in opposite directions. In early, linear stages of the instability, β is very small. As shown in Fig. 6.8, vortices move then horizontally along the chain. In later, nonlinear stages, β and therefore $|y_m|$ grows, and neighboring vortices form *vortex pairs*, as is confirmed in experiments on the nonlinear stages of the Kelvin–Helmholtz instability.

6.5   Vortex Accumulation

According to the above analysis, it seems that the Kelvin–Helmholtz instability can be understood as the limiting case of a discrete vortex row with the vortex separation tending to zero. Rosenhead (1931) in a pre-computer age numerical calculation considered twelve vortices per perturbation wavelength λ. Their subsequent evolution is shown in Fig. 6.11. The vortices become concentrated in certain spots at a distance λ. Note that the vortices in Fig. 6.11 accumulate at the points where the initial sinusoidal perturbation has maximum *negative* slope, at $x/\lambda = 0.5$ and 1.5, in contrast to Figs. 6.7 and 6.8, where they accumulate at the points of maximum positive slope. The reason is that Rosenhead assumes microscopic vortices that spin clockwise, in the mathematical negative sense (his shear flow has positive U for $\eta > 0$), whereas in the earlier figures they spin counterclockwise. This accumulation of vorticity is

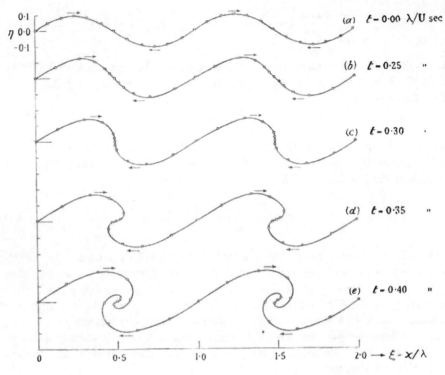

Fig. 6.11 Self-induced motion of twelve vortices after a sinusoidal perturbation. Reproduced from Rosenhead (1931)

accompanied by a roll-up of the line connecting the vortex centers, which resembles the formation of macroscopic, discrete vortices. This result is in agreement with the argument by Batchelor from the last section.

However, criticism was raised against Rosenhead's results in a key argument by Birkhoff and Fisher (1959). To express this, we follow first a calculation by Kirchhoff (1897), Lecture 20, who considers a collection of N vortices with circulation $\Gamma_i = \omega_i \, da$ (see Eq. (4.9)). The vortex centers are located at (x_i, y_i). The speed of vortex i is given by (4.11),

$$u_i = \frac{\partial \psi_i}{\partial y_i}, \qquad\qquad v_i = -\frac{\partial \psi_i}{\partial x_i}, \qquad (6.46)$$

with a streamfunction (compare with (4.12))

$$\psi_i = -\sum_{j \neq i} k_j \ln r_{ij}. \qquad (6.47)$$

Here $r_{ij} = \sqrt{(x_i - x_j)^2 + (y_i - y_j)^2}$ is the distance between vortices i and j. Multiplying both (6.46) and (6.47) with the constant k_i (no summation over i) yields

$$k_i \frac{dx_i}{dt} = \frac{\partial(k_i \psi_i)}{\partial y_i}, \qquad\qquad k_i \frac{dy_i}{dt} = -\frac{\partial(k_i \psi_i)}{\partial x_i}, \qquad (6.48)$$

where $(\dot{x}_i, \dot{y}_i)$ is the velocity of vortex i induced by all vortices with $j \neq i$. For each vortex i, $k_i \psi_i$ (from a single sum) is replaced now by a universal potential H (from a double sum), given by

$$H = -\frac{1}{2} \sum_m \sum_{n \neq m} k_m k_n \ln r_{mn}. \qquad (6.49)$$

To see this, consider the three possible, mutually exclusive combinations of index values that contribute to the double sum for a given i:

(a) $m = i$:
   Terms add up to $-\frac{1}{2} k_i \sum_{n \neq i} k_n \ln r_{in} = \frac{1}{2} k_i \psi_i$ according to (6.47).
(b) $m \neq i \neq n \neq m$:
   Then $\partial r_{mn}/\partial x_i = 0$, since $\partial x_m/\partial x_i = \partial x_n/\partial x_i = 0$; also $\partial r_{mn}/\partial y_i = 0$.
(c) $m \neq i = n$:
   Terms add up to $-\frac{1}{2} \sum_{m \neq i} k_m k_i \ln r_{mi} = -\frac{1}{2} k_i \sum_{j \neq i} k_j \ln r_{ij} = \frac{1}{2} k_i \psi_i$.

Thus, (a) and (c) together contribute $k_i \psi_i$ to H, as required. All contributions from (b) are lost in the differentiations with respect to x_i and y_i in the equations of motion (6.48). Let now $q_i = \sqrt{|k_i|}\, x_i$ and $p_i = \sqrt{|k_i|}\, y_i$. Then Eqs. (6.48) become

$$\frac{dq_i}{dt} = \frac{\partial H}{\partial p_i}, \qquad\qquad \frac{dp_i}{dt} = -\frac{\partial H}{\partial q_i}. \qquad (6.50)$$

These are Hamilton's equations, with a vortex interaction potential H. Thus, the vortex problem is a Hamiltonian dynamical system. This quite astonishing result is given in Eq. (14), p. 259 of Kirchhoff (1897). Note that in the Hamiltonian H, only vortex–vortex interactions occur. Interactions with walls will also influence the vortex motion but are not included here. In Hamiltonian dynamics, the value of the function $H(q, p)$ is a constant of motion, as is seen from

$$\frac{dH}{dt} = \sum_i \frac{\partial H}{\partial q_i} \frac{dq_i}{dt} + \sum_i \frac{\partial H}{\partial p_i} \frac{dp_i}{dt} = \sum_i \frac{\partial H}{\partial q_i} \frac{\partial H}{\partial p_i} - \sum_i \frac{\partial H}{\partial p_i} \frac{\partial H}{\partial q_i} = 0.$$

$$(6.51)$$

The argument by Birkhoff and Fisher (1959) is then as follows. When two vortices i and j approach each other, their distance r_{ij} drops and $-\ln r_{ij}$ grows. For $r_i \rightarrow r_j$, this latter term diverges. To keep H constant, this has to be compensated in either of two possible ways:

(a) an infinite number of vortices recede by finite distances or
(b) a finite number of vortices recede by infinite distances.

However, (a) is not possible for a finite number of vortices, $N < \infty$; and (b) is not possible for a finite domain in which vortices can reside. (A domain periodic in x and y is considered to be finite.)

> Since H is constant, it follows that the approach of one pair of vortices must be accompanied by the recession of other pairs. In particular, if the domain occupied by the vortices is symmetrically bounded, the negative terms in H will be bounded in magnitude. Hence, to conserve H, the positive terms must remain bounded also, and indefinitely close approach of even one vortex-pair is impossible. (Birkhoff and Fisher 1959, p. 81)

This argument suggests that the accumulation of vortices near certain points, as observed by Rosenhead (1931), may be problematic if this accumulation cannot be compensated by a corresponding recession of other vortices. However, this is a qualitative argument and gives only a hint to a possible problem. We will return to the argument in Sect. 6.10, but turn now to more profound criticism of Rosenhead's results.

The calculations by Rosenhead (1931) were repeated by many authors, starting with Birkhoff and Fisher (1959) and Hama and Burke (1960). They used more vortices per wavelength and a smaller time step. As a rather unexpected result, the vortex trajectories are then not as smooth as in the roll-up motion found by Rosenhead, but *tortuous*, i.e., irregular. More specifically,

> From these investigations of the growth of disturbances in a single sheet of vorticity one can see a tendency for concentration of vorticity in clouds, but one is cautioned not to expect detail or accurate fine structure within the cloud. (Abernathy and Kronauer 1962, p. 7)

This kind of behavior is shown in Fig. 6.12. Some researchers concluded from this and similar calculations that the limit $N \to \infty$ of vortices per perturbation wavelength is problematic, and that a discrete vortex street is not an adequate model of a continuous vortex sheet:

> Moore (1971) demonstrated that replacing the sheet by point vortices was at fault, and that using more point vortices could make matters worse, not better. It appears that there may be a fundamental difference between vortex sheets and an assembly of point vortices. (Saffman and Baker 1979, p. 104)

Or, in similar terms,

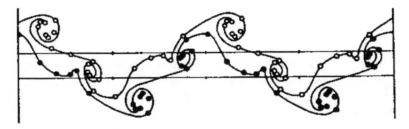

Fig. 6.12 Clouds of vorticity in a numerical calculation with a row of discrete vortices. Open circles: point vortices with clockwise rotation. Filled circles: counterclockwise rotation. The solid lines connect (certain) vortices consecutively on the basis of their ordering in the initial, straight vortex row. Reproduced from Abernathy and Kronauer (1962)

It has become widely accepted that the point-vortex approximation [i.e. finite discretization of the vortex sheet] does not converge as $N \to \infty$ and that it is inadequate for studying vortex-sheet evolution. [...] If convergence cannot be demonstrated as the discretization is refined, then conclusions about the vortex sheet are open to doubt. (Krasny 1986, p. 68)

Krasny (1986) shows that already computer roundoff errors can cause irregular motion of vortices that initially are in a straight row. But he actually succeeds to control these errors both with a numerical filtering technique and high-precision calculations, and it seems possible from the numerics to extrapolate to the limit $N \to \infty$ of a smooth vortex sheet. Thus Krasny (1986) expresses the adequacy of the approach of Rosenhead (1931).

The numerical evidence presented indicates that the point-vortex approximation converges as $N \to \infty$, up to the vortex sheet's critical time. (Krasny 1986, p. 90)

It seems thus that Rosenhead's calculation shows smooth roll-up instead of chaotic vortex motion because the number of vortices per sinusoidal perturbation wavelength is small: low resolution has usually stabilizing effects in numerical simulations, due to smoothing ('filtering') by some effective numerical viscosity. However, as will be discussed in Sect. 6.8, a continuous vortex sheet develops a singularity at some finite critical time when the sheet is still essentially flat, i.e., far from roll-up, and it is not clear how the evolution of the vortex sheet proceeds afterward.

6.6 Linear Stability Analysis

We perform now a *linear stability analysis* of the Kelvin–Helmholtz instability up to first-order terms in infinitesimal perturbations. Consider two incompressible fluids with different but constant densities, like water and oil. Both fluids are treated as being inviscid. Figure 6.13 shows the flow geometry.

The flow is independent of the Cartesian z-coordinate, and only the xy-plane is considered. The fluids at $y > 0$ and $y < 0$ have constant speeds $-U < 0$ and $U > 0$, respectively (note: positive speed at negative y). Along the x-axis, the tangential speed has a jump $2U$. If the pressure p is constant in a fluid, as are ρ, u, and v, the Euler equation is automatically fulfilled. Demanding force equilibrium at the interface of the fluids, p has the same value everywhere. This is the configuration of the undisturbed flow. The flow is assumed to be slightly disturbed, according to

$$\pm U + \tilde{u}_{\pm}(t, x, y),$$

Fig. 6.13 Flow geometry for the Kelvin–Helmholtz instability

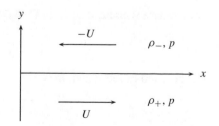

$$\tilde{v}_\pm(t, x, y),$$
$$p + \tilde{\sigma}_\pm(t, x, y),$$
$$\tilde{\eta}(t, x). \tag{6.52}$$

The velocity and pressure perturbations are $(\tilde{u}, \tilde{v})$ and $\tilde{\sigma}$, respectively. At each instant t, the function $y(x) = \tilde{\eta}(t, x)$ gives the line where the two fluids are in contact with each other. Only those terms that are linear (possibly after a series expansion of denominators) in the perturbations $\tilde{u}, \tilde{v}, \tilde{\sigma}$, and $\tilde{\eta}$ are kept. The relevant equations for the two fluids are, with $\tilde{u}_{\pm,x} = \partial \tilde{u}_\pm / \partial x$, etc.,

$$\tilde{u}_{\pm,x} + \tilde{v}_{\pm,y} = 0,$$
$$\tilde{u}_{\pm,t} \pm U\tilde{u}_{\pm,x} = -\rho_\pm^{-1}\,\tilde{\sigma}_{\pm,x},$$
$$\tilde{v}_{\pm,t} \pm U\tilde{v}_{\pm,x} = -\rho_\pm^{-1}\,\tilde{\sigma}_{\pm,y}. \tag{6.53}$$

The first equation expresses that both velocity fields are solenoidal, and the second and third equations are the horizontal and vertical Euler equations, respectively. The vertical shift of the fluid interface η is discussed later. Since the differential equations (6.53) are linear, they have solutions that are periodic in t and x,

$$\tilde{u}_\pm(t, x, y) = u_\pm(y)\, e^{i(\omega t + kx)},$$
$$\tilde{v}_\pm(t, x, y) = v_\pm(y)\, e^{i(\omega t + kx)},$$
$$\tilde{\sigma}_\pm(t, x, y) = \sigma_\pm(y)\, e^{i(\omega t + kx)}. \tag{6.54}$$

Inserting this in (6.53) gives

$$iku_\pm + v'_\pm = 0,$$
$$(\omega \pm kU)\rho_\pm u_\pm = -k\sigma_\pm,$$
$$i(\omega \pm kU)\rho_\pm v_\pm = -\sigma'_\pm, \tag{6.55}$$

where $v' = dv/dy$, etc. The first two equations yield

$$(\omega \pm kU)\rho_\pm v'_\pm = ik^2\sigma_\pm. \tag{6.56}$$

Differentiating this with respect to y, and using the third equation, gives

$$(\omega \pm kU)v''_\pm = k^2(\omega \pm kU)v_\pm, \tag{6.57}$$

or (compare with Eq. 100.23 in Chandrasekhar 1961, p. 483)

$$v''_\pm = k^2 v_\pm. \tag{6.58}$$

The solution of this equation is

$$v_\pm = ae^{ky} + be^{-ky}, \tag{6.59}$$

and since v shall fall off far away from the interface,

$$v_+ = ae^{ky}, \qquad y < 0,$$
$$v_- = be^{-ky}, \qquad y > 0,$$

(6.60)

since 'far from the interface' means $y \ll 0$ ($y \gg 0$) in the first (second) equation. This gives the correct solution of the boundary value problem since at large $|y|$ the flow shall maintain the unperturbed speeds $\pm U$. Next, the fluid interface at $\tilde{\eta}$ is considered. The kinematic boundary condition is

$$\frac{d\tilde{\eta}_\pm(t, x)}{dt} = \tilde{v}_\pm(t, x, y = \tilde{\eta}_\pm),$$

(6.61)

where d/dt is the Lagrange derivative. Therefore,

$$\frac{\partial \tilde{\eta}_\pm}{\partial t} \pm U \frac{\partial \tilde{\eta}_\pm}{\partial x} = \tilde{v}_\pm(t, x, \tilde{\eta}_\pm).$$

(6.62)

Inserting a periodic perturbation of the fluid interface, i.e.,

$$\tilde{\eta}_\pm(t, x) = \eta_\pm\, e^{i(\omega t + kx)},$$

(6.63)

where $\eta_\pm \in \mathbb{R}$, gives

$$i(\omega \pm kU)\eta_\pm = v_\pm(\tilde{\eta}_\pm).$$

(6.64)

Since v and η are real, ω or k must be complex. The fluid interface cannot split into two, thus at all times $\tilde{\eta}_+ = \tilde{\eta}_- = \tilde{\eta}$ holds, giving

$$\frac{v_+(\tilde{\eta})}{\omega + kU} = \frac{v_-(\tilde{\eta})}{\omega - kU}.$$

(6.65)

Inserting (6.60), this becomes

$$\frac{ae^{k\tilde{\eta}}}{\omega + kU} = \frac{be^{-k\tilde{\eta}}}{\omega - kU}.$$

(6.66)

The exponentials can approximately be put equal to unity, since the products $a\tilde{\eta}$ and $b\tilde{\eta}$ (from a series expansion of the exponentials) are small of second order, yielding

$$\frac{a}{\omega + kU} = \frac{b}{\omega - kU} = c,$$

(6.67)

with a new constant c. Thus (6.60) becomes finally

$$v_\pm(y) = c(\omega \pm kU)\, e^{\pm ky}.$$

(6.68)

Inserting this in the last equation in (6.55) gives

$$- i c \rho_\pm (\omega \pm kU)^2 e^{\pm ky} = \sigma'_\pm, \tag{6.69}$$

and after integration with respect to y,

$$\mp i c \rho_\pm (\omega \pm kU)^2 e^{\pm ky} = k\sigma_\pm. \tag{6.70}$$

At the interface $\tilde{\eta}$, the pressure perturbation must be continuous, $\sigma_+ = \sigma_-$. In (6.70), c and $\sigma_\pm$ are of first order, and thus one can again to first order put $e^{\pm k\tilde{\eta}} \approx 1$, which yields

$$\rho_+ (\omega + kU)^2 + \rho_- (\omega - kU)^2 = 0. \tag{6.71}$$

This gives the *dispersion relation* $\omega(k)$. To write this simply, let

$$\Omega = \frac{\omega}{kU} \tag{6.72}$$

and

$$\alpha = \frac{\rho_+ - \rho_-}{\rho_+ + \rho_-}, \tag{6.73}$$

where $-1 < \alpha < 1$. Then (6.71) becomes

$$\Omega^2 + 2\alpha\Omega + 1 = 0, \tag{6.74}$$

with solutions

$$\Omega = \frac{\omega}{kU} = -\alpha \pm i\sqrt{1 - \alpha^2}. \tag{6.75}$$

This corresponds to Eq. (101.31) in Chandrasekhar (1961, p. 484). All perturbations behave temporally like $e^{i\omega t}$. For real $k > 0$, $\omega = kU(-\alpha - i\sqrt{1 - \alpha^2})$ implies exponential growth in time. This is the Kelvin–Helmholtz instability. Note that the instability occurs for all wavenumbers $k > 0$. The growth rate $\mathrm{Im}(\omega)$, however, is largest for wavelengths $\lambda \to 0, k \to \infty$.

$\mathrm{Im}(\omega)$ vanishes for very large density contrast $|\alpha| \to 1$ of the fluids. A more interesting stabilizing agent, however, is gravity. Let the gravitational acceleration g point in the $-y$-direction, then the Euler equation for v changes to

$$\frac{dv}{dt} = -\frac{1}{\rho}\frac{\partial p}{\partial y} - g. \tag{6.76}$$

The above derivation remains unchanged up to (6.70), where continuity of pressure at the interface is imposed. The Euler equation for the unperturbed flow is now

$$0 = -\frac{1}{\rho_\pm}\frac{\partial p_\pm}{\partial y} - g, \tag{6.77}$$

or

$$p_\pm = P_\pm - \rho_\pm g y, \tag{6.78}$$

with thermal pressure P (the constant of integration) and hydrostatic pressure $-\rho g y$. For simplicity, it is assumed that P can be neglected in comparison with $\rho g y$, or is a global constant. Continuity of pressure at the interface $\tilde{\eta}$ gives then, again to first order,

$$- \rho_+ g \eta_+ + \sigma_+ = -\rho_- g \eta_- + \sigma_-, \tag{6.79}$$

where a common $e^{i(\omega t + kx)}$ cancels. The $\eta_\pm$ are expressed through (6.64) and (6.68), and $\sigma_\pm$ through (6.70). All $e^{\pm ky}$ can again approximately be put equal to 1 to first perturbation order. Then

$$\rho_+(\omega + kU)^2 + \rho_-(\omega - kU)^2 = (\rho_+ - \rho_-)gk, \tag{6.80}$$

which replaces (6.71). Introducing again Ω and α as above gives

$$\Omega^2 + 2\alpha\Omega + 1 - \frac{g\alpha}{kU^2} = 0, \tag{6.81}$$

or

$$\frac{\omega}{kU} = -\alpha \pm \sqrt{\alpha^2 - 1 + \frac{g\alpha}{kU^2}}. \tag{6.82}$$

For $\alpha < 0$, the radicand is more negative than for $g = 0$. Here $\alpha < 0$ means $\rho_+ < \rho_-$, i.e., that the heavier fluid is atop of the lighter. This is the so-called Rayleigh–Taylor instability: perturbations of the fluid interface cause the heavier fluid to fall downward through the lighter. Thereby, the system attains a stable minimum of potential energy. Note that for $U \to 0$, and still $\alpha < 0$ (and $k > 0$),

$$\omega = \pm i \sqrt{-g\alpha k}. \tag{6.83}$$

This is the pure Rayleigh–Taylor instability, without shear motion. Equation (6.83) is Eq. (92.51) in Chandrasekhar (1961), p. 435, for the case that surface tension vanishes. Finally, consider the case of a light fluid atop of a heavy one, $\alpha > 0$. Then the radicand in (6.82) is positive for

$$k < \frac{\alpha g}{(1 - \alpha^2)\, U^2} = \frac{\rho_+^2 - \rho_-^2}{\rho_+ \rho_-} \frac{g}{4U^2}, \tag{6.84}$$

see Eq. (101.34) in Chandrasekhar (1961), p. 484. Thus, a fluid stratification with light fluid atop of heavy fluid prevents the Kelvin–Helmholtz instability at long perturbation wavelengths.

6.7   The Birkhoff–Rott Equation for Vortex Sheets

As a preparation for the next section on the curvature singularity for vortex sheets, the so-called Birkhoff–Rott equation is derived here. We start by transforming Eq. (2.200) for $\vec{A}$ to the complex plane. Since our interest is in the motion of the vortex sheet itself (i.e., the interface between two shearing fluid components), the contribution from the gradient of a scalar potential ϕ to the velocity field can be dropped; thus (2.187) becomes

$$\vec{u} = \nabla \times \vec{A}. \tag{6.85}$$

With the Coulomb gauge (2.191), $\nabla \cdot \vec{A} = 0$, this yields for the vorticity,

$$\vec{\omega} = \nabla \times \vec{u} = -\Delta \vec{A}, \tag{6.86}$$

where $\Delta = \nabla \cdot \nabla$. The solution of this Poisson equation is given by

$$\vec{A}(\vec{r}) = \int \mathrm{d}^3 r' \, G(\vec{r} - \vec{r}') \, \vec{\omega}(\vec{r}'), \tag{6.87}$$

where the Green's function G is defined by $\Delta G(\vec{r}) = -\delta(\vec{r})$. In $\mathbb{R}^2$, G is given by (3.42),

$$G(\vec{r}) = \frac{1}{2\pi} \ln \frac{1}{|\vec{r}|}. \tag{6.88}$$

The vortex lines shall lie in the Cartesian xz-plane and point in the z-direction (the unit vectors are again $\hat{x}, \hat{y}, \hat{z}$). A cut with the xy-plane gives then a line $y = 0$ continuously filled with point vortices, see Fig. 6.14.

From (6.85), and in two dimensions,

$$\vec{u}(\vec{r}) = \frac{1}{2\pi} \nabla \times \iint \mathrm{d}x' \, \mathrm{d}y' \, \ln \frac{1}{|\vec{r} - \vec{r}'|} \, \vec{\omega}(\vec{r}'). \tag{6.89}$$

Fig. 6.14 Geometry for the Birkhoff–Rott equation

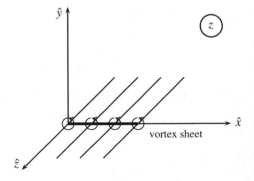

Since ∇ acts on $\vec{r}$ only, and using the product rule for derivatives of curl,

$$
\begin{aligned}
\vec{u}(\vec{r}) &= \frac{1}{2\pi} \iint dx'\, dy'\; \nabla \ln \frac{1}{|\vec{r} - \vec{r}'|} \times \vec{\omega}(\vec{r}') \\
&= -\frac{1}{2\pi} \iint dx'\, dy'\; \frac{\vec{r} - \vec{r}'}{|\vec{r} - \vec{r}'|^2} \times \vec{\omega}(\vec{r}')
\end{aligned}
\tag{6.90}
$$

is obtained. By the definitions of the circulation Γ and vorticity $\vec{\omega}$,

$$
\Gamma = \oint_{\partial A} d\vec{l} \cdot \vec{u} \stackrel{\text{Stokes}}{=} \iint_A dx\, dy\, (\nabla \times \vec{u})_z = \iint_A dx\, dy\, \omega,
\tag{6.91}
$$

with A an area and ∂A a loop in the xy-plane, and $\vec{\omega} = \omega\,\hat{z}$; thus

$$
d\Gamma = dx\, dy\, \omega.
\tag{6.92}
$$

From now on, it is assumed that vorticity ω is constant everywhere in the vortex sheet. The integral (6.90) can be rewritten using complex variables alone: first, the cross product in (6.90) is replaced by multiplication with $-i$, since $\vec{r} - \vec{r}'$ lies in the xy-plane and $\vec{\omega} = \omega\hat{z}$. Furthermore, $(x\hat{x} + y\hat{y}) \times \hat{z} = y\hat{x} - x\hat{y}$. With $x\hat{x} + y\hat{y} \equiv x + iy$, one has $y\hat{x} - x\hat{y} \equiv y - ix = -i(x + iy)$. Next, with z now the complex variable $z = x + iy$,

$$
\frac{\vec{r} - \vec{r}'}{|\vec{r} - \vec{r}'|^2} \equiv \frac{z - z'}{(z - z')(\overline{z - z'})} = \frac{1}{\overline{z - z'}}.
\tag{6.93}
$$

Identifying finally $\vec{u} \equiv \overline{w}$, with complex speed $w = u - iv$, and using (6.92), Eq. (6.90) becomes

$$
\overline{w}(z) = \frac{1}{2\pi i} \int_C \frac{d\Gamma'}{\overline{z(\Gamma')} - z}.
\tag{6.94}
$$

Here z is an arbitrary point, and $z(\Gamma')$ lies in the vortex sheet, described by the curve C. If all sheet vortices spin in the same direction, the mapping $z(\Gamma') \leftrightarrow \Gamma'$ is one-to-one, see the discussion in Sect. 2.12 after Eq. (2.132). Circulation Γ' can then be used as a Lagrange coordinate, i.e., as a continuously changing fluid property that allows to identify fluid parcels. Taking the complex conjugate of (6.94), where $\Gamma \in \mathbb{R}$, gives

$$
w(z) = -\frac{1}{2\pi i} \int_C \frac{d\Gamma'}{z(\Gamma') - z}.
\tag{6.95}
$$

This is a well-defined function for all z outside the vortex sheet. However, we are interested in the equation of motion for the sheet itself, and thus want z to lie on the curve C where the integral (6.95) becomes singular at $z(\Gamma') = z$. This is nontrivial, since the complex speed w has a finite jump Δu in its tangential component across

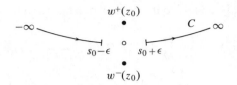

Fig. 6.15 The integral (6.96) is singular at $z_0 = z(s_0)$ marked $\circ$, with arclength s along the vortex sheet C. The Cauchy principal value $W(z_0)$ is defined by integrating *along* C up to $s_0 - \epsilon$, and starting again at $s_0 + \epsilon$. Plemelj's theorem states that $W(z_0) = (w^+(z_0) + w^-(z_0))/2$, an arithmetic mean *across* the sheet

the vortex sheet. In the following, the integration has to be performed over dz instead of $d\Gamma$, to make equations from complex function theory applicable. The increase $d\Gamma$ in circulation over a length differential dz along the curve C is $d\Gamma = \Delta u \, dz$, by definition of Γ, giving

$$w(z) = -\frac{\Delta u}{2\pi i} \int_C \frac{dz'}{z' - z}. \tag{6.96}$$

The variable z_0 shall in the following refer to locations in the vortex sheet. Then the Cauchy principal value $W(z_0)$ in z_0 for the singular integral (6.96) is defined by (see Fig. 6.15)

$$W(z_0) = -\frac{\Delta u}{2\pi i} \oint_C \frac{dz'}{z' - z_0}$$

$$= -\frac{\Delta u}{2\pi i} \lim_{\epsilon \to 0} \left(\int_{-\infty}^{s_0 - \epsilon} ds' \frac{dz'}{ds'} \frac{ds'}{z(s') - z_0} + \int_{s_0 + \epsilon}^{\infty} ds' \frac{dz'}{ds'} \frac{ds'}{z(s') - z_0} \right). \tag{6.97}$$

Here $\epsilon > 0$ is a real number, s is the arclength along C, and $z(s)$ is a parameterization of C assuming $z_0 = z(s_0)$. The vortex sheet C shall extend to infinity in both directions. For the definition (6.97), see Hurwitz and Courant (1964), p. 340.

Let t be the arclength along a curve that crosses the vortex sheet *normally* at $z_0 = z(t_0)$. This requires that the sheet has continuous curvature, else such a normal curve is not defined. (In the next section, it will be shown that the sheet actually develops a curvature singularity.) When moving along the vortex sheet C in positive direction, the part of the complex plane to the left (right) of C is referred to with an index $+$ ($-$). Since $w(z)$ from (6.96) is well defined for z outside C, the limits (see Fig. 6.15)

$$w^\pm(z_0) = \lim_{\epsilon \to 0} w(z(t_0 \pm \epsilon)) \tag{6.98}$$

can be introduced. Then the Plemelj (1908) theorem states the conditions under which the limits in (6.97) and (6.98) exist and relates the principal value from an integration *along* the vortex sheet to the velocity jump *across* the sheet, viz.,

$$w^+(z_0) + \frac{1}{2} \Delta u = W(z_0),$$

$$w^-(z_0) - \frac{1}{2} \Delta u = W(z_0), \tag{6.99}$$

see Eq. (1), p. 340 in Hurwitz and Courant (1964), where also the theorem is proved. The function $\phi(z)$ in Hurwitz and Courant corresponds to the constant $-\Delta u$ in (6.96). The requirement for Plemelj's theorem to hold is that ϕ obeys a *Hölder condition* on C,

$$|\phi(z) - \phi(z')| \le A|z - z'|^\mu, \tag{6.100}$$

with real-valued constants $A > 0$ and $0 < \mu \le 1$. This is trivially fulfilled here, since the left side in (6.100) is zero, a consequence of the assumption that vorticity ω is constant everywhere on the vortex sheet. In the general case of varying vorticity, the function $\Delta u(z) = \omega(z)\,dh$ (see Eq. (2.133)) must obey the Hölder condition (6.100).

The jump in w across the vortex sheet is $w^+ - w^- = \Delta u$, as must be. The speed of the vortex sheet is defined as arithmetic mean of w^+ and w^- (from symmetry), and obtained from (6.99),

$$w(z_0) = \frac{1}{2}\left(w^+(z_0) + w^-(z_0)\right) = W(z_0). \tag{6.101}$$

Using again $\Gamma \in \mathbb{R}$ as integration variable along C with Γ from $-\infty$ to $+\infty$ along the infinite sheet gives

$$w(z_0) = -\frac{1}{2\pi i} \int_{-\infty}^{\infty} \frac{d\Gamma'}{z(\Gamma') - z_0}. \tag{6.102}$$

The definition of fluid speed, $\dot{\vec{r}} = \vec{u}$, becomes in complex variables $\dot{z} = \overline{w}$, and (6.102) can be written as

$$\boxed{\frac{\partial \overline{z}}{\partial t}(t, \Gamma) = -\frac{1}{2\pi i} \int_{-\infty}^{\infty} \frac{d\Gamma'}{z(t, \Gamma') - z(t, \Gamma)}} \tag{6.103}$$

This is the *Birkhoff–Rott* equation (Rott 1956; Birkhoff 1962). Obviously, t is now time, not the parameter of the normal curve introduced above. The equation is also derived in Pullin (1978), p. 403, and rather briefly in Sect. 8.1 of Saffman (1995).

6.8   Curvature Singularity in Evolving Vortex Sheet

Proofs of existence and uniqueness for the motion of vortex sheets are lacking. It is not obvious that the problems are well posed and that smooth solutions (except perhaps for isolated singularities) exist for more than a finite time. (Saffman and Baker 1979, p. 98)

A wave of length λ on a plane vortex sheet of uniform circulation density or strength γ has a growth rate $\pi \gamma / \lambda$. The consequence of this pathological behaviour of short waves is that there is – on linear theory – no solution of the initial value problem for a general analytic and periodic initial disturbance which remains analytic at all subsequent times. (Moore 1979, p. 105)

Vortex sheets—similarly to point masses, point charges, strings—are mathematical idealizations, viz., fictitious lower dimensional objects in a three-dimensional world. After taking the limit of zero thickness, the tangential fluid speed is *discontinuous* across a vortex sheet. However, it is not clear whether this limit can be taken to define the vortex sheet, *before* the physical properties of such an object are analyzed. We show in this section that a major problem related to vortex sheets is the evolution of a singularity from smooth initial conditions within a finite critical time t_c. Long before the sheet rolls up into a structure resembling discrete, macroscopic vortices, the still largely flat vortex sheet develops a curvature singularity, which makes the vortex sheet nonanalytic; the subsequent evolution of the sheet remains then unclear. These are the results of a groundbreaking paper by Moore (1979), which will be followed closely in this section. The logic of the analysis is rather straightforward, although the calculation itself is occasionally quite forbidding. The initial perturbation of the sheet is an analytic sine wave of given period, corresponding to a single Fourier mode. Through nonlinear terms, which Moore succeeds to maintain throughout the calculation, the higher Fourier modes with smaller wavelengths are excited. At first, the mode amplitude at any given time drops exponentially with the mode number. However, the small-scale modes with large mode numbers grow fastest (The Kelvin–Helmholtz instability). After a *finite* critical time t_c, the mode spectrum has therefore flattened to a power law, which implies that the vortex sheet is no longer analytic and a singularity has formed. The singularity occurs in the second spatial derivative of the vortex sheet, i.e., the sheet curvature becomes infinite at a certain point, which corresponds to a cusp in the sheet. The slope (first derivative) of the sheet remains finite and continuous at this point. The point of singularity formation is the point of vortex accumulation due to vortex motion induced by neighboring vortices, i.e., the points marked with bullets in Fig. 6.8, and the centers of roll-up regions at $x/\lambda = 0.5$ and 1.5 in Fig. 6.11:

[A] straining flow drives the vorticity towards an accumulation point that becomes the point of curvature singularity for the vortex sheet. (Baker and Shelley 1990, p. 164)

Consider then a shear flow in the complex z-plane, with horizontal speed $-1/2$ for $y > 0$ and $+1/2$ for $y < 0$; the vertical speed is zero. This shear flow corresponds to a vortex sheet of uniform strength along the x-axis. The circulation between the reference point 0 and location x is, according to Eq. (2.132),

$$\Gamma = 2 \int_0^x \frac{dx'}{2} = x. \tag{6.104}$$

This is a one-to-one mapping between x and Γ, and the initial position of a fluid parcel can be written as $x(t = 0, \Gamma)$, with the Lagrange coordinate Γ (the same value of Γ refers to the same fluid parcel at all times). An initial perturbation of amplitude $0 < \epsilon < 1$ and period 2π is applied to the vortex sheet, shifting it to positions

$$z(0, \Gamma) = x + iy = x + i\epsilon \sin x = \Gamma + i\epsilon \sin \Gamma. \tag{6.105}$$

The central assumption is that $z(t, \Gamma)$ is a *real analytic* function of the real variable Γ. Here a function $f : D \subset \mathbb{R} \to \mathbb{C}$ with open interval D is called *real analytic* if for every $x_0 \in D$ there is an open neighborhood in which f can be written as a convergent power series. Then f is infinitely often differentiable and its derivatives obey a certain growth condition, see Markushevich (1965), Eq. 2.1 on p. 5. Since the initial perturbation is periodic, z can be even expanded into a Fourier series in Γ,

$$z(t, \Gamma) = \Gamma + \sum_{n=-\infty}^{\infty} A_n(t)e^{in\Gamma}, \tag{6.106}$$

assuming that z has period 2π in Γ at all times. The Fourier representation of (6.105) is

$$z(0, \Gamma) = \Gamma + \frac{\epsilon}{2}\left(e^{i\Gamma} - e^{-i\Gamma}\right). \tag{6.107}$$

Comparing this with (6.106), all $A_n(0)$ vanish, except for

$$A_{\pm 1}(0) = \pm \frac{\epsilon}{2}. \tag{6.108}$$

Eventually, the evolution of the vortex sheet will become nonlinear, and the perturbation does not remain of sinusoidal form. Thus in the course of time, all the A_n are excited. Introduce a new variable u (which is *not* the horizontal velocity component),

$$u = \Gamma' - \Gamma. \tag{6.109}$$

Then,

$$\begin{aligned} z(t, \Gamma) - z(t, \Gamma') &= \Gamma + \sum A_n e^{in\Gamma} - \Gamma' - \sum A_n e^{in\Gamma'} \\ &= -u + \sum A_n e^{in\Gamma}\left(1 - e^{inu}\right), \end{aligned} \tag{6.110}$$

and (6.103) becomes

$$\frac{\partial \bar{z}}{\partial t}(t, u) = \frac{i}{2\pi} \int_{-\infty}^{\infty} \frac{du}{u} \frac{1}{1 - S(u, \Gamma)/u}, \tag{6.111}$$

introducing the function

$$S(u, \Gamma) = \sum_{n=-\infty}^{\infty} A_n e^{in\Gamma} \left(1 - e^{inu}\right). \tag{6.112}$$

For $|S/u| < 1$, the sum formula for the geometric series can be applied,

$$\frac{\partial \bar{z}}{\partial t}(t, u) = \frac{i}{2\pi} \int_{-\infty}^{\infty} \frac{du}{u} \left(1 + \frac{S}{u} + \frac{S^2}{u^2} + \frac{S^3}{u^3} + \cdots\right). \tag{6.113}$$

Writing this out gives, with $\partial \Gamma / \partial t = 0$ (since the circulation Γ is an independent variable),

$$\sum_n \frac{d\bar{A}_n(t)}{dt} e^{-in\Gamma} = \frac{i}{2\pi} \int_{-\infty}^{\infty} \frac{du}{u} \Bigg[1 + \frac{1}{u} \sum_n A_n e^{in\Gamma} \left(1 - e^{inu}\right)$$

$$+ \frac{1}{u^2} \sum_{j,k} A_j A_k e^{i(j+k)\Gamma} \left(1 - e^{iju}\right)\left(1 - e^{iku}\right)$$

$$+ \frac{1}{u^3} \sum_{j,k,l} A_j A_k A_l e^{i(j+k+l)\Gamma} \left(1 - e^{iju}\right)\left(1 - e^{iku}\right)\left(1 - e^{ilu}\right) + \cdots \Bigg], \tag{6.114}$$

in which the sums are from $-\infty$ to ∞. The integral du/u is discussed below. Different functions I of one, two, three ... integer arguments are introduced by the formulas (with imaginary unit i)

$$I(n) = \int_{-\infty}^{\infty} \frac{du}{u^2} (1 - e^{inu}),$$

$$I(j, k) = \int_{-\infty}^{\infty} \frac{du}{u^3} (1 - e^{iju})(1 - e^{iku}),$$

$$I(j, k, l) = \int_{-\infty}^{\infty} \frac{du}{u^4} (1 - e^{iju})(1 - e^{iku})(1 - e^{ilu}), \tag{6.115}$$

and so forth. Because of the orthogonality of the sine and cosine functions, powers of $e^{in\Gamma}$ can be equated on the left- and right-hand sides of (6.114), giving

$$\frac{d\bar{A}_{-n}(t)}{dt} = \frac{i}{2\pi} \Bigg[I(n)A_n + \sum_{\substack{j,k \\ j+k=n}} I(j, k)A_j A_k + \sum_{\substack{j,k,l \\ j+k+l=n}} I(j, k, l)A_j A_k A_l + \cdots \Bigg].$$

$$\tag{6.116}$$

Note that on the left-hand side $\bar{A}_{-n}$ appears. The initial perturbation is sinusoidal, which leads to (6.108). The solution for z shall be odd at all times (i.e., unstable growth obeys the initial symmetry), and therefore

$$A_0(t) = 0,$$
$$A_{-n}(t) = -A_n(t). \tag{6.117}$$

The functions I in (6.115) can be calculated by complex contour integrals, with poles at $u = 0$. Each integral is taken along the real axis from $u \to -\infty$ to ∞. The pole at $u = 0$ is avoided by an infinitesimal semicircle in the lower half-plane. The contour is closed at ∞ in the upper half-plane, since $\lim\limits_{u \to i\infty} e^{iju} = 0$. For example,

$$\oint \frac{du}{u^4} \left(1 - e^{iju}\right)\left(1 - e^{iku}\right)\left(1 - e^{ilu}\right)$$
$$= \oint \frac{du}{u^4} (-iju)(-iku)(-ilu)$$
$$= 2\pi i (-i)^3 \, jkl, \tag{6.118}$$

where $\oint du/u = 2\pi i$ since $u = 0$ lies within the integration contour. Only the first-order Taylor terms of e^{iju}, etc., contribute to the integral; higher-order terms lead to powers u^n with $n \geq 0$ in the integral, which are analytic functions and have no poles in the complex u-plane, thus $\oint du \, u^n = 0$ for $n \geq 0$. Furthermore (where details to this rather obvious calculation can be found again in Plemelj 1908),

$$2\pi i = \oint \frac{du}{u} = \int \frac{du}{u} + \pi i + 0. \tag{6.119}$$

The three terms on the right side are, from left to right,

- integral along the real axis, with $]-\epsilon, \epsilon[$ excluded,
- infinitesimal semicircle avoiding the singularity at $u = 0$, and
- vanishing contribution from the semicircle at infinity.

Equation (6.119) thus gives

$$\int\limits_{-\infty}^{\infty} \frac{du}{u} = \pi i. \tag{6.120}$$

From (6.118) and (6.119) follows then

$$I(j, k, l) = \pi i (-i)^3 \, jkl, \tag{6.121}$$

and for general $n \geq 1$,

$$\boxed{I(j_1, \ldots, j_n) = \pi i (-i)^n \, j_1 \ldots j_n} \tag{6.122}$$

Equation (6.116) with infinite sums appearing on the right can be further simplified. The initial conditions (6.108) are $A_{\pm 1}(0) = \epsilon/2$ and $A_n = 0$ for $n \neq \pm 1$. This means that, for small times at least, (a) A_1 is of order ϵ and (b) in (6.116), the last term on the right-hand side, A_1^n, is the dominant term. Thus at small times, (6.116) implies

$$A_n(t) = \epsilon^{|n|} a_n(t) + O(\epsilon^{|n+1|}). \tag{6.123}$$

(Actually, the latter term should read $O(\epsilon^{|n+2|})$, see Eq. 2.17 in Moore). Because of (6.117), $n \geq 1$ can be assumed. Inserting (6.122) and (6.123) in (6.116) gives, to order ϵ^n,

$$2\frac{\mathrm{d}\bar{a}_n(t)}{\mathrm{d}t} = -ina_n + (-i)^2 \sum_{\substack{j,k\geq 1 \\ j+k=n}} jka_j a_k + (-i)^3 \sum_{\substack{j,k,l\geq 1 \\ j+k+l=n}} jkla_j a_k a_l + \cdots + (-i)^n a_1^n \tag{6.124}$$

Note especially the last term on the right. All sums in (6.124) run over a finite number of terms only. The $a_j, a_k, a_l, \ldots$ on the right side have $j, k, l, \ldots < n$. Thus the system can be solved recursively. This, however, does not help for the case at hand, where short scales with large n are expected to grow fastest. Thus, an asymptotic estimate for a_n as $n \to \infty$ is needed. To this end, Moore introduces a *generating function G*, defined by

$$G(t, x) = \sum_{n=1}^{\infty} a_n(t) x^n. \tag{6.125}$$

Here x and t are assumed to be real-valued variables. Then

$$x\frac{\partial G}{\partial x} = \sum_{n=1}^{\infty} na_n x^n,$$

$$\left(x\frac{\partial G}{\partial x}\right)^2 = \sum_{n=1}^{\infty} x^n \sum_{\substack{j,k\geq 1 \\ j+k=n}} jka_j a_k,$$

$$\left(x\frac{\partial G}{\partial x}\right)^3 = \sum_{n=1}^{\infty} x^n \sum_{\substack{j,k,l\geq 1 \\ j+k+l=n}} jkla_j a_k a_l \tag{6.126}$$

and so forth. Therefore,

$$2\frac{\partial \bar{G}}{\partial t} = 2\sum_{n=1}^{\infty} \frac{\mathrm{d}\bar{a}_n(t)}{\mathrm{d}t} x^n$$

$$= \sum_{n=1}^{\infty} \left[-ina_n + (-i)^2 \sum_{\substack{j,k\geq 1 \\ j+k=n}} jka_j a_k + (-i)^3 \sum_{\substack{j,k,l\geq 1 \\ j+k+l=n}} jkla_j a_k a_l + \cdots \right] x^n$$

$$= -i \sum_{n=1}^{\infty} n a_n x^n + (-i)^2 \sum_{n=1}^{\infty} x^n \sum_{\substack{j,k \geq 1 \\ j+k=n}} jk a_j a_k + (-i)^3 \sum_{n=1}^{\infty} x^n \sum_{\substack{j,k,l \geq 1 \\ j+k+l=n}} jkl a_j a_k a_l + \cdots$$

$$= -ix\frac{\partial G}{\partial x} + \left(-ix\frac{\partial G}{\partial x}\right)^2 + \left(-ix\frac{\partial G}{\partial x}\right)^3 + \cdots, \tag{6.127}$$

in which (6.124) was used. Due to the sum over n, the term $(-i)^n a_1^n$ that appeared as last term in (6.124) does no longer appear separately. It is assumed that

$$\left| x\frac{\partial G}{\partial x} \right| < 1. \tag{6.128}$$

Then the sum formula for the geometric series can be applied, giving

$$\boxed{2\frac{\partial \bar{G}}{\partial t} = \frac{-ix\dfrac{\partial G}{\partial x}}{1 + ix\dfrac{\partial G}{\partial x}}} \tag{6.129}$$

This is again (6.124), now expressed as a nonlinear partial differential equation. The initial value conditions for G are, using (6.108) and (6.123),

$$G(0, x) = a_1(0)\, x = \frac{A_1(0)}{\epsilon}\, x = \frac{x}{2}. \tag{6.130}$$

The important Eq. (6.129) can also be derived in a completely different way, as is done by Caflisch and Orellana (1986), see their Eq. (2.13). Moore (1979), in an *asymptotic* analysis, restricts himself to estimating G for large $t \gg 1$. To obtain this asymptotic behavior, return to (6.124). The solution of the linear part of this equation,

$$2\frac{d\bar{a}_n^{(1)}}{dt} = -in a_n^{(1)}, \tag{6.131}$$

is, with real-valued constants $c_n^{(1)}$,

$$a_n^{(1)}(t) = (1 + i) c_n^{(1)} e^{nt/2}. \tag{6.132}$$

Nonlinearities often lead to power-law behavior. For reasons discussed below, Moore (1979) suggests, as such,

$$a_n(t) = \frac{1 + i}{4^n} e^{nt/2} \left(c_n t^{n-1} + d_n t^{n-2} + \cdots + p_n t + q_n \right). \tag{6.133}$$

Inserting (6.133) in (6.124) gives powers $t^{n-1}, \ldots, t^0$ on the left and right sides, which cancel pairwise. All other possibilities for a power law for a_n except (6.133) can be excluded, since they lead to inconsistencies or other problems: first, if only the highest power of t occurring in (6.133), i.e., $a_n \sim e^{nt/2} t^{n-1}$, would be inserted in (6.124), then the powers of t on the left and right sides cannot balance and thus the equation is inconsistent. Second, the assumption $a_n \sim e^{nt/2} t^{n+1}$ plus lower powers of t also leads to inconsistency, since the highest power on the left side of (6.124) is t^{n+1}, but on the right side is t^{2n}. Nonlinear terms would have to cancel then, which is unlikely. Third, $a_n \sim e^{nt/2} t^n$ plus lower powers gives t^n on both sides of (6.124). Then, however, linear and quadratic terms in (6.124) would have the same temporal scaling $\sim e^{nt/2} t^n$, which would prevent the occurrence of a linear growth regime (6.132) altogether and growth would be nonlinear from the beginning, which is counterintuitive. In summary, therefore, (6.133) is the appropriate expression for $a_n(t)$. Negative powers of t could also appear, but they play no role for $t \to \infty$.

Let $t \to \infty$ (asymptotic analysis), and keep only terms in c_n and d_n in (6.133). Inserting this into (6.124) gives after short calculation

$$c_n = \bar{c}_n, \tag{6.134}$$

i.e., all the c_n are real, and

$$
\begin{aligned}
0 &= \quad n \operatorname{Im} d_n + (n-1)c_n - \sum_{\substack{j,k \geq 1 \\ j+k=n}} jk \, c_j \, c_k, \\
0 &= - \, n \operatorname{Im} d_n + (n-1)c_n,
\end{aligned}
\tag{6.135}
$$

or

$$\operatorname{Im} d_n = \frac{n-1}{n} c_n, \tag{6.136}$$

and, for $n > 1$,

$$\boxed{c_n = \frac{1}{2(n-1)} \sum_{\substack{j,k \geq 1 \\ j+k=n}} jk \, c_j \, c_k} \tag{6.137}$$

see Eq. (4.5) in Moore. The c_n are thus fully determined among themselves. We still have to obtain c_1. Equation (6.124) for $n = 1$ is

$$2\dot{a}_1 = -i a_1. \tag{6.138}$$

The last term in (6.124) applies for $n \geq 4$ only. The solution is, with initial condition $a_1(0) = \frac{1}{2}$ from (6.130),

$$a_1(t) = \frac{1}{2}\left(\cosh \frac{t}{2} + i \sinh \frac{t}{2} \right). \tag{6.139}$$

Fig. 6.16 Scaling of the Fourier coefficients $|A_n|$ with n, for $\epsilon = 10^{-3}$ and $t_c = 9.986$. The full line is a power law $A_n \sim n^{-5/2}$

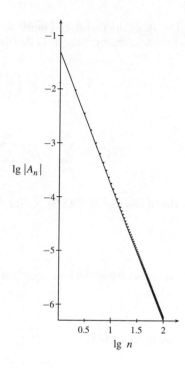

Therefore,

$$\lim_{t \to \infty} a_1(t) = \frac{1}{4}(1+i)e^{t/2}, \tag{6.140}$$

and from (6.133),

$$c_1 = 1. \tag{6.141}$$

Without the factor 4^{-n} in (6.133), $c_1 = 1/4$. Then (6.137) with $c_n \sim c_j c_{n-j}$ (for $1 \leq j < n$) suggests $c_n \sim 4^{-n}$. This is the reason why the factor 4^{-n} was included in (6.133). Figure 6.16 shows the Fourier coefficients A_n as function of n, calculated from (6.123) for $A_n(a_n)$ (only order $\epsilon^{|n|}$), (6.133) for $a_n(c_n)$ (only order $e^{nt/2}t^{n-1}$), and (6.137) for c_n. The critical time t_c used in the figure was calculated from (6.180). These Fourier components are compared with a power law $A_n \sim n^{-5/2}$, which was first suggested by Saffman. The agreement is very good.

One is interested in the Fourier coefficients A_n for large n, since small scales grow fastest. In the rest of this section, therefore, the asymptotic scaling of the A_n for $n \to \infty$ is derived, based on the first term, $c_n t^{n-1}$, in the sum in (6.133), and on the central recursion relation (6.137). A new generating function $g(x)$ is defined by

$$g(x) = \sum_{n=1}^{\infty} c_n x^n. \tag{6.142}$$

The singularity of g will allow to determine the asymptotic behavior of the c_n for $n \to \infty$. Abbreviating $g' = \mathrm{d}g/\mathrm{d}x$, (6.142) gives

$$x^2 \left(\frac{g}{x}\right)' = \sum_{n=1}^{\infty} (n-1)c_n x^n, \tag{6.143}$$

and

$$(xg')^2 = \sum_{j=1}^{\infty} jc_j x^j \sum_{k=1}^{\infty} kc_k x^k = \sum_{n=1}^{\infty} x^n \sum_{\substack{j,k \geq 1 \\ j+k=n}} jk\, c_j\, c_k. \tag{6.144}$$

Therefore, (6.137) can be written as

$$2 \left(\frac{g}{x}\right)' = g'^2. \tag{6.145}$$

This ordinary differential equation replaces (6.129). The boundary condition is, from (6.141),

$$g'(0) = 1. \tag{6.146}$$

Rewriting (6.145) as

$$g'^2 - \frac{2g'}{x} + \frac{2g}{x^2} = 0 \tag{6.147}$$

gives

$$g' = \frac{1}{x}(1 \pm \sqrt{1 - 2g}). \tag{6.148}$$

Because of (6.146), the negative root is the correct one. Introducing $u = xg'$ yields

$$(xg')^2 - 2xg' + 2g = u^2 - 2u + 2g = 0, \tag{6.149}$$

and this gives

$$g(x) = u(x) - \frac{u^2(x)}{2}. \tag{6.150}$$

From $xg' = u$ follows

$$\frac{\mathrm{d}x}{x} = \frac{\mathrm{d}g}{u} = \frac{\mathrm{d}u - u\,\mathrm{d}u}{u}, \tag{6.151}$$

or

$$\frac{\mathrm{d}x}{x} = \frac{\mathrm{d}u}{u} - \mathrm{d}u. \tag{6.152}$$

Integration gives

$$\ln x = \ln u - u + b, \tag{6.153}$$

with some constant b, and exponentiation gives

$$x = u e^{-u+b}. \tag{6.154}$$

Thus $x = 0$ is equivalent to $u = 0$. Then

$$g'(x) = \frac{u}{x} = e^{u-b}, \tag{6.155}$$

and using the boundary condition (6.146),

$$1 = g'(0) = e^{-b}, \tag{6.156}$$

thus $b = 0$, and

$$x = u e^{-u}. \tag{6.157}$$

Equations (6.150) and (6.157) are the implicit solution of (6.147). An explicit solution for $u(x)$ is not required in the following. The derivative of $x(u)$ is

$$\frac{dx}{du} = (1 - u)e^{-u}, \tag{6.158}$$

which vanishes at $u = 1$. Therefore, the function $x(u)$ is not injective near $u = 1$, and the inverse function $u(x)$ is multivalued near $x(u = 1) = e^{-1}$, which means that $x = e^{-1}$ is a branch point of $u(x)$. It is precisely this branch point that causes the curvature singularity for the vortex sheet, see Eq. (6.186). We determine in the following the behavior of g near the branch point. According to the boundary condition (6.146), the two points

$$(x, u) = (0, 0) \qquad \text{and} \qquad (x, u) = (e^{-1}, 1) \tag{6.159}$$

lie in the physically relevant branch; consider thus the real intervals $0 \leq x \leq e^{-1}$ and $0 \leq u \leq 1$. The function g is analytic in the complex x-plane except at the branch point $x = e^{-1}$. Introduce new variables y and v by

$$x = \frac{1 - y}{e},$$
$$u = 1 - v, \tag{6.160}$$

with $0 \leq y \leq 1$ and $0 \leq v \leq 1$. Inserting (6.160) in $x = u e^{-u}$ and expanding around $v \ll 1$ gives

$$1 - y = (1 - v)\left(1 + v + \frac{v^2}{2} + \frac{v^3}{6} + \frac{v^4}{24} + \cdots\right) = 1 - \frac{v^2}{2} - \frac{v^3}{3} - \frac{v^4}{8} - \cdots$$

$$(6.161)$$

to order v^4, or

$$y = \frac{v^2}{2} + \frac{v^3}{3} + \frac{v^4}{8} = \frac{v^2}{2}\left(1 + \frac{2v}{3} + \frac{v^2}{4}\right).$$

$$(6.162)$$

With $(1 + \epsilon)^{3/2} \approx 1 + \frac{3}{2}\epsilon$ follows, still to order v^4,

$$y^{3/2} = \frac{v^3}{2\sqrt{2}}(1 + v),$$

$$(6.163)$$

or

$$v^3 + v^4 = 2\sqrt{2}\, y^{3/2}.$$

$$(6.164)$$

Furthermore, again to order v^4, from (6.162),

$$v^4 = 4y^2.$$

$$(6.165)$$

Therefore,

$$g = u - \frac{u^2}{2} = 1 - v - \frac{(1 - v)^2}{2} = \frac{1}{2} - \frac{v^2}{2},$$

$$(6.166)$$

or

$$g(x) \overset{(5.162)}{=} \frac{1}{2} - y + \frac{v^3}{3} + \frac{v^4}{8}$$

$$\overset{(5.164)}{=} \frac{1}{2} - y + \frac{2\sqrt{2}}{3}\, y^{3/2} - \frac{v^4}{3} + \frac{v^4}{8}$$

$$\overset{(5.165)}{=} \frac{1}{2} - y + \frac{2\sqrt{2}}{3}\, y^{3/2} - \frac{5}{6}\, y^2 + O(y^{5/2}).$$

$$(6.167)$$

Here it is understood that $y = 1 - ex$ is inserted on the right side. The dominant singular behavior of $g(x)$ at $x = e^{-1}$ is therefore given by the term

$$\frac{2\sqrt{2}}{3}\,(1 - ex)^{3/2},$$

$$(6.168)$$

which corresponds to a branch point that connects the two Riemann sheets (because of the square root). The term $(1 - ex)^{3/2}$ will lead to a curvature singularity for the vortex sheet, since its second derivative with respect to x is $\sim(1 - ex)^{-1/2}$, which is singular at $1 - ex = 0$. The next term in (6.167) that contributes to this branch point is $(1 - ex)^{5/2}$, for which only the third derivative is singular. We therefore consider only the term in (6.168), and neglect weaker singularities. This means, near the branch point (singularity) $x = e^{-1}$, in place of the coefficients c_n for the function $g(x)$ from (6.142), new coefficients $\hat{c}_n$ defined by

$$\frac{2\sqrt{2}}{3} (1 - ex)^{3/2} = \sum_{n=0}^{\infty} \hat{c}_n x^n \tag{6.169}$$

are introduced. (Moore demonstrates that $\hat{c}_n \to c_n$ for $n \to \infty$.) The expression $(1 - ex)^{3/2}$ is expanded using the binomial theorem; an example term is

$$\frac{3 \cdot 1 \cdot 1 \cdot 3 \cdot 5 \cdot 7 \cdot 9 \cdot 11}{2 \cdot 4 \cdot 6 \cdot 8 \cdot 10 \cdot 12 \cdot 14 \cdot 16} (ex)^8. \tag{6.170}$$

Generally, the term of order n in the expansion of $(1 - ex)^{3/2}$ is

$$3 \, \frac{(2n - 5)!!}{(2n)!!} (ex)^n, \tag{6.171}$$

where $m!! = m(m - 2)(m - 4)\ldots 2$, and it is assumed that n is even. This is no restriction, since one is interested in $n \to \infty$ only. Equating powers of x gives

$$\hat{c}_n = 2\sqrt{2} \, e^n \, \frac{(2n - 5)!!}{(2n)!!} \qquad (n \to \infty). \tag{6.172}$$

The following formulas hold:

$$(2k)!! = 2^k \, k!,$$
$$(2k - 1)!! = \frac{(2k)!}{2^k \, k!}, \tag{6.173}$$

and Stirling's formula for large n is

$$n! \approx \sqrt{2\pi n} \left(\frac{n}{e}\right)^n, \qquad (n \to \infty), \tag{6.174}$$

which gives

$$\hat{c}_n = \frac{2}{\sqrt{\pi}} \, e^{n+2} \sqrt{\frac{2n - 4}{n(n - 2)}} \, \frac{(2n - 4)^{2n-4}}{2^n 2^{n-2} n^n (n - 2)^{n-2}}. \tag{6.175}$$

For large n,

$$(2n - 4)^{2n-4} \approx (2n)^{2n-4} \left(1 - \frac{4}{2n}\right)^{2n} \approx (2n)^{2n-4} \, e^{-4}, \tag{6.176}$$

and similarly $(n - 2)^{n-2} \approx n^{n-2}\, e^{-2}$, which finally yields

$$\hat{c}_n = \frac{1}{\sqrt{2\pi}}\; e^n\, n^{-5/2} \qquad\qquad (n \to \infty). \qquad\qquad (6.177)$$

Inserting this in (6.133) and neglecting orders t^{n-2} and lower yields

$$a_n(t) \approx \frac{1+i}{\sqrt{2\pi}}\; n^{-5/2}\, e^{nt/2}\, e^n\, 4^{-n}\, t^{n-1} \qquad\qquad (n \to \infty), \qquad\qquad (6.178)$$

where the approximation sign refers here and in the following to the fact that the $\hat{c}_n$ are used instead of the c_n, thus neglecting weaker singularities. Using $A_n = \epsilon^n a_n$ from (6.123) for large n yields

$$\boxed{A_n(t) \approx \frac{1+i}{\sqrt{2\pi}}\; \frac{1}{n^{5/2}\, t}\; \exp\!\big[n\big(1 + t/2 + \ln(\epsilon t/4)\big)\big]} \qquad\qquad (6.179)$$

The scaling $A_n \sim n^{-5/2}$, already found in Fig. 6.16, is a central result of Moore's analysis and leads directly to a curvature singularity for the vortex sheet, see the discussion below after Eq. (6.186). For $\epsilon \ll 1$, the argument of the exponential in (6.179) is negative for sufficiently small t because of the logarithm in (6.179). At these times, $A_n \sim n^{-5/2}\, e^{-C_t n}$, with $C_t = |1 + t/2 + \ln(\epsilon t/4)|$. This means, the A_n drop exponentially with increasing n. The critical time t_c until which this holds is given by

$$1 + t_c/2 + \ln(\epsilon t_c/4) = 0. \qquad\qquad (6.180)$$

For $\epsilon = 10^{-3}$, $t_c \approx 9.986$, which was used in Fig. 6.16. The $A_n(t)$ are the Fourier coefficients of $z(t, \Gamma)$, see (6.106). At $t = t_c$, the A_n decay no longer exponentially, but only with $A_n \sim n^{-5/2}$.

Moore (1979) applies now a theorem on the decay of the Fourier coefficients.

Theorem (Bernstein) *Let $f : [-\pi, \pi] \to \mathbb{C}$ be an integrable, periodic function. Let F_n be the Fourier coefficients of f. Then f is real analytic (cf. the definition above) if and only if the F_n decay exponentially with n.*

Thus Moore concludes (his pp. 113 and 117) that at t_c the sheet is no longer analytic and *a singularity occurs in $z(t, \Gamma)$ at a critical time t_c.*

With a singularity occurring, the assumptions made in the above analysis have to be tested:

  – Equation (6.113) assumes $|S/u| < 1$, i.e., that the A_n are sufficiently small.
  – Equation (6.123) assumes that high-order terms in ϵ can be neglected.
  – Equation (6.178) assumes that low-order terms t^{n-2} can be neglected.

Close to t_c, it is not clear whether these assumptions still hold. A partial discussion of this is given in Moore (1979), after his Eq. (5.3). Caflisch and Orellana (1986) show in a detailed convergence proof that Moore's analysis is indeed valid up to times close to t_c. They show that smooth solutions of the Birkhoff–Rott equation exist for

$$t < k |\ln \epsilon| \tag{6.181}$$

for small ϵ, where

$$\lim_{\epsilon \to 0} k = 1. \tag{6.182}$$

This is by (only) a factor of two shorter than the estimate from (6.180),

$$\lim_{\epsilon \to 0} t_c = 2 |\ln \epsilon|. \tag{6.183}$$

A central conclusion from Moore's analysis is then the following.

> The uniform smallness of the Fourier coefficients in the interval $[0, t_c]$ and the fact that even when $t = t_c$ they decay rapidly with n makes it apparent that – provided that $\epsilon^n a_n(t)$ is the dominant contribution to $A_n(t)$ – the condition $|S(u, \Gamma)/u| = O(\epsilon)$ used in the derivation is satisfied. The smallness of the Fourier coefficients means that at the instant the singularity forms the sheet is only slightly distorted. There is no sign of the rolling up of the sheet into concentrated vortices and the maximum inclination of [the] sheet to its unperturbed position is only $O(t_c^{-1})$ which is $O((\ln \epsilon^{-1})^{-1})$. (Moore 1979, p. 118)

At last, we can show that the branch point $x = e^{-1}$ of $u(x)$ (see (6.158)) corresponds to a curvature singularity in the shape of the vortex sheet. From (6.169) and (6.177) one has, for $x \to e^{-1}$,

$$\frac{2\sqrt{2}}{3} (1 - ex)^{3/2} = \frac{1}{\sqrt{2\pi}} \sum_{n=0}^{\infty} e^n \, n^{-5/2} x^n = \frac{1}{\sqrt{2\pi}} \sum_{n=0}^{\infty} \frac{(ex)^n}{n^{5/2}}, \tag{6.184}$$

or, for $y \to 1$,

$$\frac{2\sqrt{2}}{3} (1 - y)^{3/2} = \sum_{n=0}^{\infty} \frac{1}{\sqrt{2\pi}} \frac{y^n}{n^{5/2}}. \tag{6.185}$$

The time t_c of singularity formation is given by $1 + t_c/2 + \ln(\epsilon t_c/4) = 0$. At this time,

$$z(t, \Gamma) \overset{(5.106)}{=} \Gamma + \sum_{n=-\infty}^{\infty} A_n(t) e^{in\Gamma}$$

$$\overset{(A_0=0)}{=} \Gamma + \sum_{n=1}^{\infty} A_n(t) e^{in\Gamma} + \sum_{n=-1}^{-\infty} A_n(t) e^{in\Gamma}$$

$$= \Gamma + \sum_{n=1}^{\infty} A_n(t) e^{in\Gamma} + \sum_{n=1}^{\infty} A_{-n}(t) e^{-in\Gamma}$$

Fig. 6.17 Moore's curvature singularity (at the position marked with an arrow)

$$\overset{(A_{-n} = -A_n)}{=} \quad \Gamma + \sum_{n=1}^{\infty} A_n(t) e^{in\Gamma} - \sum_{n=1}^{\infty} A_n(t) e^{-in\Gamma}$$

$$\overset{(5.179)}{\approx} \quad \Gamma + \frac{1+i}{t\sqrt{2\pi}} \sum_{n=1}^{\infty} \left[\frac{e^{in\Gamma}}{n^{5/2}} - \frac{e^{-in\Gamma}}{n^{5/2}} \right]$$

$$\overset{(y_{\pm} = e^{\pm i\Gamma})}{=} \quad \Gamma + \frac{1+i}{t\sqrt{2\pi}} \sum_{n=1}^{\infty} \left[\frac{y_+^n}{n^{5/2}} - \frac{y_-^n}{n^{5/2}} \right]$$

$$\overset{(y_{\pm} \to 1)}{=} \quad \Gamma + \frac{1+i}{t} \frac{2\sqrt{2}}{3} \left[(1 - y_+)^{3/2} - (1 - y_-)^{3/2} \right]$$

$$= \quad \Gamma + \frac{1+i}{t} \frac{2\sqrt{2}}{3} \left[(1 - e^{i\Gamma})^{3/2} - (1 - e^{-i\Gamma})^{3/2} \right], \quad (6.186)$$

which is Eq. (6.5) of Moore (with a minor typo there) at the critical time; a singularity occurs at $\Gamma = 0$. Equation (6.186) is only an approximate, asymptotic formula near the singularity neglecting weaker singularities; therefore, Moore makes this formula formally exact by adding an unknown function $\psi(t, \Gamma)$ that is 'less singular.'

Figure 6.17 shows the vortex sheet at the critical time as obtained from (6.186), for one perturbation period and setting $2\sqrt{2}/(3\,t) = 0.1$; see also Eqs. (1.4) and (1.5) in Caflisch and Orellana (1989). The original sinusoidal perturbation of the sheet is still visible. The singularity occurs in the flat, central part of the figure, at $z = \Gamma = 0$. The derivative $|\mathrm{d}z/\mathrm{d}\Gamma|^{-1}$—which is proportional to the vorticity ω, see (6.92)—is finite and continuous at $z = \Gamma = 0$; however, it has a *cusp* due to the singularity, see Fig. 1 in Caflisch and Orellana (1989). For $\Gamma \approx 0$, Eq. (6.8) becomes $z \approx \Gamma + c_1 \Gamma^{3/2}$ with complex constant c_1, and thus the second derivative of z for $\Gamma \approx 0$ is $\mathrm{d}^2 z/\mathrm{d}\Gamma^2 \approx c_2 \Gamma^{-1/2}$, which is indeed singular for $z = \Gamma = 0$.

Moore concludes that because of the curvature singularity, a vortex sheet is an insufficient model for a vortex *layer* of small but finite width. Still, many papers following Moore's original work address vortex sheets with singularities as a mathematical model that may provide insights into the Kelvin–Helmholtz instability of shear flows of inviscid fluids.

6.9 Subsequent Work on Moore's Singularity

This section summarizes some of the subsequent work on Moore's vortex sheet singularity, focussing on three topics, (i) well-posedness, (ii) singularity formation, and (iii) vortex layers.

(i) *Well-Posedness*

> The singularity is caused by the nonlinear excitation of short waves which amplify rapidly. (Moore 1979, p. 119)

This amplification of small-scale perturbations is an example of an *ill-posed* problem, defined as follows. Let $f(t, x)$ be a time-dependent scalar field. Solutions of well-posed problems depend *continuously* on initial data $f(0, x)$, i.e., for all t and ϵ there is some $\delta(t, \epsilon)$ such that for all x and x',

$$|f(0, x) - f(0, x')| < \delta \quad \rightarrow \quad |f(t, x) - f(t, x')| < \epsilon. \tag{6.187}$$

An ill-posed problem is one that is not well-posed. For these definitions, see Hadamard (1902); an extensive review on ill-posed problems can be found in Kabanikhin (2008). The question thus arises whether the problem of a vortex sheet is well-posed or ill-posed.

> The linear [vortex sheet] problem [...] requires analytic initial data to be well posed and will generally be so only for a finite time. Birkhoff (1962) conjectures that the nonlinear problem with analytic initial data is well posed at least for a finite time. Richtmyer and Morton (1967) make a similar conjecture for piecewise analytic data. (Sulem et al. 1981, p. 485)

Sulem et al. (1981), using methods of functional analysis, prove that the vortex sheet problem with *analytic* initial data is well-posed. Caflisch and Orellana (1986) in an elaborate analysis find that vortex sheets are ill-posed in the *Sobolev* norms, and infinitesimally small perturbations can acquire arbitrarily large norm in arbitrarily short time. Yet, Caflisch and Orellana also demonstrate well-posedness of the vortex sheet problem in an analytic norm. Caflisch and Orellana (1989) give arguments why the latter should indeed apply to the considered idealization of vanishing viscosity and sheet thickness.

(ii) *Singularity Formation*

In the same paper, Caflisch and Orellana (1989) generalize the *real* variable Γ (circulation) to a *complex* variable, and study Moore's singularity in the complex Γ-plane, especially its motion in time toward the real Γ-axis where it becomes a physical singularity. Initially, the singularity moves away from the vortex sheet, in the purely imaginary Γ-direction. This kind of behavior is also found in other examples, see Sect. 12.8.

In a next major step, Cowley et al. (1999) derive the function ψ that determines the shape of the vortex sheet at singularity formation in the neighborhood of the singularity. This function was left unspecified in Moore (1979) and was put to zero in Fig. 6.17. Cowley et al. (1999) show that for a wide class of initial conditions, a singularity will form at which $z(\Gamma)$ shows the power-law behavior of (6.8) with power 3/2, implying a singularity in the sheet curvature. Finally, Page and Cowley (2018) give an extensive analysis of the complex Γ-plane of the vortex sheet problem, identifying branch points and branch cuts. These authors prove that for a wide range of initial conditions, curvature singularities with power 3/2—and only these—will occur on the sheet. However, even almost 40 years after Moore's work from 1979, a fundamental question is still left open:

> An important question remaining is how the vortex sheet evolves for times after the singularity has formed on the real axis. (Page and Cowley 2018, p. 202)

(iii) *Vortex Layers*

As already suggested by Moore (1979), the answer to this question may lie in recognizing that a vortex sheet (a curve) is a pathological limit of a vortex layer (a strip of finite width). If for no other reason, a vortex sheet, i.e., a shear flow with a finite jump in the tangential speed, should always broaden to some extent due to viscosity smoothing out the velocity jump.

There is a number of numerical studies that follow the temporal evolution of vortex layers. We address here a paper by Baker and Shelley (1990), which gives particularly clear and impressive results on what flow features develop in a finite-width vortex layer instead of the curvature singularity for a vortex sheet.

Baker and Shelley (1990) model the evolution of a vortex layer by following the motion of its two boundary curves: 'the motion of vortex layers depends only upon information on the boundaries themselves' (Shelley and Baker 1989, p. 39). Along each of these two curves, an initial vorticity distribution is specified, and there are two complex equations of motion for the upper and lower boundary curve, where in each equation the velocity field induced by vorticity along the two curves is summed up, similarly as in the Birkhoff–Rott equation. The boundary curves are discretized to a finite number of point vortices, and the resulting system of ordinary differential equations is solved numerically by a fourth-order difference scheme.

The initial perturbation is one period of $\sin \Gamma$ for the upper boundary curve, and one period of $- \sin \Gamma$ for the lower boundary curve; there is thus a sinusoidal perturbation both in the shape and in the thickness of the vortex layer. The results of the calculations are displayed in Fig. 6.18. Here the initial conditions for four vortex layers of different thicknesses H are shown in the four panels of the uppermost row, and their subsequent evolution from top to bottom in the four columns. The results are clear and definite:

> First, the vorticity advects to the centre [maximum thickness of the perturbed layer], causing a further thickening there due to incompressibility. Second, the vorticity in the centre quickly reforms into a roughly elliptical core with trailing arms, which subsequently wrap around the core as it revolves. (Baker and Shelley 1990, p. 175)

The vortex layer thus bulges outward instead of forming Moore's singularity. Vorticity is accumulated in roughly elliptical, rotating cores with trailing arms that wrap around these cores (the arms attach to the rotating cores in points with high curvature). The trailing arms correspond to the roll-up process expected for vortex sheets, and the cores bear resemblance to the clouds of concentrated vorticity shown in Fig. 6.12. Furthermore, these results are similar to earlier results found by Zabusky et al. (1979), Kida (1981), Pozrikidis and Higdon (1985), and Krasny (1986).

Fig. 6.18 Formation of an elliptical core of accumulated vorticity with trailing arms in the evolution of a perturbed vortex layer. Reproduced from Baker and Shelley (1990)

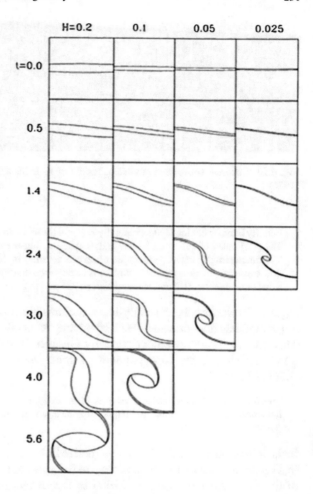

Baker and Shelley (1990) state that 'the singularity in the vortex sheet is not the limit of any type of singularity in the bounding interfaces of the vortex layer' (p. 163). The relation between vortex sheets and vortex layers remains thus rather unclear.

6.10   Why Do Large Eddies Occur in Fast Flows?

In fully developed turbulent flows with their intricate fine-grained structure, quite universally one also finds large-scale coherent structures, often in the form of large eddies. For shear flows subject to the Kelvin–Helmholtz instability, the following situation prevails:

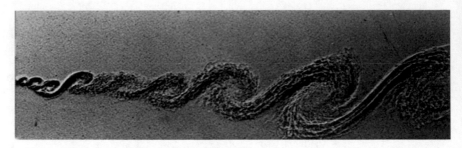

Fig. 6.19 Coherent structures in turbulent region of K–H instability. Reproduced from Konrad (1976)

> Initially laminar mixing layers eventually become turbulent in a [. . .] small-scale transition. The [. . .] onset of small-scale eddies takes place within the cores of the large structures. [. . .] interacting spanwise vortices are observed not only in the early laminar stages of the flow evolution but also farther downstream in the turbulent region, where they coexist with a fine-scale motion. (Ho and Huerre 1984, p. 385 and p. 394)

Figure 6.19 shows coherent, eddy-like structures in a turbulent flow. Why is the flow not completely fine-grained instead, consisting of small-scale, chaotic motions only? This question is addressed in a paper by Onsager (1949), showing that in turbulent flows at high energies concentrations of small vortices resulting in large vortices should occur.

> The formation of large, isolated vortices is an extremely common, yet spectacular phenomenon in unsteady flow. Its ubiquity suggests an explanation on statistical grounds. (Onsager 1949, p. 279)

In the following model, the fluid is assumed to be inviscid, the flow is two-dimensional in a square domain of area $A = \Delta x \, \Delta y$ in the xy-plane, with a large number, $n \gg 1$, of line vortices (point vortices distributed throughout the 2-D domain; the vortices lie parallel to each other, their axis is in the z-direction). Since the number of vortices is large, methods of statistical mechanics can be applied.

Each vortex is allowed to spin either clockwise or counterclockwise, and there is an equal number of vortices with either spin. The equations of motion can be written in Hamiltonian form, with an interaction Hamiltonian H from (6.49),

$$H = -\sum_{i=1}^{n} \sum_{j<i} k_i \, k_j \, \ln \sqrt{(x_i - x_j)^2 + (y_i - y_j)^2}. \tag{6.188}$$

Note that energy terms from interactions with boundaries of the finite area A are not present in (6.188). They could be included by applying image forces, as in electrostatics. For simplicity, each of the k_i is assumed to be either $+1$ or -1. For Hamiltonian systems, methods of statistical mechanics do indeed apply. To this end, consider the phase space with coordinates x_i, y_i from Sect. 6.5. Again, the unusual property of this phase space is that the canonical variables q and p are the vortex locations x and y in the plane, i.e., the phase space is a configuration space. The volume element is

$$d\Omega = dx_1 \, dy_1 \ldots dx_n \, dy_n. \tag{6.189}$$

Since each vortex can explore the full area A in the xy-plane, the total, *finite* volume available for the vortices in phase space is

$$\Omega = A^n. \tag{6.190}$$

The energy E of the system (the conserved value of H) lies in the interval from $-\infty$ to ∞. The case $E \to -\infty$ corresponds to two merging vortices of opposite sign, the case $E \to +\infty$ to two merging vortices of equal sign, see Sect. 6.5. Let $\Omega(E)$ be the phase space volume available to the system at energies $\leq E$,

$$\Omega(E) = \int\limits_{-\infty}^{E} dE \, \frac{d\Omega}{dE}, \tag{6.191}$$

with (see Fig. 6.20)

$$\Omega(-\infty) = 0,$$
$$\Omega(+\infty) = A^n. \tag{6.192}$$

Since $\Omega(E)$ is a monotonically increasing function, $\Omega' = d\Omega/dE > 0$ for all E. Since Ω has finite asymptotic values for $E \to -\infty$ and $E \to +\infty$, furthermore

$$\lim_{E \to -\infty} \Omega'(E) = 0,$$
$$\lim_{E \to +\infty} \Omega'(E) = 0. \tag{6.193}$$

Fig. 6.20 Phase space volume Ω and its first two derivatives Ω' and Ω'' with respect to the energy E of the vortex system

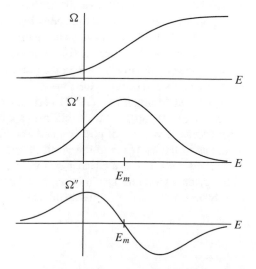

This implies that $\Omega'(E)$ has a maximum at some finite E_m, where

$$\Omega''(E_m) = 0. \tag{6.194}$$

A surface *entropy* and *temperature* can now be introduced using well-known formulas from statistical mechanics,

$$S(E) = k_B \ln \Omega'(E),$$

$$T(E) = \left(\frac{dS}{dE}\right)^{-1} = \frac{1}{k_B} \frac{\Omega'(E)}{\Omega''(E)}, \tag{6.195}$$

where k_B is the Boltzmann constant. If $\Omega'(E_m)$ is the only extremum, then $\Omega'' > 0$ for $E < E_m$, and thus

$$\begin{aligned} T > 0 & \quad \text{if } E < E_m, \\ T < 0 & \quad \text{if } E > E_m. \end{aligned} \tag{6.196}$$

This leads to the following conclusions on the behavior of the system of vortices:

> In the case $T > 0$, vortices of opposite sign will tend to approach each other. However, if $T < 0$, then vortices of the same sign will tend to cluster – preferably the strongest ones –, so as to use up excess energy at the least possible cost in terms of degrees of freedom. [...] the large compound vortices formed in this manner will remain as the only conspicuous features of the motion; because the weaker vortices, free to roam practically at random, will yield rather erratic and disorganised contributions to the flow. (Onsager 1949, p. 281, with minor modifications)

Indeed, for $E > E_m$, entropy S in (6.195) *decreases* with increasing E, see Fig. 6.20. Thus, above E_m, the 'order' in the system increases with E. This is caused by an accumulation of vortices of equal-spin direction. Large energy can be distributed in two different ways among vortices: either (a) a large number of equal-spin vortices accumulate loosely or (b) a small number of equal-spin vortices come very close. Each possible vortex configuration constitutes a so-called microstate. As usual, equal a priori probability for microstates is postulated. The number of microstates for case (a) is much higher than for case (b). Thus, at large energies, extended regions of one certain vortex spin will alternate in the plane. Or, expressed differently, large eddies ('order states') will occur in the flow.

Joyce and Montgomery (1972) performed corresponding simulations for $n = 525$ vortices. At large E, they find two spatially separated regions: one contains large excess of vortices of positive, and one of negative spin. Figure 6.21 shows the locations in the xy-plane of the vortices with negative spin (left box) and positive spin (right box). The two circular sectors in the two boxes contain half the vortices of the given sign. Thus, spatially separated concentrations of equal sign vortices do clearly occur. For further details on this statistical model, see Montgomery (1972) and

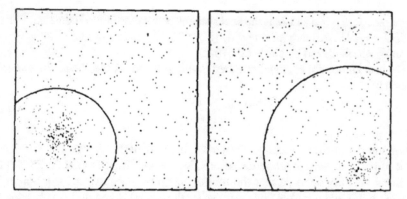

Fig. 6.21 Distribution of point vortices with negative (left box) and positive spin (right box). The circular sections contain half of the vortices of each sign. The vortices are shown separately for easy recognition, but reside in the same simulation box. Reproduced from Joyce and Montgomery (1972)

Saffman and Baker (1979). Quite clearly, a statistical model cannot (easily) explain highly organized, coherent structures as shown in Fig. 6.19. For these experiments on coherent structures in turbulent flow, see especially Brown and Roshko (1974).

6.11 Atmospheric Instability

Linear stability theory by definition leads to linear equations, and since powerful techniques exist (e.g., the Fourier and Laplace transformations) to solve such systems of equations, linear stability analysis allows for quite realistic descriptions of a complex physical situation, including the stabilizing or destabilizing tendencies of different contributing factors. In this section, shear motion and the possibility of the Kelvin–Helmholtz instability in a stratified atmosphere is briefly discussed.

> The knowledge that the tendency of gravity to keep the denser fluid below the lighter might strongly inhibit turbulence in such flows aroused interest in their stability many years ago. (Drazin 1958, p. 214)

The assumptions are that the fluid is inviscid and incompressible, the flow is in two dimensions (xz-plane), and gravity g points in the negative z-direction. The flow is stacked in horizontal layers, with an exponential decrease in density in the z-direction (barometric law, see Sect. 11.2; due to the density decreasing with height, the Rayleigh–Taylor instability does not occur) and a purely horizontal wind in the x-direction with wind speed depending on z. Specifically, Drazin (1958) assumes that

$$\rho(z) = \rho_0 \, e^{-z/H},$$
$$U = U_0 \tanh{(z/d)}, \tag{6.197}$$

where $-\infty < z < \infty$ and H is the scale height, d is an (arbitrary) constant. For $z \to \pm\infty$, $U \to \pm U_0$. Already at a few d, $|U| \approx U_0$. A similar model with $\rho = \text{const}$, however, was analyzed by Curle (1956).

The crucial nondimensional number in this problem is the *Richardson* number, defined by

$$J = \frac{gd^2}{HU_0^2}. \tag{6.198}$$

This number measures the ratio of buoyancy forces to inertial forces. For a wavenumber k of periodic *horizontal* perturbations, let $l = kd$. Then Drazin (1958) finds in his linear stability analysis that the line of marginal stability, i.e., the border line between stable and unstable flow, is given by the remarkably simple expression (see Fig. 6.22)

$$J = l^2 - l^4, \tag{6.199}$$

see his Eq. (25), and also Eq. (129) on p. 497 in Chandrasekhar (1961).

From Fig. 6.22, the flow is stable for $J > 1/4$ at all perturbation wavenumbers. This unconditional stability for $J > 1/4$ was already found by Goldstein (1931).

The second example we consider is a barometric atmosphere with a *linear* wind velocity law,

$$\rho(z) = \rho_0 \, e^{-z/H},$$
$$U = U_0 \, z/d, \tag{6.200}$$

with $z > 0$ now. A linear stability analysis leads here to confluent hypergeometric functions. At early times, exponential growth occurs for $J < 1/4$, as was the case in the first example. Case (1960) and Dyson (1960), however, find as a rather surprising result that for any H, d, and U_0, exponential growth does *not* continue to $t \to \infty$. Instead, for $t \to \infty$ perturbation amplitudes decay weakly as $1/t^\alpha$, where α lies in the range $0 < \alpha < \frac{1}{2}$.

It would, therefore, be strictly correct to say that for the distributions (6.200) extending over a semi-infinite region, the flow is stable for all Richardson numbers. (Chandrasekhar 1961, p. 494)

Fig. 6.22 Curve of marginal stability for Drazin's atmosphere

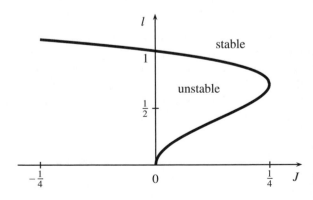

6.12   The Rayleigh Inflexion Theorem

In the foregoing sections, it was shown that a vortex sheet in an inviscid fluid, with a discrete jump in the tangential fluid speed, is subject to the Kelvin–Helmholtz instability; furthermore, atmospheric winds with a *continuously differentiable* variation of the horizontal fluid speed in the z-direction were found to be either stable or unstable. Actually, there is an important general theorem on the linear stability of shear flows with smooth velocity laws, the Rayleigh inflexion theorem, which is briefly discussed here, following Drazin and Reid (1981), p. 130.

The fluid is assumed to be inviscid and have constant density, and the velocity field is two-dimensional in the horizontal xy-plane. The flow resides in a straight canal with flow direction x, and y is the direction across the canal. Let $U(y)$ be the unperturbed flow speed, or velocity profile, in the x-direction, and the unperturbed flow speed in the y-direction is assumed to be zero.

Let u and v be velocity perturbations, derivable from a streamfunction ψ,

$$u = \partial\psi/\partial y, \qquad v = -\partial\psi/\partial x, \tag{6.201}$$

which ensures $u_x + v_y = 0$. The Fourier components $\chi(y)$ are introduced by assuming

$$\psi(t, x, y) = \chi(y)e^{ik(x-ct)}, \tag{6.202}$$

where c is some wave speed relevant to the flow and may be complex. The two Euler equations are then

$$(U + u)_t + (U + u)(U + u)_x + v(U + u)_y = -\frac{p_x}{\rho},$$

$$v_t + (U + u)v_x + vv_y = -\frac{p_y}{\rho}. \tag{6.203}$$

Taking into account that $U = U(y)$ alone, and neglecting all products of u and/or v and their derivatives, this gives, with $U' = dU/dy$,

$$u_t + Uu_x + vU' = -\frac{p_x}{\rho},$$

$$v_t + Uv_x = -\frac{p_y}{\rho}. \tag{6.204}$$

The first and second equations are differentiated with respect to y and x, respectively, which yields

$$u_{ty} + U'u_x + Uu_{xy} + v_yU' + vU'' = -\frac{p_{xy}}{\rho},$$

$$v_{tx} + Uv_{xx} = -\frac{p_{yx}}{\rho}. \tag{6.205}$$

In the first line, $U'(u_x + v_y) = 0$ since $\nabla \cdot \vec{u} = 0$. Subtracting the second from the first equation, and performing the elementary differentiations in (6.201) and (6.202) gives, with $\chi' = d\chi/dy$ and after division by $e^{ik(x-ct)}$,

$$-ikc\chi'' + ikU\chi'' - ikU''\chi - (ik)^3 c\chi + (ik)^3 U\chi = 0, \tag{6.206}$$

or

$$\chi'' - k^2\chi - \frac{U''}{U - c}\chi = 0. \tag{6.207}$$

This is *Rayleigh's stability equation*. Let $k \neq 0$, else the unperturbed speed U can be redefined accordingly. Multiplying (6.207) with the complex conjugate $\bar{\chi}$ and integrating the emerging equation from wall to wall gives

$$\int dy \left(|\chi'|^2 + k^2|\chi|^2\right) + \int dy \, \frac{U''}{U - c} \, |\chi|^2 = 0, \tag{6.208}$$

where in the first term an integration by parts was performed, using the boundary condition $v = 0$ or $\chi = 0$ at the canal walls. Assume next that the wavenumber k is real-valued, which makes the first integral in (6.208) also real-valued. Therefore, the imaginary part of the second integral must vanish,

$$\text{Im}(c) \int dy \, \frac{U''}{|U - c|^2} \, |\chi|^2 = 0. \tag{6.209}$$

Assuming then that the flow is unstable, Im $(c) > 0$, Eq. (6.209) can only hold if the second derivative U'' changes sign between the canal walls. Therefore, a *necessary* condition for fluid instability is that the unperturbed velocity law $U(y)$ has an inflexion point, $U'' = 0$. This is the Rayleigh inflexion theorem (Rayleigh 1879).

The inflexion theorem is indeed only necessary, not sufficient, and there are examples of velocity laws $U(y)$ with an inflexion point that are stable.

6.13 Kinematics of Vortex Rings

All line vortices considered so far were *straight* vortices, i.e., had a straight central axis along which vorticity approaches infinity. As such they were liable to a two-dimensional analysis in the normal plane of the axis. A *vortex ring*, by contrast, is a vortex where the central axis is bent to a circle, as is the case in a smoke ring. A vortex ring has inherently toroidal structure, and thus its mathematical description is more demanding. To make some progress, the diameter of the region in which vorticity is substantial (the diameter of a vortex tube) is assumed to be very small compared to the radius of the vortex ring.

A vortex ring in the xy-plane cannot be at rest but moves in the z-direction. The reason is that the velocity field of any segment of the ring *induces* motion of other segments in the direction of the symmetry axis.

Let the origin of the xy-plane be the center of the vortex ring. The vorticity vector is $\vec{\omega} = \omega\,\hat{\varphi}$ in cylindrical coordinates. Let $d^3r = \sigma R d\varphi$, where σ is the cross section of the ring and R its radius. The vorticity ω is assumed to be constant along the ring and over its cross section. Then the circulation Γ around the cross section of the ring is

$$\Gamma = \oint_{\partial\sigma} d\vec{l}\cdot\vec{u} = \int_{\sigma} d\vec{a}\cdot\vec{\omega} = \sigma\omega. \tag{6.210}$$

The Green's function $\vec{A}$ of the vector Poisson equation is given by (see (2.200))

$$\vec{A}(\vec{r}) = \frac{1}{4\pi}\int d^3r'\,\frac{\vec{\omega}(\vec{r}')}{|\vec{r}-\vec{r}'|} = \frac{\Gamma R}{4\pi}\int_0^{2\pi} d\varphi\,\frac{\hat{\varphi}}{|\vec{r}-\vec{r}'|}. \tag{6.211}$$

Here $|\vec{r}-\vec{r}'|$ is a function of r, r', and φ. This leads to some intricate calculations including elliptic integrals. The details can be found in Lamb (1932), p. 236 and Kotschin et al. (1954), p. 176. From $\vec{u} = \nabla\times\vec{A}$, the self-induced speed U along the z-axis is then found to be

$$U = \frac{\Gamma}{4\pi R}\left(\ln\frac{8R}{a} - \frac{1}{4}\right), \tag{6.212}$$

with ring radius R, cross section $\sigma = \pi a^2$, and circulation Γ. The pathological behavior $U \to \infty$ as $a \to 0$ is of no concern: the logarithm varies only weakly, and thus

$$U \approx \frac{\tilde{\Gamma}}{R} \tag{6.213}$$

for some constant $\tilde{\Gamma}$. From (6.213), two interesting results follow:

First, *a vortex ring cannot reach a wall normal to its propagation direction.* This result by Helmholtz (1858) is proved as follows. Consider two circular vortex rings with common symmetry axis. The vortices approach each other; each induces a velocity at the other. Clearly, the induced velocity increases the diameter of each ring. By (6.213) then, both vortex rings slow down. They meet at zero speed and infinite diameter in a midplane A. The governing equation of the problem is the Poisson equation, and its solution is uniquely determined by the boundary conditions.

Let $\vec{u}_1$ and $\vec{u}_2$ be the velocity fields of the vortices, and $\hat{n}$ the unit normal to A. For $\vec{x}\in A$, from symmetry, $\hat{n}\cdot(\vec{u}_1(t,\vec{x}) + \vec{u}_2(t,\vec{x})) = 0$ at all t. Now consider a single vortex ring approaching a fixed wall A, with direction of motion normal to the wall. The wall boundary condition is again $\hat{n}\cdot\vec{u}(t,\vec{x}) = 0$ for all $\vec{x}\in A$. The solution in one half-space limited by A is thus identical in the two cases. Therefore, a single vortex in front of a wall also slows down to normal speed zero and increases to infinite diameter.

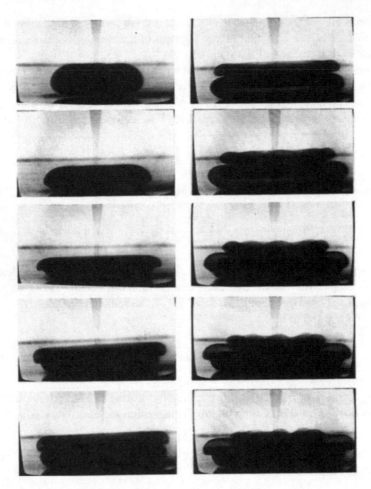

Fig. 6.23 Vortex ring impacting a plane surface, with the formation of a secondary vortex ring that develops a wavy instability at later times. Reproduced from Walker et al. (1987), Fig. 12

In an experiment, the impact of a vortex ring on a flat surface shows intriguing further details: a boundary layer can detach from the flat surface (flow separation), and form a secondary vortex ring that interacts with the primary ring. Most notably, the primary ring rebounds, and its increase in diameter is slowed down and stopped. At later times, a three-dimensional wavy instability grows on the secondary ring. For a detailed description of the processes involved, see pp. 114–117 in Walker et al. (1987). Figure 6.23 shows one of their observational sequences.

A YouTube video of two vortex rings propagating in opposite directions and colliding, thereby decreasing their speeds and increasing their diameters, can be found at https://www.youtube.com/watch?v=8iZeCo1M4Pk.

Second, *vortex rings propagating in the same direction mutually overtake each other.* This theorem is due to Gröbli (1877) and Hicks (1922). The two rings are assumed to lie on the same axis. Assume that the motion is from left to right. The left vortex induces a speed at the right vortex. The diameter of the right vortex grows, and it slows down. Similarly, the right vortex induces a speed at the left vortex. The left vortex shrinks and thus speeds up. The small and fast left vortex ring slips through the large and slow right ring. The argument repeats with exchanged roles of the rings, so that the vortices overtake each other in alternating fashion.

6.14 Curvature and Torsion

In the next section, *helical* line vortices will be considered. The expressions that are needed for the curvature and torsion of a curve are derived here. Let C be a curve in $\mathbb{R}^3$, described by the vector $\vec{r}(s)$ with arclength s. Three orthonormal unit vectors $\hat{t}$, $\hat{n}$, and $\hat{b}$, the *Frenet frame*, are introduced along C, one of them tangent ($\hat{t}$) and two normal ($\hat{n}$, $\hat{b}$) to C. Let

$$\hat{t} = d\vec{r}/ds \qquad (6.214)$$

be the unit tangent to C (with s the arclength, $|d\vec{r}| = ds$, thus $\hat{t}$ is indeed a unit vector). Define

$$\vec{n} = d\hat{t}/ds. \qquad (6.215)$$

From $\hat{t} \cdot \vec{n} = \frac{1}{2}d(\hat{t} \cdot \hat{t})/ds = 0$ follows $\vec{n} \perp \hat{t}$. From Fig. 6.24 follows, for $\alpha \to 0$,

$$\frac{ds}{R} = \frac{dt}{1}, \qquad (6.216)$$

where R is the radius of curvature of the curve C. Therefore,

$$\frac{dt}{ds} = |\vec{n}| = \frac{1}{R}. \qquad (6.217)$$

Fig. 6.24 Change of unit tangent vector $\hat{t}$ along a curve C

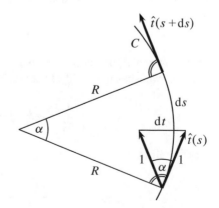

Introducing the unit *main normal* $\hat{n}$ to the curve, (6.215) and (6.216) can be written as

$$\frac{d\hat{t}}{ds} = \vec{n} = \frac{\hat{n}}{R} \tag{6.218}$$

The two unit vectors $\hat{t}$ and $\hat{n}$ define a plane, the so-called *osculating plane*. The unit *binormal* vector $\hat{b}$ is then defined by

$$\hat{b} = \hat{t} \times \hat{n}, \tag{6.219}$$

which is a unit vector since $\hat{t} \cdot \hat{n} = 0$. The change of the osculating plane along the curve is given by $\vec{q} = d\hat{b}/ds$. Let

$$|\vec{q}| = \frac{1}{Q}, \tag{6.220}$$

where Q is the *torsion radius*. The vector $\vec{q}$ points in direction $-\hat{n}$, which is seen as follows. First, $\hat{b} \cdot \vec{q} = \frac{1}{2}d(\hat{b} \cdot \hat{b})/ds = 0$. Second, $\hat{t} \cdot \hat{b} = 0$, thus $d(\hat{t} \cdot \hat{b})/ds = 0$, or

$$\hat{t} \cdot \vec{q} = -\vec{n} \cdot \hat{b} = -\frac{\hat{n} \cdot \hat{b}}{R} = 0. \tag{6.221}$$

Thus $\vec{q}$ is normal to $\hat{b}$ and $\hat{t}$, giving

$$\frac{d\hat{b}}{ds} = \vec{q} = -\frac{\hat{n}}{Q} \tag{6.222}$$

The minus sign is convention, and fixes what is meant by a positive and negative helical line (or screw) in a right-handed coordinate system. Blaschke (1945), §10 remarks that wine tendrils grow with $Q > 0$ and hop tendrils with $Q < 0$. To determine Q, use $\hat{b} = \hat{t} \times \hat{n} = R\,\hat{t} \times \vec{n}$, which gives

$$\vec{q} = R\,\hat{t} \times \vec{m} + R'\,\hat{t} \times \vec{n}, \tag{6.223}$$

where $R' = dR/ds$, and $\vec{m} = d\vec{n}/ds = d^2\hat{t}/ds^2 = \hat{t}''$. A third term, $R\hat{t}' \times \vec{n} = R\vec{n} \times \vec{n}$, vanishes in (6.223). From (6.222) follows then, with $\hat{n} \cdot (\hat{t} \times \vec{n}) = 0$,

$$\frac{1}{Q} = -\hat{n} \cdot \vec{q} = -R^2\,\vec{n} \cdot (\hat{t} \times \vec{m}), \tag{6.224}$$

and therefore

$$\frac{1}{Q} = R^2(\hat{t} \times \hat{t}') \cdot \hat{t}'' \tag{6.225}$$

This is to be supplemented by

$$\frac{1}{R} = |\hat{t}'| \tag{6.226}$$

from (6.218). Note that $R \geq 0$, whereas Q can be positive or negative. Finally, $d\hat{n}/ds$ is already fully determined by R and Q. Namely, since $\hat{t}, \hat{n}, \hat{b}$ is a right-handed orthonormal system, besides (6.219) it also holds that (cyclic permutation)

$$\hat{t} = \hat{n} \times \hat{b}, \qquad\qquad \hat{n} = \hat{b} \times \hat{t}. \tag{6.227}$$

From the latter equation,

$$\begin{aligned}
\frac{d\hat{n}}{ds} &= \frac{d\hat{b}}{ds} \times \hat{t} + \hat{b} \times \frac{d\hat{t}}{ds} \\
&= -\frac{\hat{n} \times \hat{t}}{Q} + \frac{\hat{b} \times \hat{n}}{R} \\
&= -\frac{\hat{t}}{R} + \frac{\hat{b}}{Q}.
\end{aligned} \tag{6.228}$$

Finally, new variables $\kappa = 1/R$ and $\tau = 1/Q$ are introduced, the *curvature* and *torsion* of the curve C, respectively. Equations (6.218), (6.222), and (6.228) can be written, in (formal) matrix notation, as

$$\begin{pmatrix} \hat{t} \\ \hat{n} \\ \hat{b} \end{pmatrix}' = \begin{pmatrix} 0 & \kappa & 0 \\ -\kappa & 0 & \tau \\ 0 & -\tau & 0 \end{pmatrix} \cdot \begin{pmatrix} \hat{t} \\ \hat{n} \\ \hat{b} \end{pmatrix}. \tag{6.229}$$

These three equations, together with (6.226) for κ and (6.225) for τ are termed the *Frenet formulas*.

6.15 Helical Line Vortices

After studying line vortices with straight symmetry axis (on which vorticity $\omega \to \infty$) and vortex rings with circular symmetry axis, a natural quest is for properties of line vortices with a symmetry axis that is a curve in 3-D space. The simplest example of this appears to be a *helix*; these helical line vortices were first analyzed by Betchov (1965).

Equation (6.213) for the speed of a vortex ring is also used in this case. This is a fair approximation if the torsion of the vortex line is small. Furthermore, the speed at any point on the line vortex is mainly induced by the vorticity $\vec{\omega}$ *near* this point. And up to second differential order, the line vortex near any point indeed resembles a circle segment.

Let a line vortex lie completely in the plane A and have varying radius of curvature R there. A is the osculating plane, and the self-induced speed U is normal to A. Then U varies along the line vortex, since R does. Therefore, the vortex will not stay plane but acquires torsion. We are therefore led to find the kinematic relation between curvature and torsion.

Let $\vec{x}(t, s)$ be the position vector on the helix, with arclength s and time t. (There is no danger of confusing time t and the tangent $\hat{t}$ to the helix.) Let $\dot{\vec{x}} = \partial \vec{x}/\partial t$, $\vec{x}' = \partial \vec{x}/\partial s$, and $\hat{t}' = \partial \hat{t}/\partial s$. Furthermore, let

$$k = 1/R^2 = \kappa^2, \qquad\qquad \tau = 1/Q. \qquad\qquad (6.230)$$

With $\hat{t} = \vec{x}'$, (6.225) becomes

$$k\tau = (\vec{x}' \times \vec{x}'') \cdot \vec{x}'''. \qquad\qquad (6.231)$$

The vectors $\hat{t}$ and $\hat{n}$ lie in the plane A, and $\hat{b}$ is normal to A. Written as a vector Eq. (6.213) thus becomes, for $\tilde{\Gamma} = 1$,

$$U\hat{b} = \frac{1}{R}\, \hat{t} \times \hat{n} = \hat{t} \times \vec{n}, \qquad\qquad (6.232)$$

or, with $\dot{\vec{x}} = U\hat{b}$ and $\vec{n} = \hat{t}' = \vec{x}''$,

$$\boxed{\dot{\vec{x}} = \vec{x}' \times \vec{x}''} \qquad\qquad (6.233)$$

This is an approximation only, since (6.213) was used instead of (6.212). From (6.218) follows $R^{-2} = \hat{t}' \cdot \hat{t}'$; differentiating this with respect to time and using (6.230) gives

$$\begin{aligned}
\dot{k} &= \overline{\vec{x}'' \cdot \dot{\vec{x}}''} \\
&= 2\dot{\vec{x}}'' \cdot \vec{x}'' \\
&= 2(\vec{x}' \times \vec{x}'')'' \cdot \vec{x}'' \\
&= 2(\vec{x}' \times \vec{x}''') \cdot \vec{x}'' \\
&= -2(\vec{x}' \times \vec{x}'') \cdot \vec{x}'''' \\
&= -2[(\vec{x}' \times \vec{x}'') \cdot \vec{x}''']' \\
&= -2(k\tau)'. \qquad\qquad (6.234)
\end{aligned}$$

(The scalar triple product is zero if the same vector appears twice in it.) Equation (6.234) is the first of two coupled equations for k and τ. An equation for $\dot{\tau}$ is derived along similar lines, see Betchov (1965), and reads

$$2\dot{\tau} + 4\tau\tau' = k' + \frac{k'''}{k} + \frac{k'^3}{k^3} - \frac{2k'k''}{k^2}. \qquad\qquad (6.235)$$

The greater complexity of this equation is due to the time derivative of a scalar triple product. Equations (6.234) and (6.235) are two coupled kinematic equations for k and τ. The system has the trivial solution $\dot{k} = k' = \dot{\tau} = \tau' = 0$, or

$$k(t, s) = k_0, \qquad\qquad \tau(t, s) = \tau_0, \qquad\qquad (6.236)$$

where k and τ are spatially and temporally constants. This means that *a helical line vortex with constant curvature and torsion everywhere is stationary,* i.e., a helical line vortex is a steady physical object.

Next, small perturbations are applied to such a helical vortex with constant k_0 and τ_0,

$$k(t, s) = k_0 + k_1(t, s), \qquad\qquad \tau(t, s) = \tau_0 + \tau_1(t, s). \qquad (6.237)$$

Keeping only terms that are linear in k_1 and τ_1 in (6.234) and (6.235) gives

$$\dot{k}_1 + 2\tau_0 k_1' = -2k_0 \tau_1', \qquad\qquad (6.238)$$

$$\dot{\tau}_1 + 2\tau_0 \tau_1' = \frac{1}{2}\left[k_1' + \frac{k_1'''}{k_0} \right]. \qquad\qquad (6.239)$$

On the left sides of (6.238) and (6.239), $2\tau_0$ is an advection speed, similar to the fluid speed $\vec{u}$ in the Lagrange derivative. The perturbations move with apparent speed $2\tau_0$ relative to the moving and rotating helix (for the latter two motions, see Eq. (3.3) in Betchov). In order to eliminate these advection terms from (6.238) and (6.239), Betchov (1965) introduces a new arclength $s - 2\tau_0 t$. Differentiate then (6.238)—without the term $2\tau_0 k_1'$—with respect to t, and (6.239)—without $2\tau_0 \tau_1'$—with respect to s. Subtracting the resulting equations gives the wave equation

$$\ddot{k}_1 + k_1'''' + k_0 k_1'' = 0. \qquad\qquad (6.240)$$

Differentiate next (6.238) with respect to s, and (6.239) with respect to t, both equations taken again without the $2\tau_0$ terms. Adding the equations and using (6.238) gives the wave equation

$$\ddot{\tau}_1 + \tau_1'''' + k_0 \tau_1'' = 0. \qquad\qquad (6.241)$$

A harmonic wave is assumed in (6.240) and (6.241),

$$k_1, \tau_1 \sim e^{i(ks - \sigma t)}. \qquad\qquad (6.242)$$

Inserting this gives in either case the dispersion relation

$$\sigma^2 = k^2(k^2 - k_0). \qquad\qquad (6.243)$$

The perturbations (6.242) depend on arclength s, i.e., they vary along the helical line vortex. The angular frequency σ is real, and thus $e^{i(ks-\sigma t)}$ is a wave solution, if

$$k > \sqrt{k_0}. \tag{6.244}$$

Therefore, using (6.230), only perturbations with wavelengths

$$\lambda < 2\pi R_0, \tag{6.245}$$

i.e., smaller than the circumference of the helix, can propagate in a stable way along the helical vortex. Perturbations with longer wavelength are unstable. It is clear that this simple geometric condition (6.245) is a direct reflection of the purely kinematic analysis given here.

6.16   Knotted and Linked Vortex Rings

So far we analyzed straight vortices, vortex rings in a plane, and helical vortices. As a next level of intricacy, vortex rings with knots, and two or more linked, interwoven vortex rings are treated, see Fig. 6.25. We follow a paper by Moffatt (1969)—with an interesting corrigendum in Moffatt (2017)—and consider first two vortex rings C_1 and C_2. The fluid shall be inviscid and the vector field $\vec{u}$ solenoidal. All forces shall be derivable from potentials, and the flow is barotropic, $\rho = \rho(p)$. Then the circulation $\Gamma = \oint_C d\vec{l} \cdot \vec{u}$ is conserved (the Kelvin–Helmholtz circulation theorem), $d\Gamma/dt = 0$, where the curve C moves along with the flow.

All the vorticity of the flow shall be concentrated in discrete line vortices, characterized by the vortex strength

$$\kappa = d\vec{a} \cdot \vec{\omega} = \oint_C d\vec{l} \cdot \vec{u} \tag{6.246}$$

Fig. 6.25 Two vortex rings C_1 and C_2 with winding number $L_{12} = 2$

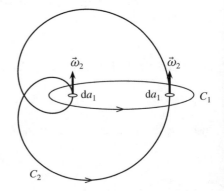

(with $\vec{\omega} = \nabla \times \vec{u}$, and κ is no longer the curvature from Sect. 6.14), where $d\vec{a} \cdot \vec{\omega}$ converges to the finite limit κ for $d\vec{a} \to 0$, see Sect. 2.11. The curve C encloses the cross section of the infinitesimal vortex tube, and κ is constant along each line vortex. The symbol Γ is reserved in the following for the circulation around a curve C for which any area A with $C = \partial A$ is *finite*.

Let vorticity $\vec{\omega}$ be equal to zero everywhere, except in two unknotted vortex rings C_1 and C_2 where it approaches infinity. In all examples considered so far, the integration curve C in the circulation Γ did loop around a line vortex. Moffatt takes the new step to identify C instead with one of the vortex rings itself, say C_1. Let A_1 be an arbitrary area that has C_1 as its boundary, $C_1 = \partial A_1$. Then

$$\Gamma_1 = \oint_{C_1} d\vec{l} \cdot \vec{u} = \int_{A_1} d\vec{a} \cdot \vec{\omega}. \qquad (6.247)$$

Again, Γ_1 is *not* the circulation of the vortex ring C_1, but the circulation along C_1 due to the vorticity $\vec{\omega}$ of the vortex ring C_2 linked with C_1. The last integral in (6.247) is a discrete sum of the constant but signed term $\kappa_2 = d\vec{a} \cdot \vec{\omega}$ (see Sect. 2.11; the sign results from the relative orientation of the area element $d\vec{a}$ of A_1 and the vorticity vector $\vec{\omega}$ along C_2). Each signed piercing of the area A_1 by the curve C_2 contributes $\pm\kappa_2$ to this sum. A *winding number* L_{12} is introduced that is zero if C_1 and C_2 are not linked, $L_{12} = \pm 1$ if C_1 and C_2 are singly linked, etc. Then

$$\Gamma_1 = L_{12}\,\kappa_2, \qquad \text{with} \qquad L_{12} \in \mathbb{Z}. \qquad (6.248)$$

Note that if C_2 or A_1 are deformed in such a way that C_2 pierces A_1 an extra of $2,4,6\ldots$ times (where C_2 is not allowed to cross through C_1), then L_{12} does *not* change, since the two members in each such pair of extra piercings have opposite sign of $\kappa_2 = d\vec{a} \cdot \vec{\omega}$. Thus the winding number L_{12} is a topological invariant. It can also be shown that $L_{12} = L_{21}$.

Before continuing, a potential problem has to be removed, the occurrence of knots in a single line vortex. A closed curve C is *unknotted* if a surface A exists with $C = \partial A$, and A does not intersect itself. For unknotted loops, Stokes' theorem applies, as in (6.247). If C is knotted, however, Stokes' theorem does not apply. The following lemma solves this problem.

Lemma *Any knotted loop can be decomposed into unknotted, linked loops.*

The proof idea (see Fig. 6.26) is to insert one or more pairs of infinitesimally close line vortices AB of strength κ and $-\kappa$ in the knot. This resolves the knot, at the cost of introducing two or more intersecting loops, which, however, pose no problem in the current analysis.

Next, (2.201) is used for the velocity field of a line vortex. The gradient term is of no importance, since for any closed loop C,

$$\oint_C d\vec{l} \cdot \nabla\phi = \oint_C d\phi = 0. \qquad (6.249)$$

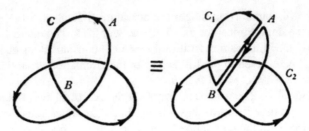

Fig. 6.26 Decomposition of a single, knotted vortex ring (left) into two linked but unknotted rings, by introducing two straight lines of opposite vorticity. Reproduced from Moffatt (1969), Fig. 5.25

In the first term in (2.201), use can be made of the fact that vortex lines are integral curves of $\vec{\omega}$, and therefore $d\vec{l}$ is parallel to $\vec{\omega}$ along a vortex line. Let V_2 be the vortex tube that becomes the closed line vortex C_2 in the limit of vanishing cross section. The velocity field $\vec{u}_2$ caused by the vortex ring C_2 (and only by C_2) is then given by

$$
\begin{aligned}
\vec{u}_2(\vec{r}) &= \frac{1}{4\pi} \int_{V_2} d^3 r' \; \vec{\omega}(\vec{r}') \times \frac{\vec{r} - \vec{r}'}{|\vec{r} - \vec{r}'|^3} \\
&= \frac{1}{4\pi} \oint_{C_2} (d\vec{l}' \cdot d\vec{a}') \; \vec{\omega}(\vec{r}') \times \frac{\vec{r} - \vec{r}'}{|\vec{r} - \vec{r}'|^3} \\
&= \frac{1}{4\pi} \oint_{C_2} (\vec{\omega}(\vec{r}') \cdot d\vec{a}') \; d\vec{l}' \times \frac{\vec{r} - \vec{r}'}{|\vec{r} - \vec{r}'|^3} \\
&= \frac{\kappa_2}{4\pi} \oint_{C_2} d\vec{l}' \times \frac{\vec{r} - \vec{r}'}{|\vec{r} - \vec{r}'|^3},
\end{aligned}
\tag{6.250}
$$

where a prime refers to quantities along the loop C_2, and $d\vec{a}'$ is the cross section of a vortex tube before taking the limit to a line vortex. Then the circulation Γ_1 along C_1 of the velocity field $\vec{u}_2$ of vortex ring C_2 is given by

$$
\begin{aligned}
\Gamma_1 = L_{12} \kappa_2 &= \oint_{C_1} d\vec{l} \cdot \vec{u}_2 \\
&= \frac{\kappa_2}{4\pi} \oint_{C_1} d\vec{l} \cdot \left[\oint_{C_2} d\vec{l}' \times \frac{\vec{r} - \vec{r}'}{|\vec{r} - \vec{r}'|^3} \right] \\
&= \frac{\kappa_2}{4\pi} \oint_{C_1} \oint_{C_2} (d\vec{l} \times d\vec{l}') \cdot \frac{\vec{r} - \vec{r}'}{|\vec{r} - \vec{r}'|^3}.
\end{aligned}
\tag{6.251}
$$

(This may be compared with the expression for inductance,

$$
L_{12} = \frac{\mu_0}{4\pi} \oint_{C_1} \oint_{C_2} \frac{d\vec{l} \cdot d\vec{l}'}{|\vec{r} - \vec{r}'|},
\tag{6.252}
$$

of two electric current loops C_1 and C_2.) The generalization of (6.251) to the circulation Γ_i caused by $n - 1$ loops linked with C_i is

$$\Gamma_i = \sum_{j \neq i} L_{ij}\, \kappa_j = \sum_{j \neq i} \oint_{C_i} d\vec{l} \cdot \vec{u}_j = \oint_{C_i} d\vec{l} \cdot \sum_{j \neq i} \vec{u}_j = \oint_{C_i} d\vec{l} \cdot \vec{u}, \qquad (6.253)$$

where $\vec{u}$ is the total velocity field obtained from all vortex rings, and it was used that along C_i, $d\vec{l} \cdot \vec{u}_i = 0$, since $d\vec{l}$ along C_i is parallel to $\vec{\omega}_i$, which is normal to $\vec{u}_i$.

Obviously, it is nearly impossible to calculate the above loop integrals except for a few simple vortex geometries. Therefore, the line integrals are replaced by a volume integral (no summation over i),

$$\begin{aligned}
\kappa_i \Gamma_i &= \kappa_i \oint_{C_i} d\vec{l} \cdot \vec{u} \\
&= \oint_{C_i} (d\vec{a} \cdot \vec{\omega})(d\vec{l} \cdot \vec{u}) \\
&= \oint_{C_i} (d\vec{a} \cdot d\vec{l})(\vec{\omega} \cdot \vec{u}) \\
&= \int_{V_i} dV \, \vec{\omega} \cdot \vec{u}, \qquad (6.254)
\end{aligned}$$

where again $d\vec{l} \parallel \vec{\omega}$ was used, and the loop C_i lies completely in the tube V_i with infinitesimal cross section. Finally, the sum over all i is performed, giving

$$I = \sum_i \kappa_i \Gamma_i = \sum_i \int_{V_i} dV \, \vec{\omega} \cdot \vec{u} = \int_V dV \, \vec{\omega} \cdot \vec{u}. \qquad (6.255)$$

Here, V is any volume that contains all the tubes V_i, and it was used that $\vec{\omega} = 0$ outside the V_i, so that the volume integral can be extended to any volume containing all V_i. On the other hand,

$$I = \sum_i \kappa_i \Gamma_i = \sum_i \sum_{j \neq i} \kappa_i \kappa_j L_{ij}. \qquad (6.256)$$

Since the integer numbers L_{ij} and the vortex strengths κ_i are constants of motion, the important result

$$\boxed{\frac{d}{dt} \int_V dV \, \vec{\omega} \cdot \vec{u} = 0} \qquad (6.257)$$

follows. The quantity I is the *knottedness* of the ensemble of linked line vortices. It is remarkable that a rather complex topological quantity like I is given by the simple volume integral in (6.257).

6.17   The Clebsch Coordinates and Knottedness

The results from the last section allow to gain a better understanding of the limitations of the Clebsch parameterization (2.217) of the velocity,

$$\vec{u} = \nabla\phi + p\nabla q, \tag{6.258}$$

which expresses the three vector components u_1, u_2, u_3 in terms of three scalar fields ϕ, p, q. Clebsch introduced (6.258) to allow for vorticity $\nabla \times \vec{u} \neq 0$ besides the potential flow $\nabla\phi$. But due to the appearance of two gradients only, it may be suspected that (6.258) is not fully general, and that vector fields $\vec{u}$ exist that cannot be written in the form (6.258).

Locally, in small neighborhoods of a point, (6.258) is indeed fully general, and the integral curves of the velocity field $\vec{u}$ allow to determine appropriate scalar fields ϕ, p, q. Globally, however, when one tries to integrate $\vec{u}$ throughout the fluid, this procedure may fail. This is demonstrated in Sect. 6 of Bretherton (1970). The key argument is quite simple: from (6.258) follows that

$$\vec{\omega} = \nabla \times \vec{u} = \nabla p \times \nabla q. \tag{6.259}$$

Therefore, integration over a volume V yields

$$
\begin{aligned}
\int_V dV \; \vec{\omega} \cdot \vec{u} &\overset{(a)}{=} \int_V dV \; (\nabla p \times \nabla q) \cdot \nabla\phi \\
&= \int_V dV \; \vec{\omega} \cdot \nabla\phi \\
&\overset{(b)}{=} \int_V dV \; \nabla \cdot (\phi\vec{\omega}) \\
&= \int_{\partial V} d\vec{a} \cdot (\phi\vec{\omega}),
\end{aligned}
\tag{6.260}
$$

where in (a) $(\nabla p \times \nabla q) \times (p\nabla q) = 0$ was used, and in (b) $\nabla \cdot (\phi\vec{\omega}) = \vec{\omega} \cdot \nabla\phi + \phi\nabla \cdot \vec{\omega}$, where $\nabla \cdot \vec{\omega} = 0$.

Consider then a collection of vortex rings fully contained in the finite volume V. Therefore, $d\vec{a} \cdot \vec{\omega} = 0$ and correspondingly $d\vec{a} \cdot (\phi\vec{\omega}) = 0$ on the whole boundary ∂V of V, and according to (6.260), the knottedness I vanishes,

$$I = \int_V dV \; \vec{\omega} \cdot \vec{u} = 0. \tag{6.261}$$

This, however, is in general *false*, and as was shown in the last section, the knottedness of an ensemble of vortex ring will in general *not* vanish. Therefore, whenever $I \neq 0$, the Clebsch parameterization (6.258) does not hold. With this, we close our considerations of ever more complex vortex geometries.

References

Abernathy, F.H., and R.E. Kronauer. 1962. The formation of vortex streets. *Journal of Fluid Mechanics* 13: 1.

Baker, G.R., and M.J. Shelley. 1990. On the connection between thin vortex layers and vortex sheets. *Journal of Fluid Mechanics* 215: 161.

Batchelor, G.K. 1967. *An introduction to fluid dynamics*. Cambridge: Cambridge University Press.

Betchov, R. 1965. On the curvature and torsion of an isolated vortex filament. *Journal of Fluid Mechanics* 22: 471.

Birkhoff, G.B. 1962. Helmholtz and Taylor instability. In *Proceedings of symposia in applied mathematics*, vol. XIII, 55. Providence: American Mathematical Society.

Birkhoff, G.B., and J. Fisher. 1959. Do vortex sheets roll up? *Rendiconti del Circolo Matematico di Palermo, Series 2*, 8: 77.

Bjerknes, V. 1898. Über einen hydrodynamischen Fundamentalsatz und seine Anwendung, etc. *Kongliga Svenska vetenskapsakademiens handlingar* 31: 1.

Blaschke, W. 1945. *Vorlesungen über Differentialgeometrie. I. Elementare Differentialgeometrie*, 4th ed. Berlin: Springer.

Bretherton, F.P. 1970. A note on Hamilton's principle for perfect fluids. *Journal of Fluid Mechanics* 44: 19.

Brown, G.L., and A. Roshko. 1974. On density effects and large structure in turbulent mixing layers. *Journal of Fluid Mechanics* 64: 775.

Caflisch, R.E., and O.F. Orellana. 1986. Long time existence for a slightly perturbed vortex sheet. *Communications on Pure and Applied Mathematics* 39: 807.

Caflisch, R.E., and O.F. Orellana. 1989. Singular solutions and ill-posedness for the evolution of vortex sheets. *SIAM Journal on Mathematical Analysis* 20: 293.

Case, K.M. 1960. Stability of an idealized atmosphere. I. Discussion of results. *The Physics of Fluids* 3: 149.

Chandrasekhar, S. 1961. *Hydrodynamic and hydromagnetic stability*. Oxford: Oxford University Press; New York: Dover (1981).

Chorin, A.J., and J.E. Marsden. 1990. *A mathematical introduction to fluid mechanics*. New York: Springer.

Cowley, S.J., G.R. Baker, and S. Tanveer. 1999. On the formation of Moore curvature singularities in vortex sheets. *Journal of Fluid Mechanics* 378: 233.

Curle, N. 1956. *Hydrodynamic stability of the laminar mixing region between parallel streams*. Unpublished Report no. 18426. London: Aeronautical Research Council.

Drazin, P.G. 1958. The stability of a shear layer in an unbounded heterogeneous inviscid fluid. *Journal of Fluid Mechanics* 4: 214.

Drazin, P.G., and W.H. Reid. 1981. *Hydrodynamic stability*. Cambridge: Cambridge University Press.

Dyson, F.J. 1960. Stability of an idealized atmosphere. II. Zeros of the confluent hypergeometric function. *The Physics of Fluids* 3: 155.

Ertel, H. 1942. Ein neuer hydrodynamischer Wirbelsatz. *Meteorologische Zeitschrift (Contributions to Atmospheric Sciences)* 59: 277.

Fließbach, T. 1996. *Mechanik*. Heidelberg: Spektrum Akademischer Verlag.

Goldstein, S. 1931. On the stability of superposed streams of fluids of different densities. *Proceedings of the Royal Society of London A* 132: 524.

Gradshteyn, I.S., and I.M. Ryzhik. 1980. *Table of integrals, series, and products*. New York: Academic Press.

Gröbli, W. 1877. Specielle Probleme über die Bewegung geradliniger paralleler Wirbelfäden. *Vierteljahrsschrift der Naturforschenden Gesellschaft in Zürich* 22/1: 37.

Hadamard, J. 1902. Sur les problèmes aux dérivées partielles et leur signification physique. *Princeton University Bulletin* 13: 49.

Hama, F.R., and E.R. Burke. 1960. *On the rolling-up of a vortex sheet*. Institute for Fluid Dynamics and Applied Mathematics, Technical Note BN-220, University of Maryland.

Helmholtz, H. 1858. Über Integrale der hydrodynamischen Gleichungen, welche den Wirbelbewe-gungen entsprechen. *Journal für die reine und angewandte Mathematik* 55: 25.

Hess, S.L. 1959. *Introduction to theoretical meteorology*. New York: Henry Holt and Company.

Hicks, W.M. 1922. On the mutual threading of vortex rings. *Proceedings of the Royal Society of London A* 102: 111.

Hill, E.L. 1951. Hamilton's principle and the conservation theorems of mathematical physics. *Reviews of Modern Physics* 23: 253.

Ho, C.-M., and P. Huerre. 1984. Perturbed free shear layers. *Annual Reviews in Fluid Mechanics* 16: 365.

Hurwitz, A., and R. Courant. 1964. *Funktionentheorie*. Berlin: Springer.

Joyce, G., and D. Montgomery. 1972. Simulation of the 'negative temperature' instability for line vortices. *Physics Letters* 39A: 371.

Kabanikhin, S.I. 2008. Definitions and examples of inverse and ill-posed problems. *Journal of Inverse and Ill-Posed Problems* 16: 317.

Kida, S. 1981. Motion of an elliptic vortex in a uniform shear flow. *Journal of the Physical Society of Japan* 50: 3517.

Kirchhoff, G. 1897. *Vorlesungen über Mechanik*, Lectures 21 and 22. Leipzig: Teubner.

Konrad, J.H. 1976. *An experimental investigation of mixing in two-dimensional turbulent shear flows with applications to diffusion-limited chemical reactions*. Internal Report CIT-8-PU. Pasadena: California Institute of Technology.

Kotschin, N.J., I.A. Kibel, and N.W. Rose. 1954. *Theoretische Hydromechanik*, vol. I. Berlin: Akademie-Verlag.

Krasny, R. 1986. A study of singularity formation in a vortex sheet by the point-vortex approxima-tion. *Journal of Fluid Mechanics* 167: 65.

Lamb, H. 1932. *Hydrodynamics*. Cambridge: Cambridge University Press; New York: Dover (1945).

Landau, L.D., and E.M. Lifshitz. 1976. *Mechanics, Course of theoretical physics*, vol. 1, 3rd ed. Amsterdam: Elsevier Butterworth-Heinemann.

Lo, S.H., P.R. Voke, and N.J. Rockliff. 2000. Three-dimensional vortices of a spatially developing plane jet. *International Journal of Fluid Dynamics* 4, Article 1.

Markushevich, A.I. 1965. *Theory of functions of a complex variable*, vol. I. Englewood Cliffs: Prentice-Hall.

McIntyre, M.E. 2015. Potential vorticity. In *Encyclopedia of atmospheric sciences*, vol. 2, ed. G.R. North, J. Pyle, and F. Zhang, 375. Amsterdam: Elsevier Scientific Publishing Company.

Moffatt, H.K. 1969. The degree of knottedness of tangled vortex lines. *Journal of Fluid Mechanics* 35: 117.

Moffatt, H.K. 2017. The degree of knottedness of tangled vortex lines – corrigendum. *Journal of Fluid Mechanics* 830: 821.

Montgomery, D. 1972. Two-dimensional vortex motion and 'negative temperatures'. *Physics Letters* 39A: 7.

Moore, D.W. 1971. *The discrete vortex approximation of a vortex sheet*. California Institute of Technology Report AFOSR-1084-69.

Moore, D.W. 1979. The spontaneous appearance of a singularity in the shape of an evolving vortex sheet. *Proceedings of the Royal Society of London A* 365: 105.

Onsager, L. 1949. Statistical hydrodynamics. *Il Nuovo Cimento* 6 (Supplemento): 279.

Page, M.A., and S.J. Cowley. 2018. On the formation of small-time curvature singularities in vortex sheets. *IMA Journal of Applied Mathematics* 83: 188.

Plemelj, J. 1908. Ein Ergänzungssatz zur Cauchyschen Integraldarstellung analytischer Funktionen, Randwerte betreffend. *Monatshefte für Mathematik und Physik* 19: 205.

Pozrikidis, C., and J.J.L. Higdon. 1985. Nonlinear Kelvin-Helmholtz instability of a finite vortex layer. *Journal of Fluid Mechanics* 157: 225.

Pullin, D.I. 1978. The large-scale structure of the unsteady self-similar rolled-up vortex sheets. *Journal of Fluid Mechanics* 88: 401.

Rayleigh, Lord. 1879. On the stability, or instability, of certain fluid motions. *Proceedings of the London Mathematical Society, Series 1*, 11: 57.

Richtmyer, R.D., and K.W. Morton. 1967. *Difference methods for initial-value problems*, 2nd ed. New York: Wiley.

Rosenhead, L. 1931. The formation of vortices from a surface of discontinuity. *Proceedings of the Royal Society of London A* 134: 170.

Rossby, C.-G. 1938. On the mutual adjustment of pressure and velocity distributions in certain simple current systems, II. *Journal of Marine Research* 1: 239.

Rott, N. 1956. Diffraction of a weak shock with vortex generation. *Journal of Fluid Mechanics* 1: 111.

Saffman, P.G. 1995. *Vortex dynamics*. Cambridge: Cambridge University Press.

Saffman, P.G., and G.R. Baker. 1979. Vortex interactions. *Annual Reviews in Fluid Mechanics* 11: 95.

Shelley, M.J., and G.R. Baker. 1989. The relation between thin vortex layers and vortex sheets. In *Mathematical aspects of vortex dynamics*, ed. R.E. Caflisch, 36. Philadelphia: SIAM.

Sulem, C., P.L. Sulem, C. Bardos, and U. Frisch. 1981. Finite time analyticity for the two and three dimensional Kelvin-Helmholtz instability. *Communications in Mathematical Physics* 80: 485.

Walker, J.D.A., C.R. Smith, A.W. Cerra, and T.L. Doligalski. 1987. The impact of a vortex ring on a wall. *Journal of Fluid Mechanics* 181: 99.

Zabusky, N.J., M.H. Hughes, and K.V. Roberts. 1979. Contour dynamics for the Euler equations in two dimensions. *Journal of Computational Physics* 30: 96.

Zhang, J., S. Childress, A. Libchaber, and M. Shelley. 2000. Flexible filaments in a flowing soap film as a model for one-dimensional flags in a two-dimensional wind. *Nature* 408: 835.

Chapter 7
Kinematics of Waves

This and the following chapters treat mechanical waves in fluids. Waves are oscillatory phenomena in time and space. Oscillations occur when two forces are in imperfect balance, so that one of them overcomes the other and one succumbs to the other periodically. (The motion of planets on ellipses instead of circles is due to an imperfect balance of gravity and centrifugal acceleration along the trajectory.) One of the two forces in imperfect balance in fluid waves is always the inertial force $dm\, \ddot{\vec{r}}$ or $\rho\, d\vec{u}/dt$ (force density) of fluid parcels. The other, material force defines the specific wave type. If the latter force is the Earth gravity, one has *water waves* per se (the old name was gravity waves), more specifically shallow water waves if the whole body of water undergoes horizontal oscillations, and free surface waves if only the uppermost layers of water near the surface are affected by the wave. If the material force is thermal pressure, one has sound waves. Water waves and sound waves are the subject of the remainder of this book, with a little digression to *capillary* and *buoyancy* waves. Waves of mixed type are also possible, if two or more material forces are in near-equilibrium with fluid inertia, e.g., in *gravo-acoustic* waves.

Section 7.1 lists basic facts about waves. A study of waves naturally starts with their *kinematics*, i.e., their basic propagation characteristics, hence Sects. 7.2 and 7.3 introduce the two fundamental wave speeds, phase speed and group speed, as the propagation speed of wave phases (e.g., wave crests or troughs) and of groups of harmonic waves with similar wavelengths, respectively. Section 7.4 introduces another fundamental concept, the wavefront and its speed. Section 7.5 contains some advanced material on the rather peculiar propagation and growth properties of waves in a wholly different example of a material force, viz., radiative waves in a continuous flow of ions accelerated by a radiative force resulting from photon absorption and/or scattering. This example will show that in extreme cases care is required in applying the very term 'wave propagation.'

© Springer Nature Switzerland AG 2019   275
A. Feldmeier, *Theoretical Fluid Dynamics*, Theoretical and Mathematical Physics,
https://doi.org/10.1007/978-3-030-31022-6_7

7.1 Wave Basics

Hydrodynamic waves are collective phenomena, with a large number of fluid parcels partaking in a macroscopically coherent excitation. For periodic excitations, the motion of a given parcel is sinusoidal in time, $\sim e^{-i\omega t}$, with angular speed ω. The simplest possible wave is a *harmonic* wave in one straight spatial dimension (coordinate x), with a spatio-temporal amplitude $\sim e^{i(kx-\omega t)}$, where k is the wavenumber. The phase $kx - \omega t$ is the position of the parcel at x and t in an excitation cycle of the wave. Neighboring parcels show a phase lag that increases with distance: a given excitation state of one parcel will appear in a neighboring parcel at a later time, or has appeared there at an earlier time. Thus the phase seems to propagate through the medium at a so-called *phase speed*.

No net flux of fluid parcels is associated with waves of very small amplitude, so-called *linear waves* (where quadratic and higher excitation terms can be neglected). Instead, the parcels oscillate on closed orbits about an equilibrium rest position, in free surface waves for example on circles, in shallow water waves in straight horizontal lines.

The force that establishes the 'elastic coupling' (in a harmonic-oscillator analogy) between neighboring parcels, and thus is responsible for the collective excitation, defines the type of the wave. In water waves, the relevant force is the weight of fluid columns and the related hydrostatic pressure. In internal gravity waves, the driving force is the buoyancy due to density differences between neighboring fluid elements. A rather unexpected wave-type was found in the 1960s in the plasma of highly rarefied ions surrounding the Earth at large distances, where the coupling responsible for collective excitations is the Coulomb force between charged particles—although particle distances are so large that direct collisions between them are extremely rare ('collisionless plasma'), see Sagdeev and Kennel (1991).

The simplest wave equation in one dimension is, with constant values of ω and k,

$$\frac{\partial^2 f}{\omega^2 \, \partial t^2} - \frac{\partial^2 f}{k^2 \, \partial x^2} = 0, \tag{7.1}$$

and it has the general solution

$$f(t, x) = F_{\pm}(\omega t \pm kx), \tag{7.2}$$

where F_+ and F_- are arbitrary (piecewise) differentiable functions. The variables ω and k are related by an algebraic *dispersion relation* of the form

$$D(\omega, k) = 0, \tag{7.3}$$

which is obtained in the derivation of the wave equation from the equations of motion. Typically, the excitation frequency ω is considered as free variable, but ω may be restricted to certain domains beyond which freely propagating waves do not exist, and k is a function of ω, not necessarily single-valued. The phase speed as the rate at which a given constant phase propagates is given by $v_p = \mathrm{d}x/\mathrm{d}t = \omega/k$. The phase speed is a function of certain flow parameters, like air temperature or water depth, and is constant for the simplest flows.

For harmonic waves $\sim e^{i(kx-\omega t)}$, the phase speed ω/k may depend then on the wavenumber k. If harmonic waves of different wavelengths propagate at different phase speeds, one has *wave dispersion*. According to the Fourier theorem, any square integrable function, especially any waveform, can be written as integral over k of functions e^{ikx} with an amplitude factor $a(k)$. If the phase speed changes with k, then the different harmonics making up the waveform will propagate at different speeds, thus their relative phases will change and their superposition, the overall waveform, will change in time.

Although there is no transport of matter in waves of very small amplitude, there is usually energy transport. The energy, however, is not transported at the phase speed, but at the so-called group speed (see Milne-Thomson 1968, p. 436 for the proof). The latter can be derived from the dispersion relation $D(\omega, k) = 0$, as is discussed in the following sections. Note that both the phase speed and the group speed are not related to the fluid speed u. Thus, despite formal similarities, the one-dimensional advection equation, $(\partial_t + u\,\partial_x)\,f = 0$, with fluid speed u, and the one-dimensional wave equation, $(\partial_t + c\,\partial_x)\,(\partial_t - c\,\partial_x)\,f = 0$, for a *constant* wave speed c, describe fundamentally different processes.

7.2 Group Speed

Harmonic waves are characterized by their wavelength and amplitude (e.g., the elevation of water). Correspondingly, there are two fundamental propagation speeds: the propagation speed of fixed wave phases like extrema or nodes and that of groups of waves of similar wavelengths, termed phase speed and group speed, respectively. The latter is introduced by Lamb as follows:

> It has often been noticed that when an isolated group of waves, of sensibly the same length, is advancing over relatively deep water, the velocity of the group as a whole is less than that of individual waves composing it. (Lamb 1932, p. 380)

This can be seen in an animation at https://en.wikipedia.org/wiki/Phase_velocity. Reynolds (1877) found that wave energy propagates at group speed. There is no physical quantity associated with a wave phase, and thus no physical quantity (except the phase itself) propagates at phase speed. Therefore, phase speeds can exceed the speed of light.

To obtain the group speed, a 'group of waves, of sensibly the same length' is needed. Stokes realized that *two* harmonic waves of equal amplitude and nearly equal wavelength are sufficient. Their superposition is given by

$$\zeta = a \sin(kx - \omega t) + a \sin(k'x - \omega't)$$

$$= 2a \, \cos\left[\frac{(k - k')x}{2} - \frac{(\omega - \omega')t}{2} \right] \, \sin\left[\frac{(k + k')x}{2} - \frac{(\omega + \omega')t}{2} \right]. \quad (7.4)$$

For $k' \to k$ and with abbreviations $dk = k - k'$ and $d\omega = \omega - \omega'$, this reads

$$\zeta = 2a \, \cos\left[\frac{1}{2}(dk \; x - d\omega \; t) \right] \, \sin(kx - \omega t). \quad (7.5)$$

The sine is the original, fast oscillation (large ω) with short wavelength (large k). The cosine, by contrast, is a slow oscillation (small $d\omega$) with large wavelength (small dk). The cosine defines a large-scale modulation of the amplitude of the sine. The two curves that connect smoothly the extrema of the sine-wave under the cosine-modulation define the *wave envelope*. This wave envelope itself is an oscillation, with the propagation speed of its nodes and extrema and its whole shape termed *group speed*, and given by

$$\boxed{v_g = \frac{d\omega}{dk}} \quad (7.6)$$

As an example, for surface waves on water it is found that (see Eq. 8.14)

$$\omega^2 = gk, \quad (7.7)$$

giving

$$v_p = \sqrt{g/k}, \quad (7.8)$$

and

$$v_g = \frac{1}{2}\sqrt{g/k}. \quad (7.9)$$

The group speed for surface waves on water is half the phase speed, independent of the wavenumber k, in agreement with the above quote from Lamb.

The next step is to consider a wave packet made up of harmonic waves of essentially the same wavelength, and derive the propagation speed of its envelope. Lamb (1904) gives a very concise summary of the results obtained then:

Assuming a disturbance

$$y = \sum C \cos(\omega t - kx + \epsilon), \quad (7.10)$$

where the summation (which may of course be replaced by an integration) embraces a series of terms in which the values of ω, and therefore also of k, vary very slightly, we remark that the phase of the typical term at time $t + \Delta t$ and place $x + \Delta x$ differs from the phase at time t and place x by the amount $\omega \Delta t - k \Delta x$. Hence if the variations of ω and k from term to term be denoted by $\delta \omega$ and δk, the change of phase will be sensibly the same for all the terms, provided

$$\delta \omega \, \Delta t - \delta k \, \Delta x = 0. \quad (7.11)$$

The group as a whole therefore travels with the velocity

$$v_g = \frac{\Delta x}{\Delta t} = \frac{d\omega}{dk}. \tag{7.12}$$

(Lamb 1904, p. 473, with symbols adapted)

Note that $\delta\omega$ and δk must be the same between any two terms in the sum (7.10), thus curvature in the dispersion relation $\omega(k)$ is not included, as is clear from the resulting equation for v_g. Note also that the constant C in (7.10) should carry an index k.

A modern version of this argument looks as follows (see Griffiths and Schroeter 2018, p. 59). Consider a wave packet, i.e., a superposition of harmonic waves,

$$\zeta(t, x) = \int_{-\infty}^{\infty} dk\, a(k)\, e^{i(kx - \omega(k)t)}. \tag{7.13}$$

The function $a(k)$ gives the initial amplitudes of waves e^{ikx} at each wavenumber k. The linear superposition principle is assumed to hold, i.e., waves of different k do not interact. Then the amplitude $a(k)$ of each wave k stays constant in time, and $a(k)$ does not depend on time. However, these waves are added at different times with different phases $e^{-i\omega t}$, which controls their constructive or destructive interference.

The essential assumption that the wave packet is made up of harmonic waves of essentially the same wavelength allows to break off the Taylor series for $\omega(k)$ at first order,

$$\omega(k) \approx \omega_0 + \omega_0' (k - k_0), \tag{7.14}$$

with $\omega_0 = \omega(k_0)$ and $\omega_0' = d\omega/dk|_{k_0}$. Inserting this in (7.13), and introducing a term $1 = e^{ik_0 x - ik_0 x}$, gives

$$\zeta(t, x) = e^{i(k_0 x - \omega_0 t)} \int_{-\infty}^{\infty} dk\, a(k)\, e^{i(k - k_0)(x - \omega_0' t)}, \tag{7.15}$$

where $a(k)$ has a narrow peak at k_0. The factor in front of the integral is the high-frequency oscillation of the harmonic wave with wavenumber k_0 (all waves have approximately the same k_0) within the envelope. The integral gives the modulation of this k_0-oscillation, i.e., the envelope of the wave packet, with a large wavelength $2\pi/(k - k_0)$ (small denominator). The integral becomes independent of time, i.e., one moves with the envelope, if $x - \omega_0' t = 0$, or

$$\frac{x}{t} = v_g = \omega_0', \tag{7.16}$$

which is again (7.6).

Finally, the asymptotic wave motion can be considered, for $x \to \infty$ and $t \to \infty$. The phase function $e^{i(kx - \omega t)}$ shows then severe changes for small changes in k and ω, if only $kx \approx \omega t \approx \pi$. For large x and t, a rather uniform random distribution of phases is thus expected, and the individual harmonic waves will show destructive interference (this argument obviously requires further substantiation). This is however *not* the case at the extrema of the phase function,

$$\frac{d}{dk}(kx - \omega t) = 0,$$ (7.17)

where contributions from a whole neighborhood of k will add up, leading to constructive interference. Waves from a *neighborhood* of k make up a group of waves of sensibly the same wavelength. This is the asymptotic wave packet, and it travels at group speed: the space–time trajectory of this constructive interference is, according to (7.17), given by

$$\frac{x}{t} = \frac{d\omega}{dk} = v_g.$$ (7.18)

This can be compared with the above case of two superimposed waves. There, the speed v_g corresponds to the speed of the extremum of the wave envelope. And this extremum is due to constructive interference between the two waves. Further considerations in this vein can be found in Chap. 5, 'The method of stationary phase', in the book by Copson (1965).

The above derivations of the group speed are slightly overcomplicated by considering harmonic waves, i.e., performing a Fourier decomposition of the wave packet. Quite interestingly, this is not necessary, as was shown by Lamb (1904). The argument can also be found, in the same words, in Lamb (1932), p. 382. Due to its conciseness, it is again quoted in full. As a preparatory step, differentiate $\omega = v_p k$ on both sides with respect to k, giving

$$\frac{d\omega}{dk} = v_p + k\frac{dv_p}{dk}.$$ (7.19)

Since $k \sim \lambda^{-1}$, $k\,d/dk = -\lambda\,d/d\lambda$, and (7.19) becomes

$$v_g = v_p - \lambda\frac{dv_p}{d\lambda}.$$ (7.20)

Then Lamb's argument runs as follows.

[If] the wave-velocity varies with the frequency, a limited initial disturbance gives rise in general to a wave-system in which the different wave-lengths, travelling with different velocities, are gradually sorted out. If we regard the wave-length λ as a function of x and t, we have

$$\frac{\partial\lambda}{\partial t} + v_g\frac{\partial\lambda}{\partial x} = 0,$$ (7.21)

since λ does not vary in the neighbourhood of a geometrical point travelling with velocity v_g; this is, in fact, the definition of v_g. Again, if we imagine another geometrical point to travel with the *waves*, we have

$$\frac{\partial\lambda}{\partial t} + v_p\frac{\partial\lambda}{\partial x} = \lambda\frac{\partial v_p}{\partial x} = \lambda\frac{dv_p}{d\lambda}\frac{\partial\lambda}{\partial x},$$ (7.22)

the second member expressing the rate at which two consecutive wave-crests are separating from one another. Combining (7.21) and (7.22), we are led, again, to the formula (7.20). (Lamb 1904, p. 474)

And the latter formula gives again $v_g = d\omega/dk$. The essential step in this argument is to identify the change in λ when moving with the crest, i.e., at v_p, with the speed at which adjacent crests at a distance λ separate from each other, i.e., the middle term in (7.22).

7.3 Kinematic Waves

Different harmonic waves $e^{i(kx-\omega t)}$ will usually have different phase speeds ω/k. Fast waves will outrun slow waves, and after some time, different λ are found at different locations, as indicated in the foregoing quote from Lamb. For *normal* dispersion, long waves propagate fast, short waves propagate slow. At large t then, long waves are found far from the wave source, and short waves stay close to the source. In a sense, wave dispersion acts as a wave filter. This idea is at the core of the following approach to group speed, adapted from Lighthill and Whitham (1955), p. 286 and Whitham (1960).

Consider the real x-axis, and assume that wave dispersion has spread individual harmonic waves $e^{i(kx-\omega t)}$ in such a way that there is a one-to-one mapping $\lambda(x)$, telling which λ can be found at which x. This requires obviously that the function $\omega(k)$ is monotonic. The function $\lambda(x)$ shall vary slowly with position,

$$|d\lambda/dx| \ll 1. \tag{7.23}$$

In a sufficiently small interval dx around x, the waveform is approximately harmonic, $\zeta = ae^{ikx}$, or

$$\zeta = \sin \frac{2\pi x}{\lambda(x)}. \tag{7.24}$$

The change in $\lambda(x)$ shall be so slow that dx contains $dN \gg 1$ crests of waves of sensibly the same wavelength. Also, dx is adjusted so as to contain an integer number of these wavelengths. Then

$$\lambda = \frac{dx}{dN}, \tag{7.25}$$

and a crest density can be introduced by

$$n(t, x) = \frac{dN}{dx} = \frac{1}{\lambda}. \tag{7.26}$$

The total number of wave crests is assumed to be conserved during motion. (This is not exactly true, since extrema can temporarily vanish when passing through the nodes of a group envelope.) Then dN in the interval dx can only change by in- and outflow of crests through the interval boundaries $x - dx/2$ and $x + dx/2$,

$$\mathrm{d}x\,\frac{\partial n(t,x)}{\partial t}$$

$$= -n\left(t, x + \frac{\mathrm{d}x}{2}\right) v_p\left(t, x + \frac{\mathrm{d}x}{2}\right) + n\left(t, x - \frac{\mathrm{d}x}{2}\right) v_p\left(t, x - \frac{\mathrm{d}x}{2}\right), \quad (7.27)$$

since, by definition, the propagation speed of crests is the phase speed v_p. Dividing both sides by $\mathrm{d}x$, this becomes

$$\frac{\partial n(t, x)}{\partial t} = -\frac{\partial[n(t, x)\, v_p(t, x)]}{\partial x}. \tag{7.28}$$

Therefore, using $n = \lambda^{-1} = (2\pi)^{-1} k$ and $v_p = \omega/k$,

$$\frac{1}{2\pi}\frac{\partial k}{\partial t} = -\frac{1}{2\pi}\frac{\partial}{\partial x}\left(k\,\frac{\omega}{k}\right), \tag{7.29}$$

or

$$\frac{\partial k}{\partial t} + \frac{\partial \omega}{\partial x} = 0. \tag{7.30}$$

Inserting the dispersion relation $\omega = \omega(k)$ gives

$$\frac{\partial k}{\partial t} + \frac{\mathrm{d}\omega}{\mathrm{d}k}\frac{\partial k}{\partial x} = 0. \tag{7.31}$$

Introducing the group speed $v_g = \mathrm{d}\omega/\mathrm{d}k$, this yields

$$\frac{\partial k}{\partial t} + v_g\frac{\partial k}{\partial x} = 0, \tag{7.32}$$

which is again (7.21), the definition of group speed: when moving at v_g, always the same λ and k are found.

7.4 The Wavefront

The conceptually important phase speed v_p was defined for harmonic waves only, as $v_p = \omega/k$. This raises the question whether a more general definition of waves and their propagation speed can be given. The points of constant phase of a wave define a surface in space, a so-called *wavefront* (known from geometrical optics and Huygens' principle), which propagates through the medium. All fluid parcels in a wavefront are in the same stage of their periodic motion. Following Herglotz (1985), p. 89 it is shown here that the motion of any surface in a fluid across which only *second* and higher derivatives of fluid properties may have *jumps* (but do not need to have jumps) propagates according to a wave equation.

To demonstrate this, let x_1, x_2, x_3 be Cartesian coordinates, and $\phi : \mathbb{R}^3 \to \mathbb{R}$ a smooth function (continuously differentiable to any order). Let, moreover, a family of surfaces in $\mathbb{R}^3$ with parameter t be defined by the equation

$$\phi(x_1, x_2, x_3) = t. \tag{7.33}$$

The change of the surface $\phi = t = \text{const}$ with time is interpreted as propagation of a wavefront. For example, for a spherical wave from a point source in the origin with propagation speed c,

$$c^{-1}(x_1^2 + x_1^2 + x_3^2)^{1/2} = t. \tag{7.34}$$

Let $f(t, x_1, x_2, x_3)$ be a hydrodynamic field, or a function thereof, and abbreviate as usual, with i and j running from 1 to 3,

$$f_j = \frac{\partial f}{\partial x_j}, \qquad \phi_j = \frac{\partial \phi}{\partial x_j}, \qquad f_t = \dot{f} = \frac{\partial f}{\partial t}. \tag{7.35}$$

A jump of any function g (f and its derivatives), when crossing the surface $\phi = t$ orthogonally, is written as

$$[g] = g^+ - g^-, \tag{7.36}$$

where g^+ and g^- are the values of g at infinitesimal distances from this surface, on a curve that crosses the surface orthogonally. Within the wavefront $\phi = t$,

$$f(t, x_1, x_2, x_3) = f(\phi(x_1, x_2, x_3), x_1, x_2, x_3). \tag{7.37}$$

The function f suffers no jump across the wavefront, $f^+ = f^-$. Thus f^+ and f^-, considered as *functions* of x_1, x_2, x_3, are identical everywhere, and so are their derivatives with respect to x_1, x_2, x_3, i.e.,

$$f_i^+ + \dot{f}^+ \phi_i = f_i^- + \dot{f}^- \phi_i, \tag{7.38}$$

or, since ϕ and its derivatives have no jumps,

$$[f_i] + [\dot{f}] \phi_i = 0. \tag{7.39}$$

Similarly, the first partial derivatives of f have no jump, which, on the basis of the same argument, yields

$$[f_{ij}] + [\dot{f}_i] \phi_j = 0, \tag{7.40}$$
$$[\dot{f}_i] + [\ddot{f}] \phi_i = 0. \tag{7.41}$$

Multiplying (7.41) with ϕ_j and subtracting this from (7.40) gives

$$\boxed{[f_{ij}] - [\ddot{f}]\phi_i\phi_j = 0}$$
(7.42)

The derivatives ϕ_i correspond to inverse speeds, which is seen as follows. Consider two surfaces $\phi = t$ and $\phi + d\phi = t + dt$ at two infinitesimally close times. Then

$$d\phi = dn\,|\nabla\phi|,$$
(7.43)

where dn is the shortest, normal distance between the two surfaces in the point under consideration (therefore the abs of the gradient suffices in Eq. 7.43). Therefore the speed of progression of the surface ϕ is, with $d\phi = dt$,

$$\frac{dn}{dt} = \frac{1}{|\nabla\phi|}.$$
(7.44)

Thus,

$$\nabla\phi = (\phi_i) = (c_i^{-1})$$
(7.45)

is the vector of inverse propagation speeds c_i^{-1} of the surface ϕ in directions x_i at any location, and therefore (7.42) is a classical wave equation. Since the wavefront is a surface of constant phase, the phase speed is $v_p = |\nabla\phi|^{-1}$. The present results can be compared with those in Sect. 12.12 where it is shown that discontinuities in second derivatives can only occur across characteristics, and indeed characteristics describe the propagation of waves.

7.5 Waves and Instability from a Radiative Force

All *forces* considered in this book are either external forces like gravity and act directly on each fluid parcel or they are short-range contact or surface forces like pressure and viscous stress and act at the interface—and only at the interface—between fluid parcels. As an exception to this, the present section discusses a force that falls off with the penetration depth into the fluid and thus *couples distant locations* in the fluid. Such a *nonlocal* force can lead to many new physical effects; in this section, some unexpected behavior in the propagation of waves and the growth of instabilities related to a nonlocal force is briefly discussed. A physical situation in which a nonlocal force occurs is the absorption of photons in bound-bound transitions of plasma ions (giving raise to spectral absorption lines), to which the photons submit their momentum and accelerate them. This physical background from radiation hydrodynamics is not further elaborated upon here, since the resulting force perturbations are still sufficiently simple to rather postulate them from a coarse intuitive understanding of the underlying physics. The foundations of radiation hydrodynamics are treated in an extensive memoir by Mihalas and Weibel-Mihalas (1984).

The flow is assumed to be a one-dimensional, planar stream of plasma with distance coordinate z, which is accelerated by some (external) radiation source at $z \to -\infty$. This results in a stationary, monotonically growing velocity law $u_0(z)$. Thermal gas pressure and viscosity are neglected, since the flow is typically accelerated to some fraction of the speed of light, and thus to much higher speeds than the thermal sound speed in the plasma.

To analyze wave propagation and unstable growth, a linear stability analysis is performed in the following. For this, only the linear response of the radiative force to velocity perturbations is needed. Specifically, it is assumed that (a) all locations $z' < z$ below some z under consideration contribute to the force perturbation at z, leading to a (nonlocal) force *integral*; (b) this contribution drops off exponentially with distance $z - z'$; and (c) the contribution of z' to the force perturbation at z is proportional to the perturbation of the velocity gradient $du(z')/dz'$ of the flow at z'. In the underlying physical model, this is caused by a drop in the plasma absorption coefficient at z' brought about by the Doppler frequency shift due to the extra acceleration of the gas if $du(z')/dz' > 0$ (an effect termed, quite intuitively, 'deshadowing'). This drop in photon absorption at z' leads to a larger remaining photon flux at z, and thus to a positive force perturbation. One has then for the linear perturbation g_1 of the radiative force due to velocity gradient perturbations $\partial u_1 / \partial z'$ at all $z' < z$,

$$g_1(t, z) = \int_{-\infty}^{z} dz' \, e^{-(z-z')} \frac{\partial u_1(t, z')}{\partial z'}. \tag{7.46}$$

This equation is derived by Owocki and Rybicki (1984) for a model of absorption of photons (from a star) in bound-bound transitions of an ion flow (a wind from this star), see their Eqs. (21) and (32). The constants $\omega_b = \chi_b$ in the latter equations are normalized to unity here.

The stationary Euler equation for the flow caused by the radiative force is, with $u_0' = du_0/dz$,

$$u_0 u_0' = g_0. \tag{7.47}$$

Despite the apparent simplicity of this equation, it should be noted that g_0 will depend on u_0 and/or u_0', and typically even in a nonlinear fashion. Small perturbations are introduced by replacing $u_0(z)$ by $u(z) = u_0(z) + u_1(t, z)$ and g_0 by $g = g_0 + g_1$. The non-stationary Euler equation $\dot{u} + uu' = g$ gives then, after subtracting (7.47) and neglecting products of perturbations,

$$\dot{u}_1 + u_0' u_1 + u_0 u_1' = g_1. \tag{7.48}$$

Furthermore, it is assumed from now on that $|u_1'/u_0'| \gg |u_1/u_0|$. This is the so-called WKB approximation, meaning that the wavelength of perturbations, $|u_1/u_1'|$, is much smaller than the lengthscale of changes in the underlying unperturbed model, $|u_0/u_0'|$. Then $u_0' u_1$ in (7.48) can be neglected as compared to $u_0 u_1'$. Consider an arbitrary but fixed location z. In a frame moving at $u_0(z)$, Eq. (7.48) becomes then Newton's second law,

$$\dot{u}_1(t, z) = g_1(t, z). \tag{7.49}$$

This is all what is required in the following. Integration by parts in (7.46) gives

$$g_1(t, z) = u_1(t, z) - \int_{-\infty}^{z} dz'\, e^{-(z-z')}\, u_1(t, z'). \tag{7.50}$$

Differentiating (7.49) with respect to z and using (7.50) gives

$$\partial_z \partial_t u_1 = \partial_z u_1 - u_1 + \int_{-\infty}^{z} dz'\, e^{-(z-z')}\, u_1(t, z'), \tag{7.51}$$

where the argument is (t, z) if not explicitly given. The last two terms in this equation are from differentiating the integral in (7.50), where z appears both in the limit of integration and in the integral kernel.

Equation (7.51) is an integro-differential equation for the velocity perturbation field u_1: the radiative force makes itself felt only through its dependence on u via the Doppler effect. That (7.51) is a purely *kinematic* equation (i.e., there appear locations and speeds only) lies at the core of some of the unexpected (or pathological) behavior to be explored in the following.

The integral in (7.51) is $u_1 - g_1$ by (7.50), or $u_1 - \dot{u}_1$ by (7.49). Thus, the problem obeys the linear partial differential equation of second order,

$$\partial_z \partial_t u_1 = \partial_z u_1 - \partial_t u_1, \tag{7.52}$$

without any nonlocal terms. Inserting wave solutions $u_1 = e^{i(kz - \omega t)}$ in (7.52) gives the dispersion relation

$$\omega = -\frac{k}{1 + ik}. \tag{7.53}$$

This is the *bridging law* of Owocki and Rybicki (1984), Eq. (28). For $k \to 0$, $\omega/k = -1$. This corresponds to *radiative waves* propagating upstream, i.e., towards smaller z, at phase speed -1, first discussed by Abbott (1980). For $k \to \infty$, on the other hand, $\omega = i$, which corresponds to exponential growth $u_1 \sim e^t$ of perturbations. This is the so-called *line-driven instability*, first suggested by Lucy and Solomon (1970) on purely physical grounds. In simple terms, if a fluid parcel at z experiences a positive velocity perturbation, it leaves the absorption shadow of intervening gas at $z' < z$, and can absorb so far unabsorbed radiation from $z \to -\infty$. This means that the fluid parcel experiences a stronger photon flux and therefore a stronger radiative force—which will *further amplify* the initial velocity perturbation, thereby lifting the parcel still further out of the absorption shadow of intervening material; thus, a self-amplifying feedback occurs, i.e., an instability. Parts of this mechanism were already suggested by Milne (1926). This instability is also found by neglecting the second term on the right in (7.50). Then (7.49) becomes $\dot{u}_1 = u_1$, leading to unstable growth $u_1 \sim e^t$. The dichotomy between waves at small k and instability at large k

is the origin of the name 'bridging law' for (7.53). At intermediate k, the radiative waves are subject to the line-driven instability.

To proceed, write (7.50) as

$$g_1(t, z) = \int_{-\infty}^{\infty} dz' \, N(z - z') \, u_1(t, z'),$$ (7.54)

with an integral kernel

$$N(z) = \delta(z) - e^{-z} \, H(z),$$ (7.55)

where H is the Heaviside function, $H(z) = 1$ for $z > 0$ and zero else. Note that N is independent of time. Its Fourier transform with respect to z is (an overbar indicates now the transform function, not the complex conjugate)

$$
\begin{aligned}
\overline{N}(k) &= \int_{-\infty}^{\infty} dz \, e^{-ikz} \, [\delta(z) - e^{-z} \, H(z)] \\
&= 1 - \int_0^{\infty} dz \, e^{-z(1+ik)} \\
&= \frac{ik}{1 + ik}.
\end{aligned}
$$ (7.56)

Equation (7.49) is then

$$\partial_t u_1(t, z) = \int_{-\infty}^{\infty} dz' \, N(z - z') \, u_1(t, z'),$$ (7.57)

and the convolution theorem gives

$$\partial_t \overline{u}_1(t, k) = \overline{N}(k) \, \overline{u}_1(t, k).$$ (7.58)

Note that no Fourier's transform is taken with respect to t. The solution of the differential equation (7.58) is

$$\overline{u}_1(t, k) = e^{\overline{N}(k) t} \, \overline{u}_1(0, k).$$ (7.59)

The convolution theorem is applied again, back to z, giving

$$u_1(t, z) = \int_{-\infty}^{\infty} dz' \, M(t, z - z') \, u_1(0, z'),$$ (7.60)

with

$$
\begin{aligned}
M(t, z) &= \frac{1}{2\pi} \int_{-\infty}^{\infty} dk \, e^{ikz} e^{\overline{N}(k) t} \\
&= \frac{1}{2\pi} \int_{-\infty}^{\infty} dk \, e^{ikz} \, e^{ikt/(1+ik)}.
\end{aligned}
$$ (7.61)

Fig. 7.1 Integration contour
in the complex s-plane for
the inverse Laplace
transform (7.63), with a
singularity at $s = 0$

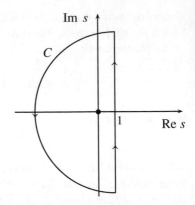

The function inside the last integral has a singularity at $1 + ik = 0$. A new complex
variable is therefore introduced,

$$s = 1 + ik, \qquad (7.62)$$

for which (7.61) becomes

$$
\begin{aligned}
M(t, z) &= e^{t-z} \frac{1}{2\pi i} \oint_C ds \; e^{sz - t/s} \\
&= e^{t-z} \frac{d}{dz} \left[\frac{1}{2\pi i} \oint_C ds \; e^{sz} \; \frac{e^{-t/s}}{s} \right],
\end{aligned}
\qquad (7.63)
$$

see equations (B10) and (B11) in Owocki and Rybicki (1986). The appropriate closed
integration contour C is easily found: $s \to \pm i\infty$ for $k \to \pm\infty$, and $s = 1$ for $k = 0$.
Thus C passes to the *right* of the singularity at $s = 0$. The integrand vanishes for
$s \to -\infty$, and C can be closed as shown in Fig. 7.1, since the integration along the
semicircle at infinity delivers no contribution to the integral.

The Laplace transform $\tilde{f}(s)$, $s \in \mathbb{C}$ of a real-valued function $f(z)$ with variable
$z \in \mathbb{R}$ is given by

$$\tilde{f}(s) = \int_0^\infty dz \; e^{-sz} f(z), \qquad (7.64)$$

see (29.1.1) in Abramowitz and Stegun (1964; A&S from now). The *inverse* Laplace
transform f of $\tilde{f}$ is given by (29.2.2) in A&S,

$$f(z) = \frac{1}{2\pi i} \int_{c-i\infty}^{c+i\infty} ds \; e^{sz} \; \tilde{f}(s). \qquad (7.65)$$

The integral in (7.63) including the factor $\frac{1}{2\pi i}$ is the inverse Laplace transform of

$$\tilde{f}(s) = \frac{e^{-t/s}}{s}, \tag{7.66}$$

where $c = 1$. From (29.3.75) in A&S, the corresponding real-valued $f(z)$ is

$$f(z) = H(z) \, J_0(2\sqrt{tz}), \tag{7.67}$$

where H occurs due to (29.1.3) in A&S. Here J_0 is the zeroth order Bessel function of the first kind. Using

$$H'(z) = \delta(z) \quad \text{and} \quad J_0'(z) = -J_1(z) \tag{7.68}$$

(see 9.1.28 in A&S for the latter), it follows that

$$\begin{aligned} M(t, z) &= e^{t-z} \frac{d}{dz} \left[H(z) \, J_0(2\sqrt{tz}) \right] \\ &= e^{t-z} \left[\delta(z) \, J_0(2\sqrt{tz}) - H(z) \, \sqrt{t/z} \, J_1(2\sqrt{tz}) \right]. \end{aligned} \tag{7.69}$$

Using $f(z) \delta(x) = f(0) \delta(z)$ and $J_0(0) = 1$, finally yields

$$\boxed{M(t, z) = e^t \, \delta(z) - e^{t-z} \, H(z) \, \sqrt{t/z} \, J_1(2\sqrt{tz})} \tag{7.70}$$

This is Eq. (21) of Owocki and Rybicki (1986), with normalized units. To be specific, $J_1(x)$ is the solution of the differential equation

$$x^2 J_1'' + x J_1' + (x^2 - 1) J_1 = 0, \tag{7.71}$$

and its Taylor series is given by (9.1.10) in A&S,

$$J_1(x) = \frac{x}{2} \sum_{k=0}^{\infty} \frac{(-x^2/4)^k}{(k+1)! \, k!}. \tag{7.72}$$

We can now analyze the significance of the singularity at $s = 0$ in (7.63). To avoid this singularity, simplify the bridging law (7.56) to

$$\overline{N}(k) = ik, \tag{7.73}$$

which holds for $k \ll 1$. Then (7.61) becomes

$$M(t, z) = \frac{1}{2\pi} \int_{-\infty}^{\infty} dk \, e^{ik(z+t)} = \delta(z + t) \tag{7.74}$$

(a standard representation of the delta function), and (7.60) becomes

$$u_1(t, z) = \int_{-\infty}^{\infty} dz'\, \delta(z - z' + t)\, u_1(0, z') = u_1(0, z + t). \qquad (7.75)$$

As already found above, this corresponds to upstream propagation of (the signal) u_1 at a speed -1 for $k \ll 1$. By contrast, however, inserting the full M from (7.70) in (7.60) yields

$$u_1(t, z) = e^t\, u_1(0, z) - \int_{-\infty}^{z} dz'\, e^{t-z+z'} \sqrt{\frac{t}{z - z'}} \; J_1\big(2\sqrt{t(z - z')}\big)\, u_1(0, z').$$

$$\qquad (7.76)$$

Only initial velocity perturbations $u_1(0, z')$ at points $z' \leq z$ contribute to $u_1(t, z)$ at later times. Thus M in (7.70) shows no upstream influence from z' to $z < z'$ at all, in contradiction to what was found in (7.75).

> The resolution of this paradox lies in the realization that information in a completely smooth (analytic) function is *not* localized, since the properties in the neighborhood of any point can be used to infer the function's value at all other points by analytic continuation. (Owocki and Rybicki 1986, p. 131)

Thus for $k \ll 1$, the kernel M in (7.70) only mimics upstream wave propagation. In fact, all what happens is a purely *local* analytic continuation. To substantiate this explanation, Figs. 7.2 and 7.3 show initial conditions for u_1 in form of a truncated Gaussian function, $u_1(0, z) = e^{-(z/100)^2}$ for $z < 0$, else $u_1(0, z) = 0$, and their subsequent evolution in time.

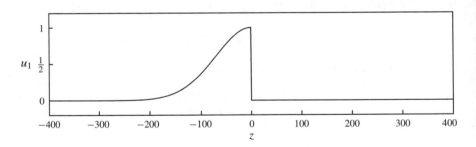

Fig. 7.2   A truncated Gaussian as initial condition

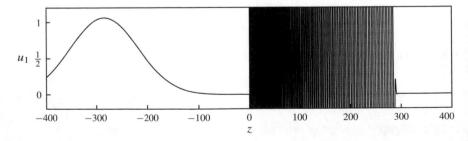

Fig. 7.3   The truncated Gaussian in the absorption model (7.49) and (7.50) at $t = 300$

Rather unexpectedly, Fig. 7.3 for $u_1(t = 300, z)$ shows the emergence of a *full* Gaussian that travels upstream to smaller z, i.e., a fake signal: viz., the 'best guess' that can be made about the signal shape from its knowledge at $z < 0$ only. One could say that the only localized, true information content of the original signal at $t = 0$, namely the truncation at $z = 0$, has vanished at $t = 300$.

> The apparent inward propagation of the pulse is due to its construction from local properties alone, without inward propagation of information. (Owocki and Rybicki 1986, p. 140)

Figure 7.3 was obtained by direct numerical convolution of M from (7.70) and u_1. This calculation is rather pathological in itself: the exponential terms in (7.76) can become very large, of order e^{300}, which is also a measure for the worst noise seen at $z > 0$ in Fig. 7.3. On the other hand, due to the oscillations of the Bessel function J_0, exponential terms may also almost perfectly cancel. To obtain an apparent signal propagation as seen in the figure for $z < 0$, accuracies down to $u_1 \ll 1$ must be reached. The calculation was thus performed with a precision of 616 decimal figures; calculations with (much) smaller numbers of significant digits lead to no sensible results. Quite clearly, therefore, the present model of a radiative force is pathological in physical terms, i.e., too strongly simplified. Despite of these shortcomings, Fig. 7.3 shows an interesting coexistence of order at $z < 0$ and 'chaos' at $z > 0$.

A major reason for the above problems is the total lack in this model of photon *scattering* in bound-bound transitions. After absorption in a bound-bound transition of an ion and release of their momenta, photons are usually re-emitted by de-excitation of the same bound-bound transition. As was first noted by Sobolev (1960), re-emission does not appear in the momentum balance for the unperturbed solution u_0, since re-emission is fore-aft symmetric and there is no net momentum transfer associated with it. Yet, re-emission appears as a dominant term in the Euler equation for first-order perturbations, as first recognized by Lucy (1984), and leads to a *line drag effect*. This renders the flow to be marginally stable, i.e., the instability $\omega = i$ for $k \to \infty$ from (7.53) vanishes. The reason is that force perturbations at $z' < z$ are completely cancelled by force perturbations at $z' > z$, i.e., due to the back-scattered radiation field: the symmetrization and cancellation of radiative force perturbations caused by back-scattering radiation is an essential ingredient to overcome the pathologies of the pure absorption model. This, however, requires that exactly one half of the radiation is scattered forward and one half backward, as was shown by Lucy (1984) and Owocki and Rybicki (1985). Equation (7.46) becomes in this case, quite plausibly,

$$g_1(t, z) = \frac{1}{2} \int_{-\infty}^{z} \mathrm{d}z' \, e^{-(z-z')} \, u_1'(t, z') + \frac{1}{2} \int_{z}^{\infty} \mathrm{d}z' \, e^{-(z'-z)} \, u_1'(t, z'). \qquad (7.77)$$

see Eq. (7) in Thomas and Feldmeier (2016). Integration by parts gives

$$g_1(t, z) = -\frac{1}{2} \int_{-\infty}^{z} \mathrm{d}z' \, e^{-(z-z')} \, u_1(t, z') + \frac{1}{2} \int_{z}^{\infty} \mathrm{d}z' \, e^{-(z'-z)} \, u_1(t, z'). \qquad (7.78)$$

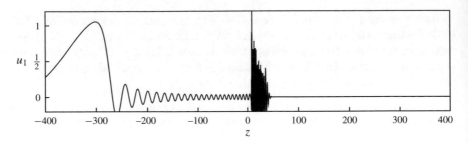

Fig. 7.4 The truncated Gaussian in the scattering model (7.49) and (7.78) at $t = 300$

The first term on the right side of (7.50), $u_1(t, z)$, does not appear here. Thus, there is indeed no line-driven instability in this scattering model. The dispersion relation is now (see Eq. 11 in Thomas and Feldmeier (2016))

$$\omega = -\frac{k}{1 + k^2}. \tag{7.79}$$

This is purely real, with no instability. For $k \to 0$, again $\omega/k = -1$.

Figure 7.4 shows $u_1(300, z)$ for this scattering model. The truncated Gaussian propagates upstream and remains truncated, the pathological behavior of the absorption model is gone. Indeed, due to back-scattered photons propagating upstream, regions $z' > z$ can now influence the point z, thereby giving physical ground for the appearance of upstream propagating radiative waves, which was not given for a radiative force from pure absorption, allowing only regions $z' < z$ to influence the point z. The deformation of the truncated Gaussian (7.4) is caused by wave dispersion according to (7.79). Similarly, the perturbations propagating downstream from $z = 0$ to $z > 0$ in (7.4) are not amplified by an instability (which is completely absent in this model), but they correspond to those wavenumbers k that propagate at maximum speed $1/8$ according to the dispersion relation (7.79). Their peak amplitude is indeed $u_1 = O(1)$ only, many orders of magnitude smaller than those of the noise in Fig. 7.3. If fractions $\eta_\pm < \frac{1}{2}$ of photons are scattered fore and back, and a fraction $1 - \eta_+ - \eta_- > 0$ vanishes (either by being re-emitted in different bound-bound transitions at other frequencies or by collisional de-excitation of the excited ion), one finds upstream propagation of radiative waves with strongly growing amplitudes due to the line-driven instability. Time-dependent numerical simulations of this nonlinear growth in presence of *wave stretching* caused by the strongly accelerating background flow can be found in Owocki et al. (1988) and Feldmeier and Thomas (2017).

To summarize this section, we have found that in an unstable flow, strictly local analytic continuation can mimic wave or signal propagation. Instead of using the phase speed or group speed from a dispersion relation then, it is safer to rely on a Green's function analysis with a strictly localized, peaked initial signal and the subsequent propagation of an outer wavefront originating from it.

References

Abbott, D.C. 1980. The theory of radiatively driven stellar winds. I. A physical interpretation. *Astrophysical Journal* 242:1183.

Abramowitz, M., and I.A. Stegun. 1964. *Handbook of mathematical functions.* New York: Dover.

Copson, E.T. 1965. *Asymptotic expansions.* Cambridge: Cambridge University Press.

Feldmeier, A., and T. Thomas. 2017. Non-linear growth of the line-driving instability. *Monthly Notices of the Royal Astronomical Society* 469: 3102.

Griffiths, D.J., and D.F. Schroeter. 2018. *Introduction to quantum mechanics.* Cambridge: Cambridge University Press.

Herglotz, G. 1985. *Vorlesungen über die Mechanik der Kontinua.* Leipzig: B.G. Teubner.

Lamb, H. 1904. On group-velocity. *Proceedings of the London Mathematical Society, Series 2,* 1: 473.

Lamb, H. 1932. *Hydrodynamics.* Cambridge: Cambridge University Press; New York: Dover (1945).

Lighthill, J., and G.B. Whitham. 1955. On kinematic waves. I. Flood movement in long rivers. *Proceedings of the Royal Society of London A* 229: 281.

Lucy, L.B. 1984. Wave amplification in line-driven winds. *Astrophysical Journal* 284: 351.

Lucy, L.B., and P.M. Solomon. 1970. Mass loss by hot stars. *Astrophysical Journal* 159: 879.

Mihalas, D., and B. Weibel-Mihalas. 1984. *Foundations of radiation hydrodynamics.* Oxford: Oxford University Press; Mineola: Dover (1999).

Milne, E.A. 1926. On the possibility of the emission of high-speed atoms from the sun and stars. *Monthly Notices of the Royal Astronomical Society* 86: 459.

Milne-Thomson, L.M. 1968. *Theoretical hydrodynamics,* 5th ed. London: Macmillan; New York: Dover (1996).

Owocki, S.P., and G.B. Rybicki. 1984. Instabilities in line-driven stellar winds. I. Dependence on perturbation wavelength. *Astrophysical Journal* 284: 337.

Owocki, S.P., and G.B. Rybicki. 1985. Instabilities in line-driven stellar winds. II. Effect of scattering. *Astrophysical Journal* 299: 265.

Owocki, S.P., and G.B. Rybicki. 1986. Instabilities in line-driven stellar winds. III. Wave propagation in the case of pure line absorption. *Astrophysical Journal* 309: 127.

Owocki, S.P., J.I. Castor, and G.B. Rybicki. 1988. Time-dependent models of radiatively driven stellar winds. I. Nonlinear evolution of instabilities for a pure absorption model. *Astrophysical Journal* 335: 914.

Reynolds, O. 1877. On the rate of progression of groups of waves and the rate at which energy is transmitted by waves. *Nature* 16: 343.

Sagdeev, R.Z., and C.F. Kennel. 1991. Collisionless shock waves. *Scientific American* 264 (4): 106.

Sobolev, V.V. 1960. *Moving envelopes of stars.* Cambridge: Harvard University Press; Russian original: Leningrad University (1947).

Thomas, T., and A. Feldmeier. 2016. Radiative waves in stellar winds with line scattering. *Monthly Notices of the Royal Astronomical Society* 460: 1923.

Whitham, G.B. 1960. A note on group velocity. *Journal of Fluid Mechanics* 9: 347.

Chapter 8
Shallow Water Waves

In shallow water waves, the water is shallow with respect to the wave, in the sense that the wavelength is (much) larger than the water depth, and the whole body of water, in every depth, is approximately equally affected by the wave. If the wave has sufficient lateral extent, the surfaces of equal wave phase (wavefronts) are vertical planes. The back and forward phase motion of water parcels is largely in the horizontal direction. Still, the wave can have substantial elevation comparable to the water height. For a linear shallow wave, fluid parcels oscillate about a rest position, and there is no absolute (drift) motion connected with the wave.

There are three main reasons for the occurrence of shallow water waves on oceans. *First*, when waves approach a sloping coast, the water height decreases until the wave affects the full water column from surface to ground. *Second*, tidal forces from the Moon and the Sun act *horizontally* on ocean water (the vertical tidal force cannot set water into motion, but merely reduces its weight by a tiny fraction). Tidal forces are nearly independent of depth, since the ocean ground is maximally eleven kilometers below the water surface (in the Mariana Trench), which is a small fraction of the Earth's radius over which the tidal force varies from its maximum at the surface to zero at the center of the Earth. Tidal waves are shallow water waves per se, since the driving material force does almost not vary over the water column in vertical direction. *Third*, tsunamis or seismic sea waves are shallow water waves, and originate in a displacement of water (columns) due to lifting or lowering of the seafloor caused by earthquakes, underwater landslides, volcanic eruptions, etc. There are also many applications of shallow water theory for rivers or canals, for example, in hydraulic jumps or tidal bores.

In the following, the basic theory of shallow water waves is developed, up to their dispersion relation. The prototypical example of tides in canals and rivers is considered, followed by examples of linear and nonlinear, breaking and non-breaking shallow water waves, most notably tidal *bores*. The final section of the chapter presents an extended calculation from an important paper by Crease (1956), demonstrating the novel possibility of waves propagating behind landmasses, as is sometimes observed on Britain's *eastern* shore during storms on the Atlantic Ocean.

© Springer Nature Switzerland AG 2019
A. Feldmeier, *Theoretical Fluid Dynamics*, Theoretical and Mathematical Physics,
https://doi.org/10.1007/978-3-030-31022-6_8

8.1   Continuity Equation

The fluid, typically water, is assumed to be inviscid, its density ρ constant, and therefore the velocity field $\vec{u}$ solenoidal, $\nabla \cdot \vec{u} = 0$. The basin is a *straight* canal (or channel or river), with the Cartesian x-coordinate pointing along the canal, and z pointing vertically upward. The canal is assumed to have *uniform* cross section, which means that its width at any fixed height z is the same at all x (the cross section could be a triangle or a parabola). Furthermore, in the following, the cross section is assumed to be *rectangular*, i.e., the canal width is the same at every height z. A generalization of some of the following material to uniform canals of arbitrary profile is given by McCowan (1892). The canal width is normalized to unity, and there shall be no cross-canal dependence of flow properties. The origin of the xz-system is put in the undisturbed water surface, since this surface is flat, whereas the canal ground may be sloped. The water height from the ground to the undisturbed water surface is $h(x) > 0$, the vertical water elevation with respect to the unperturbed surface is $\zeta(t, x)$, and thus the total water height from ground to wave surface is $h(x) + \zeta(t, x)$. The horizontal and vertical velocity components are $u(t, x, z)$ and $w(t, x, z)$, respectively, and the cross-canal speed is zero.

Let the free fluid surface be given by the equation

$$S(t, x, z) = 0, \tag{8.1}$$

with scalar function S. The normal direction $\hat{n}$ to the (momentary) fluid surface is given by

$$\hat{n} = \frac{1}{\sqrt{S_x^2 + S_z^2}} \begin{pmatrix} S_x \\ S_z \end{pmatrix}, \tag{8.2}$$

and the change in S along this normal direction $\hat{n}$ is

$$dS = |\nabla S| \, dn. \tag{8.3}$$

This gives (the trivial equation)

$$\frac{\partial S}{\partial n} = S_n = |\nabla S| = \sqrt{S_x^2 + S_z^2}. \tag{8.4}$$

Consider next the natural coordinate system for the surface S, i.e., the normal, $\hat{n}$, and tangent, $\hat{t}$, direction in any point of the surface. Following the motion of the fluid surface in time means

$$0 = dS = S_t \, dt + S_n \, dn. \tag{8.5}$$

There is no further term on the right, since the function S does not change within the surface $S = 0$. The speed of any surface point in the normal direction $\hat{n}$ is given by

$$\frac{dn}{dt} = -\frac{S_t}{S_n} = -\frac{S_t}{\sqrt{S_x^2 + S_z^2}}.$$ (8.6)

The *kinematic* boundary condition is that fluid parcels stay in S at all times, i.e., there is no transport of fluid parcels across the water surface. This is a consequence of the assumption that fluid motion is continuous in time, see Sect. 2.6. The speed of a fluid parcel in direction $\hat{n}$ is, using (8.2),

$$u_n = \vec{u} \cdot \hat{n} = \frac{u S_x + w S_z}{\sqrt{S_x^2 + S_z^2}}.$$ (8.7)

The kinematic boundary condition states that

$$\frac{dn}{dt} = u_n,$$ (8.8)

and inserting (8.6) and (8.7) yields

$$- S_t = u S_x + w S_z,$$ (8.9)

or

$$\frac{dS}{dt} = S_t + u S_x + w S_z = 0,$$ (8.10)

see Stoker (1957), p. 11, Eq. (1.4.1) or Crapper (1984), p. 19, Eq. (1.28)—who actually *starts* with this equation—and Wehausen and Laitone (1960), p. 451. With the quantities introduced above, the free fluid surface reads

$$S(t, x, z) = \zeta(t, x) - z = 0.$$ (8.11)

Inserting this in (8.10) gives

$$\frac{\partial \zeta}{\partial t} + u \frac{\partial \zeta}{\partial x} - w = 0 \quad \text{on} \quad z = \zeta.$$ (8.12)

This is the boundary condition (2.262) in Sect. 2.20, with $u = \phi_x$, $w = \phi_z$, and $\zeta = h$ (see also Stoker 1957, p. 11, Eq. 1.4.5 or Crapper 1984, p. 19, Eq. 1.31), where the latter was obtained formally from the Euler–Lagrange equation. Next, the kinematic boundary condition at the canal ground is

$$S(t, x, z) = h(x) + z = 0$$ (8.13)

(note the sign), and (8.10) gives

$$u \frac{dh}{dx} + w = 0 \quad \text{on} \quad z = -h.$$ (8.14)

The first assumption on shallow water waves is as follows.

(I) In shallow water waves, the horizontal speed is independent of height.

Therefore,

$$u = u(t, x). \tag{8.15}$$

This says that shallow water waves affect the fluid equally at all depths. Adding (8.12) and (8.14) gives then, and only then,

$$\frac{\partial \zeta(t, x)}{\partial t} + u(t, x) \frac{\partial [h(x) + \zeta(t, x)]}{\partial x} - w(t, x, \zeta) + w(t, x, -h) = 0. \tag{8.16}$$

Furthermore, integrating the equation

$$\frac{\partial u}{\partial x} + \frac{\partial w}{\partial z} = 0 \tag{8.17}$$

vertically from bottom to top gives, assuming again that $u = u(t, x)$ alone,

$$0 = \int_{-h}^{\zeta} dz \, \frac{\partial u(t, x)}{\partial x} + \int_{-h}^{\zeta} dz \, \frac{\partial w}{\partial z}$$

$$= (h + \zeta) \frac{\partial u}{\partial x} + w(t, x, \zeta) - w(t, x, -h), \tag{8.18}$$

or

$$w(t, x, \zeta) - w(t, x, -h) = -(h(x) + \zeta(t, x)) \frac{\partial u(t, x)}{\partial x}. \tag{8.19}$$

Inserting (8.19) in (8.16) yields

$$\frac{\partial \zeta}{\partial t} + u \frac{\partial (h + \zeta)}{\partial x} + (h + \zeta) \frac{\partial u}{\partial x} = 0, \tag{8.20}$$

with height $h = h(x)$, vertical elevation $\zeta = \zeta(t, x)$, and horizontal speed $u = u(t, x)$. This gives the continuity equation

$$\boxed{\frac{\partial \zeta}{\partial t} + \frac{\partial}{\partial x} [(h + \zeta) u] = 0} \tag{8.21}$$

Note that the only assumption entering here (besides continuity) is that $u = u(t, x)$ alone. Apart from this, the equation is general, and holds especially for linear and *nonlinear* waves: no linearization was performed. The somewhat simpler linear theory of shallow water waves is developed in Lamb (1932), p. 254.

8.2 The Euler Equations

The second property defining shallow water waves is the following.

(II) In shallow water waves, the vertical acceleration can be neglected.

More precisely, the vertical acceleration has to be very small as compared to the gravitational acceleration g (see McCowan 1892). It will be found that this is the case for large wavelengths $\lambda \gg h$. Still, the water elevation $\zeta(t, x)$ above or below the unperturbed height can be large (in breaking waves, bores, and solitons, it can become comparable to the unperturbed water height), and the foregoing assumption is *not* a restriction to linear waves.

> We assume [...] that the vertical acceleration is small. The equation of motion in the vertical direction then reduces to the hydrostatic equation. (Crease 1956, p. 87)

The *vertical* Euler equation in the z-direction, with gravitational acceleration g, and since the vertical acceleration is negligible, is given by

$$0 = \frac{dw}{dt} = -\frac{1}{\rho}\frac{\partial p}{\partial z} - g. \tag{8.22}$$

Integration in the $-z$-direction gives

$$p(t, x, z) = p(t, x, \zeta) + (\zeta(t, x) - z)\, g\rho, \tag{8.23}$$

where $p(t, x, \zeta)$ is the pressure at the water surface, assumed to be constant. Differentiating (8.23) with respect to x gives then

$$\boxed{\frac{\partial p}{\partial x} = g\rho\,\frac{\partial \zeta}{\partial x}} \tag{8.24}$$

Remarkably, the horizontal acceleration at any depth is given by the tilt of the water surface.

Note that whereas the vertical acceleration dw/dt is zero by assumption, the vertical speed w itself cannot vanish and, in contrast to $u = u(t, x)$, must even depend on z: else $\partial u/\partial x + \partial w/\partial z = \partial u/\partial x = 0$, and there is no wave at all but only a uniform stream $u = \text{const}$.

Next, for the *horizontal* Euler equation in the x-direction, $\partial u/\partial z = 0$ since $u = u(t, x)$, and therefore

$$\frac{\partial u}{\partial t} + u\frac{\partial u}{\partial x} = -\frac{1}{\rho}\frac{\partial p}{\partial x}. \tag{8.25}$$

Inserting (8.24), this becomes

$$\boxed{\frac{\partial u}{\partial t} + u \frac{\partial u}{\partial x} = -g \frac{\partial \zeta}{\partial x}} \tag{8.26}$$

Equations (8.21) and (8.26) are the one-dimensional equations for shallow water waves in a straight canal, for the hydrodynamic fields $\zeta(t, x)$ and $u(t, x)$. For later use, the horizontal displacement $\xi(t, x)$ of fluid parcels is introduced,

$$\frac{\partial \xi}{\partial t} = u. \tag{8.27}$$

8.3   Wave Equation for Linear Water Waves

Linear water waves are defined by the assumptions that the elevation ζ is very small compared to the water height, $\zeta \ll h$, and that the inertia term $u \, \partial u/\partial x$ in the horizontal Euler equation can be neglected. Equations (8.21) and (8.26) then become

$$\frac{\partial \zeta}{\partial t} = -\frac{\partial}{\partial x}(hu), \tag{8.28}$$

$$\frac{\partial u}{\partial t} = -g \frac{\partial \zeta}{\partial x}. \tag{8.29}$$

Eliminating ζ gives

$$\ddot{u} - g \, (hu)'' = 0, \tag{8.30}$$

where a prime indicates differentiation along the canal. Assuming constant canal depth, $h = \text{const}$,

$$\ddot{u} - ghu'' = 0. \tag{8.31}$$

This is the wave equation for the horizontal speed u. A wave equation for the horizontal displacement ξ is derived along similar lines: with $u = \dot{\xi}$, Eq. (8.28) becomes, for $h = \text{const}$,

$$\dot{\zeta} = -h \dot{\xi}'. \tag{8.32}$$

Integration yields the linearized continuity equation,

$$\zeta = -h \xi'. \tag{8.33}$$

The linearized Euler equation (8.29) gives, with (8.27),

$$\ddot{\xi} = -g\zeta'. \tag{8.34}$$

Eliminating ζ from the last two equations gives the wave equation for ξ,

$$\boxed{\frac{\partial^2 \xi}{\partial t^2} - gh \frac{\partial \xi^2}{\partial x^2} = 0} \tag{8.35}$$

The phase speed in both (8.31) and (8.35) is

$$\boxed{c = \sqrt{gh}} \tag{8.36}$$

Note that there is no factor of two as in the free-fall law, $v = \sqrt{2gh}$, and that the group speed equals the phase speed.

The conditions can be quantified now under which the vertical acceleration is indeed negligible, which for shallow water waves is assumption (II). A full wavelength λ of a wave with speed c will pass through a fluid parcel in time λ/c. The parcel will be accelerated vertically by this wave by $\approx \zeta c^2/\lambda^2$. The horizontal acceleration, by contrast, is $g \, \partial \zeta/\partial x$, which is of order $g\zeta/\lambda$. The vertical acceleration is therefore negligible if

$$\frac{\zeta c^2}{\lambda^2} \ll \frac{g\zeta}{\lambda}, \tag{8.37}$$

and upon inserting for the wave speed $c^2 = gh$,

$$\lambda \gg h. \tag{8.38}$$

Thus the vertical acceleration can be neglected compared to the horizontal acceleration if the wavelength λ is very large compared to the water depth.

The speed of fluid parcels hit by a linear water wave is much smaller than the wave speed itself. To see this, it is sufficient to consider horizontal motion only (see below). For linear waves,

$$u \frac{\partial u}{\partial x} \ll \frac{\partial u}{\partial t}. \tag{8.39}$$

For changes within the wave, $\partial/\partial t = c \, \partial/\partial x$ holds, or

$$\frac{\partial u}{\partial t} = c \frac{\partial u}{\partial x}. \tag{8.40}$$

Comparison with (8.39) gives

$$u \ll c, \tag{8.41}$$

thus the parcel speed in a wave is much smaller than the wave speed.

The general solution of the wave Eq. (8.35) can easily be verified to be given by

$$\xi(t, x) = \xi_-(x - ct) + \xi_+(x + ct), \tag{8.42}$$

where ξ_- and ξ_+ are arbitrary differentiable functions of one argument. The functions ξ_- and ξ_+ describe waveforms that propagate to larger and smaller x, respectively, without change of shape. The special case of a plane wave solution is given by

$$\xi(t, x) = ae^{i(kx-\omega t)}, \tag{8.43}$$

with $\omega = ck$. From (8.27) and (8.42) follows

$$\frac{u}{c} = \frac{1}{c} \frac{\partial \xi}{\partial t} = -\xi'_- + \xi'_+, \tag{8.44}$$

where $\xi'_\pm = d\xi_\pm(y)/dy$ assuming $\xi_\pm = \xi_\pm(y)$. From (8.33) and (8.42) follows

$$\frac{\zeta}{h} = -\frac{\partial \xi}{\partial x} = -\xi'_- - \xi'_+. \tag{8.45}$$

Adding and subtracting the last two equations gives

$$\xi'_- = -\frac{1}{2}\left(\frac{u}{c} + \frac{\zeta}{h}\right), \tag{8.46}$$

$$\xi'_+ = \frac{1}{2}\left(\frac{u}{c} - \frac{\zeta}{h}\right). \tag{8.47}$$

For waves propagating to larger x, $\xi_+ = 0$, thus $\xi'_+ = 0$, and from (8.47),

$$\frac{u}{c} = \frac{\zeta}{h}. \tag{8.48}$$

For waves propagating to smaller x, $\xi_- = 0$, thus $\xi'_- = 0$, and (8.46) gives

$$\frac{u}{c} = -\frac{\zeta}{h}. \tag{8.49}$$

Thus $u \ll c$ implies $|\zeta| \ll h$, and therefore for linear shallow water waves,

$$\boxed{|\zeta| \ll h \ll \lambda} \tag{8.50}$$

which confirms that the motion is essentially horizontal.

In the following section, shallow water waves resulting from a tidal force are studied. For this, a new horizontal force f is added to the horizontal equation of motion. Then (8.34) becomes

$$\frac{\partial^2 \xi}{\partial t^2} = -g\frac{\partial \zeta}{\partial x} + f. \tag{8.51}$$

Inserting (8.33) gives, for linear waves and constant h, the *forced* wave equation

$$\frac{\partial^2 \xi}{\partial t^2} - gh \frac{\partial^2 \xi}{\partial x^2} = f. \tag{8.52}$$

8.4 Tides in Canals

Observed tides can vary strongly from harbor to harbor, and the analytical theory of tides on a two-dimensional, rotating Earth with landmasses is rather limited in its predictive power, so that one has to retreat to numerical calculations and semi-empirical approaches. In the following, therefore, only a simple theory of one-dimensional tides in straight canals is developed. The first example, however, treats a lake or sea of arbitrary shape and gives a quite trivial, yet illuminating estimate for the height of tides from horizontal lunar gravity. Note, again, that tides on the Earth result from horizontal gravitational pull f of the Moon on water. Its vertical pull can be neglected, since it only reduces the water weight by a very tiny amount, and leads to no observable dynamical consequences.

The tidal force is the difference between the lunar (or solar) gravity at the Earth's center and its surface. From Newton's law of gravity, roughly $f/g \approx 10^{-7}$, and this already allows to estimate tidal heights. Let a lake have diameter l (in arbitrary direction), as shown in Fig. 8.1.

The water surface tilts uniformly until it is normal to the force $f\hat{x} - g\hat{z}$ at every point. Thus for the tidal height ζ, from elementary geometry,

$$\frac{2\zeta}{l} = \frac{f}{g}. \tag{8.53}$$

Note that the tidal height ζ is independent of the water depth. For a lake with $l = 100\,\text{km}$, this gives $\zeta \approx 5\,\text{mm}$, which is small but measurable. For example, tides on lake Geneva with a height of $\zeta \approx 2\,\text{mm}$ are observed. Inserting (somewhat arbitrarily, regarding the assumed planar geometry) for l the radius of the Earth gives $\zeta \approx 50\,\text{cm}$ as an estimate for the height of tides on (open) oceans, which compares well with a more accurate estimate of $\zeta \approx 70\,\text{cm}$.

Fig. 8.1 Force balance in equilibrium tides

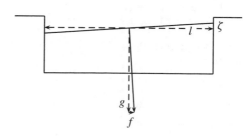

From now on, the progression of tides as shallow water waves that are caused by the apparent motion of the Moon (i.e., by the daily rotation of the Earth) is studied; either in straight canals or in one-dimensional 'strips' on the open sea. Tides caused by the Sun are not treated, although . . .

> . . . there are some places where the tides are regulated by the sun as much as they are by the moon, or even more. (Whewell 1848, p. 8; cited in Aldersey-Williams 2016, p. 222)

The tidal force per mass f is calculated from a tidal potential V; we skip here the derivation of V, which can be found in Lamb (1932), p. 358, Proudman (1953), p. 6, or Feldmeier (2013), p. 92, resulting in

$$V = \gamma a \left(\frac{1}{3} - \cos^2 \Psi \right),$$ (8.54)

where

$$\gamma = \frac{3}{2} \frac{m}{M} \frac{a^3}{d^3} \frac{GM}{a^2} \approx 8.6 \times 10^{-8} \, g.$$ (8.55)

Here $M/m \approx 81$ is the mass ratio of the Earth and Moon (only lunar tides are considered in the following) and $d/a \approx 60$ is the ratio of the Earth–Moon distance d to the radius a of the Earth. The independent variable in the potential is Ψ, the *zenith angle* of the Moon, which is to be defined shortly. As stated above, only tidal waves in straight, one-dimensional canals and rivers are considered. To simplify matters further, the Moon is assumed to be orbiting in the equatorial plane of the Earth throughout.

The first example is that of a tidal wave in a canal or river *along the equator*, see Fig. 8.2. Here, E is an arbitrary but fixed meridian, P is a fluid parcel, and m is the Moon. The angle φ is the azimuth of P, measured from E (eastward counts positive), and Ψ_0 is zenith angle of the Moon at E. Note that the zenith angle at E is measured with respect to the center O of the Earth, i.e., it is the angle between Om and OE, and *not* the angle between Em and EE'. Finally, $\Psi = \Psi_0 + \varphi$ is the zenith angle of the Moon at P.

Fig. 8.2 Equatorial cut through the Earth. The Moon m is in equatorial orbit around the Earth, the canal or river is along the equator. E : fixed meridian, P : fluid parcel, Ψ_0 : zenith angle of the Moon at E

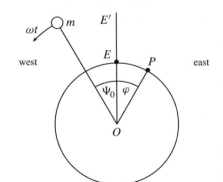

The Moon travels westward at an apparent angular speed $\omega = 2\pi/d_l$, where d_l is the length of a lunar day. Since always the same lunar hemisphere is seen from the Earth, this is given by (with 1 month = 27.3 Earth days d)

$$d_l = \left(1 + \tfrac{1}{27.3}\right) d = 24\,\text{h}\,50\,\text{m}. \tag{8.56}$$

Then, with $\omega > 0$,

$$\Psi = \omega t + \varphi. \tag{8.57}$$

The tidal acceleration per mass at P is westward (negative). From (8.54) for the tidal potential,

$$f = -\frac{\partial V}{a\,\partial\Psi} = -2\gamma \sin\Psi \, \cos\Psi = -\gamma \sin(2\Psi). \tag{8.58}$$

The phase 2Ψ evidences the fact that there are *two* tidal bulges on Earth, with a separation of 180°. With this f, Eq. (8.52) for the horizontal displacement becomes

$$\frac{\partial^2 \xi}{\partial t^2} - gh\,\frac{\partial^2 \xi}{a^2\,\partial\varphi^2} = -\gamma \sin(2(\omega t + \varphi)). \tag{8.59}$$

This is the equation of a *forced* harmonic oscillator and has the well-known solution for the long-term amplitude (see Lamb 1932, p. 268),

$$\xi = -\frac{1}{4}\,\frac{\gamma a^2}{c^2 - \omega^2 a^2}\,\sin(2(\omega t + \varphi)) \tag{8.60}$$

(with $c^2 = gh$), which is easily checked by insertion. The minus sign in front of the right side is central in the following. The continuity equation (8.33) is $\zeta = -(h/a)\,(\partial\xi/\partial\varphi)$ and gives as tidal elevation

$$\zeta = \frac{a\gamma}{2g}\,\frac{c^2}{c^2 - \omega^2 a^2}\,\cos(2(\omega t + \varphi)). \tag{8.61}$$

For water directly below the Moon, $\Psi = 0$, the elevation becomes

$$\frac{\zeta}{a} = \frac{\gamma}{2g}\,\frac{1}{1 - \omega^2 a^2/gh}. \tag{8.62}$$

The (inverse) parameter in the denominator of the right side has the numerical value

$$\frac{gh}{\omega^2 a^2} \approx 300\,\frac{h}{a}. \tag{8.63}$$

The radius of the Earth is $a \approx 6000\,\text{km}$, and thus the denominator in (8.62) changes sign at $h \approx 20$ km. The maximum ocean depth on Earth is 11 km, implying $\omega^2 a^2/gh > 1$ for all oceans. Therefore, $\zeta < 0$ directly below the Moon, which is termed an *inverted* tide. Tidal inversion corresponds to a phase shift π of a fast-driven oscillator: the apparent motion of the Moon (i.e., the rotation of the Earth) is faster than propagation of shallow water waves in the oceans. The prediction of inverted tides was an important success of the dynamical theory of tides (i.e., tides treated as tidal waves), as compared to the older equilibrium theory of tides, which does not account for propagation effects (and phases) and can only give direct tides (high water below the Moon). Finally, from (8.60), the horizontal displacement of fluid parcels in tides is ≈ 40 m for canals and rivers and can reach ≈ 70 m for the deepest oceans. By contrast, the wavelength λ for the longest tidal waves on oceans can reach planetary scale.

The second example of tidal motion in this section is for a tidal wave in a canal or river *parallel to the equator*, i.e., at a constant colatitude ϑ, see Fig. 8.3. The azimuthal distance between fluid parcel and the Moon is again $\omega t + \varphi$. The normalized lengths 1, $\sin \vartheta$, $\cos \vartheta$, $\sin \Psi$, and $\cos \Psi$ are measured from the origin O of the Earth. The line x is normal to Om, therefore

$$\cos(\omega t + \varphi) = \frac{\cos \Psi}{\sin \vartheta}. \tag{8.64}$$

Thus, for the angle Ψ in the tidal potential $V(\Psi)$,

$$\cos \Psi = \sin \vartheta \, \cos(\omega t + \varphi). \tag{8.65}$$

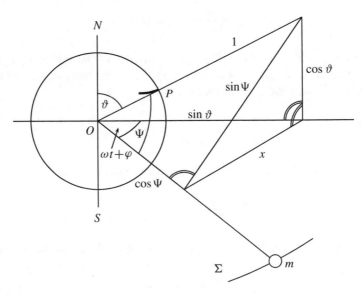

Fig. 8.3 Tides in a canal or river of constant colatitude, with the Moon m in the equatorial plane Σ of the Earth. Bold line: canal, P: fluid parcel

The length element along the canal is $dl = a \sin\vartheta \, d\varphi$. Therefore, the acceleration along the canal is

$$f = -\frac{\partial V}{a \sin\vartheta \; \partial\varphi} = -\gamma \; \sin\vartheta \; \sin(2(\omega t + \varphi)). \tag{8.66}$$

The vertical displacement ζ is now found to be (see Lamb 1932, p. 269, Eq. 3)

$$\zeta = \frac{a\gamma}{2g} \frac{c^2 \sin^2 \vartheta}{c^2 - \omega^2 a^2 \sin^2 \vartheta} \; \cos(2(\omega t + \varphi)). \tag{8.67}$$

For all rivers and canals, $c \ll \omega a$. Thus, in an equatorial belt where $|\sin\vartheta| > c/\omega a$, tides are inverted. However, close to the poles, $\sin\vartheta \to 0$, and $c^2 - \omega^2 a^2 \sin^2 \vartheta > 0$. Canals and rivers within these polar caps show therefore *direct* tides.

The third example is a tidal wave in a canal or river *along a meridian*. Here, Lamb (1932), p. 271 finds that if the Moon is in the meridian of the canal, tides are inverted in an equatorial belt of colatitudes $\frac{\pi}{4} < \vartheta < \frac{3\pi}{4}$, and tides are direct in the polar caps $\vartheta < \frac{\pi}{4}$ and $\vartheta > \frac{3\pi}{4}$. If the Moon, however, is at one of the two points at the horizon that lie perpendicular to the canal, then the tides are direct in the equatorial belt and inverted in the polar caps.

We close this section with a rather peculiar example of a 'river tide.' Figure 8.4 shows a strong tidal current in the Ormhullet in the Barents Sea. The explanation of this phenomenon is as follows:

> The reason for the strong tidal current in the Heley and Freeman Sounds is that the tide on the eastern side of the Barents Island is delayed with 4–6 hours compared to the tide in Storfjorden on the western side of the island. […] The tide at two locations, the Henchel islands and Cape Bessel, which are located respectively on the west and east side of the Heley Sound, have almost opposite phase (i.e., 180° phase lag). This means that when it is high water on the eastern side it is low water on the western side and vice versa. At certain times up to one meter difference is measured in water level between the two recording stations. This large difference in sea level drives the sea as a river through Ormhullet (the Wormhole), a narrow side branch of the Heley Sound. (Gjevik 2009, p. 181; English translation provided by B. Gjevik, priv. comm.)

Fig. 8.4 Tidal current in the Ormhullet near the Heley Sound between the Barents Island and Nordaustlandet near Spitzbergen.
Photograph: Prof. Johan Ludvig Sollid, University of Oslo

8.5   Cotidal Lines and Amphidromic Points

The theory of two-dimensional tides in oceans and large lakes is rather complex and not treated here; but we take a brief look at two interesting topics that are central to the theory, *cotidal lines* and *amphidromic points*. Cotidal lines are lines on an ocean or lake that have the same *phase*—usually the highest elevation or deepest suppression of a tidal wave—at one time. This obviously refers to one universal time, often Greenwich standard time, and, in the case of extended seas, not to local time zones. Furthermore, note that if the cotidal line is defined by either high or low water, then the absolute height of the tidal wave will usually change along a cotidal line. Lines of constant elevation, by contrast, are termed *co-range lines*. Tidal waves consist of different harmonic constituents, and each of them has its own system of cotidal lines.

Cotidal lines show the progression of the phase of a tidal wave, see Fig. 8.5. The roman numerals in the figure are Greenwich standard time.

> These lines [give] a graphic impression of the tide as a single majestic wave, sweeping progressively from New Zealand through the Southern Ocean and then deflecting through the South Atlantic and on up through the North Atlantic toward the Arctic. (Aldersey-Williams 2016, p. 222)

Rather surprisingly, cotidal lines over the full period range of a tidal wave (i.e., roughly twelve hours for the main lunar tide) can merge in one stationary point, as is shown in Fig. 8.6 to occur between the southeast coast of Britain and the coast of Holland. The interpretation is quite clear: the tidal wave (its phase) propagates around this point, which leads to the name *amphidromic point* (Greek amphi = around, as in amphitheater; dromic = running, as in hippodrome). To *fully* account for such a

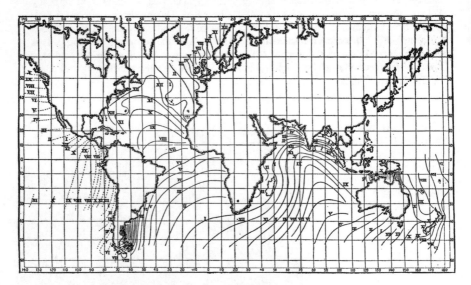

Fig. 8.5 Cotidal lines without amphidromic points. Reproduced from Darwin (1902)

circular motion, besides the tidal force one has to include (i) the Coriolis acceleration due to the rotation of the Earth, (ii) the propagation of waves along landmasses, and (iii) the influence of the sea depth on the speed of the wave propagation.

Amphidromic points are tidal *nodes*, i.e., there is no vertical elevation or suppression at them; the difference between high water and low water, called the tidal range, is zero at amphidromic points and increases with distance from them along cotidal lines (up to some point). Still, there can be tidal currents across amphidromic points, since high water may occur on one of their sides, and low water on the opposite side. This, for example, was one of the early obstacles for a proper understanding of amphidromic points, as some authors thought there could be no tidal current across them.

There were strong objections raised against the occurrence of circular motion near amphidromic points. The first description of amphidromic points by Whewell (1836) was rejected by eminent fluid dynamicist George Airy:

> This circumstance [of two approaching tidal waves, one propagating from north to south, one from south to north] is so strange, that Mr. Whewell, in order to explain it, has had recourse to the supposition of a revolving tide in the German Ocean [the North Sea], in which the tide-wave would run as on the circumference of a wheel, the line of high water at any instant being in the position of a spoke of the wheel. Although our mathematical acquaintance with the motion of extended waters is small, we have little hesitation in pronouncing this to be impossible. (Airy 1845, p. 376, art. 525)

Airy's objection became the leading tradition in the nineteenth century and was renewed by Darwin (1902), a son of Charles Darwin and one of the founders of the theory of tidal friction; for Darwin's cotidal lines, there are no amphidromic points in either the Atlantic or the Pacific Ocean, see Fig. 8.5. Actually, the figure from Darwin's 1902 article is a reproduction of plate 6 from Airy's famous treatise on

Fig. 8.6 Amphidromic points in the North Sea. Reproduced from Proudman and Doodson (1924)

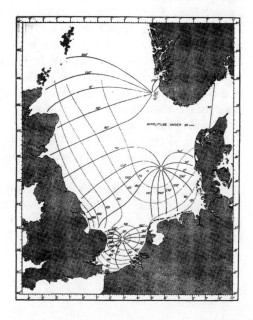

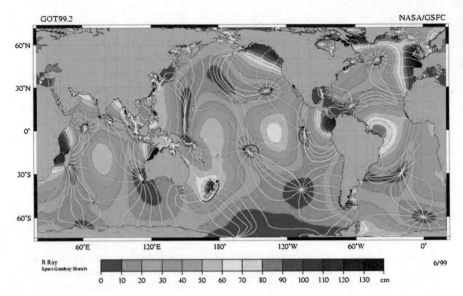

Fig. 8.7 'The M2 tidal constituent [the principal lunar semidiurnal tide with period 12.42 h]. Amplitude is indicated by color, and the white lines are cotidal differing by 1 h. The curved arcs around the amphidromic points show the direction of the tides, each indicating a synchronized 6 h period.' Image: R. Ray, NASA Goddard Space Flight Center, Jet Propulsion Laboratory, Scientific Visualization Studio, https://svs.gsfc.nasa.gov/stories/topex/tides.html

tides and waves from 1845. For comparison, Fig. 8.7 shows a modern world map with cotidal lines and amphidromic points.

This persisting rejection of amphidromic points is more surprising as their principal mechanism is quite easily understood. As Thorade (1941), p. 74 shows, the superposition of two orthogonal standing plane waves in a square basin, each with a wavelength twice the side length a of the basin, and with a temporal phase lag of $\pi/2$ between the two waves, gives rise to an amphidromic point at the center of the basin and cotidal lines radiating out from this point (like 'spokes of a wheel'). A standing wave in the x-direction is given by $\sin(kx)\sin(\omega t)$, and the total normalized elevation ζ from two orthogonal standing waves with phase lag $\pi/2$ is, with the center of the basin as origin, i.e., $x, y \in [-\pi/2, \pi/2]$ for a side length $a = \pi$ of the basin, and choosing $\omega = 1$,

$$\zeta(t, x, y) = \sin x \, \sin t - \sin y \, \cos t. \tag{8.68}$$

Here $\zeta(t, 0, 0) = 0$ at all times, thus the center of the basin is an amphidromic point. The steepest gradient of ζ occurs at the node of the wave. The integral curves through the normal directions of this gradient are the locations of wave nodes and, with a phase lag of $t = \pi/2$, also of extrema of ζ. From $\nabla\zeta \cdot d\vec{r} = 0$, these nodal lines are given by the differential equation

$$dx \ \cos x \ \sin t = dy \ \cos y \ \cos t. \tag{8.69}$$

Therefore, the curves $y(x)$ of wave nodes (or, correspondingly, tidal extrema) at a given time t are given by

$$y = \sin^{-1}(\tan t \ \sin x), \tag{8.70}$$

and are shown in Fig. 8.8.

The velocity field of the wave is obtained from the linearized Euler equation (8.26), which is readily generalized to two dimensions, $\dot{\vec{u}} = -g \ \nabla \zeta$, giving

$$\vec{u} = -g(\cos x \ \cos t, \cos y \ \sin t). \tag{8.71}$$

The velocity $\vec{u} \sim -(\cos x, \cos y)$ at phase $t = \pi/4$ is shown in Fig. 8.9, and there is clearly a current through the amphidromic point. The current changes direction by 360° within the period of the two standing waves (this is an example of a *rotary* velocity field, see Sect. 8.12). A rather similar example of amphidromic points and cotidal lines, however, in an infinitely long canal and resulting from the superposition of the two Kelvin waves traveling in opposite directions, is given in Eq. (12.12), p. 58 and Fig. 12.4, p. 60 of Hutter et al. (2012).

A detailed historical account of cotidal lines and amphidromic points is given in Chap. 9 of Cartwright (1999). A full treatment of amphidromic points in lakes can be found in Hutter et al. (2012), especially Chaps. 12, 14, and 22. An early analytical calculation of cotidal lines and amphidromic points in a basin with wave reflection was performed by Taylor (1922), see Fig. 8.10.

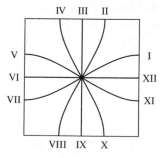

Fig. 8.8 Cotidal lines for the wave node $\zeta = 0$ of two standing waves (8.68) in a basin, with amphidromic point at the center. The Roman numerals are the time for an assumed wave period of 12 h (cf. Thorade 1941, Fig. 48, p. 76)

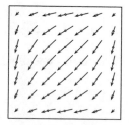

Fig. 8.9 Velocity field (8.71) for two superimposed standing waves at phase $t = \pi/4$ (cf. Thorade 1941, Fig. 49, p. 77)

Fig. 8.10 Amphidromic
point and cotidal lines in a
straight semi-infinite canal.
Reproduced from Taylor
(1922), parts of Fig. 1

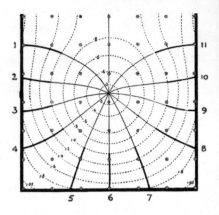

8.6   Waves of Finite Amplitude

So far, the assumption was made that the vertical elevation ζ in a shallow water
wave is very small compared to the depth h of the undisturbed water, $\zeta \ll h$. This
assumption is dropped now. Parts of the wave profile that are highly elevated will
then propagate faster than parts less elevated. This leads to a distortion of the wave
profile in course of time. The solution of the differential equation for the vertical
elevation $\zeta(t, x)$ of the wave profile can be given in the form of an implicit algebraic
equation, which can be further solved in terms of a power series.

Consider again an infinite, straight, uniform, and rectangular canal of constant
depth h and normalized width unity. The unperturbed water shall flow at a constant
speed $u_0 = -b$ in the canal. The speed of shallow water waves for small elevations
was found above to be $c = \sqrt{gh}$. This enters the following equations as a constant
parameter. The equations of continuity (8.21) and motion (8.26) are again

$$\frac{\partial \zeta}{\partial t} + \frac{\partial}{\partial x}[(h + \zeta)\,u] = 0, \tag{8.72}$$

$$\frac{\partial u}{\partial t} + u\,\frac{\partial u}{\partial x} + g\,\frac{\partial \zeta}{\partial x} = 0. \tag{8.73}$$

The following calculation is adapted from Proudman (1958), with elements from
Lamb (1932), §187. First, the continuity equation is multiplied by $\sqrt{g/(h + \zeta)}$.
Subtracting the resulting equation from the equation of motion gives

$$\frac{\partial u}{\partial t} - \sqrt{\frac{g}{h + \zeta}}\,\frac{\partial \zeta}{\partial t} + \left[u - \sqrt{g(h + \zeta)}\right]\left[\frac{\partial u}{\partial x} - \sqrt{\frac{g}{h + \zeta}}\,\frac{\partial \zeta}{\partial x}\right] = 0. \tag{8.74}$$

A sufficient condition for this to hold is that

$$\frac{\partial u}{\partial t} - \sqrt{\frac{g}{h + \zeta}} \frac{\partial \zeta}{\partial t} = 0, \tag{8.75}$$

$$\frac{\partial u}{\partial x} - \sqrt{\frac{g}{h + \zeta}} \frac{\partial \zeta}{\partial x} = 0. \tag{8.76}$$

These equations may be integrated to give

$$u = 2\sqrt{g(h + \zeta)} + \text{const.} \tag{8.77}$$

To fix the constant and to keep here the undisturbed solution $\zeta = 0$ and $u = -b$ of (8.72) and (8.73), let

$$u + b = 2\sqrt{g(h + \zeta)} - 2c, \tag{8.78}$$

or

$$\frac{u + b}{c} = 2\sqrt{1 + \frac{\zeta}{h}} - 2, \tag{8.79}$$

from which it follows that

$$\frac{\zeta}{h} = \frac{u + b}{c} + \left(\frac{u + b}{2c}\right)^2. \tag{8.80}$$

Inserting this in the equation of motion (8.73) gives, since $c = \text{const}$,

$$\boxed{\frac{\partial u}{\partial t} + U(u(t, x)) \frac{\partial u}{\partial x} = 0} \tag{8.81}$$

with

$$\boxed{U = c + \frac{b}{2} + \frac{3u}{2}} \tag{8.82}$$

Remarkably, (8.81) together with (8.82) is a nonlinear differential equation for u alone. The derivation so far is an example of Riemann's method of characteristics, by which coupled quasi-linear differential equations can be decoupled. A partial differential equation is quasi-linear if it is linear in all its derivatives. Aspects of Riemann's method will be encountered again in Sect. 8.8 on dam break and 8.10 on bores, and will be systematically developed to a certain degree in Sects. 12.9–12.12.

Inserting alternatively (8.80) in the continuity equation (8.72) gives again (8.81). Replacing then $\dot{u}$ and u' by $\dot{\zeta}$ and ζ' using (8.75) and (8.76) gives

$$\boxed{\frac{\partial \zeta}{\partial t} + U(u(t, x)) \frac{\partial \zeta}{\partial x} = 0} \tag{8.83}$$

From (8.79), furthermore, one may deduce by simple manipulation

$$c + \frac{b}{2} + \frac{3u}{2} = 3c\sqrt{1 + \frac{\zeta}{h}} - 2c - b. \tag{8.84}$$

Using this in (8.83) gives a nonlinear differential equation for ζ alone,

$$\frac{\partial \zeta}{\partial t} + \left(3c\sqrt{1 + \frac{\zeta}{h}} - 2c - b\right)\frac{\partial \zeta}{\partial x} = 0, \tag{8.85}$$

which, however, will not further be followed. The nonlinear partial differential equations (8.81) and (8.83) have the obvious solutions

$$\frac{\zeta(t, x)}{h} = F(t - x/U), \tag{8.86}$$

$$\frac{u(t, x) + b}{c} = G(t - x/U), \tag{8.87}$$

with (at first) arbitrary functions F and G. This is analogous to the general solution (8.42) of the linear wave Eq. (8.35). Note, however, that U from (8.82) depends on the fluid speed u itself and is not a constant wave speed. Furthermore, solutions to the first-order advection-type Eqs. (8.81) and (8.83) have only one propagation direction, whereas the one-dimensional wave equation has two.

Specifically, F is a continuously differentiable, otherwise arbitrary function of one argument, where $F(-x/U)$ is the initial wave profile at time 0, which is assumed to be known (e.g., corresponds to some incoming tidal wave in the canal). The function G is then *not* arbitrary, but determined by (8.79) as

$$G(x) = 2\sqrt{1 + F(x)} - 2. \tag{8.88}$$

The important Eq. (8.86) with the specific U from (8.82) was already derived by McCowan (1892) using a different approach, see his equations (8′) and (11′). McCowan's method is also interesting in that it starts from the assumption $\zeta(t, x) = F(x - a(t, x)t)$ and derives the continuity and Euler equation for shallow water waves alongside the determination of $a(t, x)$ as function of $u(t, x)$ (and b and c).

Proudman (1958) solves the algebraic equations (8.86) and (8.87) by a power series for ζ and u in terms of the variables x and t. For new independent variables X and T defined by

$$X = \frac{x}{c - b}, \qquad\qquad T = t - \frac{x}{c - b}, \tag{8.89}$$

he obtains the series

$$\frac{\zeta(t,x)}{h} = F(T) + \sum_{n=1}^{\infty} X^n f_n(T),$$

$$\frac{u(t,x)+b}{c} = 2\sqrt{1+F(T)} - 2 + \sum_{n=1}^{\infty} X^n g_n(T). \tag{8.90}$$

Here $f_n(T)$ and $g_n(T)$ are given as follows; let

$$h(T) = \frac{\sqrt{1+F(T)} - 1}{\sqrt{1+F(T)} - b/(3c) - 2/3}. \tag{8.91}$$

Then, with $F'(T) = dF/dT$,

$$f_n(T) = \frac{1}{n!} \frac{d^{n-1}}{dT^{n-1}} \left[h(T)^n \, F'(T) \right],$$

$$g_n(T) = \frac{1}{n!} \frac{d^{n-1}}{dT^{n-1}} \left[h(T)^n \, \frac{F'(T)}{\sqrt{1+F(T)}} \right]. \tag{8.92}$$

This rather peculiar appearance of a series expansion is a direct consequence of Lagrange's inversion theorem on the expansion of a function that is defined implicitly (which is a form of the inverse function theorem).

Theorem (Lagrange's inversion theorem) *Let f be a given function and y a solution of the equation*

$$y = x + f(y). \tag{8.93}$$

Then y is given by the series

$$y = x + \sum_{n=1}^{\infty} \frac{1}{n!} \frac{d^{n-1}}{dx^{n-1}} f(x)^n. \tag{8.94}$$

For a proof as well as references to the history of the theorem, see Grossman (2005).

Proudman (1958) obtains explicit expressions up to fifth order in the above expansions; terms up to third order were already obtained by Airy (1845).

8.7 Nonlinear Tides in an Estuary

Figure 8.11 shows a photograph of the estuary of the Rio de la Plata, Fig. 8.12 a sketch of a simple *estuary*. An *estuary* is a water basin of roughly rectangular or conical shape, and it forms when a river enters the open sea, and the latter shows

Fig. 8.11 Rio de la Plata estuary. Image source: Wikipedia entry 'Estuary' at https://en.wikipedia.org/wiki/Estuary. Original image: Earth Sciences and Image Analysis Laboratory, NASA Johnson Space Center

Fig. 8.12 Geometry of a simple estuary

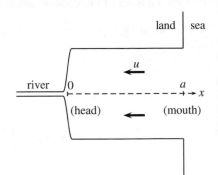

significant tides that enter the estuary. If tides on the sea are small, rivers enter the sea in *deltas*, not in estuaries.

In this section, the tidal wave elevation in an estuary is calculated up to second order in a perturbation expansion, including products of small quantities, i.e., nonlinear terms. The calculation follows Proudman (1957). The estuary shall have constant width, and the undisturbed water has constant depth h. A one-dimensional frictionless flow is assumed, with coordinate x along the estuary. There is no velocity component across the estuary. The *head* of the estuary where the river enters is at $x = 0$, the *mouth* where it is open to the sea is at $x = a$. For simplicity, the head of the estuary is assumed to be closed (no river). The tidal wave entering the estuary is assumed to be a shallow water wave.

The continuity and Euler equations are (8.21) and (8.26),

$$\frac{\partial \zeta}{\partial t} + \frac{\partial}{\partial x}[(h + \zeta)u] = 0, \tag{8.95}$$

$$\frac{\partial u}{\partial t} + u\frac{\partial u}{\partial x} + g\frac{\partial \zeta}{\partial x} = 0. \tag{8.96}$$

In *perturbation theory*, series expansions are performed of the unknowns,

$$\zeta = \zeta_0 + \epsilon\zeta_1 + \epsilon^2\zeta_2 + \cdots,$$
$$u = u_0 + \epsilon u_1 + \epsilon^2 u_2 + \cdots, \tag{8.97}$$

with a small parameter $\epsilon < 1$, and it is assumed that

(a) for all $n \geq 0$, neither ζ_n nor u_n depend on ϵ,
(b) ϵ depends continuously on parameters (here g, h) of the problem.

A perturbation expansion is meaningful if the above series converges quickly, and already the first few terms give a reasonable approximation to ζ and u. With the expansions (8.97), Eqs. (8.95) and (8.96) take the general form $\sum_{n=0}^{\infty} \epsilon^n S_n = 0$, where the S_n are obtained from the differential equations. From (a) and (b) it follows that

$$S_n = 0, \qquad n = 1, 2, 3, \ldots \tag{8.98}$$

For the unperturbed estuary, $u_0 = \zeta_0 = 0$ is assumed due to the closed head. Inserting (8.97) into (8.95) and (8.96) gives

$$\epsilon(\dot{\zeta}_1 + hu_1') + \epsilon^2(\dot{\zeta}_2 + hu_2' + (\zeta_1 u_1)') + O(\epsilon^3) = 0, \tag{8.99}$$

$$\epsilon(\dot{u}_1 + g\zeta_1') + \epsilon^2(\dot{u}_2 + u_1 u_1' + g\zeta_2') + O(\epsilon^3) = 0, \tag{8.100}$$

where $O(\epsilon^3)$ are terms of order ϵ^3 and higher, and not considered here. Putting each order of ϵ separately to zero, four equations result,

$$\dot{\zeta}_1 + hu_1' = 0, \tag{8.101}$$
$$\dot{u}_1 + g\zeta_1' = 0,$$
$$\dot{\zeta}_2 + hu_2' + (\zeta_1 u_1)' = 0, \tag{8.102}$$
$$\dot{u}_2 + u_1 u_1' + g\zeta_2' = 0.$$

Cross-differentiation gives, with the constant parameter $c^2 = gh$,

$$\ddot{\zeta}_1 - c^2\zeta_1'' = 0,$$
$$\ddot{u}_1 - c^2 u_1'' = 0, \tag{8.103}$$

$$\ddot{u}_2 - c^2 u_2'' = g(\zeta_1 u_1)'' - \tfrac{1}{2} \overline{\dot{u}_1^2}{}'. \tag{8.104}$$

The continuity equation (8.102) is kept as second-order equation for ζ_2. Note that ζ_2 does not enter (8.104) for u_2, which can therefore be solved directly. The solution of (8.103) is, with arbitrary functions f and g,

$$u_1(t, x)/c = -f(t + x/c) + g(t - x/c), \tag{8.105}$$

where $f(t + x/c)$ describes a wave propagating from mouth to head of the estuary. According to Fig. 8.12, the wave speed is negative, which is accounted for by the minus sign in (8.105). The boundary condition at the head of the estuary is

$$u(t, 0) = 0. \tag{8.106}$$

To first order then, $f(t) = g(t)$ for all t, thus f and g are the same function,

$$u_1(t, x)/c = -f(t + x/c) + f(t - x/c). \tag{8.107}$$

The function f is left unspecified in the following and shall describe the arbitrary tidal waveform that enters the mouth of the estuary. To first order, the continuity equation (8.101) is $\dot{\zeta}_1 = -hu_1'$, and hence after differentiation and a trivial integration with wave speed $c = \text{const}$,

$$\zeta_1(t, x)/h = f(t + x/c) + f(t - x/c). \tag{8.108}$$

Proudman (1957) introduces new independent variables by

$$n_+ = t + \frac{x}{c}, \qquad\qquad n_- = t - \frac{x}{c}. \tag{8.109}$$

Then $t = (n_+ + n_-)/2$ and $x = c(n_+ - n_-)/2$, and

$$\frac{\partial}{\partial t} = \frac{\partial}{\partial n_+} + \frac{\partial}{\partial n_-}, \tag{8.110}$$

$$c\frac{\partial}{\partial x} = \frac{\partial}{\partial n_+} - \frac{\partial}{\partial n_-}, \tag{8.111}$$

$$\frac{\partial^2}{\partial t^2} - c^2 \frac{\partial^2}{\partial x^2} = 4\frac{\partial^2}{\partial n_+ \partial n_-}. \tag{8.112}$$

The following abbreviations are introduced: $\partial_\pm = \partial/\partial n_\pm$ and $f_\pm = f(n_\pm)$ and $f_\pm' = df(n_\pm)/dn_\pm$. Note that f_+' means differentiation with respect to n_+, but f_-' differentiation with respect to n_-. Equations (8.107) and (8.108) read then

$$\zeta_1/h = f_+ + f_-,$$
$$u_1/c = -f_+ + f_-. \tag{8.113}$$

The remaining calculation is the solution of the two second-order equations. Equation (8.104) becomes, in the variables $n_\pm$,

$$4\,\partial_+\partial_-\,u_2 = \frac{1}{h}\,(\partial_+ - \partial_-)^2(\zeta_1 u_1) - \frac{1}{2c}\,(\partial_+^2 - \partial_-^2)u_1^2. \tag{8.114}$$

Inserting (8.113) gives

$$\frac{4}{c}\,\partial_+\partial_-\,u_2 = -(\partial_+ - \partial_-)^2(f_+^2 - f_-^2) - \frac{1}{2}\,(\partial_+^2 - \partial_-^2)(f_+ - f_-)^2, \tag{8.115}$$

see Eq. (18) in Proudman (1957). With $\partial_\pm f_\mp = 0$, this becomes

$$\partial_+\partial_-\,\frac{u_2}{c} = -\frac{3}{8}\,f_+^{2\prime\prime} + \frac{3}{8}\,f_-^{2\prime\prime} + \frac{1}{4}\,f_+^{\prime\prime}f_- - \frac{1}{4}\,f_+f_-^{\prime\prime}. \tag{8.116}$$

Due to the symmetry of the right side with respect to f_+ and f_-, and in order to obey the boundary condition $u_2 = 0$ at the head of the estuary, where $x = 0$ or $n_+ = n_- = t$, a solution for u_2 of the form

$$\frac{u_2(n_+, n_-)}{c} = \Phi(n_+, n_-) - \Phi(n_-, n_+) \tag{8.117}$$

is looked for, where the function Φ is to be determined. Inserting this in (8.116) gives two equations (since n_+ and n_- are independent variables),

$$\partial_+\partial_-\Phi(n_-, n_+) = \partial_+^2\left(\frac{3}{8}\,f_+^2 - \frac{1}{4}\,f_+f_-\right), \tag{8.118}$$

$$\partial_-\partial_+\Phi(n_+, n_-) = \partial_-^2\left(\frac{3}{8}\,f_-^2 - \frac{1}{4}\,f_+f_-\right). \tag{8.119}$$

Integrating once gives

$$\partial_-\Phi(n_-, n_+) = \partial_+\left(\frac{3}{8}\,f_+^2 - \frac{1}{4}\,f_+f_-\right) + b_-, \tag{8.120}$$

$$\partial_+\Phi(n_+, n_-) = \partial_-\left(\frac{3}{8}\,f_-^2 - \frac{1}{4}\,f_+f_-\right) + b_+, \tag{8.121}$$

where $b_\pm = b(n_\pm)$. Clearly, the same function b must occur in the two lines, however, with different arguments n_- and n_+. Next, in (8.120) an integration over n_- (or actually n'_-) is performed from n_+ to n_-; and in (8.121) an integration over n_+ from n_- to n_+. At the head $n_+ = n_-$ of the estuary, the following simple boundary conditions are applied:

$$\Phi(n_+, n_+) = \Phi(n_-, n_-) = 0. \tag{8.122}$$

This gives

$$\Phi(n_-, n_+) = \frac{3}{8} f_+^{2'}(n_- - n_+) - \frac{1}{4} f_+'(F_- - F_+) + B_- - B_+, \tag{8.123}$$

$$\Phi(n_+, n_-) = \frac{3}{8} f_-^{2'}(n_+ - n_-) - \frac{1}{4} f_-'(F_+ - F_-) + B_+ - B_-, \tag{8.124}$$

where $F' = f$ and $B' = b$. Inserting (8.123) and (8.124) in (8.117) gives

$$\frac{u_2}{c} = \frac{3}{8} (f_+^{2'} + f_-^{2'})(n_+ - n_-) - \frac{1}{4} (f_+' + f_-')(F_+ - F_-) + 2B_+ - 2B_-. \tag{8.125}$$

Thus at the head $n_+ = n_-$ of the estuary, the boundary condition $u_2(t, 0) = 0$ is indeed fulfilled, as was intended with (8.117). Therefore, (8.125) is the solution of the forced wave equation (8.104), with u_1 from (8.107) and ζ_1 from (8.108).

Next, the continuity equation (8.102) for ζ_2 is solved. The prime in (8.102) is $\partial/\partial x$ and is rewritten with $\partial/\partial n_\pm$ using (8.111) as

$$\frac{\dot{\zeta}_2}{h} = -u_2' - \frac{1}{h} (\zeta_1 u_1)'$$
$$= -\frac{1}{c} \frac{\partial u_2}{\partial n_+} + \frac{1}{c} \frac{\partial u_2}{\partial n_-} - \frac{1}{ch} \frac{\partial(\zeta_1 u_1)}{\partial n_+} + \frac{1}{ch} \frac{\partial(\zeta_1 u_1)}{\partial n_-}. \tag{8.126}$$

Inserting ζ_1 and u_1 from (8.113), and u_2/c from (8.125) gives

$$\frac{\dot{\zeta}_2}{h} = \frac{1}{4} \left(f_+^{2'} + f_-^{2'} \right) - \frac{3}{8} \left(f_+^{2''} - f_-^{2''} \right) (n_+ - n_-) + \frac{1}{4} (f_+ + f_-) (f_+' + f_-')$$
$$+ \frac{1}{4} (f_+'' - f_-'') (F_+ - F_-) - 2b_+ - 2b_-. \tag{8.127}$$

This can be integrated over time to give ζ_2, where $n_+ - n_- = 2x/c$ can be taken out of the integral. Instead of performing this calculation, we note that, using $\partial n_\pm/\partial t = 1$,

$$\frac{\partial}{\partial t} \left[(f_+ + f_-)^2 + (f_+' - f_-')(F_+ - F_-) \right]$$
$$= 2(f_+ + f_-)(f_+' + f_-') + (f_+'' - f_-'')(F_+ - F_-) + (f_+' - f_-')(f_+ - f_-)$$
$$= f_+^{2'} + f_-^{2'} + (f_+' + f_-')(f_+ + f_-) + (f_+'' - f_-'')(F_+ - F_-). \tag{8.128}$$

The whole last line appears in (8.127). The first line in (8.128) and the remaining terms in (8.127) give then

$$\frac{\zeta_2}{h} = \frac{1}{4}(f_+ + f_-)^2 - \frac{3x}{4c}(f_+^{2\prime} - f_-^{2\prime}) + \frac{1}{4}(f_+' - f_-')(F_+ - F_-) - 2B_+ - 2B_-,$$

(8.129)

see Eq. (25) in Proudman (1957).

The final and in a sense crucial step in the calculation is to determine the unknown function B from the boundary condition at the mouth of the estuary. Proudman states,

> In this paper, a prescribed incident wave is taken to progress up the estuary [to the head], and the remaining part of the motion is taken to reduce, at the mouth, to a wave traveling down the estuary [away from it] [...] The condition for this is $\zeta_2/h = u_2/c$ [our symbols], where $x = a$, and this will be used as the determining condition in this paper. (Proudman 1957, p. 372, 374)

This is an early example of the use of *nonreflecting* boundary conditions, often used nowadays in numerical simulations on finite spatial grids. If waves propagate through the grid and reach a boundary, they should be able to leave the calculational domain. However, the boundary may reflect parts of the wave back into the interior, where they can lead to spurious results. A classical paper on this subject is by Hedstrom (1979). From (8.97), (8.107), and (8.108), with $u_0 = \zeta_0 = 0$,

$$\frac{\zeta}{h} = \epsilon(f_+ + f_-) \quad + \epsilon^2 \frac{\zeta_2}{h},$$

(8.130)

$$\frac{u}{c} = \epsilon(-f_+ + f_-) + \epsilon^2 \frac{u_2}{c}.$$

(8.131)

The term $f_+ = f(t + x/c)$ corresponds to a wave propagating from mouth to head, to smaller x, and $f_- = f(t - x/c)$ to a wave propagating from head to mouth, to larger x. The relative sign of their amplitudes in (8.130) and (8.131) agrees with (8.48) and (8.49): for propagation to larger x, i.e., for terms in f_-, one has $u/c = \zeta/h$; for propagation to smaller x, i.e., f_+, $u/c = -\zeta/h$.

Proudman uses as boundary condition (his Eq. 12) at the mouth $x = a$ of the estuary that the incident wave is fully specified by the function f, i.e., by the terms ϵf_+ in (8.130) and $-\epsilon f_+$ in (8.131) corresponding to propagation toward the head. At the mouth, the two terms $+\epsilon f_-$ in (8.130) and (8.131) *plus* the terms $\epsilon^2 \zeta_2/h$ and $\epsilon^2 u_2/c$ at $x = a$ (shall) correspond to waves propagating away from the estuary. In order that there is no component from second-order terms at the mouth that propagate into the estuary, the condition

$$\frac{\zeta_2(t, a)}{h} = \frac{u_2(t, a)}{c} = q(t)$$

(8.132)

must hold, with abbreviation $q(t)$, so that the wave leaving the estuary at its mouth $x = a$ has amplitudes $\zeta/h = u/c = \epsilon f_- + \epsilon^2 q$.

Inserting in (8.132) u_2/c from (8.125) and ζ_2/h from (8.129) at $x = a$ yields

$$B\left(t + \frac{a}{c}\right) = \frac{1}{16}\left[f\left(t + \frac{a}{c}\right) + f\left(t - \frac{a}{c}\right)\right]^2 - \frac{3a}{8c} f^{2\prime}\left(t + \frac{a}{c}\right)$$

$$+ \frac{1}{8} f'\left(t + \frac{a}{c}\right)\left[F\left(t + \frac{a}{c}\right) - F\left(t - \frac{a}{c}\right)\right]. \quad (8.133)$$

This equation holds for all t; therefore

$$B(t) = \frac{1}{16}\left[f(t) + f\left(t - \frac{2a}{c}\right)\right]^2 - \frac{3a}{8c} f^{2\prime}(t) + \frac{1}{8} f'(t)\left[F(t) - F\left(t - \frac{2a}{c}\right)\right].$$
$$(8.134)$$

Note that here for the first time, as a consequence of the boundary condition at the mouth, an argument $t - 2a/c$ appears, corresponding to the time it takes a wave to travel up and down the estuary. The $B_\pm$ are obtained by replacing t by $t \pm x/c$. Inserting this in (8.129) gives finally as solution for the second-order elevation

$$\frac{\zeta_2(t, x)}{h} = \frac{1}{8} f^2\left(t + \frac{x}{c}\right) + \frac{1}{8} f^2\left(t - \frac{x}{c}\right) + \frac{1}{2} f\left(t + \frac{x}{c}\right)f\left(t - \frac{x}{c}\right)$$

$$- \frac{1}{8} f^2\left(t + \frac{x}{c} - \frac{2a}{c}\right) - \frac{1}{8} f^2\left(t - \frac{x}{c} - \frac{2a}{c}\right)$$

$$- \frac{1}{4} f\left(t + \frac{x}{c}\right)f\left(t + \frac{x}{c} - \frac{2a}{c}\right) - \frac{1}{4} f\left(t - \frac{x}{c}\right)f\left(t - \frac{x}{c} - \frac{2a}{c}\right)$$

$$+ \frac{3}{4}\frac{a - x}{c} f^{2\prime}\left(t + \frac{x}{c}\right) + \frac{3}{4}\frac{a + x}{c} f^{2\prime}\left(t - \frac{x}{c}\right)$$

$$- \frac{1}{4} f'\left(t + \frac{x}{c}\right)\left[F\left(t - \frac{x}{c}\right) - F\left(t + \frac{x}{c} - \frac{2a}{c}\right)\right]$$

$$- \frac{1}{4} f'\left(t - \frac{x}{c}\right)\left[F\left(t + \frac{x}{c}\right) - F\left(t - \frac{x}{c} - \frac{2a}{c}\right)\right]. \quad (8.135)$$

This is Eq. (27) in Proudman (1957), without the viscosity terms. The corresponding formula for u_2/h is also readily obtained, but not quoted here. From (8.130), the total elevation at the head of the estuary, $x = 0$, is

$$\frac{\zeta(t, 0)}{h} = 2\epsilon f(t) + \epsilon^2\left\{\frac{3}{4} f^2(t) - \frac{1}{4} f^2\left(t - \frac{2a}{c}\right) - \frac{1}{2} f(t)f\left(t - \frac{2a}{c}\right)\right.$$

$$\left. + \frac{3a}{2c} f^{2\prime}(t) - \frac{1}{2} f'(t)\left[F(t) - F\left(t - \frac{2a}{c}\right)\right]\right\}. \quad (8.136)$$

The origin of the terms $f(t - 2a/c)$ is as follows: a wave propagating through the estuary and reaching its mouth at time $t - a/c$ must already upon its start at the head at time $t - 2a/c$ obey certain amplitude relations, in order that the boundary condition (8.132) is obeyed; and this 'maintenance' of the boundary condition at the mouth at $t - a/c$ makes itself felt at the head at t, after a wave propagation time. Thus, in second order there is a coupling of the elevation at the head at time t to that at time $t - 2a/c$. In third and higher order perturbation theory, there are then 'higher' retardation terms of this sort. A similar argument applies to the expression $F(t) - F(t - 2a/c)$, which is the integral of $f(t')$ from time $t - 2a/c$ to time t.

To draw some physical conclusions from (8.136), Proudman (1957) assumes that f grows monotonically in the interval $-2a/c < t < 0$. It follows then (see his Eqs. 29 to 31) that in second order, *the maximum elevation at the head of the estuary is larger and earlier* than to first order. This is one of the major conclusions of the paper.

Proudman (1957) gives more examples of observational consequences on tides in estuaries (including also friction and the length of the estuary) due to second-order effects. Still more material on tides in estuaries can be found in Proudman (1955).

8.8 Similarity Solution: Dam Break

A similarity solution is a solution of the one-dimensional hydrodynamic equations of motion that does not depend on time and space as separate variables, but only on a combination thereof, the so-called *similarity variable*, e.g., $\eta = qr/t^\alpha$, where q is a constant (often making η dimensionless) and the *similarity exponent* α a positive real number (for $\alpha = 0$ one has a stationary, not a similarity solution). With all hydrodynamic fields depending on a single parameter η only, the equations of motion are reduced from partial differential equations in t and r (or x) to ordinary differential equations in η, for which many methods of solution exist. A similarity solution, say $U(\eta)$, has the same shape at all times. With respect to the spatial variable r (or x), the solution at later times is merely an expanded or contracted version of that at earlier times. A typical example of an expansion process is the explosion from a point source, an example for a contraction process is the implosion of a gas bubble in a liquid to a point. There is thus no real change and evolution in similarity solutions, and they only occur in rather simple, featureless physical processes. The small number of physical parameters in a problem is an indication of the presence of a similarity solution. The idea of similarity solutions was introduced by Sedov (1946) and Taylor (1950) in the theory of point blasts.

If the similarity exponent α is known beforehand (either by assumption, by dimensional analysis, from the conservation laws, or from the boundary conditions), the similarity problem is termed to be of the first type; if, by contrast, α is determined as part of the solution (e.g., is fixed in such a way that the solution of a differential equation passes through a critical point, in order to obey certain boundary conditions), the similarity problem is of the second type. Often (not universally), the similarity exponent is a rational number for the first type and an irrational number (e.g., obtained by some iteration procedure) for problems of the second type. For this distinction of types, see Zel'dovich and Raizer (2002), p. 792. For processes with expansion, clearly $\alpha > 0$. For processes with contraction, one keeps $\alpha > 0$, but lets the process evolve from time $t \to -\infty$ to end at time $t = 0$, at which the solution vanishes in a point (see Zel'dovich and Raizer 2002, p. 796).

In this section, we consider a very simple similarity problem connected with shallow water equations, the problem of a dam where the vertical wall holding up the water is completely destroyed at time $t = 0$, and the subsequent motion of water is studied. This section follows Stoker (1957), p. 202 and Stoker (1948).

Starting point are again the dynamical Eqs. (8.21) and (8.26) for water motion in a one-dimensional, straight canal,

$$\dot{u} + uu' = -g\zeta', \tag{8.137}$$

$$\dot{\zeta} + [(h + \zeta)\, u]' = 0, \tag{8.138}$$

where a prime indicates differentiation with respect to x. These equations can be rewritten by introducing a wave speed c defined by

$$c = \sqrt{g(h + \zeta)}. \tag{8.139}$$

Note that the former derivation of $c = \sqrt{gh}$ assumed small height perturbations ζ. The spatial and temporal derivatives of c are

$$c' = g(h + \zeta)'/2c, \tag{8.140}$$

$$\dot{c} = g\dot{\zeta}/2c. \tag{8.141}$$

Inserting (8.140) in (8.137) gives

$$\dot{u} + uu' + 2cc' - gh' = 0. \tag{8.142}$$

From now on, a flat canal bed is assumed, $h' = 0$; then the Euler equation becomes

$$\boxed{\dot{u} + uu' + 2cc' = 0} \tag{8.143}$$

Inserting (8.141) in the continuity equation (8.138) gives

$$2c\dot{c}/g + 2ucc'/g + (h + \zeta)\, u' = 0. \tag{8.144}$$

Using (8.139) in the last term yields

$$\boxed{2\dot{c} + 2uc' + cu' = 0} \tag{8.145}$$

Equations (8.143) and (8.145) describe shallow water motion in a flat canal. Note that no linearization was performed in deriving them. The equations have constant-state solutions $u = \text{const}$ and $c = \text{const}$ (both in space and time). The idea is to look for some 'middle' solution to (8.143) and (8.145) that continuously connects two different constant states that reside (at all times) at small x and at large x. Since the adjacent states are constant, one may assume that they are connected by a simple similarity solution, which depends on one variable only; the following calculation shows that this is indeed the case. The solution connecting two constant states and depending only on a single variable is called a *simple wave*. This idea is developed more systematically in Sect. 12.12. As the single variable, consider a combination of x and t of the simplest possible form,

$$\boxed{\eta = \frac{x}{Vt}} \tag{8.146}$$

Here V is some speed to be determined in the following, and the similarity exponent is assumed to be $\alpha = 1$. This is so to make η dimensionless. The space and time differentials, expressed in η, become

$$\frac{\partial}{\partial x} = \frac{\partial \eta}{\partial x} \frac{d}{d\eta} = \frac{\eta}{x} \frac{d}{d\eta},$$

$$\frac{\partial}{\partial t} = \frac{\partial \eta}{\partial t} \frac{d}{d\eta} = -\frac{\eta}{t} \frac{d}{d\eta}, \tag{8.147}$$

and the hydrodynamic fields shall take the form

$$u(t, x) = U(\eta),$$
$$c(t, x) = C(\eta), \tag{8.148}$$

with functions $U(\eta)$ and $C(\eta)$ to be determined. Then (8.143) and (8.145) become

$$(U - \eta V)\, U' + 2CC' = 0, \tag{8.149}$$
$$2(U - \eta V)\, C' + CU' = 0, \tag{8.150}$$

where $U' = dU/d\eta$ and $C' = dC/d\eta$. Written in matrix notation,

$$\begin{pmatrix} U - \eta V & 2C \\ C & 2(U - \eta V) \end{pmatrix} \cdot \begin{pmatrix} U' \\ C' \end{pmatrix} = \begin{pmatrix} 0 \\ 0 \end{pmatrix}. \tag{8.151}$$

The determinant of the matrix must vanish, hence $(U - \eta V)^2 = C^2$, or

$$U(\eta) = \eta V \pm C(\eta). \tag{8.152}$$

Inserting this into (8.149) gives, with $U' = V \pm C'$,

$$V = \mp 3C'. \tag{8.153}$$

Since V is assumed to be constant, so is C', thus $C(\eta)$ is a *linear* function,

$$C(\eta) = \eta C' + D = \mp \frac{1}{3}\eta V + D, \tag{8.154}$$

with some constant D. According to (8.152), $U(\eta)$ is also linear,

$$U = \eta V \pm C = \eta V \pm \left(\mp \frac{1}{3} \eta V + D \right) = \frac{2}{3} \eta V \pm D. \qquad (8.155)$$

Rewriting this in terms of the fields u and c in the variables t and x gives

$$u(t, x) = \frac{2}{3} \frac{x}{t} \pm D, \qquad (8.156)$$

$$c(t, x) = \mp \frac{1}{3} \frac{x}{t} + D. \qquad (8.157)$$

Therefore, the similarity solution connecting two constant states (to its left and right) has a linear velocity law and a linear law for the wave speed c. The latter determines, through $c = \sqrt{g(h + \zeta)}$, the height of the water, which is thus a quadratic function of x. Finally, the constant D is determined from the two constant states to which the similarity solution connects.

To proceed, we consider the dam break as a specific example where two constant states—the water in the dam and the dry land outside—are connected by the simple wave given by (8.156) and (8.157). A planar, vertical concrete wall shall be located at $x = 0$. Water of constant height h is located at $x > 0$. At time $t = 0$, the wall is suddenly removed (destroyed), and water flows toward $x < 0$. Figure 8.13 shows a space–time diagram of the kinematics.

At time τ, the water has reached the point p on the line OA. The 'signal' about the occurrence of a dam break (in form of one end of the simple wave) propagates into the reservoir. At time τ, it has reached the point q on the line OB. The region pq, i.e., the simple wave, is more specifically called a *rarefaction wave*, and obeys (8.156) and (8.157). Quite clearly, the point q should propagate at the undisturbed wave speed $c_0 = \sqrt{gh}$, since it propagates into the original, unperturbed water of height h. Since the point q propagates at constant speed, the line OB is straight. This allows to fix the integration constant D: choose the *lower* sign in (8.156) and (8.157). Then for $x/t = c_0$,

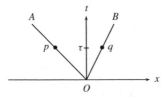

Fig. 8.13 Kinematics of the dam break problem. The dam wall is at $x = 0$. Water is right of the line OA: at time τ, outflowing water has reached the point p on OA. Water beyond point q on OB is still unaffected by dam break, and has unperturbed height h

$$u_q = \frac{2}{3} c_0 - D = 0, \tag{8.158}$$

$$c_q = \frac{1}{3} c_0 + D = c_0, \tag{8.159}$$

from which $D = 2c_0/3$, and thus

$$u(t, x) = \frac{2}{3} \left(\frac{x}{t} - c_0 \right), \tag{8.160}$$

$$c(t, x) = \frac{1}{3} \left(\frac{x}{t} + 2c_0 \right). \tag{8.161}$$

This gives

$$u - 2c = -2c_0 = \text{const.} \tag{8.162}$$

At the point p on the line OA, the water height is zero, thus the wave speed is $c_p = 0$, and (8.161) gives

$$x_p = -2c_0 t. \tag{8.163}$$

Either directly or from inserting x_p in (8.160), one obtains

$$u_p = -2c_0. \tag{8.164}$$

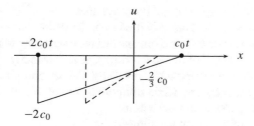

Fig. 8.14 Water speed in the dam break problem, at two instants

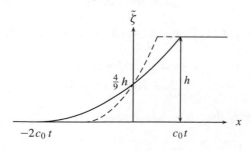

Fig. 8.15 Water height in the dam break problem, at two instants

Thus, OA is also a straight line. Figure 8.14 shows the linear velocity law $u(t, x)$ at two instants in time.

The water height ζ is obtained from $c = \sqrt{g(h + \zeta)} = \sqrt{g\tilde{\zeta}}$, where h is a constant parameter, the undisturbed water height behind the dam, and the new height variable $\tilde{\zeta} = h + \zeta$ is measured directly from the flat ground. The water height $\tilde{\zeta}$ is shown in Fig. 8.15 and takes the form of a parabola, according to (8.161) and $c = \sqrt{2g\tilde{\zeta}}$. Note that at $x = c_0 t$, a discontinuity occurs in the slope $\partial \tilde{\zeta}/\partial x$ of the water surface. Discontinuities in *slopes* of hydrodynamic fields are called *weak* discontinuities, and it can be proved that they propagate at the local wave speed, which is here the shallow water speed c_0, see (8.159).

Finally, from (8.160) one has for $x = 0$ that

$$\frac{u}{c_0} = -\frac{2}{3} \quad \text{and} \quad \frac{\tilde{\zeta}}{h} = \frac{4}{9}. \tag{8.165}$$

Both quantities do not depend on time, and thus the water speed and the water height at the original location of the broken wall are at all times constant!

8.9 Non-breaking Waves

When a wave propagates from the open ocean to the shore, the water height decreases (except for bumps in the ground), and the wave speed $c = \sqrt{g(h + \zeta)}$ also reduces. Faster parts of the wave further out may then overtake slower parts closer in. Usually, this leads to *breaking* of waves, which means that the crest of the wave becomes faster than parts of the wave ahead of it, and overturns, as is constantly observed on ocean coasts. In this section, an unexpected result by Carrier and Greenspan (1958) is derived, showing that breaking, while ubiquitous, is not *necessary*, and that there may be non-breaking waves on slopy shores.

A *constant slope* of the ground is allowed for,

$$gh' = m = \text{const}, \tag{8.166}$$

and the dynamical Eqs. (8.142) and (8.145) for shallow water waves take the form

$$\dot{u} + uu' + 2cc' - m = 0,$$
$$2\dot{c} + 2uc' + cu' = 0. \tag{8.167}$$

Adding and subtracting these equations gives

$$\boxed{\left[\frac{\partial}{\partial t} + (u \pm c) \frac{\partial}{\partial x} \right] (u \pm 2c - mt) = 0} \tag{8.168}$$

These are two partial differential equations of first order for the unknown fields $u(t, x)$ and $c(t, x)$, which can be rewritten as four ordinary differential equations. To this end, we introduce new variables $\eta_\pm$ (these are not similarity variables), replacing x and t, by writing (8.168) as

$$\boxed{\frac{\partial}{\partial \eta_\pm}(u \pm 2c - mt) = 0}\tag{8.169}$$

where

$$\frac{\partial}{\partial \eta_\pm} = \frac{\partial}{\partial t} + (u \pm c)\frac{\partial}{\partial x}.\tag{8.170}$$

This gives, since x and t are independent variables,

$$\frac{\partial x}{\partial \eta_\pm} = u \pm c, \qquad\qquad \frac{\partial t}{\partial \eta_\pm} = 1,\tag{8.171}$$

and therefore

$$\boxed{\frac{\partial x}{\partial \eta_\pm} - (u \pm c)\frac{\partial t}{\partial \eta_\pm} = 0}\tag{8.172}$$

The four ordinary differential equations (8.169) and (8.172) are solved in the remainder of this section. The quantity $u \pm 2c - gh't$ is constant along curves $x(t)$ for which $dx/dt = u \pm c$, or

$$0 = du \pm 2dc - g\,\frac{dh}{dx}\,dt$$
$$= du \pm 2dc - g\,\frac{dh}{dx}\,dx\,\frac{dt}{dx}$$
$$= du \pm 2dc - \frac{g}{u \pm c}\frac{dh}{dx}.\tag{8.173}$$

The wave shall propagate from $x < 0$ toward $x = 0$, where $x = 0$ corresponds to the coastline $h(0) = 0$ for unperturbed water. For simplicity, units with $g = 1$ are assumed, and the ground shall have a positive slope of $45°$,

$$gh = -x, \qquad\qquad gh' = m = -1.\tag{8.174}$$

Although the slope is positive, minus signs appear in these equations since h is the positive depth of water (at $x < 0$) and h decreases (water gets shallower) with $x < 0$ increasing toward $x = 0$. Then,

$$c^2 = g(h + \zeta) = \zeta - x.\tag{8.175}$$

The first two ordinary differential equations (8.169) are readily solved,

$$u + 2c + t = \quad \eta_-,$$
$$u - 2c + t = - \eta_+, \tag{8.176}$$

since η_+ and η_- are independent variables. Here the integration 'constants' on the right sides are chosen by Carrier and Greenspan (1958) in such a way that the following calculation is simplified. Subtracting and adding Eqs. (8.176) yields

$$c = \frac{\eta_- + \eta_+}{4} =: \frac{\sigma}{4}, \tag{8.177}$$

$$u + t = \frac{\eta_- - \eta_+}{2} =: \frac{\tau}{2}, \tag{8.178}$$

where from now on new variables σ and τ are used in place of $\eta_\pm$,

$$\sigma = \eta_- + \eta_+, \qquad\qquad \tau = \eta_- - \eta_+. \tag{8.179}$$

From (8.177) and (8.178),

$$\frac{\partial \sigma}{\partial \eta_-} = \frac{\partial \sigma}{\partial \eta_+} = \frac{\partial \tau}{\partial \eta_-} = 1, \qquad\qquad \frac{\partial \tau}{\partial \eta_+} = -1. \tag{8.180}$$

We now turn to the remaining two ordinary differential equations (8.172), and apply the chain rule; this leads to

$$0 = \frac{\partial x}{\partial \sigma}\frac{\partial \sigma}{\partial \eta_+} + \frac{\partial x}{\partial \tau}\frac{\partial \tau}{\partial \eta_+} - (u+c)\frac{\partial t}{\partial \sigma}\frac{\partial \sigma}{\partial \eta_+} - (u+c)\frac{\partial t}{\partial \tau}\frac{\partial \tau}{\partial \eta_+}$$
$$= \frac{\partial x}{\partial \sigma} - \frac{\partial x}{\partial \tau} - (u+c)\frac{\partial t}{\partial \sigma} + (u+c)\frac{\partial t}{\partial \tau}, \tag{8.181}$$

and

$$0 = \frac{\partial x}{\partial \sigma}\frac{\partial \sigma}{\partial \eta_-} + \frac{\partial x}{\partial \tau}\frac{\partial \tau}{\partial \eta_-} - (u-c)\frac{\partial t}{\partial \sigma}\frac{\partial \sigma}{\partial \eta_-} - (u-c)\frac{\partial t}{\partial \tau}\frac{\partial \tau}{\partial \eta_-}$$
$$= \frac{\partial x}{\partial \sigma} + \frac{\partial x}{\partial \tau} - (u-c)\frac{\partial t}{\partial \sigma} - (u-c)\frac{\partial t}{\partial \tau}. \tag{8.182}$$

Adding and subtracting these equations yields

$$\frac{\partial x}{\partial \sigma} - u\frac{\partial t}{\partial \sigma} + c\frac{\partial t}{\partial \tau} = 0, \tag{8.183}$$

$$\frac{\partial x}{\partial \tau} + c\frac{\partial t}{\partial \sigma} - u\frac{\partial t}{\partial \tau} = 0. \tag{8.184}$$

These equations are the transformation of (8.172) from variables η_+ and η_- to variables σ and τ. For an arbitrary function $a(\tau, \sigma)$, employ the abbreviations $a_\tau = \partial a/\partial \tau$ and $a_\sigma = \partial a/\partial \sigma$. Then differentiate (8.183) with respect to τ and

(8.184) with respect to σ, which gives, after subtraction of the resulting equations,

$$-u_\tau t_\sigma + c_\tau t_\tau + ct_{\tau\tau} - c_\sigma t_\sigma - ct_{\sigma\sigma} + u_\sigma t_\tau = 0. \tag{8.185}$$

Equations (8.177) and (8.178) give

$$c_\sigma = \frac{1}{4}, \qquad c_\tau = 0, \qquad u_\sigma = -t_\sigma, \qquad u_\tau = \frac{1}{2} - t_\tau. \tag{8.186}$$

Inserting this into (8.185) and using $c = \sigma/4$ yields

$$\sigma(t_{\tau\tau} - t_{\sigma\sigma}) - 3t_\sigma = 0. \tag{8.187}$$

Using again $t = \dfrac{\tau}{2} - u$ and $\tau_{\tau\tau} = \tau_{\sigma\sigma} = \tau_\sigma = 0$, one arrives at

$$\boxed{\sigma(u_{\tau\tau} - u_{\sigma\sigma}) - 3u_\sigma = 0} \tag{8.188}$$

This partial differential equation can be further simplified by introducing a new field $\phi = \phi(\tau, \sigma)$ via

$$u = \frac{\phi_\sigma}{\sigma}. \tag{8.189}$$

Then,

$$u_\sigma = \frac{\phi_{\sigma\sigma}}{\sigma} - \frac{\phi_\sigma}{\sigma^2},$$

$$u_{\sigma\sigma} = \frac{\phi_{\sigma\sigma\sigma}}{\sigma} - \frac{2\phi_{\sigma\sigma}}{\sigma^2} + \frac{2\phi_\sigma}{\sigma^3}. \tag{8.190}$$

Inserting this in (8.188) gives

$$\begin{aligned}
0 &= \sigma\left(\frac{\phi_{\sigma\tau\tau}}{\sigma} - \frac{\phi_{\sigma\sigma\sigma}}{\sigma} + \frac{2\phi_{\sigma\sigma}}{\sigma^2} - \frac{2\phi_\sigma}{\sigma^3}\right) - \frac{3\phi_{\sigma\sigma}}{\sigma} + \frac{3\phi_\sigma}{\sigma^2}\\
&= \phi_{\tau\tau\sigma} - \phi_{\sigma\sigma\sigma} - \frac{\phi_{\sigma\sigma}}{\sigma} + \frac{\phi_\sigma}{\sigma^2}\\
&= \frac{\partial}{\partial\sigma}\left[\phi_{\tau\tau} - \phi_{\sigma\sigma} - \frac{\phi_\sigma}{\sigma}\right].
\end{aligned} \tag{8.191}$$

Therefore,

$$\phi_{\sigma\sigma} + \frac{\phi_\sigma}{\sigma} - \phi_{\tau\tau} = \text{const} = 0, \tag{8.192}$$

since ϕ and $\phi + \text{const}$ gives the same u. Thus,

$$\sigma\phi_{\sigma\sigma} + \phi_\sigma - \sigma\phi_{\tau\tau} = 0, \tag{8.193}$$

or

$$\boxed{(\sigma\phi_\sigma)_\sigma - \sigma\phi_{\tau\tau} = 0} \tag{8.194}$$

Both (8.188) and (8.194) are linear equations of second order, and each of them can serve to replace the two ordinary differential equations (8.172) of first order. As usual, the latter are slightly more restrictive, since a differentiation is applied to derive the second-order equations. Equation (8.194) is readily solved. For this, assume a factorization

$$\phi(\tau, \sigma) = \phi_0(\sigma)\,\phi_1(\tau), \tag{8.195}$$

which gives

$$\sigma\phi_{0\sigma\sigma}\phi_1 + \phi_{0\sigma}\phi_1 - \sigma\phi_0\phi_{1\tau\tau} = 0. \tag{8.196}$$

If we now select

$$\phi_{1\tau\tau} = -\phi_1, \tag{8.197}$$

then (8.196) becomes the Bessel differential equation,

$$\sigma^2\phi_{0\sigma\sigma} + \sigma\phi_{0\sigma} + \sigma^2\phi_0 = 0. \tag{8.198}$$

This restriction is sufficient to derive the relevant, non-breaking solutions. Equation (8.197) is solved by $\phi_1 = a_1 \cos\tau$ (neglecting a trivial phase), and (8.198) is solved by the Bessel function $\phi_0 = a_0 J_0(\sigma)$. Therefore,

$$\boxed{\phi(\tau, \sigma) = a\,J_0(\sigma)\,\cos\tau} \tag{8.199}$$

Finally, the water elevation ζ is expressed using the velocity 'potential' ϕ. First,

$$0 \overset{(7.183)}{=} x_\sigma - u t_\sigma + c t_\tau$$
$$\overset{(7.178)}{=} x_\sigma + u u_\sigma + c\left(\frac{1}{2} - \frac{\phi_{\sigma\tau}}{\sigma}\right)$$
$$\overset{(7.177)}{=} x_\sigma + \frac{(u^2)_\sigma}{2} + \frac{\sigma}{8} - \frac{\phi_{\tau\sigma}}{4}. \tag{8.200}$$

After integration with respect to σ,

$$x = \frac{\phi_\tau}{4} - \frac{\sigma^2}{16} - \frac{u^2}{2}, \tag{8.201}$$

where an integration constant was set to zero (corresponding to a shift of the origin). Then, from (8.175),

$$\zeta = x + c^2 = \frac{\phi_\tau}{4} - \frac{(\phi_\sigma)^2}{2\sigma^2}, \tag{8.202}$$

using (8.177) and (8.189) in the last equation. With $dJ_0(\sigma)/d\sigma = -J_1(\sigma)$, one arrives at

$$u = \frac{\phi_\sigma}{\sigma} = -\frac{a}{\sigma} \, J_1(\sigma) \, \cos \tau, \tag{8.203}$$

$$\zeta = -\frac{a}{4} \, J_0(\sigma) \, \sin \tau - \frac{a^2}{2\sigma^2} \, J_1^2(\sigma) \, \cos^2 \tau, \tag{8.204}$$

$$t = \frac{\tau}{2} - u,$$

$$x = -\frac{\sigma^2}{16} + \zeta.$$

This completely specifies the solution $u(t, x)$, $\zeta(t, x)$ implicitly in terms of functions $u(\tau, \sigma)$, $\zeta(\tau, \sigma)$ (we keep the same names) and $t(\tau, u(\tau, \sigma))$ and $x(\zeta(\tau, \sigma), \sigma)$.

For example, let first $\tau = \pi/2$ and then $\tau = 3\pi/2$. In both cases, $\cos \tau = 0$, thus $u = 0$ and $t = \tau/2$ is a constant. Therefore, the momentary wave profiles $\zeta(x)$ at times $t = \pi/4$ and $3\pi/4$ are given in parametric form by

$$\zeta = \mp \frac{a}{4} \, J_0(\sigma),$$

$$x = \mp \frac{a}{4} \, J_0(\sigma) - \frac{\sigma^2}{16}, \tag{8.205}$$

where $\sigma \in \mathbb{R}^+$. Figure 8.16 shows these two wave profiles for $a = 1$. At $t = \pi/4$, the elevation at the shore is a minimum, at $t = 3\pi/4$ a maximum, and the wave *climbs the sloping beach without breaking*.

The surface wave shown in Fig. 8.16 is a *standing* wave, as is clear from the factorization of the periodicities with respect to time and space variables in (8.203) and (8.204), resulting from the *assumed* factorization in (8.195). The physical origin of this standing wave is the reflection of the incoming wave at the shore, resulting in an outward propagating wave.

The wave amplitude $a = 1$ is a limiting amplitude, since it gives infinite slope at the shore. For $a < 1$, the wave slope at the shore is finite, whereas for $a > 1$, a *bore* must be included in the wave (Carrier and Greenspan 1958).

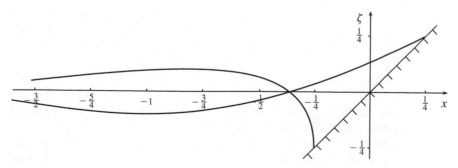

Fig. 8.16 Standing, non-breaking wave reaching a shore at a 45°-slope (Carrier and Greenspan 1958)

A follow-up analysis on breaking and non-breaking waves at shores is given by Tuck and Hwang (1972), and their Fig. 5 for nonlinear solutions can be compared to Fig. 8.16. Hibberd and Peregrine (1979) calculate numerical solutions for run-up on shores. For example, they find a landward-facing bore in the backwash of a wave.

8.10   Bores

When physical phenomena are described by differential equations, one is led to expect smooth solutions, differentiable at least to the order of the relevant differential equation. This, however, need not be the case. Physical processes in $\mathbb{R}^3$ can show surfaces of discontinuity for physical fields, corresponding to solutions that are only piecewise differentiable, where two or more spatial regions with (vastly) different physical conditions come into contact. In fluids, this happens when a wave source (an external force) moves through the fluid at larger speed than the local wave speed. Then no adaption of the fluid to the 'appearance' of the force is possible (there is no precursor), and the force and the motion it causes appear suddenly. Examples for this are the Mach cones of supersonic projectiles and jet airplanes. Another example is the discontinuous front ahead of a piston that is driven at supersonic speed into a tube filled with a gas.

Rather surprisingly, and only discovered in the late nineteenth century, such surfaces of discontinuity need not be transient phenomena and smooth out or decay quickly, but can be permanent features with physical properties and characteristics of their own, like the propagation speed of the discontinuity surface, the jump conditions for discontinuous quantities, and so forth. These properties are deduced by returning from the differential equations for the process to the original conservation laws in integral form, from which the differential equations were derived—since there is no problem in integrating across a finite discontinuity. (In reality, the contact surface is a thin layer in which physical processes often irrelevant in the adjacent domains lead to adaption from one side of the surface to the other, by viscosity, heat transfer, radiative cooling, etc.)

In this section, we study jumps in the water height of a tidal wave. As observed in Sect. 8.5, the rotation of the Earth close to or at the surface is faster than the propagation speed of shallow water waves in oceans and rivers; thus the Moon or Sun (as origin of the external tidal force) moves at 'supercritical' speed relative to water, and discontinuities may occur. They are observed on certain rivers in the form of so-called hydraulic jumps, which are sudden jumps in the height of the water. A moving hydraulic jump caused by the lunar or solar tidal force on rivers is called a *bore*. Figures 8.17 and 8.18 show examples of bores.

It is important to realize (i) that bores can indeed be described as jumps, i.e., there is no need to consider the microphysics in the bore transition to derive the bore speed; (ii) that fluid passes from one side of the bore to the other, i.e., there is fluid flow across the bore, and the fluid itself undergoes a transition in the bore (a fluid parcel traverses zero distance in a bore, thus the transition in a bore is instantaneous;

Fig. 8.17 Tidal bore on a
channel in an estuary near
Silverdale, Britain.
Photograph: Arnold Price,
https://www.geograph.org.
uk/photo/324581

Fig. 8.18 Tidal bore in the
Petitcodiac River in Canada,
circa 1902. Image source:
Centre d'études acadiennes

similarly, a bore adjusts instantaneously to changes in its surroundings); and (iii) that
bores propagate relative to the fluid ahead and behind it, even if they reside at a fixed
position in a certain reference frame.

The fluid is again assumed to be inviscid and to have constant density $\rho = 1$, the
flow is a two-dimensional canal flow obeying the shallow water equations (horizontal
propagation; vertical elevation). We work in the rest frame of the bore, which may be
an accelerating, i.e., non-inertial frame. To emphasize this choice of frame, boldface
letters are used for the fluid speed **u** relative to the bore ('boldface = bore frame').

The physical conditions in the region immediately behind the bore, the so-called
post-bore region, are fully determined by the conditions immediately ahead of the
bore, the *pre-bore* conditions. Thus, it is sufficient to consider an infinitesimally thin
fluid layer passing through the bore with constant density and velocity. Since this
passage is instantaneous, all time derivatives can be neglected in the bore frame.
Then mass and momentum conservation fixes the jump in water height across the
bore and the speed at which the bore itself propagates.

Let X and $\underline{X}$ be the pre- and post-bore values, respectively, of a quantity X. Mass conservation and $\rho = 1$ imply that $dV = d\underline{V}$ for the volume of a fluid parcel, or

$$\boxed{(\zeta + h)\,\mathbf{u} = (\underline{\zeta} + h)\,\underline{\mathbf{u}}} \qquad (8.206)$$

Here $h(x)$ is the unperturbed water height at the bore, and $\mathbf{u}$ and $\underline{\mathbf{u}}$ are the pre- and post-bore speeds relative to the bore. Let $\mathbf{u} > 0$ and $\underline{\mathbf{u}} > 0$, i.e., water crosses the bore from left to right. Consider a narrow, vertical strip that contains the full bore jump. The hydrostatic pressure law (8.23) in Sect. 8.2 is, for vanishing pressure at the water surface and $\rho = 1$,

$$p(t, x, z) = g(\zeta(t, x) - z). \qquad (8.207)$$

Integrating this in vertical direction from ground to water surface gives

$$\int_{-h}^{\zeta} dz\, p = g \int_{-h}^{\zeta} dz\,(\zeta - z) = \frac{1}{2} g(\zeta + h)^2. \qquad (8.208)$$

The depth-integrated momentum equation in the narrow strip containing the bore then takes the form

$$\boxed{(\zeta + h)\,\mathbf{u}^2 + \frac{1}{2} g(\zeta + h)^2 = (\underline{\zeta} + h)\,\underline{\mathbf{u}}^2 + \frac{1}{2} g(\underline{\zeta} + h)^2} \qquad (8.209)$$

The signs of the pressure terms are a result of the fact that negative pressure gradients give positive acceleration. Furthermore, in (8.209) it was used that the horizontal speed $\mathbf{u}$ (and $\underline{\mathbf{u}}$) of a shallow water wave does not depend on height. The two Eqs. (8.206) and (8.209) fully determine the bore. For these equations, see also Stoker (1957), p. 316.

Next, the wave energy is considered. Mechanical energy is *not* conserved in bores, since heat Q is necessarily generated in them. Thus, the second law of thermodynamics applies to bores, and actually fixes the unique direction in which a bore must be passed. Fluid parcels passing a bore in the opposite direction would violate the second law. Energy splits into kinetic and potential energy, pressure work, and heat. The vertical integral of wave energy over the bore height reads

$$\int\limits_{-h}^{\zeta} dz\left(\frac{1}{2} \mathbf{u}^3 + g(z + h)\,\mathbf{u} + p\mathbf{u}\right) + Q = \int\limits_{-h}^{\zeta} dz\left(\frac{1}{2} \underline{\mathbf{u}}^3 + g(z + h)\,\underline{\mathbf{u}} + \underline{p}\,\underline{\mathbf{u}}\right) + \underline{Q}. \qquad (8.210)$$

Inserting here the hydrostatic pressure from (8.207) gives

$$\boxed{\frac{1}{2}(\zeta + h)\,\mathbf{u}^3 + g\mathbf{u}\,(\zeta + h)^2 + Q = \frac{1}{2}(\underline{\zeta} + h)\,\underline{\mathbf{u}}^3 + g\underline{\mathbf{u}}\,(\underline{\zeta} + h)^2 + \underline{Q}} \qquad (8.211)$$

Equations (8.206), (8.209), and (8.211) determine the post-bore quantities $\underline{\mathbf{u}}, \underline{\zeta}, \underline{Q}$ for given pre-bore quantities $\mathbf{u}, \zeta, Q$.

Let

$$N = \zeta + h. \tag{8.212}$$

Then the conservation laws (8.206), (8.209), and (8.211) become

$$\mathbf{u}N = \underline{\mathbf{u}}\underline{N} =: j, \tag{8.213}$$

$$\mathbf{u}^2 N + \frac{1}{2} g N^2 = \underline{\mathbf{u}}^2 \underline{N} + \frac{1}{2} g \underline{N}^2, \tag{8.214}$$

$$\frac{1}{2} \mathbf{u}^3 N + g \mathbf{u} N^2 + Q = \frac{1}{2} \underline{\mathbf{u}}^3 \underline{N} + g \underline{\mathbf{u}} \underline{N}^2 + \underline{Q}. \tag{8.215}$$

(We refrain from writing j in boldface type.) From (8.214) follows, using (8.213),

$$(\mathbf{u} - \underline{\mathbf{u}})j = \mathbf{u} \left(1 - \frac{N}{\underline{N}}\right) j = \frac{1}{2} g(\underline{N}^2 - N^2). \tag{8.216}$$

Therefore,

$$\mathbf{u} = \frac{g}{2j}(\underline{N} + N)\underline{N}. \tag{8.217}$$

Using again (8.213), this gives

$$\underline{\mathbf{u}} = \frac{g}{2j}(\underline{N} + N)N. \tag{8.218}$$

One thus has

$$N\mathbf{u} = \underline{N}\underline{\mathbf{u}} = j = \frac{g}{2j}(\underline{N} + N)\underline{N}N. \tag{8.219}$$

The heat difference is calculated using (8.215),

$$
\begin{aligned}
\underline{Q} - Q &= j\left(\frac{1}{2}\mathbf{u}^2 - \frac{1}{2}\underline{\mathbf{u}}^2 + gN - g\underline{N}\right) \\
&= j\left(\frac{g^2}{8j^2}(\underline{N} + N)^2(\underline{N}^2 - N^2) + gN - g\underline{N}\right) \\
&= j\left(\frac{g^2}{8j^2}(\underline{N} + N)^3 - g\right)(\underline{N} - N) \\
&\overset{(7.219)}{=} jg\left(\frac{(\underline{N} + N)^2}{4\underline{N}N} - 1\right)(\underline{N} - N) \\
&= jg\frac{(\underline{N} - N)^3}{4\underline{N}N},
\end{aligned}
\tag{8.220}
$$

which is Eq. (10.6.16) in Stoker (1957), p. 319. From the second law of thermodynamics, the heat increases in the transition,

$$\underline{Q} - Q > 0. \tag{8.221}$$

Since $j > 0$ by assumption, Eq. (8.220) implies that $\underline{N} > N$, and thus

$$\boxed{\underline{\zeta} > \zeta} \tag{8.222}$$

Therefore, the pre-bore height $h + \zeta$ is *smaller* than the post-bore height $h + \underline{\zeta}$, and
water is *lifted upward* in a bore transition. This is a consequence of $\underline{\mathbf{u}} < \mathbf{u}$ and volume
conservation: water piles up behind the bore. Correspondingly, the post-bore speed
$\underline{\mathbf{u}}$ relative to the bore is *smaller* than the pre-bore speed $\mathbf{u}$.

The above calculations were first performed by Rayleigh (1914). Heat generation
in bores is also confirmed empirically, since real bores show turbulence in the post-
bore domain. In small turbulent eddies, mechanical energy is indeed converted to
heat.

Bore Speed

Next, the speed of the bore will be derived. The speeds $\mathbf{u}$ and $\underline{\mathbf{u}}$ above are relative to
the speed of the bore jump. We assume now that the bore propagates into *unperturbed*
water *at rest*, e.g., water at a shore,

$$\zeta = 0, \qquad\qquad u = 0. \tag{8.223}$$

This defines an inertial frame, in which U shall be the speed of the bore jump and $\underline{u}$
the speed of water behind the bore. Then

$$\mathbf{u} = u - U = -U, \qquad\qquad \underline{\mathbf{u}} = \underline{u} - U. \tag{8.224}$$

From (8.217) and (8.218) follows

$$\mathbf{u}\underline{\mathbf{u}} = -U(\underline{u} - U) = \frac{g^2}{4j^2}(\underline{N} + N)^2 \underline{N}N = \frac{g}{2}(\underline{N} + N), \tag{8.225}$$

where (8.219) was used. The continuity equation (8.213) is in this case

$$-UN = (\underline{u} - U)\underline{N}. \tag{8.226}$$

Inserting this in (8.225) gives the bore speed relative to the unperturbed water that enters the bore,

$$U = \sqrt{\frac{g\,(2h + \underline{\zeta})(h + \underline{\zeta})}{2h}} \tag{8.227}$$

where (8.212) has been used. Once U is known, (8.226) gives the post-bore speed,

$$\underline{u} = \frac{\underline{N} - N}{\underline{N}}\,U, \tag{8.228}$$

or, by definition of N and since $\zeta = 0$,

$$\underline{u} = \frac{\underline{\zeta}}{h + \underline{\zeta}}\,U \tag{8.229}$$

Water enters the bore at a speed

$$\mathbf{u} = u - U = 0 - U < 0, \tag{8.230}$$

and leaves the bore at a slower speed,

$$\underline{\mathbf{u}} = \underline{u} - U = -\frac{h}{h + \underline{\zeta}}\,U < 0. \tag{8.231}$$

Furthermore, one has the obvious inequality

$$c = \sqrt{gh} < \sqrt{\frac{g\,(2h + \underline{\zeta})(h + \underline{\zeta})}{2h}} = U = |\mathbf{u}|. \tag{8.232}$$

Therefore, the bore propagates into unperturbed water faster than a linear wave (the linear wave is regained for a bore of vanishing height, $\underline{\zeta} \to 0$). On the other hand, the bore recedes from water left behind at a speed $-\underline{\mathbf{u}} = U - \underline{u} > 0$, where

$$|\underline{\mathbf{u}}| = \frac{h}{h + \underline{\zeta}}\,U = \sqrt{\frac{gh(2h + \underline{\zeta})}{2(h + \underline{\zeta})}} < \sqrt{g(h + \underline{\zeta})} = \underline{c}. \tag{8.233}$$

Since $|\underline{\mathbf{u}}|$ is smaller than the post-bore wave speed $\underline{c}$, linear waves can propagate throughout the whole post-bore region, especially they can reach the bore although the latter is receding. We emphasize that $u < U$ and $\underline{u} < U$ according to (8.229): the bore speed U is larger than *both* the pre- and post-bore speeds; this reflects the fact that the bore is a *transition*.

Fig. 8.19 Bore (full) and
post-bore wave (dashed)

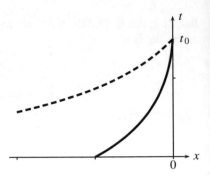

Bores at Shores

The relations derived above allow to prove that a bore cannot reach a coastline at finite
jump height, but that the jump must decrease to zero when approaching the coast.
For simplicity, the shore is again assumed to have constant slope, $gh' = m = $ const.
As found above, a linear shallow water wave behind the bore propagating toward the
bore will indeed reach the bore jump, which recedes at a speed *below* the post-bore
wave speed from the post-bore water. Figure 8.19 shows this situation, where a bore
propagates toward a coast and is overtaken by a linear wave. A subscript zero shall
indicate the value of a quantity at the shore $x = 0$. The following arguments are taken
from Keller et al. (1960), p. 305.

Lemma *The post-bore speed at a shore is finite, $\underline{u}_0 < \infty$.*

Proof by contradiction. Assume the opposite, $\underline{u}_0 \to \infty$. Then $U_0 > 0$ by (8.229),
and the bore reaches the shore at a finite time $t_0 < \infty$. Consider the unique curve
$x(t)$ in an xt-diagram (cf. Fig. 8.19) that *meets* the bore at time t_0 at $x = 0$, and has
slope $dx/dt = \underline{u} + \underline{c}$ in each of its points (propagation at local wave speed $\underline{c}$ in water
of local speed $\underline{u}$). According to (8.168) in Sect. 8.9, $\underline{u} + 2\underline{c} + |m|t$ is constant along
this curve $x(t)$ (here $m < 0$ according to 8.174). Continuity at $x = 0$ and $\underline{u}_0 \to \infty$
imply $\underline{u} + 2\underline{c} + |m|t \to \infty$ along the whole curve $x(t)$. For $t < t_0 < \infty$, therefore
$\underline{u} \to \infty$ or $\underline{c} \to \infty$. However, $\underline{u}$ and $\underline{c}$ are finite for $x < 0$ (namely, for $x < 0$ one
has $h > 0$ and $\underline{\zeta} < \infty$, thus $U < \infty$ by (8.227), and hence $\underline{u} < \infty$ and $\underline{c} < \infty$). This
is a contradiction, and thus the assumption is wrong and $\underline{u}_0 < \infty$.   □

Lemma *Any bore jump must vanish at a shore, $\underline{\zeta}_0 = 0$.*

Proof by contradiction. Assume the opposite, $\underline{\zeta}_0 > 0$. Then (8.227) and $h_0 = 0$
imply $U_0 = \underline{\zeta}_0 (g/2h_0)^{1/2} \to \infty$, and (8.229) implies $\underline{u}_0 \to \infty$. Contradiction to the
foregoing lemma, and thus the assumption is wrong and $\underline{\zeta}_0 = 0$.   □

Shen and Meyer (1963) study in more detail what happens, when a bore reaches a shore and vanishes. They find that the speed of the shore line $h(t, x) = 0$ has a *singular jump* upon arrival of the bore.

> It is in the nature of a singularity [...] that it dominates the solution over a whole region, and thus firm indications concerning the internal structure of run-up and back-wash [...] will be obtained. (Shen and Meyer 1963, p. 115)

Shen and Meyer find that a whole *water sheet* of run-up and backwash forms near the coastline. A new, secondary bore forms at the shore and moves seaward, in opposite direction to the primary bore. The calculations are somewhat involved, and we do not go into them here.

8.11 The Poincaré and Kelvin Waves

On the rotating Earth, two types of shallow water waves occur that are subject to the Coriolis force, the *Poincaré* and *Kelvin* waves. The Poincaré waves propagate freely in the open ocean as plane waves, or they can be radiated from some origin as cylindrical waves. Furthermore, they can be reflected at coasts. The Kelvin waves, by contrast, propagate in narrow strips or belts along coastlines, with no velocity component normal to the coastline, and one can think of this strip as a waveguide. The amplitude of a Kelvin wave falls off exponentially with distance from the coast.

In the next section, some basic facts about the Poincaré and Kelvin waves are used. Due to their importance in oceanography, an extensive literature exists on both wave types, but only a few key concepts are covered here. Books that deal at length with the Poincaré and Kelvin waves are LeBlond and Mysak (1978) and Hutter et al. (2012).

The fluid shall be inviscid and have constant density, and we consider again shallow water waves, with no dependence on the height coordinate. The flow domain is a Cartesian xy-plane tangent to the Earth surface at some point. The relevant fluid fields are the elevation $\zeta(t, x, y)$ and the two horizontal velocity components $u(t, x, y)$ and $v(t, x, y)$ in the x- and y-directions, respectively. In this and the next section, only linear waves with infinitesimal elevations $\zeta \ll h$ are treated, with all quadratic terms (e.g., uu') and higher order terms neglected.

Moving fluid parcels experience a Coriolis acceleration $\vec{g}_C$ (an apparent force) on the rotating Earth, given by

$$\vec{g}_C = -2\vec{\Omega} \times \vec{u}, \tag{8.234}$$

where $\vec{\Omega}$ is the vector of angular velocity of the Earth pointing from the South Pole to the North Pole.

A derivation of the Coriolis force can be found in most textbooks on mechanics; its basic scaling is obtained as follows (see, e.g., Mortimer 1975, p. 33). A force-free point mass m shall move radially outward at constant speed u from the center of a disk rotating counterclockwise with constant angular speed $\Omega > 0$. The distance traversed in radial direction in time dt is $u \, dt$. For a polar coordinate system fixed

Fig. 8.20 Tangential plane
on the rotating Earth

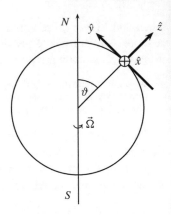

to the rotating disk, the azimuth of m changes in dt by $d\varphi = \Omega \, dt$. Attributing this azimuthal motion to an acceleration g_C by an apparent force, one has

$$\Omega \, dt = d\varphi = \frac{\text{arclength}}{\text{radius}} = \frac{\frac{1}{2} g_C \, dt^2}{u \, dt}, \tag{8.235}$$

or

$$g_C = 2\Omega u. \tag{8.236}$$

In returning to the full expression (8.234) for $\vec{g}_C$, only the component of $\vec{\Omega}$ that is normal to the local xy-plane of (wave) motion leads to horizontal displacements within this plane. This vector component is given by (see Fig. 8.20)

$$\vec{g}_{C,z} = -2\Omega \cos \vartheta \, \hat{z} \times (u\hat{x} + v\hat{y}), \tag{8.237}$$

with colatitude ϑ and angular speed $\Omega = 2\pi/\text{day}$ of the Earth. Introducing the abbreviation

$$\gamma = 2\Omega \cos \vartheta, \tag{8.238}$$

this gives

$$\vec{g}_{C,z} = \gamma (v\hat{x} - u\hat{y}). \tag{8.239}$$

The Coriolis parameter γ is assumed here to be constant, which is termed the f-plane approximation. While still considering flows in an xy-plane tangent to the Earth, one can include a first order correction for the height dependence of the Coriolis force, $\gamma = \gamma_0 + \beta y$, with

$$\beta = -\frac{1}{R} \frac{\partial \gamma}{\partial \vartheta} = \frac{2\Omega}{R} \sin \vartheta, \tag{8.240}$$

where R is the Earth's radius, and the minus sign accounts for $dy = -R d\vartheta$. This is termed the beta plane approximation, and β itself is termed the Rossby parameter. Rossby waves (or planetary waves), which are atmospheric and oceanic wave phenomena with the largest known lengthscales, do indeed occur only for nonzero β (i.e., there are no Rossby waves on a rotating cylinder). Furthermore, a finite β allows to study waves near the equator that are influenced by the Coriolis force, where $g_C = 0$ exactly on the equator.

For an ocean of depth h, the propagation speed of shallow water waves is $\sqrt{gh}$. Thus, shallow water waves should be affected by the Coriolis effects if their extension is comparable to the *Rossby radius*,

$$L_R = \frac{\sqrt{gh}}{\gamma}. \tag{8.241}$$

At a latitude of $45°$, $\gamma \approx 1.0 \times 10^{-4}$ s^{-1}. Thus for an ocean of depth $h = 1$ km, the Rossby radius is $L_R \approx 10^3$ km.

The Euler equation (8.26) in Sect. 8.3 gives the acceleration in the x-direction, and the corresponding equation for the y-direction is obtained by replacing u by v and x by y. Furthermore, the continuity equation (8.21) is readily generalized to flow in both the x- and y-directions. Adding the Coriolis force (8.239), the *linearized* equations of motion become

$$\begin{aligned} u_t - \gamma v &= -g\zeta_x, \\ v_t + \gamma u &= -g\zeta_y, \\ u_x + v_y &= -h^{-1}\zeta_t. \end{aligned} \tag{8.242}$$

These are the central equations for the present and the next section.

The Kelvin Waves

The Kelvin waves propagate parallel to coastlines. In the northern hemisphere, they propagate in such a way that the coastline lies to the right of the propagation direction. If the fluid velocity is perturbed and attains a small component away from the coastline, the Coriolis acceleration $-\gamma \hat{z} \times \vec{u}$ points to the right side of $\vec{u}$, and thus *opposes* the original perturbation. The equations of motion allow for an equilibrium solution where the velocity vector is parallel to the coast. Since the propagation direction of the wave and the direction of the oscillating velocity vector field $\vec{u}$ (which, for water waves, is what the fields $\vec{E}$ and $\vec{B}$ are for electromagnetic waves) are both parallel to the coastline, the velocity field $\vec{u}$ of a Kelvin wave is called *longitudinal*.

To obtain this solution, assume that the region $y < 0$ is occupied by land, and the Kelvin wave propagates in the x-direction in the region $y > 0$. Assuming $v = 0$ right from the outset, the equations of motion (8.242) become

$$u_t = -g\zeta_x, \tag{8.243}$$
$$\gamma u = -g\zeta_y, \tag{8.244}$$
$$u_x = -h^{-1}\zeta_t. \tag{8.245}$$

Assuming that u can be factorized as

$$u(t, x, y) = f(x - ct)\,\tilde{f}(y),$$ (8.246)

the continuity equation (8.245) gives

$$\zeta(t, x, y) = \frac{h}{c}\,u(t, x, y),$$ (8.247)

the Euler equation (8.243) in x-direction yields

$$c^2 = gh,$$ (8.248)

and the Euler equation (8.244) in the y-direction gives

$$\gamma u = -\frac{gh}{c}\,u_y,$$ (8.249)

or

$$\gamma \tilde{f}(y) = -c\,\frac{\mathrm{d}\tilde{f}}{\mathrm{d}y},$$ (8.250)

and therefore

$$u(t, x, y) = f(x - ct)\,e^{-\gamma y/c}.$$ (8.251)

Both the acceleration γu from the Coriolis force and the acceleration $-g\zeta_y$ from hydrostatic pressure stand normal to the coast, the former points toward the coast, the latter away from it. The two forces are in equilibrium according to (8.244). The Kelvin wave propagates force-free in the x-direction, i.e., according to the dispersion-free law $c = \sqrt{gh} = $ const of shallow water waves in water of constant depth. Therefore, any arbitrary waveform f is maintained in the course of time. In the y-direction, the elevation and horizontal speed fall off exponentially with distance from the coast, with a lengthscale c/γ, called the Rossby radius.

The Poincaré Waves

The Poincaré waves are oscillations in the open ocean, and their velocity field is not restricted to lie parallel to a coastline. More than that, the velocity field is not even restricted to be parallel to the propagation direction of the wave. The Poincaré waves are not longitudinal, but have a *rotary* velocity field, i.e., a 'rotation of current direction in a particular phase relationship with the wave' (Mortimer 1975, p. 70), which reflects the presence of the Coriolis force. Even for a plane wave, the velocity vector $\vec{u}$ rotates in the full xy-plane in space and time. Diagrams of this interesting wave kinematics can be found in Mortimer (1975) [Figs. 47, 54, 59, 60] and Hutter et al. (2012).

> Instead of setting in one direction for a period of six hours and in the opposite direction during the following period of six hours, the tidal current offshore changes its direction continually, so that in a period of about twelve and a half hours it will have set in all directions of the compass. This type of current is called a rotary current in distinction from the reversing type of current found in inland tidal waters. (Le Lacheur 1924, p. 282)

To describe a Poincaré wave, assume a plane wave propagating in the x-direction and let y point as usual to the left. The wave amplitude depends on x only, not on y, i.e., the wave is uniform in the y-direction. Assuming harmonic waves, the following is easily seen to be a solution of (8.242),

$$u = C \cos(kx - \omega t), \tag{8.252}$$

$$v = \frac{\gamma}{\omega} C \sin(kx - \omega t), \tag{8.253}$$

$$\zeta = \frac{kh}{\omega} C \cos(kx - \omega t), \tag{8.254}$$

if the different parameters and constants obey the dispersion relation

$$\omega^2 - ghk^2 - \gamma^2 = 0. \tag{8.255}$$

This, in contrast to the Kelvin waves, includes a Coriolis term (see Proudman 1953, p. 263). Equations (8.252) and (8.253) describe a rotary velocity field.

Wave Equation

To derive more systematically a shallow water wave equation in the presence of the Coriolis force, the equations of motion (8.242) are differentiated as follows:

$$-u_{tx} + \gamma v_x = g\zeta_{xx},$$
$$-v_{ty} - \gamma u_y = g\zeta_{yy},$$
$$u_{xt} + v_{yt} = -h^{-1}\zeta_{tt}. \tag{8.256}$$

Adding these three equations gives

$$\zeta_{xx} + \zeta_{yy} - \frac{\zeta_{tt}}{gh} = \frac{\gamma}{g}(v_x - u_y). \tag{8.257}$$

Then (8.242) are differentiated to give

$$-u_{ty} + \gamma v_y = g\zeta_{xy},$$
$$v_{tx} + \gamma u_x = -g\zeta_{yx},$$
$$-\gamma u_x - \gamma v_y = \frac{\gamma}{h}\zeta_t. \tag{8.258}$$

Adding the equations yields

$$v_{tx} - u_{ty} = \frac{\gamma}{h}\zeta_t. \tag{8.259}$$

Integrating this equation with respect to t, one has

$$v_x - u_y = \frac{\gamma}{h}\zeta. \tag{8.260}$$

The integration constant is 0, since both sides vanish for $\zeta = u = v = 0$. Inserting (8.260) in (8.257) gives

$$\zeta_{xx} + \zeta_{yy} - \frac{1}{gh}(\zeta_{tt} + \gamma^2\zeta) = 0. \tag{8.261}$$

This is the general wave equation in the presence of a Coriolis acceleration. For harmonic waves $\sim e^{i(kx-\omega t)}$, this gives again the dispersion relation (8.255).

8.12 Wave Behind a Barrier

The Sommerfeld diffraction problem in a rotating fluid was first considered by Crease (1956) for the case of normal incidence. Modeling the British Isles by a semi-infinite vertical barrier, he showed that the amplitude of the Kelvin wave generated in the North Sea by the diffraction of a Poincaré wave incident from the North Atlantic could exceed the amplitude of the incident wave. This coastal amplification was proposed by Crease as a possible explanation for the origin of some of the large-amplitude storm surges causing flooding on the eastern British coast. (LeBlond and Mysak 1978, p. 267)

If a plane, longitudinal wave, for example, sound, hits a semi-infinite screen at normal incidence, a part of the wave may be diffracted at the boundary and propagate behind the screen, dying out with distance from the boundary. In acoustics, this is a classical result derived by Sommerfeld (1896), see also Lamb (1932). In a seminal paper, Crease (1956) demonstrates that something drastically different can happen if a plane *rotary* wave—see the discussion of the Poincaré waves in the last section— hits a semi-infinite planar obstacle at normal incidence:

[...] one might expect that after the waves have passed the barrier their transverse velocity components act as a source for waves propagating into the region behind the barrier. (Crease 1956, p. 86)

Crease proves, using the Wiener–Hopf technique, that this intuition is correct and that the incoming, rotary Poincaré wave is the origin, behind the screen, of a Kelvin wave that propagates in a strip along the backside of the obstacle, perpendicular to the incident wave. The wave amplitude is constant in the propagation direction (because of energy conservation) and decays exponentially with distance from the barrier. More than that, this Kelvin wave can have a larger amplitude than the incoming Poincaré wave, which Crease makes responsible for certain storm surges on the eastern British coast, following a Poincaré wave from the North Atlantic that enters the North Sea across the northern tip of Britain.

The basic flow geometry is shown in Fig. 8.21. A straight semi-infinite barrier runs from $x = 0$ to $x \to \infty$. A plane shallow water wave $\sim e^{i(ky-\omega t)}$ approaches the barrier from negative y in the positive y-direction at normal incidence. The whole system is rotating in a counterclockwise manner at angular speed $\gamma/2$, see (8.238). This leads to a Kelvin wave propagating behind the barrier. (If the barrier extends

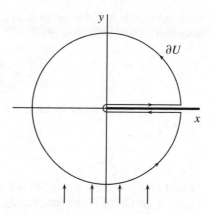

Fig. 8.21 Plane wave at normal incidence on a straight, semi-infinite barrier

from $x \to -\infty$ to $x = 0$, the Kelvin wave will instead propagate down the front of the barrier; thus closed tidal currents may occur around islands in mid-ocean.) The vertical wave elevation shall be harmonic in time,

$$\zeta(t, x, y) = e^{-i\omega t} \, \bar{\zeta}(x, y). \tag{8.262}$$

The wave Eq. (8.261) for the Poincaré waves becomes then,

$$\zeta_{xx} + \zeta_{yy} + \frac{\omega^2 - \gamma^2}{gh} \zeta = 0. \tag{8.263}$$

Abbreviating

$$k^2 = \frac{\omega^2 - \gamma^2}{gh}, \tag{8.264}$$

which corresponds to the dispersion relation (8.255) for plane waves, one has

$$\bar{\zeta}_{xx} + \bar{\zeta}_{yy} + k^2 \bar{\zeta} = 0. \tag{8.265}$$

In order that the wavenumber k is real, i.e., the incoming wave is not evanescent, let

$$\omega > \gamma, \tag{8.266}$$

and thus

$$p = \frac{\gamma}{\omega} < 1, \tag{8.267}$$

which is a central parameter in the following. From now on, we write again $\zeta(x, y)$ instead of $\bar{\zeta}(x, y)$.

First, the Green's function for the Helmholtz equation (8.265) is determined. The fluid is excited by a point-like perturbation. Since the water is homogeneous and isotropic, azimuthal symmetry is assumed for the solution, thus $\zeta = \zeta(r)$ in polar coordinates, and (8.265) reads

$$\frac{d^2\zeta}{dr^2} + \frac{1}{r}\frac{d\zeta}{dr} + k^2\zeta = 0. \tag{8.268}$$

Introducing an auxiliary variable $s = kr$, this becomes (still keeping the function name ζ)

$$s^2\frac{d^2\zeta}{ds^2} + s\frac{d\zeta}{ds} + s^2\zeta = 0, \tag{8.269}$$

see Eq. (4.19) in Sect. 4.3. In the following, $k > 0$ is assumed. The solutions to this equation are (i) the Bessel function J_0, (ii) the Weber function Y_0, and (iii) the Hankel function H_0, which are given by (replacing s by kr again)

$$J_0(kr) = \frac{2}{\pi}\int_0^\infty dt \ \sin(kr\cosh t), \tag{8.270}$$

$$Y_0(kr) = -\frac{2}{\pi}\int_0^\infty dt \ \cos(kr\cosh t), \tag{8.271}$$

$$H_0(kz) = J_0(kz) + iY_0(kz) = \frac{1}{i\pi}\int_{-\infty}^\infty dt \ e^{ikz\cosh t}, \tag{8.272}$$

see Eq. (9.1.23) in Abramowitz and Stegun (1964; A&S hereafter). The functions J_0 and Y_0 are real-valued functions of the real variable kr, and H_0 is a complex function of the complex variable kz. The Green's function for the differential equation (8.265) is defined by

$$G_{xx} + G_{yy} + k^2G = -\delta(x - x')\delta(y - y'), \tag{8.273}$$

where the minus sign on the right side is convention. The delta functions describe a point-like perturbation at (x', y'). Instead of deriving G, we only check that the following G is the solution of (8.273),

$$G(x - x', y - y') = \frac{i}{4}H_0\big(k\sqrt{(x - x')^2 + (y - y')^2}\big), \tag{8.274}$$

where $r = \sqrt{(x - x')^2 + (y - y')^2}$. For $x \neq x'$ or $y \neq y'$ one has $\delta(x - x')\delta(y - y') = 0$, and the right side of (8.273) is zero; hence, $H_0(kr)$ is indeed a solution (with azimuthal symmetry) of (8.273), as it is of (8.269). For $x \to x'$ and $y \to y'$, on the other hand, consider an infinitesimal neighborhood of (x', y'). Let $x' = y' = 0$ without loss of generality, and $x, y \to 0$. Inserting (8.274) in (8.273) and integrating over a disk D with radius r centered at $(0, 0)$ gives

$$\lim_{r \to 0} \frac{i}{4}\int_D da \ (\Delta H_0(kr) + k^2H_0(kr)) \overset{!}{=} -1. \tag{8.275}$$

Using $\Delta = \nabla \cdot \nabla$ and applying Gauss' theorem in the plane, one has

$$\lim_{r \to 0} \left(\frac{i}{4} \oint_{\partial D} dl\, \hat{r} \cdot \nabla H_0(kr) + \frac{i}{4} \int_D da\, k^2 H_0(kr) \right) \stackrel{!}{=} -1, \tag{8.276}$$

where dl is the arclength along the circular boundary of the disk D. In the limit $r \to 0$, according to equation A&S (9.1.8),

$$H_0(kr) = \frac{2i}{\pi} \ln(kr). \tag{8.277}$$

The area integral in (8.276) contains the term $da\, H_0 = dr\, rd\varphi\, H_0$ and vanishes since $r \ln r \to 0$ for $r \to 0$. Using $\nabla H_0(kr) = \hat{r}\, k\, dH_0(kr)/d(kr)$, one obtains from the line integral in (8.276),

$$\lim_{r \to 0} \frac{i}{4} 2\pi rk \frac{2i}{\pi kr} \stackrel{!}{=} -1, \tag{8.278}$$

which is indeed correct, and thus (8.274) is the required Green's function.

The boundary conditions at the barrier are $v = 0$ and $\dot{v} = 0$ (no normal speed). Inserting this in (8.242) and assuming $u \sim e^{-i\omega t}$ gives

$$i\omega u = g\zeta_x \quad \text{and} \quad \gamma u = -g\zeta_y, \tag{8.279}$$

hence, with $p = \gamma/\omega$,

$$\zeta_y = ip\zeta_x \quad \text{for} \quad x > 0,\, y = 0. \tag{8.280}$$

Green's integral theorem in the plane is, for some domain U and twice continuously differentiable functions f and g,

$$\int_U dx\, dy\, (f\Delta g - g\Delta f) = \oint_{\partial U} dl\, \left(f \frac{\partial g}{\partial n} - g \frac{\partial f}{\partial n} \right). \tag{8.281}$$

Here, the differential $\partial/\partial n$ is taken in the normal direction pointing away from U. Let $f = \zeta$ and $g = G$, and replace x, y by x', y'. Let $(x, y) \in U$. Then, using (8.265) and (8.273), Eq. (8.281) takes the form

$$\int_U dx'\, dy'\, (\zeta\Delta G - G\Delta\zeta)$$
$$= \int_U dx'\, dy'\, (-\zeta k^2 G - \zeta(x', y')\delta(x - x')\delta(y - y') + Gk^2\zeta)$$
$$= -\zeta(x, y). \tag{8.282}$$

A sign change is performed on the right-hand side of (8.281), so $\partial/\partial n$ becomes the differential in the direction of the interior of U, where ζ is calculated from the boundary conditions on ∂U. Thus,

$$\zeta(x, y) = \oint_{\partial U} dl \left(\zeta \frac{\partial G}{\partial n} - G \frac{\partial \zeta}{\partial n} \right). \tag{8.283}$$

The contour ∂U shall be a circle centered at the origin and with radius tending to infinity. Furthermore, ∂U shall be indented so as to exclude the barrier, see Fig. 8.21. For $r \to \infty$, all *outgoing* waves are assumed to have died away. Then

$$\zeta(x, y) = \int_0^\infty dx' \left([\zeta] \frac{\partial G}{\partial y'} - G \left[\frac{\partial \zeta}{\partial y'} \right] \right)_{y'=0} + ae^{iky}, \tag{8.284}$$

where, for $y = 0$, the discrete jumps are defined by

$$[\zeta(x)] = \lim_{\epsilon \to 0} \left(\zeta(x, \epsilon) - \zeta(x, -\epsilon) \right), \tag{8.285}$$

$$\left[\frac{\partial \zeta}{\partial y}(x) \right] = \lim_{\epsilon \to 0} \left(\frac{\partial \zeta}{\partial y}(x, \epsilon) - \frac{\partial \zeta}{\partial y}(x, -\epsilon) \right). \tag{8.286}$$

It is assumed here that G and $\partial G / \partial y$ are continuous across the barrier. Note the sign change when differentiating in the $+y$- and $-y$-directions, respectively. The boundary condition (8.280) gives

$$[\zeta_y] = ip[\zeta_x], \tag{8.287}$$

and with this, (8.284) turns into

$$\zeta(x, y) = \int_0^\infty dx' \left([\zeta] \frac{\partial G}{\partial y'} - ipG \left[\frac{\partial \zeta}{\partial x'} \right] \right)_{y'=0} + ae^{iky}. \tag{8.288}$$

After integration by parts, this becomes

$$\zeta(x, y) = \int_0^\infty dx' \, [\zeta(x')] \left(\frac{\partial G(x - x', y - y')}{\partial y'} + ip \frac{\partial G(x - x', y - y')}{\partial x'} \right)_{y'=0} + ae^{iky} \tag{8.289}$$

which is Eq. (11) in Crease (1956). For the following technicalities, we also refer to a classic paper by Karp (1950), on which Crease (1956) largely builds. In the integration by parts, it was assumed that the function $[\zeta] G$ vanishes in both the points $(0, 0)$ and $(\infty, 0)$. This is reasonable since ζ should be continuous at the origin. Furthermore, at infinity, $[\zeta]$ is bounded, and $G \to 0$. Equation (8.289) is an *integral equation* that is to be solved for the unknown function

$$[\zeta(x > 0)], \tag{8.290}$$

i.e., the difference in water height across the barrier. Crease applies the differential $\partial_y - ip\partial_x$ to (8.289), then restricts x and y to the semi-infinite line $x > 0$, $y = 0$ along which the boundary condition (8.280) applies, giving

$$
0 = \int_0^\infty dx' \, [\zeta(x')] \left\{ \left(\frac{\partial}{\partial y} - ip \frac{\partial}{\partial x} \right) \left(\frac{\partial}{\partial y'} + ip \frac{\partial}{\partial x'} \right) G \right\}_{y=y'=0} + aik
$$

(8.291)

Note that the integral is still a function of x. Equation (8.291) is a first-order integral equation of the form

$$
g(x) = \int_0^\infty dx' \, K(x - x') \, f(x') + h(x),
$$

(8.292)

with kernel (a linear operator)

$$
K(x - x') = \left\{ \left(\frac{\partial}{\partial y} - ip \frac{\partial}{\partial x} \right) \left(\frac{\partial}{\partial y'} + ip \frac{\partial}{\partial x'} \right) G \right\}_{y=y'=0}
$$

(8.293)

and $h(x) = aik = $ const. The solution of (8.292) is obtained using the Wiener–Hopf method (the original paper is by Wiener and Hopf 1931), which allows to solve wave problems with semi-infinite barriers, as was first demonstrated by Schwinger (1946) and Copson (1946). For the key idea, which is also developed below, see Karp (1950), Stoker (1957), p. 143 and Noble (1958), p. 36. Applications of the Wiener–Hopf method to diffraction theory are discussed in Flügge (1961).

With a lower bound of integration $x' \to -\infty$, (8.291) would be a convolution. The restriction to $x' \geq 0$ is central to the Wiener–Hopf method:

> In [Fourier integrals] $g(x)$ is given for all x, and $f(x)$ has to be found for all x; here [in the Wiener-Hopf method] $g(x)$ is given for $x > 0$ and $f(x)$ is identically zero for $x < 0$, and we have to find $g(x)$ for $x < 0$ and $f(x)$ for $x > 0$. (Copson 1946, p. 24)

Thus the Wiener–Hopf equation allows to determine two different functions on two different half-axes, where both functions are known on the complementary half-axes. One says that the Wiener–Hopf equation allows to solve for two *half-unknowns*, see Hutter and Olunloyo (1980).

First, the integral equation (8.291) is extended to the full axis $-\infty < x' < \infty$. Using the abbreviations f, g, h from (8.292), Crease (1956) assumes in his Eq. (13),

$$
\left. \begin{aligned} f(x) &= [\zeta(x)] \\ g(x) &= 0 \\ h(x) &= iak \end{aligned} \right\} \quad x > 0,
$$

(8.294)

and on the physically irrelevant negative half-axis,

$$\left.\begin{array}{l} f(x) = 0 \\ g(x) = q(x) \\ h(x) = 0 \end{array}\right\} \quad x < 0. \tag{8.295}$$

The unknown functions $[\zeta]$ for $x > 0$ and q for $x < 0$ are then determined by the integral equation (note the range of x'),

$$g(x) = h(x) + \int_{-\infty}^{\infty} dx' \, K(x - x') \, f(x'). \tag{8.296}$$

On both sides of this equation, an integration $\int_{-\infty}^{\infty} dx \, e^{-ilx}$ is performed. Then, according to the convolution theorem for the Fourier transforms (which can be applied now since the integration is over the full x-axis),

$$\bar{g}(l) = \bar{h}(l) + \bar{K}(l)\bar{f}(l), \tag{8.297}$$

where $\bar{f}(l) = \int_{-\infty}^{\infty} dx \, e^{-ilx} \, f(x)$ is the Fourier transform of $f(x)$ (not the complex conjugate). Four transforms (i) to (iv) are performed in the following for the functions h and f, the kernel K, and the function g, respectively. To simplify the calculations, the factor $\frac{1}{2\pi}$ is applied in the back-transform from $\bar{f}(l)$ to $f(x)$, etc.

(i) The first Fourier transform is

$$\bar{h}(l) = \int_{-\infty}^{\infty} dx \, e^{-ilx} \, h(x) = iak \int_{0}^{\infty} dx \, e^{-ilx}. \tag{8.298}$$

The restriction of the integral to $x \geq 0$ is due to $h(x < 0) = 0$. The wavenumber l shall have an infinitesimal negative imaginary part,

$$\text{Im} \, l < 0. \tag{8.299}$$

Then e^{-ilx} goes quickly to zero as $x \to \infty$, and the upper integration bound $x \to \infty$ in (8.298) does not contribute to the integral, giving

$$\bar{h}(l) = \frac{ak}{l}. \tag{8.300}$$

This function is analytic on the lower half-plane $\text{Im} \, l < 0$.

(ii) For the function $f(x) = [\zeta(x)]$ for $x > 0$, one expects a Kelvin wave behind the barrier, $f(x) = O(e^{ik_1 x})$, with unknown wavenumber k_1. For

$$\text{Im} \, k_1 > 0, \tag{8.301}$$

$f(x)$ is bounded as $x \to \infty$. The Fourier transform is, with $f(x < 0) = 0$,

$$\bar{f}(l) = \int_{0}^{\infty} dx \, e^{-ilx} \, f(x). \tag{8.302}$$

If

$$\text{Im } l < 0,$$ (8.303)

then $e^{-ilx} \to 0$ for $x \to \infty$, and thus $\bar{f}(l)$ is analytic on the lower half-plane $\text{Im } l < 0$. Furthermore, $\bar{f}(l)$ is also *bounded* for $\text{Im } l < 0$, which is a central assumption in the Liouville theorem applied below.

(iii) The Fourier transform of the kernel (8.293) requires a little more effort. Let $r = \sqrt{(x - x_0)^2 + (y - y_0)^2}$. First, consider the case of no rotation (no Coriolis force), $p = 0$; the corresponding kernel is termed K_0. Then,

$$
\begin{aligned}
K_0(x - x') &= \frac{\partial}{\partial y} \frac{\partial}{\partial y'} G(x - x', y - y') \bigg|_{y=y'=0} \\
&= \frac{i}{4} \frac{\partial}{\partial y} \frac{\partial}{\partial y'} \left(J_0(kr) + i Y_0(kr) \right) \bigg|_{y=y'=0} \\
&\stackrel{(a)}{=} \frac{ik}{4} \frac{\partial}{\partial y} \left(\frac{y - y'}{r} \left(J_1(kr) + i Y_1(kr) \right) \right) \bigg|_{y=y'=0} \\
&\stackrel{(b)}{=} \frac{ik}{4r} \left[J_1(kr) + i Y_1(kr) \right]_{y=y'=0}.
\end{aligned}
$$ (8.304)

In (a), $J_0' = -J_1$ and $Y_0' = -Y_1$ was used. In (b), terms in $\partial J_1/\partial y$ and $\partial Y_1/\partial y$ drop out due to a factor $y - y' = 0$. For the same reason, the derivative of the denominator r drops out. Therefore,

$$K_0(x) = \frac{ik}{4} \frac{J_1(k|x|) + i Y_1(k|x|)}{|x|},$$ (8.305)

where $|x|$ occurs because of $r > 0$. This is an even function of x, thus e^{ilx} in the Fourier transform of K_0 can be replaced by $\cos(lx)$,

$$
\begin{aligned}
\bar{K}_0(l) = \int_{-\infty}^{\infty} dx\, e^{-ilx} K_0(x) &= \frac{ik}{4} \int_{-\infty}^{\infty} dx\, \cos(lx) \frac{J_1(k|x| + i Y_1(k|x|)}{|x|} \\
&= \frac{ik}{2} \int_0^{\infty} dx\, \cos(lx) \frac{J_1(kx) + i Y_1(kx)}{x}.
\end{aligned}
$$ (8.306)

These integrals can be written in terms of H_1, but most integral tables list J_1 and Y_1: for the first integral, Erdélyi (1954), p. 44, Eq. (9) applies,

$$
\int_0^{\infty} dx\, \cos(xl) \frac{J_1(kx)}{x} =
\begin{cases}
\dfrac{\sqrt{\pi}}{2\Gamma(3/2)} \sqrt{k^2 - l^2} & \text{if } l < k \\
0 & \text{if } l > k.
\end{cases}
$$ (8.307)

The same result is obtained from A&S (11.4.36), p. 487. For the second integral in (8.305), use A&S (9.1.5), $Y_1(x) = -Y_{-1}(x)$. Then Erdelyi (1954), p. 47, Eq. (30) can be used with $\nu = -1$,

$$\int_0^\infty dx \, \cos(xl) \, \frac{Y_1(kx)}{x} = \begin{cases} 0 & \text{if } l < k \\ \dfrac{\sqrt{\pi}}{2\Gamma(3/2)} \, \sqrt{l^2 - k^2} & \text{if } l > k. \end{cases} \qquad (8.308)$$

(These cosine integrals of J_n and Y_n for $n = 1, 2, 3, \ldots$ are calculated using the theorem of residues. An example of such a calculation is given in Copson 1946, p. 26.) Using $\Gamma(3/2) = \sqrt{\pi}/2$ (see A&S 9.1.20), one has for the Fourier transform of K_0, for all l,

$$\boxed{\bar{K}_0(l) = \frac{i}{2} \sqrt{k^2 - l^2}} \qquad (8.309)$$

This is given without derivation in Stoker (1957), Eq. (5.6.13) on p. 144; and it agrees with the last equation on p. 90 of Crease (1956) for $p = 0$.

For the full kernel from (8.293) including rotation, one has

$$K(x - x') = \frac{i}{4} \left[(\partial_y - ip\partial_x)(\partial_{y'} + ip\partial_{x'})(J_0(kr) + iY_0(kr)) \right]_{y=y'=0}. \qquad (8.310)$$

A short calculation similar to the one above gives

$$K(x) = \frac{ik}{4} \frac{J_1(k|x|) + iY_1(k|x|)}{|x|} - \frac{ip^2 k^2}{4} \left(J_0''(k|x|) + iY_0''(k|x|) \right). \qquad (8.311)$$

The Bessel differential equation (8.269) gives, for $J_0(s)$ with $s = kr$,

$$J_0''(s) = -J_0 - \frac{J_0'}{s} = -J_0 + \frac{J_1}{s}, \qquad (8.312)$$

and the same equation holds with J replaced by Y. Thus (note that $1 - p^2 > 0$ by assumption)

$$K(x) = \frac{ik}{4} (1 - p^2) \frac{J_1(k|x|) + iY_1(k|x|)}{|x|} + \frac{ip^2 k^2}{4} \left(J_0(k|x|) + iY_0(k|x|) \right). \qquad (8.313)$$

Therefore, using (8.309) and that K is an even function of x,

$$\bar{K}(l) = \int_{-\infty}^\infty dx \, e^{-ilx} \, K(x)$$

$$= \frac{i}{2} (1 - p^2) \sqrt{k^2 - l^2} + \frac{ip^2 k^2}{2} \int_0^\infty dx \, \cos(lx) \left(J_0(kx) + iY_0(kx) \right). \qquad (8.314)$$

For the J_0 integral, one has from Erdelyi, p. 43, Eq. (1),

$$\int_0^\infty dx \; \cos(lx) \; J_0(kx) = \begin{cases} \dfrac{1}{\sqrt{k^2 - l^2}} & \text{if } l < k \\ 0 & \text{if } l > k. \end{cases} \tag{8.315}$$

And for the Y_0 integral from Erdelyi, p. 47, Eq. (28),

$$\int_0^\infty dx \; \cos(lx) \; Y_0(kx) = \begin{cases} 0 & \text{if } l < k \\ -\dfrac{1}{\sqrt{l^2 - k^2}} & \text{if } l > k. \end{cases} \tag{8.316}$$

Altogether therefore, for arbitrary l,

$$\bar{K}(l) = \frac{i}{2} \, (1 - p^2)\sqrt{k^2 - l^2} + \frac{i}{2} \, \frac{p^2 k^2}{\sqrt{k^2 - l^2}}, \tag{8.317}$$

or

$$\boxed{\bar{K}(l) = \frac{i}{2} \, \frac{k^2 - l^2(1 - p^2)}{\sqrt{k^2 - l^2}}} \tag{8.318}$$

This is the equation at the end of p. 90 in Crease (1956). The objective in the following is to factorize $\bar{K}$ as

$$\bar{K}(l) = L_+(l) \, L_-(l), \tag{8.319}$$

where

$$L_+ \sim \frac{1}{\sqrt{k + l}}, \qquad L_- \sim \frac{1}{\sqrt{k - l}}. \tag{8.320}$$

As is discussed below, the Wiener–Hopf method requires that the functions L_+, L_- and $\bar{f}, \bar{g}, \bar{h}$ have a common strip in the complex l-plane (i.e., for all values of Re l) on which they all are analytic. To avoid a zero in the denominator and have a common strip for L_+ and L_- in the complex l-plane, choose

$$\begin{aligned} \text{for } L_+: & \qquad \text{Im } l > -\text{Im } k, \\ \text{for } L_-: & \qquad \text{Im } l < \text{Im } k. \end{aligned} \tag{8.321}$$

Then the kernel K is analytic on the strip $-\text{Im } k < \text{Im } l < \text{Im } k$ (the wavenumber k of the incoming plane wave is a constant parameter), see Fig. 8.22.

(iv) Finally, the function $g(x)$ is nonvanishing for $x < 0$ only, where, according to (8.294) to (8.296),

$$g(x) = \int_0^\infty dx' \, K(x - x') \, f(x'). \tag{8.322}$$

The kernel K consists of second spatial derivatives of $H_0(kr)$. We require that $g(x)$ remains bounded for $x \to -\infty$, and thus for $r \to |x| \to \infty$. From (9.2.3) in A&S,

$$\lim_{r \to \infty} H_0(kr) = (1 - i) \frac{e^{ikr}}{\sqrt{\pi kr}}. \tag{8.323}$$

For $x \to -\infty$, the integral in the Fourier transform $\bar{g}(l)$ reads

$$\int_{-\infty}^{\infty} dx \, e^{-ilx} \, e^{ik|x|} \ldots \tag{8.324}$$

This integral is well defined for $x \to -\infty$, and thus $\bar{g}(l)$ is analytic, if the argument of the exponential has a (small) negative real part,

$$\text{Re}\,(-ilx + ik|x|) < 0,$$

or $\quad -i^2 \, \text{Im}(l)x + i^2 \, \text{Im}(k)|x| < 0,$

or $\quad \text{Im}\,l + \text{Im}\,k > 0. \tag{8.325}$

Furthermore, $\bar{g}(l)$ is also *bounded* for $\text{Im}\,l > -\text{Im}\,k$, which is used in applying the Liouville theorem below.

This finishes the calculation of the Fourier transforms of h, f, K, g (an explicit expressing for $\bar{g}$ is not needed). Figure 8.22 shows the domains of analyticity for these functions in the complex l-plane, with a common strip $-\text{Im}(k) < \text{Im}(l) < 0$ in which all the functions are analytic. A nonvanishing value of $\text{Im}(k)$ is needed to obtain this common strip of analyticity:

> The Wiener-Hopf method [...] employs the artificial device of assuming a positive imaginary part for the wavenumber k. This brings with it [...] that while the primary wave dies out as $y \to +\infty$, it becomes exponentially infinite as $y \to -\infty$. (Stoker 1957, p. 145, with x replaced by y)

We return to Eq. (8.297), $\bar{g}(l) = \bar{h}(l) + \bar{K}(l)\bar{f}(l)$. This equation is rearranged in such a way that the left side of the new equation is analytic for $\text{Im}\,l > -\text{Im}\,k$, and the right side for $\text{Im}\,l < 0$.

The functions $\bar{g}$ and $\bar{f}$ shall still appear on the left and right side of the new equation, respectively, as this is in accordance with the demanded domains of analyticity,

Fig. 8.22 Domain of analyticity of the Fourier transforms in the complex l-plane. The wavenumber of the incident wave is k and is assumed to have a small positive imaginary part $\text{Im}\,k$

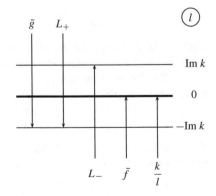

see Fig. 8.22. To gain some feeling for the procedure, the case $p = 0$ of no rotation is considered again first. Here (8.297) becomes

$$\bar{g}(l) = \frac{ak}{l} + \frac{i}{2} \sqrt{k+l} \sqrt{k-l} \; \bar{f}(l). \tag{8.326}$$

The function $\bar{g}$ is analytic for Im $l > -$Im k, thus (8.326) is written as

$$\frac{\bar{g}(l)}{\sqrt{k+l}} - \frac{ak}{l\sqrt{k+l}} = \frac{i}{2} \sqrt{k-l} \; \bar{f}(l). \tag{8.327}$$

However, the second term on the left, $-ak/(l\sqrt{k+l})$ is analytic only for Im $l > 0$. To make it analytic for Im $l > -$Im k, a term $a\sqrt{k}/l$ is added on both sides of the equation, giving

$$\frac{\bar{g}(l)}{\sqrt{k+l}} - \frac{ak}{l} \left(\frac{1}{\sqrt{k+l}} - \frac{1}{\sqrt{k}} \right) = \frac{a\sqrt{k}}{l} + \frac{i}{2} \sqrt{k-l} \; \bar{f}(l). \tag{8.328}$$

When $1/\sqrt{k+l}$ is expanded in a Taylor series around $l = 0$, one obtains for the term in brackets,

$$\frac{ak}{l} \left(\frac{1}{\sqrt{k+l}} - \frac{1}{\sqrt{k}} \right) = \frac{ak}{l\sqrt{k}} \left(-\frac{l}{2k} + \frac{3l^2}{8k^2} + \cdots \right). \tag{8.329}$$

The right side of this equation and therefore also the left side has no singularity at $l = 0$. Thus the left side of (8.328) is indeed analytic for Im $l > -$Im k, and the right side of (8.328) is analytic for Im $l < 0$ due to the functions $1/l$ and $\bar{f}$ (see Fig. 8.22), as was intended.

With the full kernel for $p \neq 0$ one proceeds analogously. Equation (8.297) is, with the Fourier transform (8.318) of the kernel,

$$\bar{g}(l) = \frac{ak}{l} + \frac{i}{2} \frac{k+l\sqrt{1-p^2}}{\sqrt{k+l}} \frac{k-l\sqrt{1-p^2}}{\sqrt{k-l}} \; \bar{f}(l). \tag{8.330}$$

This is rearranged as

$$\frac{\sqrt{k+l}}{k+l\sqrt{1-p^2}} \bar{g}(l) - \frac{ak}{l} \left(\frac{\sqrt{k+l}}{k+l\sqrt{1-p^2}} - \frac{1}{\sqrt{k}} \right) = \frac{a\sqrt{k}}{l} + \frac{i}{2} \frac{k-l\sqrt{1-p^2}}{\sqrt{k-l}} \; \bar{f}(l), \tag{8.331}$$

which is equation (17) in Crease (where his $a\sqrt{k}/\alpha$ on the left should read ak/α). The Taylor series expansion to first order around $l = 0$ gives for the term in brackets,

$$\frac{ak}{l}\left(\frac{\sqrt{k+l}}{k+l\sqrt{1-p^2}} - \frac{1}{\sqrt{k}}\right) = \frac{a}{\sqrt{k}}\left(\frac{1}{2} - \sqrt{1-p^2}\right) + O(l), \qquad (8.332)$$

which has no singularity at $l = 0$. The left side of (8.331) is therefore analytic for

$$\operatorname{Im} l > -\frac{\operatorname{Im} k}{\sqrt{1-p^2}}, \qquad (8.333)$$

and since $0 \le p \le 1$, it is also analytic in the smaller domain $\operatorname{Im} l > -\operatorname{Im} k$. The right side of (8.331) is analytic for $\operatorname{Im} l < 0$ due to the functions $1/l$ and $\bar{f}$ (see Fig. 8.22), again as was intended. This leads to the following important conclusion:

> The left-hand side of Eq. (8.331) is regular [analytic] for $\operatorname{Im} l > -\operatorname{Im} k$, and the right-hand side for $\operatorname{Im} l < 0$. Therefore, as both sides are regular in the strip $0 > \operatorname{Im} l > -\operatorname{Im} k$, they define a function $E(l)$ which is regular over the entire l-plane by analytic continuation. (Crease 1956, p. 91, with our symbols and equation numbers)

The last few words are further emphasized in the following quote:

> Each side of the Eq. (8.331) furnishes the analytic continuation of the function defined by the other side. (Stoker 1957, p. 145)

Here 'analytic continuation' refers to a rather general principle of complex function theory, by which analytic functions that agree on certain domains define an analytic function on the union of these domains:

> An analytic function $f(z)$ defined in a region [domain] Ω will constitute a *function element*, denoted by (f, Ω), and a *global analytic function* will appear as a collection of function elements which are related to each other in a prescribed manner. Two function elements (f_1, Ω_1) and (f_2, Ω_2) are said to be *direct analytic continuations* of each other if $\Omega_1 \cap \Omega_2$ is nonempty and $f_1(z) = f_2(z)$ in $\Omega_1 \cap \Omega_2$. More specifically, (f_2, Ω_2) is called a direct analytic continuation of (f_1, Ω_1) to the region Ω_2. (Ahlfors 1966, p. 275)

The underlying idea is that analytic functions are infinitely often differentiable, and fully determined by their values—and thus their differentials of any order—in an arbitrarily small domain, which allows an expansion of the function in a Taylor series, i.e., a power series.

We return to Eq. (8.331), which defines a complex function $E(l)$ that is analytic on the whole complex l-plane. On the left side of (8.331), defining $E(l)$ for $\operatorname{Im} l > -\operatorname{Im} k$, the function $\bar{g}(l)$ is *bounded* for $\operatorname{Im} l > -\operatorname{Im} k$ according to item (iv) above; and on the right side of (8.331), defining $E(l)$ for $\operatorname{Im} l < 0$, the function $\bar{f}(l)$ is *bounded* for $\operatorname{Im} l < 0$ according to (ii) above. The highest power of l for $l \to \infty$ on the left side (i.e., in the upper half-plane of l) of (8.331) is thus $l^{-1/2}$, and for the right side (i.e., the lower half-plane of l) is $l^{1/2}$. Overall, the function $E(l)$ for $|l| \to \infty$ is therefore $O(|l|^{1/2})$,

$$|E(l)| \le M \, |l|^{1/2} \qquad (8.334)$$

for sufficiently large l. This function is subject to the generalized Liouville theorem of complex function theory.

Theorem (Generalized Liouville theorem) *Let $f(z)$ be a single-valued complex function that is analytic for all $z \in \mathbb{C}$, and for which*

$$|f(z)| \leq M\,|z|^n \tag{8.335}$$

holds for all $z \in \mathbb{C}$, with some constant $M > 0$ and $n \in \mathbb{N}$. Then $f(z)$ is a polynomial of degree $\leq n$.

The proof is a mild generalization (an exercise in Markushevich 1965, p. 367) to that of the classical Liouville theorem of complex function theory (see Appendix A), stating that a bounded entire function is constant, which itself is a direct consequence of the Cauchy integral formula.

For $E(l)$, the smallest $n \in \mathbb{N}$ that bounds E according to (8.335) is $n = 1$. From the generalized Liouville theorem, E is thus a polynomial of order $n = 1$ or 0. However, if E would be a polynomial of order one, it would violate (8.334) for sufficiently large l, where the l-term would dominate over the $l^{1/2}$-term. Thus the function E must be a polynomial of order zero, that is, a constant! Since the right side of (8.331) scales as $l^{-1/2}$, for $|l| \to \infty$ in the upper half-plane one has $E \to 0$. Therefore

$$E(l) = 0 \tag{8.336}$$

for all l, which gives the central conclusion:

Both the left and right sides of (8.331) *vanish.*

This gives one equation for $\bar{f}$ and one for $\bar{g}$, i.e., one common equation for $\bar{f}$ and $\bar{g}$ has been separated into two equations. Putting the right side of (8.331) to zero gives as solution for the Fourier transform of the jump across the barrier,

$$\boxed{\bar{f}(l) = \frac{2ia\sqrt{k}\sqrt{k-l}}{l(k-l\sqrt{1-p^2})}} \tag{8.337}$$

which is Eq. (18) in Crease (1956). This allows to calculate the jump across the barrier from the integral equation (8.289) for $\zeta(x, y)$. The kernel in (8.289) is

$$M(x - x', y) = [(\partial_{y'} + ip\partial_{x'})\,G(x - x', y - y')]_{y'=0}$$
$$= \frac{ik}{4} \frac{y + ip(x - x')}{\sqrt{(x - x')^2 + y^2}}\,H_1\!\left(k\sqrt{(x - x')^2 + y^2}\right). \tag{8.338}$$

The Fourier transform of the first term in this sum, $iky H_1(kr)/4r$, is, with $x' = 0$,

$$\bar{M}_1(l) \overset{(a)}{=} \frac{iky}{2} \int_0^\infty dx\ \cos(lx)\ \frac{H_1\left(k\sqrt{x^2+y^2}\right)}{\sqrt{x^2+y^2}}$$

$$\overset{(b)}{=} \frac{iy}{2}\ \sqrt{k^2-l^2}\ \sqrt{\frac{\pi}{2}}\ \frac{H_{1/2}\left(|y|\sqrt{k^2-l^2}\right)}{\sqrt{|y|}\ (k^2-l^2)^{1/4}}$$

$$\overset{(c)}{=} \frac{iy}{2}\ \sqrt{k^2-l^2}\ h_0\!\left(|y|\sqrt{k^2-l^2}\right) \tag{8.339}$$

$$\overset{(d)}{=} \frac{y}{2}\ \sqrt{k^2-l^2}\ \left(i j_0\!\left(|y|\sqrt{k^2-l^2}\right) - y_0\!\left(|y|\sqrt{k^2-l^2}\right)\right)$$

$$\overset{(e)}{=} \frac{y}{2}\ \left(i\ \sin\!\left(|y|\sqrt{k^2-l^2}\right) + \cos\!\left(|y|\sqrt{k^2-l^2}\right)\right)$$

$$= \frac{y}{2|y|}\ e^{i|y|\sqrt{k^2-l^2}}.$$

(a) $H_1(kr)/r$ is even with respect to x, thus use $\cos(lx)$ instead of e^{-ilx}.
(b) See Erdelyi (1954), p. 56, formula (42), with erratum on p. xvi.
(c) h_0 is a spherical Bessel function of the third kind, A&S (10.1.1).
(d) $y_0(|y|)$ should cause no confusion:
      y_0 is a spherical Bessel function, $|y|$ the Cartesian coordinate.
(e) See A&S (10.1.11) and (10.1.12) on p. 438.

For the second sum term in (8.338), put $H_1 = J_1 + iY_1$. Since $x J_1(kr)/r$ is odd with respect to x, the Fourier term e^{-ilx} can be replaced by $-i\sin(lx)$, giving for the Fourier transform,

$$\bar{M}_{\mathrm{II},J_1}(l) = \frac{ikp}{2} \int_0^\infty dx\ \sin(lx)\ \frac{x}{\sqrt{x^2+y^2}}\ J_1\!\left(k\sqrt{x^2+y^2}\right)$$

$$= \begin{cases} \dfrac{ip}{2}\ \dfrac{l}{\sqrt{k^2-l^2}}\ \cos\!\left(|y|\sqrt{k^2-l^2}\right) & \text{if } l < k \\[2mm] 0 & \text{if } l > k, \end{cases} \tag{8.340}$$

using Erdelyi (1954), p. 111, (35) in the case distinction of the second equation sign. For the Fourier transform $\bar{M}_{\mathrm{II},Y_1}(l)$ of the second sum term in (8.338) with iY_1 entering instead of J_1, one has only to replace $\cos(|y|\sqrt{k^2-l^2})$ in (8.340) by $i\sin(|y|\sqrt{k^2-l^2})$. Thus, finally, for $\bar{M} = \bar{M}_\mathrm{I} + \bar{M}_\mathrm{II}$,

$$\bar{M}(l) = \frac{1}{2}\left(\frac{y}{|y|} + \frac{ipl}{\sqrt{k^2-l^2}}\right) e^{i|y|\sqrt{k^2-l^2}}. \tag{8.341}$$

Replacing in (8.289) the jump $[\zeta](x)$ by the function $f(x)$ with the understanding that $f(x) = 0$ for $x < 0$, the lower integration bound can be shifted from 0 to $-\infty$. Then, with the Fourier convolution theorem, one has

$$\zeta(x, y) = \int_{-\infty}^{\infty} dx' \, M(x - x') \, f(x') + ae^{iky}$$

$$= \frac{1}{2\pi} \int_{-\infty}^{\infty} dl \, e^{ilx} \, \bar{M}(l) \, \bar{f}(l) + ae^{iky}, \tag{8.342}$$

and one arrives at a key result of Crease (1956), his Eq. (23),

$$\zeta(x, y) = \frac{ia}{2\pi} \int_{-\infty}^{\infty} dl \, \frac{\sqrt{k}\sqrt{k - l}}{l\left(k - l\sqrt{1 - p^2}\right)} \left(\frac{y}{|y|} + \frac{ipl}{\sqrt{k^2 - l^2}}\right) e^{i\left(lx + |y|\sqrt{k^2 - l^2}\right)} + ae^{iky} \tag{8.343}$$

This has four singularities on the real l-axis, at $l = -k, 0, k$, and $k/\sqrt{1 - p^2}$ (where $l = -k$ and k are branch points). The integration contour can be closed along a semicircle at infinity in the upper l-half-plane, on which e^{ikl} is zero; the semicircle gives then no contribution to the integral. To have the four singularities enclosed by the resulting closed contour (else the integral vanishes according to Cauchy's theorem), the integration from $l = -\infty$ to $l = \infty$ is carried out along a line shifted slightly *below* the real l-axis.

Crease (1956) proceeds by first extracting from (8.343) an elevation ζ_p that is due to the Coriolis terms alone (i.e., he subtracts from (8.343) the corresponding expression for $p = 0$), and transforming the integral for ζ_p 'into more recognizable forms' by using advanced techniques developed in applications of the Wiener–Hopf method to diffraction theory (see Eqs. 24–32 in Crease 1956; see also Copson 1946 and the book by Noble 1958). The calculation is somewhat oblique and leads to the asymptotic expression for ζ_p given in Eq. (33) of Crease (named ζ_3 there). It is somewhat odd that Crease does not mention a much simpler approach to derive this equation, which is used in the following. Let

$$s = \sqrt{1 - p^2}, \tag{8.344}$$

where $0 \le s \le 1$. From the four singularities of $\zeta(x, y)$ in (8.342), only the last one,

$$l = \frac{k}{s}, \tag{8.345}$$

contains the Coriolis term p. We can calculate ζ from this singularity *alone* by performing an integral along a circle of infinitesimal radius $\epsilon \to 0$ centered at k/s in the l-plane (i.e., calculate the residue; note that no further extraction of the Coriolis-induced part of the wave elevation is then necessary),

$$l = \frac{k}{s} + \epsilon e^{i\varphi}. \tag{8.346}$$

Then $dl = i\epsilon e^{i\varphi} d\varphi$ and $k - ls = s\epsilon e^{i\varphi}$ and $\sqrt{k^2 - l^2} = ipk/s$, and (8.343) gives for $y > 0$ behind the barrier,

$$\zeta_p = \frac{ia}{2\pi} \, 2\pi \, \frac{i\epsilon e^{i\varphi} \sqrt{k} \sqrt{k} \sqrt{1-s^{-1}}}{ks^{-1} \, s\epsilon e^{i\varphi}} \left(1 + \frac{ipk/s}{ipk/s}\right) \, e^{ikx/s} e^{-pky/s}. \qquad (8.347)$$

Note especially the appearance of a factor of two from the terms in the bracket; correspondingly, $\zeta_p = 0$ ahead of the barrier where $y/|y| = -1$. Shifting an overall minus sign to the oscillating term $e^{ikx/s}$ leads to

$$\zeta_p = 2ia \, \sqrt{\frac{1-s}{s}} \, e^{ikx/s} e^{-pky/s}. \qquad (8.348)$$

Finally, using $k^2 = (\omega^2 - \gamma^2)/gh$ (Eq. 8.264), $p = \gamma/\omega$ (Eq. 8.267), and $s^2 = 1 - p^2$, one obtains from (8.348)

$$\boxed{\zeta_p(x, y) = 2ia \left(\frac{\omega}{(\omega^2 - \gamma^2)^{1/2}} - 1\right)^{1/2} \exp\left[\frac{i\omega x - \gamma y}{\sqrt{gh}}\right]} \qquad (8.349)$$

This is the asymptotic equation (33) in Crease (1956) for a *Kelvin wave* behind the barrier, at small $y > 0$ and large x (derived there from direct calculation of the integral in 8.342). The wave has a resonant amplitude for $\omega \to \gamma$: for a Coriolis parameter $p > 3/5$, giving $s > 4/5$ and $(s^{-1} - 1)^{1/2} > 1/2$, the Kelvin wave amplitude $> 2a/2$ is indeed larger than the amplitude a of the incoming wave at normal incidence to the barrier!

A follow-up paper by Crease (1958) considers the case of *two* parallel barriers, one roughly corresponding to Britain and the other to Denmark plus Norway. Empirical facts about surges can be found in Schmitz (1965). Some simplifications to the above calculation are due to Kapoulitsas (1977) and Chambers (1954, 1964). Chambers follows Lamb in using parabolic coordinates, see also Williams (1964). For later applications of the method, see Manton et al. (1970) and Crighton and Leppington (1974), the latter authors applying it to a vortex sheet.

Hutter and Olunloyo (1980) employ the Wiener–Hopf technique to calculate velocity deviations from a mean Poiseuille flow as well as the normal and shear pressures at the bottom and top of an infinitely long, tilted slab of ice, meant to model a glacier atop of rock, which is partly sliding over and partly adhering to its bed, see Fig. 8.23. This is a novel example for the use of different types of boundary conditions (Neumann vs. Dirichlet) not just at different boundaries (e.g., at the bottom of an ocean and at its free surface), but at different portions of the same boundary. More specifically, Hutter and Olunloyo assume perfect sliding, $u > 0$ and $\partial u/\partial z = 0$, of temperate ice at $x < 0$ (the Neumann boundary conditions), and adherence, $u = 0$ and $\partial u/\partial z \neq 0$ (the Dirichlet boundary conditions) of cold ice at $x > 0$. (The same boundary conditions as on the bed are also applied at the top of the ice slab, so that the flow becomes a canal flow symmetric about the x-axis.) This pair of half-unknowns, u and $\partial u/\partial z$, can be determined along the whole boundary by the Wiener–Hopf method. The entire function E, defined on the whole complex plane by analytic continuation from two half-planes overlapping in a parallel strip of finite

Fig. 8.23 Tilted slab of ice between two walls, with perfect slip, Neumann's boundary conditions for $x < 0$, and no-slip, Dirichlet's boundary conditions for $x > 0$

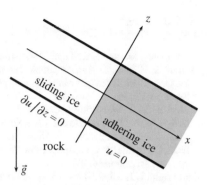

width 'must be set to zero to ensure an integrable shear stress at the point of discontinuity' (Hutter and Olunloyo, p. 392). This argument replaces in the present example the above application of the generalized Liouville theorem. With this integrable pressure, total forces remain finite even in the presence of a pressure singularity: a central result from this analysis is that the basal shear stress and pressure exhibit a singularity $x^{-1/2}$ at the point $x = 0$ of the change in boundary conditions. An interesting physical implication is that these 'stress singularities could account for the high gravel concentration frequently found near the bottom of ice boreholes' (Hutter and Olunloyo 1980, p. 385) by 'tear[ing] off portions of the rock base' (p. 386).

This same problem is taken up again by Barcilon and MacAyeal (1993) in an elaborate mathematical analysis using the Wiener–Hopf method (their Sects. 5 and 6), who find, partially in contrast to Hutter et al. (1980), that the transition in streamwise velocity near the location of change in boundary conditions at the bed is smooth, see their Figs. 7 and 8. As an important new result, they find a dip of 20% in the ice elevation near the transition in boundary conditions (their Fig. 5c), which is a strong empirical indicator for such changes taking place. Despite all mathematical efforts, Barcilon and MacAyeal (1993) conclude that the basal pressure singularity $p \to -\infty$ (not the shear stress singularity $\tau \to \infty$) violates basic mathematical assumptions of the model (e.g., that $p \to -\infty$ is of first order and small compared to the zeroth-order pressure), so that 'the problem, as formulated here, remains open and unsolved' (Barcilon and MacAyeal, p. 184).

References

Abramowitz, M., and I.A. Stegun. 1964. *Handbook of mathematical functions*. New York: Dover.

Ahlfors, L.V. 1966. *Complex analysis*, 2nd ed. New York: McGraw-Hill.

Airy, G.B. 1845. Tides and waves. In *Encyclopaedia metropolitana*, vol. 5, 241. London.

Aldersey-Williams, H. 2016. *The tide: The science and stories behind the greatest force on Earth*. New York: W. W. Norton & Company.

Barcilon, V., and D.R. MacAyeal. 1993. Steady flow of a viscous ice stream across a no-slip/free-slip transition at the bed. *Journal of Glaciology* 39: 167.

Carrier, G.F., and H.P. Greenspan. 1958. Water waves of finite amplitude on a sloping beach. *Journal of Fluid Mechanics* 4: 97.

Cartwright, D.E. 1999. *Tides – A scientific history*. Cambridge: Cambridge University Press.

Chambers, Ll.G. 1954. Diffraction by a half-plane. *Proceedings of the Edinburgh Mathematical Society* 10: 92.

Chambers, Ll.G. 1964. Long waves on a rotating Earth in the presence of a semi-infinite barrier. *Proceedings of the Edinburgh Mathematical Society* 14: 25.

Copson, E.T. 1946. On an integral equation arising in the theory of diffraction. *Quarterly Journal of Mathematics* 17: 19.

Crapper, G.D. 1984. *Introduction to water waves*. Hemel: Ellis Horwood Ltd.

Crease, J. 1956. Long waves on a rotating Earth in the presence of a semi-infinite barrier. *Journal of Fluid Mechanics* 1: 86.

Crease, J. 1958. The propagation of long waves into a semi-infinite channel in a rotating system. *Journal of Fluid Mechanics* 4: 306.

Crighton, D.G., and F.G. Leppington. 1974. Radiation properties of the semi-infinite vortex sheet: The initial-value problem. *Journal of Fluid Mechanics* 64: 393.

Darwin, G.H. 1902. Tides. *Encyclopaedia Britannica*, 9th ed. https://www.1902encyclopedia.com/T/TID/tides.html.

Erdélyi, A. (ed.). 1954. *Tables of integral transforms*, vol. II. New York: McGraw-Hill.

Feldmeier, A. 2013. *Theoretische Mechanik*. Berlin: Springer.

Flügge, S. (ed.). 1961. *Crystal optics, diffraction, encyclopedia of physics*, vol. XXV/1. Berlin: Springer.

Gjevik, B. 2009. *Flo og fjære langs kysten av Norge og Svalbard* (in Norwegian). Oslo: Farleia Forlag.

Grossman, N. 2005. A C^∞ Lagrange inversion theorem. *The American Mathematical Monthly* 112: 512.

Hedstrom, G.W. 1979. Nonreflecting boundary conditions for nonlinear hyperbolic systems. *Journal of Computational Physics* 30: 222.

Hibberd, S., and D.H. Peregrine. 1979. Surf and run-up on a beach: A uniform bore. *Journal of Fluid Mechanics* 95: 323.

Hutter, K., and V.O.S. Olunloyo. 1980. On the distribution of stress and velocity in an ice strip, which is partly sliding over and partly adhering to its bed, by using a Newtonian viscous approximation. *Proceedings of the Royal Society of London A* 373: 385.

Hutter, K., Y. Wang, and I.P. Chubarenko. 2012. *Physics of lakes: Lakes as oscillators*, vol. 2. Berlin: Springer.

Kapoulitsas, G.M. 1977. Diffraction of long waves by a semi-infinite vertical barrier on a rotating Earth. *International Journal of Theoretical Physics* 16: 763.

Karp, S.N. 1950. Wiener-Hopf techniques and mixed boundary value problems. *Communications on Pure and Applied Mathematics* 3: 201.

Keller, H., D. Levine, and G. Whitham. 1960. Motion of a bore over a sloping beach. *Journal of Fluid Mechanics* 7: 302.

Lamb, H. 1932. *Hydrodynamics*. Cambridge: Cambridge University Press; New York: Dover (1945).

LeBlond, P.H., and L.A. Mysak. 1978. *Waves in the ocean*. Elsevier oceanography series 20. Amsterdam: Elsevier Scientific Publishing Company.

Le Lacheur, E.A. 1924. Tidal currents in the open sea: Subsurface tidal currents at Nantucket Shoals Light Vessel. *Geographical Review* 14: 282.

Manton, M.J., L.A. Mysak, and R.E. McGorman. 1970. The diffraction of internal waves by a semi-infinite barrier. *Journal of Fluid Mechanics* 43: 165.

Markushevich, A.I. 1965. *Theory of functions of a complex variable*, vol. I. Englewood Cliffs: Prentice-Hall.

McCowan, J. 1892. On the theory of long waves and its application to the tidal phenomena of rivers and estuaries. *Philosophical Magazine Series 5*, 33: 250.

Mortimer, C.H. 1975. *Environmental status of the Lake Michigan region. Vol. 2. Physical limnology of Lake Michigan. Part 1. Physical characteristics of Lake Michigan and its responses to applied forces*, ANL/ES-40 Vol. 2 Environmental and Earth Sciences, National Technical Information Service, U. S. Department of Commerce, Springfield, Virginia. Available as pdf-file.

Noble, B. 1958. *The Wiener-Hopf technique*. London: Pergamon Press.

Proudman, J. 1953. *Dynamical oceanography*. London: Methuen & Co.

Proudman, J. 1955. The propagation of tide and surge in an estuary. *Proceedings of the Royal Society of London A* 231: 8.

Proudman, J. 1957. Oscillations of tide and surge in an estuary of finite length. *Journal of Fluid Mechanics* 2: 371.

Proudman, J. 1958. On the series that represent tides and surges in an estuary. *Journal of Fluid Mechanics* 3: 411.

Proudman, J., and A.T. Doodson. 1924. The principal constituent of the tides of the North Sea. *Philosophical Transactions of the Royal Society of London A* 224: 185.

Rayleigh, Lord. 1914. On the theory of long waves and bores. *Proceedings of the Royal Society of London A* 90: 324.

Schmitz, H.P. 1965. Modellrechnungen zur deep water surge Entwicklung – das external surge Problem. *Deutsche Hydrographische Zeitschrift* 18: 49.

Schwinger, J.S. 1946. *Fourier transform solution of integral equations*. M. I. T. Radiation Laboratory Report.

Sedov, L.I. 1946. The movement of air in a strong explosion. *Doklady Akademii nauk SSSR* 52: 17.

Shen, M.C., and R.E. Meyer. 1963. Climb of a bore on a beach. III. Run-up. *Journal of Fluid Mechanics* 16: 113

Sommerfeld, A. 1896. Mathematische Theorie der Diffraction. *Mathematische Annalen* 47: 317.

Stoker, J.J. 1948. The formation of breakers and bores. The theory of nonlinear wave propagation in shallow water and open channels. *Communications on Pure and Applied Mathematics* 1: 1.

Stoker, J.J. 1957. *Water waves. The mathematical theory with applications*. New York: Interscience Publishers.

Taylor, G.I. 1922. Tidal oscillations in gulfs and rectangular basins. *Proceedings of the London Mathematical Society, Series 2*, 20: 148.

Taylor, G.I. 1950. The formation of a blast wave by a very intense explosion. *Proceedings of the Royal Society of London A* 201: 159.

Thorade, H. 1941. *Ebbe und Flut. Ihre Entstehung und ihre Wandlungen*. Berlin: Springer.

Tuck, E.O., and L.-S. Hwang. 1972. Long wave generation on a sloping beach. *Journal of Fluid Mechanics* 51: 449.

Wehausen, J.V., and E.V. Laitone. 1960. Surface waves. In *Handbuch der Physik = Encyclopedia of Physics*, vol. 3/9, Strömungsmechanik 3, ed. S. Flügge and C. Truesdell, 446. Berlin: Springer.

Whewell, W. 1836. Researches on the tides. Sixth series. On the results of an extensive system of tide observations made on the coasts of Europe and America in June 1835. *Philosophical Transactions of the Royal Society of London* 126: 289.

Whewell, W. 1848. The Bakerian Lecture. – Researches on the tides. Thirteenth series. On the tides of the pacific, and on the diurnal inequality. *Philosophical Transactions of the Royal Society of London* 138: 1.

Wiener, N., and E. Hopf. 1931. Über eine Klasse singulärer Integralgleichungen. *Sitzungsberichte der preussischen Akademie der Wissenschaften, Physikalisch-Mathematische Klasse* XXXI: 696.

Williams, W.E. 1964. Note on the scattering of long waves in a rotating system. *Journal of Fluid Mechanics* 20: 115.

Zel'dovich, Ya.B., and Yu.P. Raizer. 1967. *Physics of shock waves and high-temperature hydrodynamic phenomena*, vol. 2. New York: Academic Press; Mineola: Dover (2002).

Chapter 9
Free Surface Waves

This chapter treats waves that affect only relatively thin layers at the surface of a body of water, called *free surface waves*, and which therefore are essentially complementary to shallow water waves. Free surface waves have a vertical penetration depth into the water that corresponds roughly to their horizontal wavelength. The external mechanical force responsible for these waves is again gravity. On oceans, free surface waves are mostly excited by winds.

The main interest is in the *shape* of the wave, nowadays especially in the shape of nonlinear waves and eventually the universal shape of the *steepest* wave before breaking takes place. This steepest wave is called the *Stokes wave*, and its profile has at its highest point a sharp corner of 120° opening angle. This was already suggested in the nineteenth century by Stokes on physical grounds; but only Krasovskii (1961) could prove rigorously, using methods of nonlinear functional analysis:

> At each given magnitude of $0 < h/\lambda \le \infty$ [mean water depth h; wavelength λ] there exist waves of the steady-state kind, for which the maximum angle of inclination of the tangent to the profile of the free boundary takes any value from the range $(0, \pi/6)$. (Krasovskii 1961, p. 1012)

The supremum of the latter interval corresponds to a full inner opening angle of 120° at the wave crest. Another variant of the proof was given later by Keady and Norbury (1978).

Free surface waves pose an interesting boundary value problem, since the shape of the upper water surface, on which boundary conditions are specified, is not known in advance, but has to be found. A standard technique going back to Stokes is not to solve for the complex potential $\Phi(z)$ of the wave, but to consider instead $z(\Phi)$ with Φ as the independent variable and the waveform z as the unknown function. As already used in Chap. 5 on jets, wakes, and cavities, the streamfunction ψ is constant along the free fluid surface, and the latter can thus be specified by $\psi = 0$ and a curve parameter ϕ, the velocity potential. The two-dimensional wave problem is then reduced to that of finding an analytic mapping from a rectangle in the Φ-plane to a simply connected domain in the z-plane, with one boundary representing the waveform.

© Springer Nature Switzerland AG 2019

A. Feldmeier, *Theoretical Fluid Dynamics*, Theoretical and Mathematical Physics,
https://doi.org/10.1007/978-3-030-31022-6_9

First in this chapter, the dispersion relation for surface water waves is derived, which, by contrast to that for shallow water waves, shows *dispersion*. Therefore, a wave packet will spread out in time, and a central question is for the shape of the water surface after a point-like perturbation, i.e., for the Green's function of the differential equation for surface water waves. After the derivation of this Green's function, some relatively elementary wave propagation problems are treated, followed by a rather superficial treatment (without existence proof) of the steepest wave, the Stokes wave. The closing section gives one of the very few known examples of an exact nonlinear wave solution, the Crapper wave, which holds for capillary waves of (very) short wavelength, for which surface tension due to the curvature of the water surface becomes important.

9.1 Dispersion Relation

The fluid is assumed to be inviscid and to have constant density. A straight canal with constant depth h and constant rectangular profile shall point in the x-direction. The z-coordinate points vertically upward, with $z = 0$ at the unperturbed water level. The wave elevation above and below $z = 0$ is $\zeta(t, x, z)$, and $u(t, x, z)$ and $w(t, x, z)$ are the speeds in the x- and z-directions, respectively. The hydrodynamic fields ζ, u, and w do not depend on the cross-canal coordinate y, and there is no cross-canal speed v.

The equation of motion for the fluid is $\Delta\phi = 0$, with a velocity potential $\phi(t, x, z)$. Two boundary conditions apply on the free surface: first, the Bernoulli equation (2.208) is

$$\phi_t + \tfrac{1}{2}\phi_x^2 + \tfrac{1}{2}\phi_z^2 + g\zeta = 0. \tag{9.1}$$

This is a dynamic boundary condition and holds at the free surface

$$S(t, x, z) = \zeta(t, x) - z = 0. \tag{9.2}$$

The pressure on S is assumed to be constant or zero. Second, the kinematic boundary condition (8.10) on S is

$$0 = \frac{\mathrm{d}S}{\mathrm{d}t} = \left(\frac{\partial}{\partial t} + u\frac{\partial}{\partial x} + w\frac{\partial}{\partial z}\right)\left(\zeta(t, x) - z\right). \tag{9.3}$$

With $u = \phi_x$ and $w = \phi_z$, this yields

$$0 = \zeta_t + \phi_x\,\zeta_x - \phi_z. \tag{9.4}$$

For *linear* waves, ϕ and ζ are small, and all quadratic terms can be neglected. Then the boundary conditions (9.1) and (9.4) on the free water surface ζ become

$$g\zeta + \phi_t = 0,$$
$$\zeta_t - \phi_z = 0. \tag{9.5}$$

An important new idea is that, to first order in perturbations, (9.5) can be assumed to hold at the *unperturbed* water surface, $z = 0$, instead of at $z = \zeta$, which is unknown. Indeed, performing the Taylor series expansions of the terms ζ and ϕ in (9.5) about $z = 0$, the linear terms in these expansions lead to the expressions $\zeta_z\zeta$, $\zeta_{tz}\zeta$, $\phi_{tz}\zeta$, and $\phi_{zz}\zeta$, which are all of second order and thus negligible. Surface wave equations that are correct to second order in perturbations are given in Stoker (1957), p. 20.

Furthermore, no differentiation with respect to x appears in the system (9.5). Thus, these equations also hold for three-dimensional waves, where $\phi = \phi(t, x, y, z)$ and $\zeta = \zeta(t, x, y)$.

Eliminating ζ in (9.5) gives the free surface boundary condition

$$\boxed{[\phi_{tt} + g\phi_z]_{z=0} = 0} \tag{9.6}$$

An alternative derivation of this equation is given in Lighthill (1978), p. 207. Next, at the lower boundary, $z = -h = $ const, the boundary condition is that the vertical speed must vanish,

$$\phi_z|_{z=-h} = 0. \tag{9.7}$$

Once a solution for ϕ is found, the elevation ζ is given by the first equation in (9.5), $\zeta = -\phi_t/g$. Assuming time-periodic waves, one has

$$\phi(t, x, z) = e^{i\omega t} \chi(x, z). \tag{9.8}$$

The fluid equation $\Delta\phi = 0$ becomes $\Delta\chi = 0$, and a solution of the Laplace equation $\chi_{xx} + \chi_{zz} = 0$ is given by

$$\chi(x, z) = ae^{ikx} e^{\pm kz}. \tag{9.9}$$

To obey the lower boundary condition (9.7), a linear combination of these two solutions is used,

$$\chi(x, z) = \frac{a}{2} e^{ikx} \left(e^{k(z+h)} + e^{-k(z+h)}\right) = ae^{ikx} \cosh(k(z + h)). \tag{9.10}$$

Then, at the free surface $z = 0$,

$$\phi_{tt}(t, x, 0) = -\omega^2 e^{i\omega t} ae^{ikx} \cosh(kh),$$
$$\phi_z(t, x, z)|_{z=0} = e^{i\omega t} ake^{ikx} \sinh(kh). \tag{9.11}$$

The boundary condition (9.6) at $z = 0$ gives therefore

$$-\omega^2 \cosh(kh) + gk \sinh(kh) = 0, \tag{9.12}$$

or

$$\boxed{\omega^2 = gk \tanh(kh) \overset{h \to \infty}{\longrightarrow} gk} \tag{9.13}$$

The phase speed $c = \omega/k$ is then

$$\boxed{c = \sqrt{\frac{g}{k} \tanh(kh)} \overset{h \to \infty}{\longrightarrow} \sqrt{\frac{g}{k}}} \tag{9.14}$$

The phase speed of surface waves depends on their wavelength, and therefore *free surface waves are dispersive*. Although rather different assumptions were made in the derivation of shallow and free surface waves, the present Eq. (9.13) allows to regain (8.36) for shallow water waves: for very long wavelengths, $kh \ll 1$ and $\tanh(kh) \to kh$, giving as before,

$$c = \sqrt{gh}.$$

The real part of the velocity potential is

$$\phi(t, x, z) = a \cos(\omega t + kx)\ \cosh(k(z+h)), \tag{9.15}$$

and is used in the following calculations. Let δx and δz be the displacements of fluid parcels. Then

$$\dot{\overline{\delta x}} = u = \phi_x = -ak \sin(\omega t + kx)\ \cosh(k(z+h)),$$
$$\dot{\overline{\delta z}} = v = \phi_z =\ \ \ ak \cos(\omega t + kx)\ \sinh(k(z+h)). \tag{9.16}$$

Let

$$A(z) = ak\omega^{-1} \cosh(k(z+h)),$$
$$B(z) = ak\omega^{-1} \sinh(k(z+h)). \tag{9.17}$$

Then, from (9.16),

$$\frac{(\delta x)^2}{A(z)^2} + \frac{(\delta z)^2}{B(z)^2} = 1. \tag{9.18}$$

(To obtain the square of δx and δz, it was necessary to use the real part of ϕ alone.) At each height z, this is the equation of an ellipse. On the ground, this ellipse becomes a horizontal straight line, since $\sinh 0 = 0$. For deep water, $h \to \infty$, one has $A(z) = B(z) = ake^{kz}$ with $z < 0$. Then (9.18) describes circles, with radius decreasing as $e^{-k|z|}$ with depth. Figure 9.1 shows the parcel trajectories in shallow, intermediate, and deep water. More details on this are found in Stoker (1957), p. 47 and Sommerfeld (1957), p. 165.

Fig. 9.1 Parcel trajectories
in shallow, intermediate, and
deep water

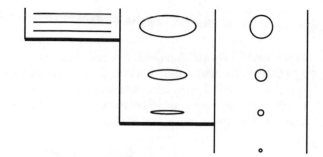

9.2 Sudden Impulse

The last section showed that the phase speed of free surface waves depends on their
wavelength. Therefore, a wave packet consisting of a spectrum of harmonics will
disperse in course of time. A fundamental question is then, how the water elevation
looks at later time, when at time $t = 0$ a sudden and point-like pressure impulse is
applied to a flat surface? First solutions to this problem were obtained by Cauchy,
Poisson, and Fourier.

The fluid is again assumed to be inviscid and to have constant density. The flow is
in a straight canal of infinite length and infinite depth, and with a rectangular profile.
The impulse is applied with equal strength across the whole canal.

To find the initial conditions for the pressure impulse, the Bernoulli equation (9.1)
is written as

$$\phi_t + \tfrac{1}{2}\phi_x^2 + \tfrac{1}{2}\phi_z^2 + g\zeta + \frac{p}{\rho} = 0. \tag{9.19}$$

Keeping again only first-order terms, one has at the free surface $z = 0$,

$$\phi_t(t, x, 0) + g\zeta(t, x, 0) + \frac{p(t, x, 0)}{\rho} = 0. \tag{9.20}$$

The pressure perturbation is of the form

$$p(t, x, 0) = I\,\delta(t)\,\delta(x), \tag{9.21}$$

where I is a finite number. Inserting this into (9.20), and integrating over a short time
interval τ, one has

$$\phi(\tau, x, 0) - \phi(0, x, 0) = -\frac{I}{\rho}\delta(x) - g\int_0^\tau dt\,\zeta(t, x). \tag{9.22}$$

At $t = 0$, the surface is undisturbed and there is no motion, thus one can assume
$\phi(0, x, 0) = 0$. The elevation ζ is assumed to be finite (this is actually not true at
$x = 0$), hence the last term in (9.22) can be made arbitrarily small for $\tau \to 0$, and
the equation becomes

$$\phi(0_+, x, 0) = -\frac{I}{\rho} \, \delta(x). \tag{9.23}$$

This is Eq. (6.1.6) in Stoker (1957), p. 150. Next, instead of integrating the Bernoulli equation (9.20) over a time interval τ, we take directly the limit $t \to 0_+$ in this equation. One can then assume that the elevation ζ is still zero, as at the initial time $t = 0$; for the pressure term, it is assumed that the perturbation $I \, \delta(x) \, \delta(t)$ has already died out. This yields, again to first order,

$$\phi_t(0_+, x, 0) = 0. \tag{9.24}$$

The further treatment follows that in Stoker (1957), p. 156. One looks for a velocity potential $\phi(t, x, z)$ that satisfies, for $-\infty < x < \infty$,

$$\phi_{xx} + \phi_{zz} = 0, \qquad\qquad -\infty < z < 0, \qquad\qquad \text{(I)}$$
$$\phi_z = 0, \qquad\qquad z \to -\infty \qquad\qquad \text{(II)}$$
$$\phi_{tt} + g\phi_z = 0, \qquad\qquad z = 0, \qquad\qquad \text{(III)}$$
$$\phi_t(0_+, x, 0) = 0, \qquad\qquad\qquad \text{(IV)}$$
$$\phi(0_+, x, 0) = -I\delta(x)/\rho. \qquad\qquad\qquad \text{(V)}$$

Here (I) is the equation of motion, (II) and (III) are boundary conditions, and (IV) and (V) are initial conditions for a sudden impulse. Equations (I) and (II) hold for $t \geq 0$, whereas (III)–(V) hold for $t > 0$.

A Fourier transform of (I)–(V) with respect to x is performed, with

$$\phi(t, x, z) = \int_{-\infty}^{\infty} dk \, e^{ikx} \bar{\phi}(t, k, z), \tag{9.25}$$

$$\bar{\phi}(t, k, z) = \frac{1}{2\pi} \int_{-\infty}^{\infty} dx' \, e^{-ikx'} \phi(t, x', z). \tag{9.26}$$

From this one obtains, interchanging integration and differentiation,

$$\overline{\phi_x} = ik\bar{\phi}. \tag{9.27}$$

Furthermore, from (9.26) the Fourier transform of $\delta(x)$ is evidently

$$\bar{\delta}(k) = \frac{1}{2\pi}. \tag{9.28}$$

Both sides in (I) are multiplied by $\frac{1}{2\pi} e^{-ikx}$ and integrated over dx. Using then (9.27) and $\overline{\phi_z} = \bar{\phi}_z$ (there is no Fourier's transform with respect to z), one finds

$$-k^2\bar{\phi}(t, k, z) + \bar{\phi}_{zz}(t, k, z) = 0. \tag{Ī}$$

This is an ordinary differential equation, with solution

$$\bar{\phi}(t, k, z) = A(t, k)\, e^{|k|z} + B(t, k)\, e^{-|k|z}. \tag{9.29}$$

Equation (II) implies, after transformation from ϕ to $\bar{\phi}$, that

$$B = 0. \tag{ĪĪ}$$

Equation (III) remains the same when going from ϕ to $\bar{\phi}$,

$$\bar{\phi}_{tt} + g\bar{\phi}_z = 0 \tag{9.30}$$

for $z = 0$, or

$$A_{tt} + g|k|A = 0, \tag{ĪĪĪ}$$

which is again an ordinary differential equation. Equation (IV) becomes after the Fourier transformation,

$$A_t(0_+, k) = 0. \tag{ĪV}$$

The solution of (ĪĪĪ) and (ĪV) is

$$A(t, k) = a(k)\, \cos\!\left(t\sqrt{g|k|}\right). \tag{9.31}$$

The Fourier transformation of (V) gives

$$\bar{\phi}(0_+, k, 0) = -\frac{I}{2\pi\rho} = a(k). \tag{V̄}$$

Collecting together and using (9.25), one has

$$\phi(t, x, z) = -\frac{I}{2\pi\rho} \int_{-\infty}^{\infty} dk\, e^{ikx}\, e^{|k|z}\, \cos\!\left(t\sqrt{g|k|}\right). \tag{9.32}$$

Keeping only the real part gives

$$\phi(t, x, z) = -\frac{I}{\pi\rho} \int_0^{\infty} dk\, \cos(kx)\, e^{kz}\, \cos\!\left(t\sqrt{gk}\right). \tag{9.33}$$

This integral converges for all $z < 0$ because of the exponential e^{kz}. The free fluid surface is then given by (9.5),

$$\zeta(t, x) = -\frac{I}{\pi\rho\sqrt{g}} \lim_{z \to 0_-} \int_0^{\infty} dk\, \sqrt{k}\, \cos(kx)\, e^{kz}\, \sin\!\left(t\sqrt{gk}\right). \tag{9.34}$$

The limit $z \to 0_-$ is left here since the integral diverges for $z = 0$.

So far, a flat water surface was assumed at $t = 0$, at which time a sudden and localized pressure impulse is applied to the surface. A slightly different case is considered now, in which the surface is initially ($t \le 0$) elevated at the point $x = 0$ (e.g., by applying an underpressure between two parallel metal sheets put across the canal and brought into the fluid), and then released. No (further) impulse is exerted. The initial conditions are then

$$\phi(0_+, x, 0) = 0, \tag{IV$'$}$$

$$\phi_t(0_+, x, 0) = -g\,J\delta(x). \tag{V$'$}$$

The first line follows from (9.23) by putting $I = 0$, and the second line follows from (9.20) by putting $p = 0$. The initial elevation of the surface in an interval of width $\epsilon \to 0$ centered at $x = 0$ is J/ϵ.

The discussion of Eqs. (I) to (III) remains unchanged. Equation (IV$'$) becomes after the Fourier transformation

$$A(0_+, k) = 0. \tag{$\overline{\text{IV}'}$}$$

The solution of $(\overline{\text{III}})$ and $(\overline{\text{IV}}')$ is now

$$A(t, k) = a(k)\,\sin\!\big(t\sqrt{g|k|}\big). \tag{9.35}$$

Equation (V$'$) becomes after the Fourier transformation

$$\bar{\phi}_t(0_+, k, 0) = -\frac{gJ}{2\pi} = A_t(0_+, k) = a(k)\sqrt{g|k|}, \tag{$\overline{\text{V}'}$}$$

or

$$a(k) = -\frac{\sqrt{g}\,J}{2\pi\sqrt{|k|}}. \tag{9.36}$$

By the corresponding steps as in the derivation of (9.32), one obtains now

$$\phi(t, x, z) = -\frac{\sqrt{g}\,J}{2\pi}\int_{-\infty}^{\infty}\frac{dk}{\sqrt{|k|}}\,e^{ikx}\,e^{|k|z}\,\sin\!\big(t\sqrt{g|k|}\big). \tag{9.37}$$

And keeping only the real part,

$$\phi(t, x, z) = -\frac{\sqrt{g}\,J}{\pi}\int_{0}^{\infty}\frac{dk}{\sqrt{k}}\,\cos(kx)\,e^{kz}\,\sin\!\big(t\sqrt{gk}\big). \tag{9.38}$$

The free fluid surface is again given by (9.5),

$$\boxed{\;\zeta(t, x) = \frac{J}{\pi}\,\lim_{z\to 0_-}\int_{0}^{\infty}dk\,\cos(kx)\,e^{kz}\,\cos\!\big(t\sqrt{gk}\big)\;} \tag{9.39}$$

This important result, obtained from a rather formal analysis, can also be derived in a semi-intuitive way, as is demonstrated by Lamb (1932), p. 384. Lamb starts from the observation that a single Fourier mode with wavenumber k corresponds to a harmonic wave that is periodic in time with an angular frequency $\omega = \sqrt{gk}$, and therefore one has

$$\bar{\zeta}(t, k) = \bar{\zeta}(0, k) \, \cos(\omega t). \tag{9.40}$$

Using the real part of (9.26), this gives

$$\int dx' \, \zeta(t, x') \cos(kx') = \int dx' \, \zeta(0, x') \cos(kx') \, \cos(\omega t). \tag{9.41}$$

Starting then with the general Fourier formula, one obtains

$$\begin{aligned}
\zeta(t, x) &= \frac{1}{\pi} \int_0^\infty dk \int_{-\infty}^\infty dx' \, \zeta(t, x') \, \cos(k(x - x')) \\
&= \frac{1}{\pi} \int_0^\infty dk \, \cos(\omega t) \int_{-\infty}^\infty dx' \, \zeta(0, x') \, \cos(k(x - x')) \\
&= \frac{J}{\pi} \int_0^\infty dk \, \cos(t\sqrt{gk}) \int_{-\infty}^\infty dx' \, \delta(x') \, \cos(k(x - x')) \\
&= \frac{J}{\pi} \int_0^\infty dk \, \cos(t\sqrt{gk}) \, \cos(kx),
\end{aligned} \tag{9.42}$$

where $\zeta(0, x) = J \, \delta(x)$ was used. This is again (9.39), except for the limit $z \to 0_-$.

In a final step, the integrals (9.34) and (9.39) for ζ can be expanded into power series. For the calculational details, we refer to Lamb (1932), p. 385 and p. 389, and to Kotschin et al. (1954), p. 396. For (9.34), Lamb gives the expansion

$$\zeta(t, x) = \frac{It}{\pi \rho x^2} \left[1 - \frac{3}{1 \cdot 3 \cdot 5} \left(\frac{gt^2}{2x} \right)^2 + \frac{5}{1 \cdot 3 \cdot 5 \cdot 7 \cdot 9} \left(\frac{gt^2}{2x} \right)^4 - \cdots \right], \tag{9.43}$$

and for (9.39),

$$\zeta(t, x) = \frac{J}{\pi x} \left[\frac{gt^2}{2x} - \frac{1}{1 \cdot 3 \cdot 5} \left(\frac{gt^2}{2x} \right)^3 + \frac{1}{1 \cdot 3 \cdot 5 \cdot 7 \cdot 9} \left(\frac{gt^2}{2x} \right)^5 - \cdots \right]. \tag{9.44}$$

The water surface $\zeta(t, x)$ from (9.44) is shown in Fig. 9.2. To avoid curve crowding, the inner interval $0 < x < 0.011$ is not shown. The distance between maxima grows with increasing x. The reason for this is that waves with larger λ propagate faster and thus reach larger distances earlier than short waves. This is also the reason for the decrease of oscillation amplitude with x: at any given time, the number of waves that have reached a certain x decreases with increasing x.

Figure 9.3 shows the elevation ζ at fixed location x_0 as a function of time. (To obtain good numerical accuracy as $t \to 2.78$, we added 138 terms of Eq. 9.44 with 35-digit precision.) Waves with decreasing λ pass x_0; the reason is again that the

Fig. 9.2 Water elevation ζ
from (9.44) after elevation
$\zeta = J \, \delta(x)$ at $t = 0$.
Parameters are $\frac{1}{2} g t^2 = 1$ and
$J = 0.0004$

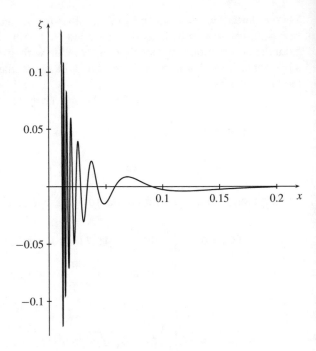

phase speed grows with wavelength, and thus long waves arrive first at a given point
and short waves later.

The above series expansions depend on t and x only in the dimensionless combination $\frac{1}{2} g t^2 / x$. This can be understood as follows (for this argument, see Stoker
1957, p. 168). Only two different dimensional constants occur in the present problem,
the acceleration g and either the pulse strength I or the initial elevation J. Both I
and J control only the absolute height of the elevation and can be dropped under the
assumption that the water elevation is correspondingly normalized; then this surface
is given by some function $\zeta(\xi(t, x, g))$, where ξ is a dimensionless combination of
x, t, and g, i.e., $\xi = g t^2 / x$, as is indeed found in (9.43) and (9.44). Any phase φ of
ζ (e.g., a node or an extremum of ζ) has then a certain constant value of $g t^2 / x$, and
φ moves at a speed $dx/dt \sim t \sim \sqrt{x}$, as is shown in Fig. 9.2.

In determining the response of a water surface to an initial point-like impulse or
point-like elevation, we have essentially solved the one-dimensional *Green's function*
problem for free surface waves on water in a straight canal of infinite depth.

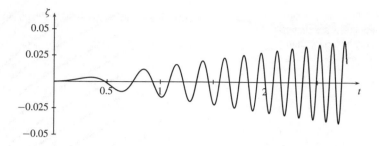

Fig. 9.3 Similar to Fig. 9.2, but now ζ at fixed location $x_0 = 0.1$ in course of time

9.3 Refraction and Breaking at a Coast

Water waves are ideally suited for studying wave refraction (and diffraction); Fig. 9.4 shows refraction of water waves at a shallow coast. The following analysis, adapted from a paper by Longuet-Higgins (1956), addresses different aspects of refraction of water waves approaching a coast from the open sea.

When a plane harmonic wave with a given wavelength and propagation direction approaches a coast, then (a) its wavelength becomes shorter and (b) its direction of propagation becomes more normal to the coast. Both effects are a simple consequence of the dispersion relation for water waves. Regarding (a), if a wave runs up a coast, the wave slows down and therefore its wavelength, which is the distance passed per period, becomes shorter. Regarding (b), if a broad wavefront approaches a coast at some intermediate angle of propagation, sections of the wave that are close to the coast propagate slower than sections that are further out. Thus, the propagation direction of the wave turns and becomes more normal to the coastline, and the wavefront becomes more parallel to the coastline.

Fig. 9.4 Wave refraction near San Juanico, Baja California, Mexico. Photograph: Google Earth

Fig. 9.5 Single plane water
wave (only crests are shown)
approaching a sloping shore
(right vertical line)

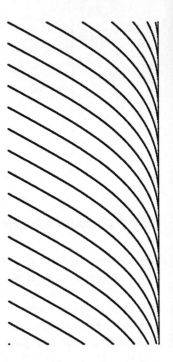

Both effects are clearly visible in Fig. 9.4. Using the dispersion relation (9.13) for
free surface waves at a shore with constant slope, they are also easily demonstrated
for a plane harmonic wave approaching a coastline, see Fig. 9.5.

The quantitative expression for the change in wave direction toward the coast is
Snell's law, which is derived, for water waves as for any other wave type, as follows
(see, e.g., Stoneley 1935). The dispersion relation (9.13) for free surface waves was
derived by assuming constant water depth h. To account for a sloping shore, the
latter is replaced by a sequence of small horizontal steps in the ground, and in the
end a continuum limit is taken. The step in height is from deeper to shallower water,
and thus the phase speed of the wave gets smaller at the step. Huygens' principle is
applied, by which every point on a wavefront (e.g., a crest) at time t is the source
of a spherical wave, and the new wavefront at time $t + dt$ is the superposition of all
these spherical waves (constructive and destructive interference). Assume then that
from point A in Fig. 9.6 at the edge of the wavefront, a spherical wave needs time
dt to reach point B on the step, $\overline{AB} = v_p\,dt$. From point C on the step, the parts of
the spherical wave propagating toward the coast propagate with phase speed v_p' over
a distance $\overline{CD} = v_p'\,dt$. The tangent from B to the circle with radius $\overline{CD}$ about C
gives the new wavefront. Before deriving Snell's law, we check for the correctness
of this construction, by showing that from any point E on the wavefront, the wave
propagates in time dt to point F lying on the straight line BD, where q and q' are
parallel to AB and CD, respectively.

Fig. 9.6 Wave refraction at a step (vertical line)

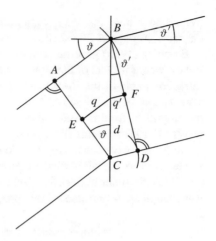

Indeed, with $\overline{BC} = h$,

$$\frac{v_p\, dt}{h} = \frac{q}{d}, \qquad\qquad \frac{v_p'\, dt}{h} = \frac{q'}{h-d}, \qquad (9.45)$$

and therefore the sum of the travel times along q and q' is

$$\frac{q}{v_p} + \frac{q'}{v_p'} = \frac{d}{h}\, dt + \frac{h-d}{h}\, dt = dt. \qquad (9.46)$$

Snell's law is then obtained from

$$\sin\vartheta = \frac{v_p\, dt}{h}, \qquad\qquad \sin\vartheta' = \frac{v_p'\, dt}{h}, \qquad (9.47)$$

or

$$\frac{\sin\vartheta}{v_p} = \frac{\sin\vartheta'}{v_p'}, \qquad (9.48)$$

or, with $v_p = \omega/k$ and $v_p' = \omega/k'$, since ω is not changed by refraction at the step,

$$k\,\sin\vartheta = k'\,\sin\vartheta'. \qquad (9.49)$$

Decomposing the wavenumber vector $\vec{k}$ into components $k_\parallel$ and $k_\perp$ parallel and normal to the step, respectively, one has $\sin\vartheta = k_\parallel/k$ and $\sin\vartheta' = k_\parallel'/k'$, and Snell's law takes the form

$$\boxed{k_\parallel = k_\parallel'} \qquad (9.50)$$

Remarkably, thus, the wavelength component $\lambda_\parallel$ is a constant of motion, although the wave propagates into shallower water: the decrease in λ affects only $\lambda_\perp$.

Stoneley (1935) generalizes these considerations to the refraction of a group of waves, where instead of the phase speed v_p one has to use the group speed v_p. The refraction angle of the so-called amplitude front of the wave group is then not the same as that of the phase front considered above, which has the consequence 'that the amplitude front is arranged *en échelon* [staggered]' (Stoneley p. 362), i.e., the wavefront looks roughly like the cross section of a Fresnel lense.

So far, we have considered on two occasions the superposition of two harmonic waves, first in the discussion of group speed in Sect. 7.2, then in setting up an amphidromic point in Sect. 8.5. Here now, two harmonic plane waves of the same wavelength are superimposed that propagate in slightly different directions,

$$\zeta = \sin[(\vec{k} - d\vec{k}) \cdot \vec{x} - \omega t] + \sin[(\vec{k} + d\vec{k}) \cdot \vec{x} - \omega t], \tag{9.51}$$

where $\vec{x}$ and $\vec{k}$ are vectors in the plane of unperturbed water, $\zeta = 0$. With the trigonometric addition formula, (9.51) becomes

$$\zeta = 2 \sin(\vec{k} \cdot \vec{x} - \omega t) \, \cos(d\vec{k} \cdot \vec{x}). \tag{9.52}$$

Since the magnitude of the wavenumbers $\vec{k} \pm d\vec{k}$ is the same for the two waves, $d\vec{k}$ stands normal to $\vec{k}$. Thus, the cosine term in (9.52) describes a sinusoidal modulation of the plane wave normal to its propagation direction, with a wavelength

$$\lambda_\perp = \frac{2\pi}{dk} = \frac{2\pi}{k \, d\vartheta}, \tag{9.53}$$

where $k = |\vec{k}|$ and $d\vartheta = dk/k$. Here ϑ is the angle between $\vec{k}$ and the normal direction to the straight coastline. This lateral extent of a wave is termed its *crest length*; within a crest length, the wave amplitude changes over its full range in a direction normal to the propagation direction.

Lemma *If two water waves approaching a sloping shore have the same wavelength but slightly different directions of propagation, then the waves get collimated toward the shore, i.e., become more parallel to each other.*

Note According to (9.53), this makes the crest length of the superimposed waves *larger* when approaching the coastline.

Proof Consider the two waves in a wavenumber diagram, where $k_\perp$ and $k_\parallel$ are the components of $\vec{k}$ normal and parallel to the straight shoreline, respectively. At a large distance from the coast, the incident plane waves shall have wavenumbers $\vec{k}_0$ and $\vec{k}_0'$. Since $k_0 = k_0'$ by assumption, both wavenumbers lie on a common circle about the origin in the wavenumber diagram, see Fig. 9.7. The angle $d\vartheta_0$ between

Fig. 9.7 Wave refraction in a wavenumber diagram

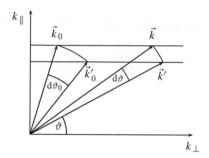

the vectors $\vec{k}_0$ and $\vec{k}'_0$ is a measure of the difference in propagation directions of the two waves. The projections to the $k_\perp$- and $k_\parallel$-axes give these two components for the waves. Especially, the component $k_\parallel$ will not be influenced by the approach to the coastline (Snell's law) and is a constant. Accordingly, two horizontal lines are drawn in the wavenumber diagram through the endpoints of the vectors $\vec{k}_0$ and $\vec{k}'_0$. The components $k_\perp$ get *larger* when the waves approach the coast (due to the shortening of their wavelength). Thus, at a later time the two waves are represented by vectors $\vec{k}$ and $\vec{k}'$ in the diagram. Since $k_0 = k'_0$, also $k = k'$, and both wavenumbers lie again on a circle about the origin. The diagram shows that $d\vartheta < d\vartheta_0$ for $k > k_0$, i.e., the two waves become more unidirectional, and they get collimated. $\qquad\square$

This effect is shown in Fig. 9.8, where two plane waves of identical wavelengths and amplitudes, however, with slightly different propagation directions, approach a coast. In this figure, amplitudes that lie above 80% of the maximum amplitude are marked with a dot (the dots lie dense enough to make up the stripes seen in the figure).

The complementary case of two waves with the same propagation direction but slightly different wavenumbers k and k' leads to the opposite conclusion that the crest length gets smaller with proximity to the coast (still, λ gets smaller and the propagation direction more normal to the coast). The reason is that waves with different λ get refracted by different angles (similar to light rays in a prism), since the phase speed (9.14) of free surface waves depends on their wavelength. Thus, starting with $d\vartheta_0 = 0$, a nonvanishing interval $d\vartheta$ of propagation directions is established and grows with proximity to the coast, which means that the crest length *decreases*.

Instead of discussing this in a wavenumber diagram as in Fig. 9.7, one can derive both effects, the widening and narrowing of the interval of propagation angles, by simple analysis. Let dk and $d\vartheta$ be infinitesimal intervals of wavenumbers and propagation angles, respectively, about some mean values k and ϑ, of waves approaching a shore from the open sea. Quantities on the open sea are referred to by a subscript 0, and for simplicity $\omega_0 = \sqrt{g/k_0}$ is assumed for free surface waves on very deep water, whereas close to the shore (no subscript), shallow water waves with $\omega = \sqrt{gh}$ are assumed. The wave period is not affected by refraction, $\omega = \omega_0$ and $d\omega = d\omega_0$, thus

Fig. 9.8 Superposition of
two water waves with
identical wavenumbers k, but
slightly different propagation
directions $\hat{k}$ and $\hat{k}'$
approaching a coast

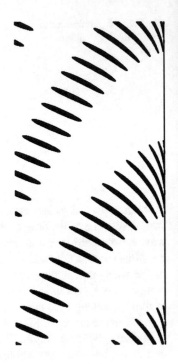

$$\frac{k_0}{k}\frac{dk}{dk_0} = \frac{d\omega_0/dk_0}{d\omega/dk}\frac{\omega/k}{\omega_0/k_0} = \frac{v_{0g}}{v_g}\frac{v_p}{v_{0p}} = \frac{1}{2}, \qquad (9.54)$$

where v_g and v_p are the group and phase speeds, respectively, with $v_{0g} = v_{0p}/2$ for free surface waves on deep water, and $v_g = v_p$ for shallow water waves. Therefore,

$$\frac{dk}{k} = \frac{k_0}{k}\frac{dk}{dk_0}\frac{dk_0}{k_0} = \frac{1}{2}\frac{dk_0}{k_0}. \qquad (9.55)$$

Snell's law is $k_\parallel = k_{0\parallel}$, or

$$k\sin\vartheta = k_0 \sin\vartheta_0. \qquad (9.56)$$

Forming the total differential on both sides gives

$$dk\,\sin\vartheta + k\cos\vartheta\,d\vartheta = dk_0\,\sin\vartheta_0 + k_0\cos\vartheta_0\,d\vartheta_0. \qquad (9.57)$$

Dividing by (9.56) and using (9.55) yields

$$\cot\vartheta\,d\vartheta = \cot\vartheta_0\,d\vartheta_0 + \frac{dk_0}{2k_0}. \qquad (9.58)$$

For the case of incoming waves with identical k_0, one has $dk_0 = 0$. The angle ϑ of $\vec{k}$ with the normal direction to the shore decreases with proximity to the shore;

thus, the cotangent grows and $d\vartheta < d\vartheta_0$ from (9.58), the waves get collimated. In the complementary case of waves with identical propagation direction, $d\vartheta_0 = 0$ and $dk_0 > 0$, one has $d\vartheta > 0$, the waves get de-collimated when approaching the shore. Longuet-Higgins (1956) suggests that the first effect should usually dominate, and waves get collimated.

Waves at a beach are often found to have a rather large crest length, which is usually *not* the case for wind-induced waves on the open sea. The latter appear short-crested, although long-crested waves of small amplitude may also be present. The increase in crest length due to wave refraction near a beach is only one possible reason for the observed long-crested waves. Another reason was suggested by Jeffreys (1924). He uses a method of *successive approximation* (see also Chap. 10, where successive approximation plays a major role) from Lamb (1932), p. 280, which goes back to Airy and which comprises a simple iteration step for the fluid equations: one uses the solutions of the first-order equations in order to obtain second-order corrections and thus better solutions. More specifically, the plane wave solutions of the linearized fluid equations are used to evaluate the quadratic terms uu_x, vu_y, etc., that appear in the full wave equations. The resulting algebraic expressions are inserted for the quadratic terms in the fluid equations, but the linear terms of these differential equations are left as they are. Thus, one still has the linear equations, now, however, with inhomogeneous, quadratic, algebraic terms.

The calculation proceeds as follows: the exact equations of motion for shallow water waves (only these are treated here) for a sea of constant depth h are (see Eqs. 8.21 and 8.26; with obvious generalizations to two space dimensions),

$$\frac{\partial u}{\partial t} + g\frac{\partial \zeta}{\partial x} = -u\frac{\partial u}{\partial x} - v\frac{\partial u}{\partial y},$$

$$\frac{\partial v}{\partial t} + g\frac{\partial \zeta}{\partial y} = -u\frac{\partial v}{\partial x} - v\frac{\partial v}{\partial y},$$

$$\frac{\partial \zeta}{\partial t} + h\frac{\partial u}{\partial x} + h\frac{\partial v}{\partial y} = -\frac{\partial(u\zeta)}{\partial x} - \frac{\partial(v\zeta)}{\partial y}. \tag{9.59}$$

The linearized equations are obtained when all right sides, which contain quadratic terms in the fields u, v, ζ, are set to zero. From the linearized equations follows, e.g., the wave equation for ζ,

$$\frac{\partial^2 \zeta}{\partial t^2} - gh\left(\frac{\partial^2 \zeta}{\partial x^2} + \frac{\partial^2 \zeta}{\partial y^2}\right) = 0, \tag{9.60}$$

and the solutions to these linear equations are given by

$$\zeta(t, x, y) = a \, \cos(kx - \omega t) \, \cos(ly),$$

$$u(t, x, y) = \frac{gka}{\omega} \, \cos(kx - \omega t) \, \cos(ly),$$

$$v(t, x, y) = -\frac{gla}{\omega} \, \sin(kx - \omega t) \, \sin(ly), \tag{9.61}$$

provided that k, l, ω obey the dispersion relation

$$\omega^2 = gh(k^2 + l^2). \tag{9.62}$$

Airy's method of successive approximation is to calculate the quadratic terms on the right sides of (9.59) using the linear solutions (9.61); this yields

$$-u \frac{\partial u}{\partial x} - v \frac{\partial u}{\partial y} = \frac{g^2 ka^2}{4\omega^2} \, \sin(2kx - 2\omega t)\left[k^2(1 + \cos(2ly)) - l^2(1 - \cos(2ly))\right],$$

$$-u \frac{\partial v}{\partial x} - v \frac{\partial v}{\partial y} = \frac{g^2 la^2}{4\omega^2} \left[k^2(1 + \cos(2kx - 2\omega t)) - l^2(1 - \cos(2kx - 2\omega t))\right] \sin(2ly),$$

$$-\frac{\partial(u\zeta)}{\partial x} - \frac{\partial(v\zeta)}{\partial y} = \frac{ga^2}{2\omega} \, \sin(2kx - 2\omega t)\left[k^2(1 + \cos(2ly)) + l^2 \cos(2ly))\right], \tag{9.63}$$

and to insert these expressions on the right side of the exact equations (9.59), to obtain linear equations for the corrections to the harmonic first-order solution. A minor technicality has to be observed at this point: one is interested only in those correction terms that include sine and cosine terms of *both* $2kx - 2\omega t$ and $2ly$; the other terms do not correspond to (small) corrections of the first-order plane harmonic wave (9.61). Dropping the latter terms, the resulting *inhomogeneous* linear equations for shallow water waves are

$$\frac{\partial u}{\partial t} + g \frac{\partial \zeta}{\partial x} = \frac{gka^2}{4h} \, \sin(2kx - 2\omega t) \, \cos(2ly),$$

$$\frac{\partial v}{\partial t} + g \frac{\partial \zeta}{\partial y} = \frac{gla^2}{4h} \, \cos(2kx - 2\omega t) \, \sin(2ly), \tag{9.64}$$

$$\frac{\partial \zeta}{\partial t} + h \frac{\partial u}{\partial x} + h \frac{\partial v}{\partial y} = \frac{\omega a^2}{2h} \, \sin(2kx - 2\omega t) \, \cos(2ly).$$

Eliminating u and v, one finds the forced wave equation

$$\frac{\partial^2 \zeta}{\partial t^2} - gh \left(\frac{\partial^2 \zeta}{\partial x^2} + \frac{\partial^2 \zeta}{\partial y^2} \right) = -\frac{3}{2} \frac{\omega^2 a^2}{h} \, \cos(2kx - 2\omega t) \, \cos(2ly). \tag{9.65}$$

The general solution of this is the sum of the general solution of the linear, homogeneous wave equation (9.60), i.e., the harmonic plane wave (9.61), and of a particular solution of the inhomogeneous equation (9.65), viz.,

$$\zeta = a \, \cos(kx - \omega t) \, \cos(ly) + \frac{3}{8} \frac{\omega^2 a^2}{kgh^2} \, x \, \sin(2kx - 2\omega t) \, \cos(2ly). \qquad (9.66)$$

The appearance of a factor x in front of the sine function is central to the following argument. (The origin $x = 0$ is placed at a location where the amplitude of the wave is still sufficiently small for the wave to be fully linear). When the second-order term, which grows with x, becomes of the same magnitude as the first-order term, the wavefront has become so steep as to break—and the wave is destroyed. Our interest is mainly in long-crested waves, $l \leq k$. It is plausible to assume that for waves generated on the open sea by winds in a rather turbulent process, waves with short crest length have also short wavelength λ ('isotropy'). Thus, k and l can be assumed to be of the same order, and ω^2/k^2 is of the order of gh, see (9.62). This implies that the ratio of the forefactor of the second-order term and the first-order term scales like xka/h. Breaking occurs when xka/h is of order unity, and thus a large wavenumber k and a large wave amplitude a lead to early breaking and wave destruction, before the wave reaches the beach. Waves with large crest length have small wavenumber k, and they are hardly found on the open sea, i.e., have small amplitude a there (being the largest coherent structures in the turbulent sea, they are rarely excited). Both their small k and a favor late breaking of these waves.

> Hence as a compound wave advances in shallow water the short-crested constituents tend to annihilate themselves by breaking, while the long-crested wave does not break until all the shorter waves have disappeared. (Jeffreys 1924, p. 47)

Therefore, the long-crested waves become dominant near the coastline.

9.4 Waves in a Nonuniform Stream

The last section showed how in a fluid of changing depth, the propagation direction of waves can get changed by refraction. This section gives an even more striking example for the action now of a fluid *flow* on the waves propagating through it. Following a paper by Longuet-Higgins and Stewart (1961), it is shown that a one-dimensional flow with speed $U + U'x$, in which a wave propagates in the direction x of the flow, can transfer energy to the wave.

The following calculations are rather straightforward in terms of the involved differential equations and their solutions, but require some algebraic effort regarding a perturbation expansion to second order, which describes the interaction of the flow with a wave.

The fluid is assumed to have constant density and to be inviscid, the flow is solenoidal, irrotational, and one-dimensional in a straight canal of rectangular profile with no cross-canal variations. Cartesian coordinates are used, with x pointing in the flow direction, z pointing vertically upward, and the origin is in the unperturbed fluid surface. The horizontal flow is assumed to have a nonuniform, linear velocity law $u(x) = U + U'x$, where U and U' are independent, nonvanishing constants. There is no dependence of the horizontal flow on z, i.e., there is no shear. Without any further velocity component, the continuity equation for solenoidal flow would be violated, $0 \neq U' = \partial u / \partial x = 0$. As a remedy, a vertical flow speed $w(x, z)$ (a so-called vertical upwelling) is introduced, to enforce $\partial u / \partial x + \partial w / \partial z = 0$. The flow carries a free surface wave, which is assumed to be linear.

The horizontal flow velocity $U + U'x$ is assumed to be a *weakly* varying function of x,

$$\frac{U'}{\omega} \ll 1, \tag{9.67}$$

with angular frequency ω of the wave. Using $\omega / k = v_p$, with v_p the phase speed of the wave, this can be written as

$$\lambda U' \ll 2\pi v_p. \tag{9.68}$$

Thus, over a wavelength λ the change in the horizontal speed of the nonuniform flow is small compared to the phase speed of the wave. For irrotational flow, a velocity potential ϕ can be introduced, $\vec{u} = \nabla \phi$, and since the flow is solenoidal, $\nabla \cdot \vec{u} = 0$, one has

$$\Delta \phi = 0. \tag{9.69}$$

The unknown functions are then the vertical elevation $\zeta(t, x)$ of the wave and the velocity potential $\phi(t, x, z)$. The lower boundary of the fluid is at $z \to -\infty$, where no wave motion remains. The kinematic boundary condition at the upper, free surface is given by Eq. (8.12),

$$\left[\frac{\partial \zeta}{\partial t} + u \frac{\partial \zeta}{\partial x} - w \right]_{z=\zeta} = 0, \tag{9.70}$$

or

$$\frac{\partial \zeta}{\partial t} + \left[\frac{\partial \phi}{\partial x} \frac{\partial \zeta}{\partial x} - \frac{\partial \phi}{\partial z} \right]_{z=\zeta} = 0. \tag{9.71}$$

The dynamic boundary condition at the free surface is pressure constancy. The Bernoulli equation is (see Sect. 2.17)

$$\left[\frac{\partial \phi}{\partial t} + \frac{1}{2} \nabla \phi \cdot \nabla \phi \right]_{z=\zeta} + g\zeta = C(t), \tag{9.72}$$

where the constant p/ρ is absorbed into $C(t)$, the Bernoulli 'constant,' which may depend on time. The form ζ of the free wave surface is not known; therefore, as in Sect. 9.1 after Eq. (9.5), in Eqs. (9.71) and (9.72) the boundary line $z = \zeta$ is replaced by the unperturbed water surface $z = 0$. Since second-order terms are relevant now, the Taylor expansion about $z = 0$ has to be carried out to first order (the first term in the expansion is of first order, and thus first order is also sufficient in the Taylor series). One has

$$\frac{\partial \zeta}{\partial t} + \left[\frac{\partial \phi}{\partial x} \frac{\partial \zeta}{\partial x} - \frac{\partial \phi}{\partial z} \right]_{z=0} + \zeta \left[\frac{\partial}{\partial z} \left(\frac{\partial \phi}{\partial x} \frac{\partial \zeta}{\partial x} - \frac{\partial \phi}{\partial z} \right) \right]_{z=0} + \ldots = 0, \quad (9.73)$$

$$g\zeta + \left[\frac{\partial \phi}{\partial t} + \frac{1}{2} \nabla \phi \cdot \nabla \phi \right]_{z=0} + \zeta \left[\frac{\partial}{\partial z} \left(\frac{\partial \phi}{\partial t} + \frac{1}{2} \nabla \phi \cdot \nabla \phi \right) \right]_{z=0} + \ldots = C.$$

$$(9.74)$$

The core idea is to assume a perturbation expansion of the form

$$\phi = Ux + \epsilon_a \phi_a + \epsilon_b \phi_b + \epsilon_a \epsilon_b \phi_c, \qquad (9.75)$$

$$\zeta = \qquad \epsilon_a \zeta_a + \epsilon_b \zeta_b + \epsilon_a \epsilon_b \zeta_c. \qquad (9.76)$$

Here $\epsilon_a \ll 1$ and $\epsilon_b \ll 1$ are small constant numbers, $\phi_a(x, z)$ is the potential for the unperturbed, nonuniform flow (for the assumed linear velocity law, ϕ_a is a quadratic function in x), $\phi_b(t, x, z)$ is the potential for a surface wave on deep water, and $\phi_c(t, x, z)$ is the interaction potential between flow and wave. The small number ϵ_a will be found to depend on the velocity gradient U', see (9.92). Terms in $\epsilon_a^2 \phi_a^2$ and $\epsilon_b^2 \phi_b^2$ are of no interest, and therefore neglected. For the nonuniform flow, a velocity field $(u_a(x), w_a(x, z))$ is assumed. Then

$$0 = \frac{\partial u_a}{\partial z} = \frac{\partial^2 \phi_a}{\partial z \, \partial x}, \qquad (9.77)$$

and therefore, to the considered order,

$$\nabla \phi = U\hat{x} + \epsilon_a \nabla \phi_a + \epsilon_b \nabla \phi_b + \epsilon_a \epsilon_b \nabla \phi_c, \qquad (9.78)$$

$$\frac{\partial \zeta}{\partial t} = \epsilon_b \frac{\partial \zeta_b}{\partial t} + \epsilon_a \epsilon_b \frac{\partial \zeta_c}{\partial t}, \qquad (9.79)$$

where $\partial \zeta_a / \partial t = 0$ since the unperturbed nonuniform flow is stationary. This gives, neglecting terms in ϕ_a^2 and ϕ_b^2,

$$\frac{1}{2} \vec{u} \cdot \vec{u} = \frac{1}{2} U^2 + U \left(\epsilon_a \frac{\partial \phi_a}{\partial x} + \epsilon_b \frac{\partial \phi_b}{\partial x} \right) + \epsilon_a \epsilon_b \left(U \frac{\partial \phi_c}{\partial x} + \frac{\partial \phi_a}{\partial x} \frac{\partial \phi_b}{\partial x} + \frac{\partial \phi_a}{\partial z} \frac{\partial \phi_b}{\partial z} \right).$$

$$(9.80)$$

Inserting (9.75)–(9.80) in (9.73) yields, to the considered order and without terms in ϕ_a^2 and ϕ_b^2,

$$\epsilon_a \left[\quad U \frac{\partial \zeta_a}{\partial x} - \frac{\partial \phi_a}{\partial z} \right]_{z=0}$$

$$+ \epsilon_b \left[\frac{\partial \zeta_b}{\partial t} + U \frac{\partial \zeta_b}{\partial x} - \frac{\partial \phi_b}{\partial z} \right]_{z=0} \tag{9.81}$$

$$+ \epsilon_a \epsilon_b \left[\frac{\partial \zeta_c}{\partial t} + U \frac{\partial \zeta_c}{\partial x} - \frac{\partial \phi_c}{\partial z} + \frac{\partial \phi_a}{\partial x} \frac{\partial \zeta_b}{\partial x} + \frac{\partial \phi_b}{\partial x} \frac{\partial \zeta_a}{\partial x} - \zeta_a \frac{\partial^2 \phi_b}{\partial z^2} - \zeta_b \frac{\partial^2 \phi_a}{\partial z^2} \right]_{z=0} = 0.$$

Inserting (9.75)–(9.80) in (9.74) gives

$$\frac{1}{2} U^2 + \epsilon_a \left[g\zeta_a \quad \quad + U \frac{\partial \phi_a}{\partial x} \right]_{z=0}$$

$$+ \epsilon_b \left[g\zeta_b + \frac{\partial \phi_b}{\partial t} + U \frac{\partial \phi_b}{\partial x} \right]_{z=0} \tag{9.82}$$

$$+ \epsilon_a \epsilon_b \left[g\zeta_c + \frac{\partial \phi_c}{\partial t} + U \frac{\partial \phi_c}{\partial x} + \frac{\partial \phi_a}{\partial x} \frac{\partial \phi_b}{\partial x} + \frac{\partial \phi_a}{\partial z} \frac{\partial \phi_b}{\partial z} + \zeta_a \left(\frac{\partial}{\partial t} + U \frac{\partial}{\partial x} \right) \frac{\partial \phi_b}{\partial z} \right]_{z=0} = C.$$

In the last line, the equations $\partial^2 \phi_a/(\partial t\, \partial z) = 0$ and $\partial^2 \phi_a/(\partial x\, \partial z) = 0$ were used (stationary flow and 9.77). From (9.82), the unperturbed nonuniform flow obeys $C = \frac{1}{2} U^2$. In order to fulfill (9.81) and (9.82), each term in square brackets is assumed to vanish individually. This is only a sufficient, not a necessary condition, since three terms in each equation are assumed to vanish, but there are only two independent parameters ϵ_a and ϵ_b. One obtains from (9.81) and (9.82) three equation pairs to the order considered:

(I) For the nonuniform flow potential ϕ_a and elevation ζ_a,

$$\left. \begin{array}{l} U \dfrac{\partial \zeta_a}{\partial x} - \dfrac{\partial \phi_a}{\partial z} = 0 \\[2mm] g\zeta_a + U \dfrac{\partial \phi_a}{\partial x} = 0 \end{array} \right\} \quad \text{at } z = 0. \tag{9.83}$$

(II) For the wave potential ϕ_b and elevation ζ_b,

$$\left. \begin{array}{l} \left(\dfrac{\partial}{\partial t} + U \dfrac{\partial}{\partial x} \right) \zeta_b - \dfrac{\partial \phi_b}{\partial z} = 0 \\[3mm] g\zeta_b + \left(\dfrac{\partial}{\partial t} + U \dfrac{\partial}{\partial x} \right) \phi_b = 0 \end{array} \right\} \quad \text{at } z = 0. \tag{9.84}$$

(III) For the interaction potential ϕ_c and elevation ζ_c,

$$\left. \begin{array}{l} \left(\dfrac{\partial}{\partial t} + U \dfrac{\partial}{\partial x} \right) \zeta_c - \dfrac{\partial \phi_c}{\partial z} + \dfrac{\partial \phi_a}{\partial x} \dfrac{\partial \zeta_b}{\partial x} + \dfrac{\partial \phi_b}{\partial x} \dfrac{\partial \zeta_a}{\partial x} - \zeta_a \dfrac{\partial^2 \phi_b}{\partial z^2} - \zeta_b \dfrac{\partial^2 \phi_a}{\partial z^2} = 0 \\[3mm] g\zeta_c + \left(\dfrac{\partial}{\partial t} + U \dfrac{\partial}{\partial x} \right) \phi_c + \dfrac{\partial \phi_a}{\partial x} \dfrac{\partial \phi_b}{\partial x} + \dfrac{\partial \phi_a}{\partial z} \dfrac{\partial \phi_b}{\partial z} + \zeta_a \left(\dfrac{\partial}{\partial t} + U \dfrac{\partial}{\partial x} \right) \dfrac{\partial \phi_b}{\partial z} = 0 \end{array} \right\} \quad z = 0. \tag{9.85}$$

Regarding (I) for ϕ_a and ζ_a, cross-eliminating ζ_a in (9.83) gives

$$\left[U^2 \frac{\partial^2 \phi_a}{\partial x^2} + g \frac{\partial \phi_a}{\partial z} \right]_{z=0} = 0. \tag{9.86}$$

The flow speed at $x = 0$ shall be U, thus

$$\left. \frac{\partial \phi_a}{\partial x} \right|_{x=0} = 0. \tag{9.87}$$

Furthermore, ϕ_a has to obey the Laplace equation,

$$\frac{\partial^2 \phi_a}{\partial x^2} + \frac{\partial^2 \phi_a}{\partial z^2} = 0. \tag{9.88}$$

The solution to the last three equations is

$$\phi_a = v_p\, k(x^2 - z^2) - \frac{2U^2}{v_p} z, \tag{9.89}$$

with an arbitrary but constant wavenumber k (see below), and with v_p the phase speed of surface water waves on very deep water (obtained from the free boundary condition 9.86),

$$v_p = \sqrt{g/k}. \tag{9.90}$$

Note that v_p is a constant. The steepness U' of the nonuniform flow is also determined by ϕ_a, by

$$\epsilon_a \frac{\partial^2 \phi_a}{\partial x^2} = U', \tag{9.91}$$

which, using (9.89), fixes the constant ϵ_a to

$$\epsilon_a = \frac{U'}{2v_p k}. \tag{9.92}$$

According to (9.68), indeed $\epsilon_a \ll 1$. For the following calculations, a new constant is introduced,

$$\gamma = \frac{U}{v_p}. \tag{9.93}$$

The change in mean level of the nonuniform flow, ζ_a, can be found from (9.83), which yields

$$\zeta_a(x) = -2\gamma x. \tag{9.94}$$

Regarding (II) for ϕ_b and ζ_b, cross-eliminating ζ_b in (9.84) gives

$$\left[\left(\frac{\partial}{\partial t} + U \frac{\partial}{\partial x}\right)^2 \phi_b + g \frac{\partial \phi_b}{\partial z}\right]_{z=0} = 0. \tag{9.95}$$

For a harmonic wave that propagates in x-direction and decays with depth $z < 0$, let

$$\phi_b = b e^{kz+ikx-i\omega t}, \tag{9.96}$$

which satisfies the Laplace equation. The same wavenumber k as in (9.89) is chosen, as is required by the interaction term $\sim \epsilon_a \epsilon_b$. From (9.95) then follows

$$(\omega - Uk)^2 = gk. \tag{9.97}$$

With v_p from (9.90),

$$\omega = k(U + v_p), \tag{9.98}$$

where the sign of the root was chosen so that $\omega = v_p k$ for $U = 0$. The elevation ζ_b follows from (9.84), with $z = 0$ in this equation, together with (9.96), (9.97), and $v_p = \sqrt{g/k}$, giving

$$\zeta_b(t, x) = \frac{i}{v_p} \phi_b(t, x, 0). \tag{9.99}$$

Regarding (III) for ϕ_c and ζ_c, cross-eliminating ζ_c in (9.85) yields

$$\frac{1}{g}\left(\frac{\partial}{\partial t} + U \frac{\partial}{\partial x}\right)^2 \phi_c + \frac{\partial \phi_c}{\partial z} = \tag{9.100}$$

$$= 2\left(\frac{\partial \phi_a}{\partial x}\frac{\partial \zeta_b}{\partial x} + \frac{\partial \phi_b}{\partial x}\frac{\partial \zeta_a}{\partial x} + k\zeta_b\frac{\partial \phi_a}{\partial z}\right) - \zeta_b\frac{\partial^2 \phi_a}{\partial z^2} \qquad (z = 0),$$

where (9.77), (9.83) from (I), and (9.84) from (II) were used, together with $\partial \phi_b/\partial z = k\phi_b$ from (9.96) and $\partial \phi_a/\partial t = \partial \zeta_a/\partial t = \partial \zeta/\partial z = 0$. Inserting the solutions for ϕ_a, ζ_a, ϕ_b, and ζ_b given in (9.89), (9.94), (9.96), and (9.99), respectively, the right side of (9.100) becomes

$$[2ik(1 - 2\gamma - 2\gamma^2) - 4k^2 x]\, \phi_b(t, x, 0). \tag{9.101}$$

As a trial solution, Longuet-Higgins and Stewart (1961) assume, with constant wavenumbers $l, m \in \mathbb{R}$,

$$\phi_c(t, x, z) = i(lq + m^2 q^2)\, \phi_b(t, x, z), \qquad \text{where} \qquad q = z + ix. \tag{9.102}$$

Then the left side of (9.100) becomes, after some elementary rearrangements using (9.84) and (9.95) (as well as 9.93, 9.96, and 9.99), and with special care on the second-derivative operator $(\partial_t + U\partial_x)^2$,

$$i[(1 + 2\gamma)(l + 2im^2x) - 2\gamma^2 m^2/k] \, \phi_b(t, x, 0). \tag{9.103}$$

On equating (9.101) and (9.103), one obtains

$$l = 2k \, \frac{1 - 4\gamma^2 - 4\gamma^3}{(1 + 2\gamma)^2}, \tag{9.104}$$

$$m^2 = \frac{2k^2}{1 + 2\gamma}. \tag{9.105}$$

Inserting (9.84), (9.89), (9.90), (9.93), (9.94), (9.96), and (9.99) in (9.85) gives, using $q = z + ix$,

$$g\zeta_c(t, x) = [\gamma l + 2\gamma^2 k - ix(2k^2 + 2\gamma k^2 + kl - 2\gamma m^2) + km^2 x^2] \, v_p \, \phi_b(t, x, 0). \tag{9.106}$$

This finishes the calculation of explicit solutions for ϕ_a, ϕ_b, ϕ_c, ζ_a, ζ_b, ζ_c. The total water elevation in the wave is obtained from (9.99) and (9.106) to be

$$\epsilon_b \zeta_b + \epsilon_a \epsilon_b \zeta_c = \epsilon_b \zeta_b \left[1 - i\epsilon_a \left(\frac{\gamma l}{k} + 2\gamma^2 + m^2 x^2 \right) - 2\epsilon_a k x \left(1 + \gamma + \frac{l}{2k} - \frac{\gamma m^2}{k^2} \right) \right]. \tag{9.107}$$

(Note that in Eqs. 2.34 and 2.35 of Longuet-Higgins and Stewart, 2γ should read $2\gamma^2$.) For arbitrary terms A and B, the following formula is correct to first order in ϵ_a,

$$1 - i\epsilon_a A - \epsilon_a B \approx (1 - i\epsilon_a A)(1 - \epsilon_a B) \approx e^{-i\epsilon_a A}(1 - \epsilon_a B). \tag{9.108}$$

Therefore, (9.107) can be written as, to order $\epsilon_a \epsilon_b$ and using again (9.96) and (9.99) as well as (9.98),

$$\epsilon_b \zeta_b + \epsilon_a \epsilon_b \zeta_c = i\epsilon_b \, \frac{b}{v_p} \left[1 - 2\epsilon_a k x \left(1 + \gamma + \frac{l}{2k} - \frac{\gamma m^2}{k^2} \right) \right] \times$$
$$\times \exp \left\{ i \left[kx - k(U + v_p)t - \epsilon_a \left(\frac{\gamma l}{k} + 2\gamma^2 + m^2 x^2 \right) \right] \right\}. \tag{9.109}$$

This is a wave whose amplitude and wavenumber change slowly (since $\epsilon_a \ll 1$) with x. All the above formulas are valid near $x = 0$ only, where deviations from b and k are small. The exponential is expanded in a Taylor series,

$$e^{if(x)} e^{-i\omega t} = e^{i\{f(0) + df/dx|_{x=0} \, x + \cdots\}} e^{-i\omega t}. \tag{9.110}$$

Neglecting a constant phase as well as the terms of higher order, the local wavenumber $\tilde{k}(x)$ at x is then given by $\partial f/\partial x$, or

$$\tilde{k}(x) = k - 2\epsilon_a m^2 x. \tag{9.111}$$

The relative change in wavenumber near $x = 0$ is thus, to the perturbation order considered,

$$\frac{1}{k}\frac{\partial \tilde{k}(x)}{\partial x}\bigg|_{x=0} = -\frac{2\epsilon_a m^2}{k} \overset{(8.105)}{=} -\frac{4\epsilon_a k}{1 + 2\gamma}. \tag{9.112}$$

Inserting $\epsilon_a = U'/(2v_p k)$ from (9.92), one has

$$\boxed{\frac{1}{k}\frac{\partial \tilde{k}(x)}{\partial x}\bigg|_{x=0} = -\frac{2U'}{v_p + 2U}} \tag{9.113}$$

The amplitude $\tilde{b}(x)$ of the wave is from (9.109)

$$\epsilon_b \frac{\tilde{b}(x)}{v_p} = \epsilon_b \frac{b}{v_p}\left[1 - 2\epsilon_a kx\left(1 + \gamma + \frac{l}{2k} - \frac{\gamma m^2}{k^2}\right)\right]. \tag{9.114}$$

The relative change in amplitude near $x = 0$ is, to the order considered,

$$\frac{1}{b}\frac{\partial \tilde{b}(x)}{\partial x}\bigg|_{x=0} = -2\epsilon_a k\left(1 + \gamma + \frac{l}{2k} - \frac{\gamma m^2}{k^2}\right) = -2\epsilon_a k\frac{2 + 3\gamma}{(1 + 2\gamma)^2}, \tag{9.115}$$

in which (9.104) and (9.105) have been used. Applying again (9.92) gives

$$\boxed{\frac{1}{b}\frac{\partial \tilde{b}(x)}{\partial x}\bigg|_{x=0} = -\frac{2 + 3\gamma}{(1 + 2\gamma)^2}\frac{U'}{v_p}} \tag{9.116}$$

Equations (9.113) and (9.116) are the central results, see Eqs. (2.38) and (2.41) in Longuet-Higgins and Stewart (1961). An alternative derivation from the conservation laws of the flow is given by Whitham (1962).

Actually, as is to be expected, (9.113) can be derived directly from wave kinematics, see Eqs. (3.1) to (3.7) in Longuet-Higgins and Stewart (1961). For this, let $\tilde{b}, \tilde{k}, \tilde{v}_p, \tilde{U}$ be the local values of b, k, v_p, U at some location $x \neq 0$. Note that now functions $\tilde{U}$ and $\tilde{v}_p$ are introduced, which was not necessary in the derivations above. The wave speed relative to a fixed position x is $\tilde{v}_p(x) + \tilde{U}(x)$. The angular frequency $\omega = k(v_p + U)$ of the wave is the same everywhere, therefore

$$\tilde{k}(\tilde{v}_p + \tilde{U}) = k(v_p + U), \tag{9.117}$$

and

$$\frac{k}{\tilde{k}} = \frac{\tilde{v}_p + \tilde{U}}{v_p + U}.$$

(9.118)

With $\gamma = U/c$ (there is no $\tilde{\gamma}$), $v_p = \sqrt{g/k}$, and $\tilde{v}_p = \sqrt{g/\tilde{k}}$ for waves on deep water, one obtains

$$\frac{\tilde{v}_p^2}{v_p^2} = \frac{\tilde{v}_p + \tilde{U}}{v_p + U} = \frac{1}{1+\gamma}\left(\frac{\tilde{v}_p}{v_p} + \frac{\tilde{U}}{v_p}\right).$$

(9.119)

Differentiating this with respect to x gives

$$2\,\frac{\tilde{v}_p}{v_p}\,\frac{\mathrm{d}\tilde{v}_p}{\mathrm{d}x} = \frac{1}{1+\gamma}\left(\frac{\mathrm{d}\tilde{v}_p}{\mathrm{d}x} + U'\right),$$

(9.120)

where $\mathrm{d}\tilde{U}/\mathrm{d}x = U'$ for a linear velocity law. At $x = 0$, this becomes

$$\frac{\mathrm{d}\tilde{v}_p}{\mathrm{d}x} = \frac{U'}{1+2\gamma}.$$

(9.121)

Since $\tilde{k}\tilde{v}_p^2 = g = \mathrm{const}$, one has, still at $x = 0$,

$$\frac{1}{k}\frac{\mathrm{d}\tilde{k}}{\mathrm{d}x} = -\frac{2}{v_p}\frac{\mathrm{d}\tilde{v}_p}{\mathrm{d}x} = -\frac{2U'}{v_p(1+2\gamma)},$$

(9.122)

which is indeed (9.113).

The intuition behind (9.116) is not so obvious. To give a physical interpretation of this result, Longuet-Higgins and Stewart start with the equation

$$\boxed{\frac{\mathrm{d}}{\mathrm{d}x}\left[\tilde{E}(x)\left(\frac{\tilde{v}_p(x)}{2} + \tilde{U}(x)\right)\right] + \frac{\tilde{E}(x)}{2}U' = 0}$$

(9.123)

where $\tilde{v}_p/2 = \tilde{v}_g$ is the group speed for waves on very deep water. Equation (9.123) has the form of an energy transfer equation. Carrying out the differentiations in (9.123) at $x = 0$, where $U = \gamma v_p$, and using (9.121), one obtains

$$\frac{\mathrm{d}\tilde{E}(x)}{\mathrm{d}x}\frac{v_p}{2}(1+2\gamma) + \frac{\tilde{E}(x)}{2}\left(3 + \frac{1}{1+2\gamma}\right)U' = 0.$$

(9.124)

Therefore, with $E = \tilde{E}(x = 0)$ as for the quantities above,

$$\frac{1}{E}\frac{\mathrm{d}\tilde{E}}{\partial x}\bigg|_{x=0} = -\frac{4+6\gamma}{(1+2\gamma)^2}\frac{U'}{v_p}.$$

(9.125)

The energy of a free surface wave is proportional to the wave elevation (the wave amplitude) *squared*. This is obvious for the kinetic energy, which scales as $\frac{1}{2} \nabla \phi_b \cdot \nabla \phi_b \sim b^2$, see (9.96) and (9.99). For the potential energy, one argues as follows (see Kotschin et al. 1954, p. 422). Let $\zeta(x)$ be the elevation at x within an infinitesimal interval dx. This corresponds to a fluid mass (per length in cross-canal direction) of $\rho \zeta(x) \, dx$. The center of mass of the strip $\zeta \, dx$ is at $\zeta/2$, and thus the potential energy of this strip is $\frac{1}{2} \rho g \zeta^2 \, dx$, and the potential energy W per wavelength of a harmonic wave $b \cos(kx)$ (put $t = 0$) is

$$W = \frac{1}{2} \rho g \int_0^\lambda dx \; \zeta(x) = \frac{1}{2} \rho g b^2 \int_0^\lambda dx \; \cos^2(kx) = \frac{\rho g b^2}{4k}. \qquad (9.126)$$

Thus, the total wave energy is proportional to the wave amplitude squared, and Eq. (9.125) implies for the amplitude b,

$$\frac{1}{b} \frac{d\tilde{b}}{dx}\bigg|_{x=0} = -\frac{2 + 3\gamma}{(1 + 2\gamma)^2} \frac{U'}{v_p}, \qquad (9.127)$$

which is (9.116)! Therefore, the energy transfer equation (9.123) is consistent with the increase in wave amplitude according to (9.116). This is a somewhat unexpected result, since the usual balance argument should rather lead to an expression of the form

$$\frac{d}{dx} \left[\tilde{E}(x) \left(\frac{\tilde{v}_p(x)}{2} + \tilde{U}(x) \right) + \frac{\tilde{E}(x)}{2} U' \right] = 0, \qquad (9.128)$$

where the latter $\tilde{E}/2$ gets differentiated too. Whitham (1962) gives (rather nontrivial) intuitive arguments, why indeed (9.123) is the correct form. Equation (9.123) resolves a long-lasting uncertainty about the correct equation for the energy transfer from a mean flow to a wave. According to this equation, the flow does work at a rate $\frac{1}{2} E U'$ per unit distance on the wave. The factor $E/2$ is the simplest manifestation of a new rank-two tensor, the so-called radiation stress. Details on this can be found in Longuet-Higgins and Stewart (1960) and (1961).

A physical situation to which the above analysis applies is the change observed in wind-induced ocean waves—and thus in the appearance of the sea surface altogether—that run in the opposing direction into a tidal stream. As described by Unna (1942), when waves meet a 'foul' stream, their wavelength and speed are reduced, the wave energy crowds up, the wave height increases, and the sea steepens. This should lead to enhanced wave breaking and destruction. This idea underlies the operation of a bubble breakwater, also known as bubble curtain or pneumatic barrier, patented in 1907 by Philip Brasher. Air bubbles created by air supply from compressors at land rise from a perforated pipe at the seabed. They carry the surrounding water with them and create a vertical flow, which near the water surface splits into two horizontal streams, one toward the coast and one toward the sea. The latter opposes incoming waves from the sea and protects the coast (that this is the main working

Fig. 9.9 The Big Bubble Curtain at the construction site 'Wikinger,' May 2016. Photograph: Hydrotechnik Lübeck GmbH

mechanism of a breakwater became clear in the 1930s). The expenditure is, however, immense, so that the method is nowadays rather used to protect sea life from sound waves from offshore drill and ram work for wind parks, by sound refraction at the bubble curtain itself, see Fig. 9.9.

The above analysis of the energy transfer from a mean flow to a wave is the first step toward an analysis of wave–wave interaction, e.g., when short waves 'ride' upon long waves. The following quote gives an example of such an energy transfer from long to short waves.

> When short wind-generated gravity waves lose energy by breaking (or other dissipative processes) near the crests of longer waves, the loss is supplied partly by the longer wave because of the second-order radiation-stress interaction. (Phillips 1963, p. 321)

9.5 The Stokes Wave

In an incompressible, inviscid fluid in irrotational motion, the Stokes wave is the (presumably unique) free surface wave with a maximum ratio of wave height to wavelength. The Stokes wave is spatially periodic and has a temporally constant form. As wave of maximum elevation, the Stokes wave is inherently nonlinear. The limitation of the elevation is brought about by a unique feature of the Stokes wave, a *cusp* at its crest (a point with a discontinuous change in the slope of the waveform), with a total inner opening angle of 120°, the so-called Stokes angle, see Fig. 9.10. Figure 9.11 shows an approximation to the Stokes waves in water.

Stokes in 1880 gave a famous heuristic argument for the occurrence of a sharp corner with opening angle of 120° at the crest. The fluid is assumed to be inviscid, to have constant density, and the flow is irrotational in two dimensions; the flow geometry is shown in Fig. 9.12. Then the first part of Stokes' argument runs as follows.

> Reduce the wave motion to steady motion by superposing a velocity equal and opposite to that of propagation. Then a particle [fluid parcel] at the surface may be thought of as gliding along a fixed smooth curve [...] On arriving at a crest the particle must be momentarily at rest, and on passing it must be ultimately in the condition of a particle starting from rest down an inclined or vertical plane. Hence the velocity must vary ultimately as the square root of the distance from the crest. (Stokes 1880, p. 226)

(The acceleration near the wave crest is $-g/2$, see Longuet-Higgins 1963.) In linear waves, fluid parcels merely oscillate with a speed $u \ll c$ about a rest position. For nonlinear waves, by contrast, a nonvanishing parcel flux exists; an extreme case is that of a bore transition, where the mass flux j was calculated in Eq. (8.219). Stokes' novel assumption is that in the cusp at the crest, the parcel speed equals the wave speed,

$$u = c. \tag{9.129}$$

The argument is that for $u < c$ at the crest, the steepest wave is not yet reached, while for $u > c$, parcels in the cusp overtake the crest and the wave breaks, excluding $u > c$ for a *stationary* waveform. The velocity profile with a cusp where $u = c$ is the extreme waveform before breaking occurs.

Approximating the vicinity of the cusp by its non-horizontal tangents, the fluid parcel, from its stopping point relative to the wave at this highest point, starts to slide down an inclined plane (like in Galileo's experiments). In the wave frame with $c = 0$, one has then $u \sim \sqrt{h}$ or

$$u \sim \sqrt{r}, \tag{9.130}$$

where r is the distance from the cusp when the latter is made the origin of a polar coordinate system, see Fig. 9.12.

Fig. 9.10 The Stokes angle $2\alpha = 120°$

Fig. 9.11 Surfaces wave in Manhattan beach, Los Angeles County. Photograph: Eino Mustonen on Wikipedia, https://en. wikipedia.org/wiki/File: Manhattan_beach_wave.JPG

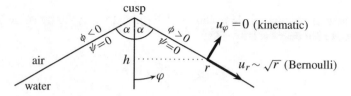

Fig. 9.12 Geometry and kinematics near the highest point of the Stokes wave

The second, older part of the argument goes back to Stokes' first paper on nonlinear waves from 1847. Let 2α be the full inner opening angle of the symmetric wave profile at the crest. Along the wave profile, the streamfunction ψ in the complex potential $\Phi = \phi + i\psi$ is constant, and this constant is set again zero. The real line $-\infty < \phi < \infty$ is mapped to the corner in the z-plane with opening angle 2α, see Fig. 9.12. This is achieved by a function

$$z = e^{i\beta}\phi^q \tag{9.131}$$

with some rational number $q < 1$ and $\beta \in \mathbb{R}$. To fix this function, introduce polar coordinates in the complex z-plane with the origin in the cusp and φ counted from the downward vertical for symmetry reasons. The real half-axis $\phi_+ > 0$ of the Φ-plane is mapped to the tangent in the z-plane with $\varphi = \alpha$, lying right of the cusp, and the real half-axis $\phi_- < 0$ is mapped to the tangent with $\varphi = -\alpha$ left of the cusp. Then, with $z = re^{i\varphi} = re^{\pm i\alpha}$ and abbreviating $p = 1/q$,

$$\begin{aligned}\phi_+ &= e^{-i\beta}\left(re^{+i\alpha}\right)^p > 0 &\rightarrow&& \alpha p - \beta &= 0,\\ \phi_- &= e^{-i\beta}\left(re^{-i\alpha}\right)^p < 0 &\rightarrow&& -\alpha p - \beta &= -\pi,\end{aligned} \tag{9.132}$$

giving

$$\beta = \frac{\pi}{2}, \qquad \alpha p = \frac{\pi}{2}. \tag{9.133}$$

This yields for the complex potential Φ, up to some positive real normalization factor (note that $e^{-i\pi/2} = -i$),

$$\Phi = -iz^p, \tag{9.134}$$

which is then real-valued along the waveform near the crest. Since the complex function $\Phi = -iz^p$ is analytic except at $z = 0$, the velocity potential ϕ is a harmonic function and thus obeys the Laplace equation, which is the equation of motion for stationary, incompressible, irritation flows of an inviscid fluid, as is assumed here.

With $\psi = 0$ on the wave profile near the crest, the kinematic boundary condition that there is no flow normal to the wave profile is already fulfilled. Indeed, with $\phi = \text{Re}\,\Phi = r^p \sin(\varphi p)$, one has

$$u_\varphi(r, \pm\alpha) = \frac{1}{r}\frac{\partial\phi}{\partial\varphi}\bigg|_{\varphi=\pm\alpha} = pr^{p-1}\cos(\pm\alpha p) \overset{(8.133)}{=} pr^{p-1}\cos\left(\pm\frac{\pi}{2}\right) = 0. \tag{9.135}$$

Finally, the half-opening angle α can be determined using the dynamic boundary condition (9.130) near the cusp: from

$$r^{1/2} \sim u_r = \frac{\partial \phi}{\partial r} \sim r^{p-1} \tag{9.136}$$

follows

$$p = \frac{3}{2}, \tag{9.137}$$

and thus, with $\alpha p = \pi/2$,

$$\boxed{\alpha = \frac{\pi}{3}} \tag{9.138}$$

The total inner angle of the wave profile at the cusp is therefore $2\alpha = 120°$. Furthermore, (9.134) gives then

$$\boxed{z = i \mathbb{O}^{2/3}} \tag{9.139}$$

which will be discussed in the next section.   □

In a modern version, this derivation can be put as follows: the free fluid surface is a streamline along which $\psi = $ const. The Bernoulli equation along the free wave surface of the stationary flow is, with constant p and ρ,

$$\frac{1}{2} u^2 + gh = \text{const}, \tag{9.140}$$

where u is the total fluid speed and h the vertical distance from the cusp (not from the unperturbed water surface). At the cusp itself, according to Stokes, $h = u = 0$, thus the constant on the right side is zero. The crest is assumed to be a symmetric cusp, where two curves meet whose tangents each make an angle α with the vertical. From Fig. 9.12, $\cos \alpha = -h/r$ (since h is negative) on the tangents meeting at the cusp, and

$$u = \sqrt{2gr \cos \alpha}, \tag{9.141}$$

i.e., $u \sim \sqrt{r}$ as above. In Eq. (4.18), the absolute fluid speed between two straight walls was derived to be

$$u = \frac{\pi a}{2\alpha} r^{\frac{\pi}{2\alpha} - 1}. \tag{9.142}$$

The variable r must appear with the same power in (9.141) and (9.142), from which it follows that

$$\frac{1}{2} = \frac{\pi}{2\alpha} - 1, \tag{9.143}$$

giving again $\alpha = \pi/3$.

There is a large number of papers addressing the Stokes wave. In the sections below, only a few properties of this wave are studied. For further topics, we refer to the following references: Amick et al. (1982) *prove* the existence of a cusp at the crest of the highest wave. Trajectories of fluid parcels in a Stokes wave are given in Fig. 6 of Longuet-Higgins (1979). A theory of 'spilling breakers' can be found in Longuet-Higgins and Turner (1974), and experiments on breaking waves are discussed in Duncan (1981) and Liu (2016). Further mathematical properties of the Stokes wave are discussed in Fraenkel (2007), and Schwartz and Fenton (1982) give a general review on strongly nonlinear water waves including the Stokes wave.

The Longuet-Higgins Waveform

Since an exact solution for the Stokes wave is (so far) not available, one is interested in a simple, closed solution that approximates the waveform well. An influential paper is by Michell (1893), who calculates the form of a wave with a 120° crest. This is done by a series expansion up to $\approx 1\%$ accuracy. A good and simple approximation to this Stokes–Michell wave is given by the following expression:

$$z = -\ln \cos x, \tag{9.144}$$

where appropriate lengthscales are chosen. This is accurate to within 3%. The derivation of (9.144) by Longuet-Higgins (1973) is quite original, and therefore reproduced here. The fluid is assumed again to propagate with wave speed $-c$ in the opposite direction of the wave, making the latter stationary. The boundary condition at the ground for the complex speed potential is

$$\Phi = -cx \quad \text{for} \quad z \to -\infty. \tag{9.145}$$

Consider the variable transformation from the complex variable $x + iz$ (the real Cartesian coordinates x and z should not lead to confusion with the usual complex variable z) to a complex variable ζ, given by

$$\zeta = \zeta_0 \, e^{-ik(x+iz)}, \tag{9.146}$$

where $k \in \mathbb{R}$ and ζ_0 is a constant complex number. Then the (horizontal) line $z \to -\infty$ of infinite depth becomes the single point $\zeta = 0$. As $\zeta \to 0$, one has, for fixed $z < 0$,

$$\Phi \to -\frac{ic}{k} \ln \zeta. \tag{9.147}$$

The unperturbed water surface $z = 0$ becomes

$$\zeta = \zeta_0 \, e^{-ikx}. \tag{9.148}$$

For $x \in \mathbb{R}$ one has $|e^{-ikx}| = 1$, and the undisturbed water surface is given by

Fig. 9.13 Hexagon in the complex ζ-plane

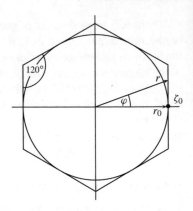

$$|\zeta| = |\zeta_0| = r_0. \tag{9.149}$$

In the complex ζ-plane, this is a circle with radius r_0 centered at the origin. The undisturbed water fills the region $|\zeta| \leq r_0$. We are interested in the waveform near this circle. Longuet-Higgins (1973) uses that the steepest waves have $120°$ cusps at their crests. The mapping (9.146) is differentiable, hence, conformal. The Stokes waveforms should therefore 'wrap' around the circle and maintain their $120°$ cusps. The geometrical figure in the ζ-plane must therefore have hexagonal symmetry, with six corners of $120°$ angle (i.e., $6 \times 60° = 360°$ turning angle). Hexagonal symmetry means that a total of six full Stokes' waves, each of wavelength λ, are present over the full circle, giving in (9.148),

$$2\pi/k = 6\lambda. \tag{9.150}$$

The approximation of Longuet-Higgins (1973) is then to assume that the Stokes waveform in the ζ-plane *is a hexagon*, i.e., the waveform from crest to crest is given by straight lines in the ζ-plane. For these, in polar representation $\zeta = r e^{i\varphi}$,

$$\cos \varphi = \frac{r_0}{r} \quad \text{for} \quad -\frac{\pi}{6} \leq \varphi \leq \frac{\pi}{6}. \tag{9.151}$$

Choosing $\zeta_0 = r_0$ as in Fig. 9.13, the conformal transformation (9.146) becomes

$$\zeta = r_0 \, e^{kz} e^{-ikx}. \tag{9.152}$$

Comparing this with the polar representation $\zeta = r e^{i\varphi}$ gives

$$e^{kz} = \frac{r}{r_0}, \tag{9.153}$$

$$e^{-ikx} = e^{i\varphi}. \tag{9.154}$$

Inserting the hexagon equation (9.151) gives

Fig. 9.14 The steepest wave $z = -\frac{3}{\pi} \ln \cos \frac{\pi x}{3}$

$$kz = -\ln \cos\varphi,$$

$$kx = -\varphi, \tag{9.155}$$

and therefore

$$kz = -\ln \cos(kx). \tag{9.156}$$

Using $k = \pi/3\lambda$ and defining $\tilde{x} = x/\lambda$ and $\tilde{z} = z/\lambda$, one obtains

$$\boxed{\tilde{z} = -\frac{3}{\pi} \ln \cos \frac{\pi \tilde{x}}{3}} \tag{9.157}$$

with $-\frac{1}{2} \leq \tilde{x} \leq \frac{1}{2}$. This wave profile is shown in Fig. 9.14. The above derivation can also be found in Crapper (1984), p. 48.

9.6 The Stokes Singularity

After the cusped cavity (Kolscher 1940; see Sect. 5.8) and the curvature singularity of a vortex sheet (Moore 1979; see Sect. 6.8), the cusp in the steepest surface wave (Stokes 1880) is the third major hydrodynamic singularity (discontinuous bore transitions and shock fronts are no singularities). As exceptional points of smooth solutions of differential equations, singularities may contain information about large-scale topological properties of a solution space.

The crest angle of 120° of the Stokes wave corresponds to a singularity of order $\frac{2}{3}$ in complex function theory, $z \sim \Phi^{2/3}$, see (9.139). Streamlines below the Stokes wave profile $\psi = 0$ have no singularity but are everywhere smooth. Since $\psi = $ const, these streamlines also correspond to possible profiles of free surface waves (one can 'peel-off' the Stokes wave to obtain lower lying nonlinear profiles). However, singularities can still occur in the fluid-free space *above* the Stokes wave. This unphysical domain is accessible via analytic continuation.

As with Moore's curvature singularity, one can doubt the viability of the model assumptions and avoid the singularity by including more physics: no Moore's singularity occurs if vortex layers (caused by viscous broadening) are considered instead of vortex sheets; and in experiments, the Stokes singularity is hardly observable due to capillary waves caused by surface tension near the cusp. On the other hand, 'the dynamics [of a system] is determined by poles/branch cuts in the complex plane'

(Dyachenko et al. 2013, p. 675), and from singularities, asymptotic wave solutions near the Stokes wave and better numerical algorithms can be obtained (see Crew and Trinh 2016, p. 259). Most notably,

> In theory, if we know the detailed locations and scalings of the set of singularities, we can subtract any undesired behaviour from the solution and extend the radius of convergence. (Crew and Trinh 2016, p. 259)

In the following, an unexpected result by Grant (1973a) is derived, that the non-steepest complex solutions of the boundary value problem for surface water waves have singularities of order $\frac{1}{2}$ in the fluid-free space above the Stokes profile:

> In general, the structure near the corner [the cusp of the Stokes wave] is considerably more complicated than has been assumed in the past. Instead of one order $\frac{2}{3}$ singularity above the fluid, there are several order $\frac{1}{2}$ singularities which coalesce at maximum amplitude. (Grant 1973a, p. 258, order of sentences reversed)

Such singularities of order $\frac{1}{2}$ in the fluid-free space were also found numerically by Schwartz (1972, 1974) and later by Williams (1981, 1985).

The fluid is again incompressible and inviscid, and the flow velocity field is solenoidal, irrotational, stationary, and two dimensional in water of infinite depth. The x-coordinate is in the flow direction, but the vertical z-coordinate is now replaced by an y-coordinate, to make the complex variable $z = x + iy$ available. Furthermore, $c = g = 1$ is assumed for the wave speed c and the gravitational acceleration g. The unperturbed flow at large depth shall move with wave speed $-c$, and waves propagate with $+c$ and thus are stationary. Since the flow is irrotational, one has $w = d\Phi/dz$, with complex speed potential $\Phi = \phi + i\psi$. At the surface $y = \eta$ of the wave, $\psi = 0$ is assumed.

The boundary conditions and the equation of motion are

$$\frac{1}{2} (\nabla \phi)^2 + y = 0 \qquad \text{for} \quad y = \eta, \tag{9.158}$$

$$\phi_x \eta_x - \phi_y = 0 \qquad \text{for} \quad y = \eta, \tag{9.159}$$

$$\phi_{xx} + \phi_{yy} = 0 \qquad \text{for} \quad y < \eta, \tag{9.160}$$

$$\phi_x = -1, \ \phi_y = 0 \qquad \text{for} \quad y \to -\infty. \tag{9.161}$$

Equation (9.158) is the Bernoulli equation, i.e., the dynamic boundary condition, cf. (9.72). With the constant in the Bernoulli equation set to zero, the cusp of the Stokes wave is at $(x, y) = (0, 0)$, since $\nabla \phi = 0$ there, according to the Stokes hypothesis about the breaking limit. In the following, the straight tangent lines with opening angle 120° near the crest are considered. The unperturbed water surface has $\nabla \phi = (u, v) = (-1, 0)$, thus $y = -1/2$. Equation (9.159) is the kinematic boundary condition, cf. (9.71), that the fluid speed is tangent to the surface and the surface is a streamline,

$$\frac{d\eta}{dx} = \frac{\phi_y}{\phi_x}. \tag{9.162}$$

A central trick is to consider the function $z(\Phi)$ instead of $\Phi(z)$:

While preparing his collected papers in 1880, Stokes had occasion to reconsider the method and discovered that the complexity of the calculations could be greatly reduced by using the velocity potential and stream function, rather than the space co-ordinates, as the independent quantities. (Schwartz 1974, p. 553)

Since the wave height [...] is one of the unknowns, the equations are transcendentally nonlinear. This can be improved by a transformation of the fluid volume [in the xy-plane] to a fixed, known domain [in the $\phi\psi$-plane]. The nonlinearity becomes 'only' polynomial. (Grant 1973a, p. 258)

(Similarly, note that in Chap. 5 on jets, wakes, and cavities, z and Φ were assumed to be functions of a common, real parameter t). With this, (9.161) becomes

$$z = -\Phi \qquad \text{for} \qquad \psi \to -\infty, \tag{9.163}$$

and the Bernoulli equation (9.158), which is central to the following, becomes for $\psi = 0$,

$$\frac{1}{2}(\nabla\phi)^2 = \frac{1}{2}\left|\frac{d\Phi}{dz}\right|^2 = \frac{1}{2}\left|\frac{dz}{d\Phi}\right|^{-2} = -y = -\text{Im } z, \tag{9.164}$$

or

$$-2\left[\text{Im } z \left|\frac{dz}{d\Phi}\right|^2\right]_{\psi=0} = 1 \tag{9.165}$$

Order of the Stokes Singularity

The surface of the Stokes wave is given by the complex function $z(\phi)$ of a real argument ϕ, with $\psi = 0$. By assumption, the Stokes singularity at the surface is the only singularity of the Stokes wave, and the region below the surface of the wave is free of singularities. Then the function $z(\Phi)$ has no singularity in the whole domain $\psi < 0$, and $z(\Phi)$ is analytic for $\psi < 0$, and its derivative does not vanish. The origin of the Φ-plane is chosen to lie in the Stokes singularity; it also lies at $z = 0$, so one can put

$$z = a\Phi^\mu \tag{9.166}$$

at the crest, mapping the 180° 'corner' on the straight line $\psi = 0$ in the Φ-plane to the 120° corner in the z-plane. The mapping (9.166) maps an angle α in the point $\Phi = 0$ to an angle $\mu\alpha$ in the point $z = 0$, therefore

$$\boxed{\mu = 2/3} \tag{9.167}$$

as in (9.139).

Order of the Non-Stokes Singularities

Waves with smaller elevation than the highest wave have no singularity at the crest or anywhere else in the fluid. Instead, when moving away 'downward' from the Stokes wave with a cusp at the crest to waves with smaller elevation, the singularity 'detaches' from the crest and moves upward along the vertical imaginary axis in the complex z-plane, into the unphysical, fluid-free region. This is demonstrated in numerical calculations by Schwartz (1974) and Longuet-Higgins and Fox (1978). Higher the location y of the singularity, the weaker the wave. Schwartz (1974) shows analytically that the singularities move to $y \to +\infty$ for the weakest, linear waves.

It came as a big surprise when Grant (1973a) showed that all these off-crest singularities have order $\frac{1}{2}$, instead of order $\frac{2}{3}$ for the Stokes wave. The situation is shown in Fig. 9.15, taken from Longuet-Higgins and Fox (1978). Some streamline with $\psi > 0$ above the free surface with $\psi = 0$ reaches the singular point above the Stokes singularity (which would lie at the origin) at a rather steep slope, at an angle of $45°$. Note that the order of the singularity changes discontinuously, in an infinitesimal neighborhood of the Stokes cusp, from $\frac{2}{3}$ to $\frac{1}{2}$. It may not be completely superfluous to remark that what will be observed in nature for the almost-highest waves is a near approximation to a Stokes profile with a $120°$ cusp, and not an indication of the $90°$ angle of a Grant singularity.

Let the surface of waves below the Stokes wave be defined again by $\psi = 0$, i.e., the origin of the Φ-plane is shifted to the wave crest. Then the Bernoulli equation (9.165) holds for all waves, since the Bernoulli constant is a global constant for irrotational flow. Grant (1973a) performs an analytic continuation from the wave surface $\psi = 0$ into the domain $\psi > 0$, to find the order of singularities located there.

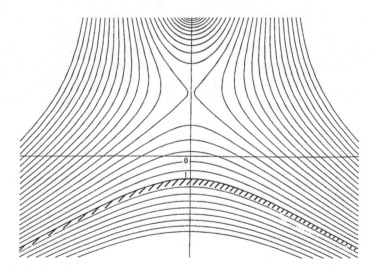

Fig. 9.15 Sketch of streamlines in the complex z-plane, for an almost-highest wave (hatched region). A singularity of order $\frac{1}{2}$ occurs in the upper, fluid-free half-plane. Streamlines in its neighborhood are hyperbolas. Reproduced from Longuet-Higgins and Fox (1978), Fig. 11

For this, he introduces a new function $Z(\mathbb{O})$ by

$$Z(\mathbb{O}) = \overline{z(\overline{\mathbb{O}})}. \tag{9.168}$$

The function $z(\mathbb{O})$ is analytic in the lower half-plane of $\mathbb{O}$, thus $Z(\mathbb{O})$ is analytic in the upper half-plane of $\mathbb{O}$ (where $z(\mathbb{O})$ has its singularities). According to Ahlfors (1966), p. 170, 'the proof of this statement consists in trivial verifications.' Note that this is not Schwarz's reflection principle, which would require that $z(\mathbb{O})$ is real-valued on a finite interval of the real ϕ-axis, which is not the case here. Similarly, $dz/d\mathbb{O} \neq 0$ in the lower and $dZ/d\mathbb{O} \neq 0$ in the upper half-plane of $\mathbb{O}$. Then

$$Z(\mathbb{O}) = Z(\phi) = \overline{z(\overline{\phi})} = \overline{z(\phi)} = \overline{z(\mathbb{O})} \qquad \text{for} \qquad \psi = 0. \tag{9.169}$$

Therefore,

$$-2 \operatorname{Im} z = iz - i\overline{z} = iz - iZ \qquad \text{for} \qquad \psi = 0. \tag{9.170}$$

Next, since z is analytic, one has $dz/d\mathbb{O} = \partial z/\partial \phi$, and

$$\overline{dz/d\mathbb{O}} = \frac{\partial \overline{z}}{\partial \phi} = \frac{\partial Z}{\partial \phi} = \frac{dZ}{d\mathbb{O}} \qquad \text{for} \qquad \psi = 0, \tag{9.171}$$

where in the last equation, analyticity of Z was used. Therefore (9.165) becomes

$$\left[\left(iz(\mathbb{O}) - iZ(\mathbb{O}) \right) \frac{dz}{d\mathbb{O}} \frac{dZ}{d\mathbb{O}} \right]_{\psi=0} = 1. \tag{9.172}$$

The effort of introducing a new complex function $Z(\mathbb{O})$ could be avoided on the straight line $\psi = 0$, which allows direct reduction of (9.172) to the real variable ϕ. The central idea of Grant (1973a), however, is to perform analytic continuation from this line $\psi = 0$ into the unphysical, fluid-free domain $\psi = 0$, in which Z is analytic.

The principle of analytic continuation was introduced in Sect. 8.12. Here now, analytic continuation is performed along a curve, from some starting point on the line $\psi = 0$ to an end point in the half-plane $\psi > 0$. The idea is to cover the given curve with a finite number of overlapping open disks D_i that serve as the domains Ω_i in the function elements (z_i, Ω_i) in the quote from Ahlfors in Sect. 8.12. The monodromy theorem from Appendix A ensures that for a *simply connected* domain—as the upper half-plane $\psi > 0$ is—the analytic continuation of $z(\mathbb{O})$ along *all possible* curves gives a unique analytic function $z(\mathbb{O})$ in this half-plane. In order that analytic continuation can be performed, it must be assumed that $z(\mathbb{O})$ is analytic at each point of the line $\psi = 0$, i.e., in an infinitesimal open disk about each point of $\psi = 0$ (see Sect. 3.10), giving the first function element.

The procedure of finding singularities of (9.172) is then easy and straightforward: dropping the restriction $\psi = 0$, Eq. (9.172) is an equation for the unknown function $z(\mathbb{O})$ of the complex argument $\mathbb{O}$ in the upper half-plane $\psi > 0$ in terms of the known analytic function $Z(\Phi)$. Therefore, consider the equation (see Eq. 4.1 in Grant 1973a)

$$\boxed{\left(iz(\Phi) - iZ(\Phi)\right) \frac{\mathrm{d}z}{\mathrm{d}\Phi} \frac{\mathrm{d}Z}{\mathrm{d}\Phi} = 1} \tag{9.173}$$

for $\psi > 0$. 'From the principle of analytic continuation, (9.173) is valid for any Φ' (Tanveer 1991, p. 144, symbols adapted). As a side remark, we briefly demonstrate how to avoid introducing the function Z, following Crew and Trinh (2016). From postulating symmetry of the wave about the crest at $x = 0$, one has

$$\bar{z}(\phi) = x(\phi) - iy(\phi) \stackrel{(*)}{=} -x(-\phi) - iy(-\phi) = -z(-\phi), \tag{9.174}$$

where at $(*)$ the symmetry of the wave profile was used. Then the Bernoulli equation (9.165) on the wave profile $\psi = 0$ becomes

$$\left(iz(\phi) + iz(-\phi)\right) \frac{\mathrm{d}z(\phi)}{\mathrm{d}\phi} \frac{\mathrm{d}z(-\phi)}{\mathrm{d}\phi} = -1. \tag{9.175}$$

In a cautious step, Crew and Trinh (2016) introduce a *complex* variable ϕ, but give it again the name Φ:

> The resultant equation (9.175) can then be used as a prescription for continuing into the rest of the complex ϕ-plane. We write the free-surface potential as $\phi = \phi_r + i\phi_c$ and, abusing notation somewhat, we relabel $\phi \mapsto \Phi \in \mathbb{C}$. (Crew and Trinh 2016, p. 262, symbols adapted)

Dyachenko et al. (2013) and Lushnikov (2016) proceed even more cautiously, by introducing a new auxiliary complex variable f (resp. w) and functions $z(f)$ and $\Phi(f)$, similarly to the procedure in Chap. 5 on jets, wakes, and cavities. In the following, we stay with Φ, and furthermore use the function $Z(\Phi)$ in order to emphasize the two different domains of analyticity of z and Z.

To simplify the algebra near a singularity in the upper half-plane of Φ, the origin $\Phi = 0$ is shifted into the singularity. Neglecting higher powers of Φ for $\Phi \to 0$, we may then approximate

$$z = a_0 + a_\alpha \Phi^\alpha, $$
$$Z = b_0 + b_1 \Phi. \tag{9.176}$$

The latter equation holds since Z is analytic in the upper Φ-half-plane, i.e., can be expanded in a power series (integer exponents). For the exponent α, one can assume $0 < \alpha < 1$, to obtain a singularity in the derivative $\mathrm{d}z/\mathrm{d}\Phi$; for this argument, see Crew and Trinh (2016) after their Eq. (3.4). Then (9.173) becomes

$$\frac{a_0 - b_0}{\Phi^{1-\alpha}} + a_\alpha \Phi^{2\alpha-1} - b_1 \Phi^\alpha = \frac{1}{i\alpha a_\alpha b_1}. \tag{9.177}$$

For $\textcircled{1} \to 0$ at the singularity, the first term has to vanish, since else the equation cannot hold (a_α and b_1 are both nonzero). Thus $a_0 = b_0$. However, as $\textcircled{1} \to 0$, also $b_1 \textcircled{1}^\alpha \to 0$ since $\alpha > 0$. Therefore,

$$\lim_{\textcircled{1} \to 0} \textcircled{1}^{2\alpha-1} = \frac{1}{i\alpha a_\alpha^2 b_1}. \tag{9.178}$$

The right side is finite, and thus the left side must be finite too. This is only possible if

$$\boxed{\alpha = 1/2} \tag{9.179}$$

when the left side becomes 1, therefore

$$a_\alpha^2 = -\frac{2i}{b_1}. \tag{9.180}$$

Thus the coefficients a_0 and a_α of $z(\textcircled{1})$ near a singularity are determined by the coefficients b_0 and b_1 of the analytic function $Z(\textcircled{1})$. The notable result, however, is (9.179) for the order of the singularity:

As an equation for $z(\textcircled{1})$, (9.173) admits only singularities of order $\frac{1}{2}$. (Grant 1973a, p. 261)

The same result $\alpha = \frac{1}{2}$ is also obtained numerically by Schwartz (1972, 1974). Tanveer (1991) shows that all off-crest singularities of surface water waves must be of square root type, i.e., have order $\frac{n}{2}$ with $n \in \mathbb{N}$, by bringing the Bernoulli equation (9.173) into a form (his Eq. 2.12), for which Painlevé proved that all singularities are of this type. As Grant (1973a) observes, singularities of order $\frac{1}{2}$ must coalesce near the highest amplitude to form a singularity of order $\frac{2}{3}$. It is not without irony that Rankine (1865) suggested $\alpha = \frac{1}{2}$ for the highest wave, which was corrected by Stokes (1880) to $\alpha = \frac{2}{3}$.

Crew and Trinh (2016) obtain the structure of the Riemann surface established by these singularities, i.e., the square root branch points. The structure is rather involved, and Lushnikov (2016) finds an infinite number of Riemann's sheets connected by the branch points. Thus a 'countably infinite number of distinct singularities exist on other branches of the solution, and [...] these singularities coalesce as Stokes' highest wave is approached' (Crew and Trinh 2016, p. 256).

The streamlines near a singularity $\textcircled{1} = z^2$ are given by the equation $\psi = \text{Im}(z^2) = 2xy = \text{const}$, which gives hyperbolas, see Fig. 9.15.

In this and the foregoing section, the progressive Stokes waves were treated. The corner angle of 120° was a result of the Stokes assumption, $u = c$, at the crest before breaking. Quite a different physical behavior can occur in the *standing* Stokes wave, where Grant (1973b) finds a crest angle of 90° instead of 120°.

9.7 The Crapper Wave

The highest inviscid surface water wave, the Stokes wave, has a broad and shallow trough and a cusped crest. In this section, we consider progressive surface water waves that have an absolute short lengthscale and are subject to surface tension. For these so-called capillary waves or ripples, an *exact, nonlinear* solution is available, the so-called Crapper wave. This has a form quite in contrast to the Stokes wave (one reason is that gravity plays no role), with an extended rounded crest and narrow troughs terminating in a singularity. Capillary waves are ubiquitous in lakes and the sea; Fig. 9.16 gives an example of their appearance.

> Capillary wave, small, free, surface-water wave with such a short wavelength that its restoring force is the water's surface tension, which causes the wave to have a rounded crest and a V-shaped trough. The maximum wavelength of a capillary wave is 1.73 centimetres (0.68 in.); longer waves are controlled by gravity and are appropriately termed gravity waves. Encyclopedia Britannica, entry 'Capillary wave' (oceanography), at https://www.britannica.com/science/capillary-wave.

The linear dispersion relation for capillary waves is derived in Kotschin et al. (1954), p. 409 from linear stability analysis. An early reference to capillary waves is Wilton (1915), who calculates their waveform including a narrow and deep trough, however, without singularity, from an expansion of the waveform in a power series. We follow here the classic paper by Crapper (1957) on capillary waves of finite amplitude.

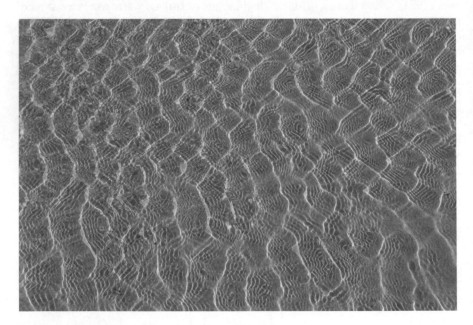

Fig. 9.16 Capillary waves. Photograph: Meinhard Spitzer

Crapper presents a finite-amplitude capillary wave that is that rarest of objects—an exact
solution to a nonlinear problem. It has some bizarre properties with important implications.
(Kinsman 1965, p. 258)

The fluid shall again have constant density and be inviscid. The velocity field $\vec{u}$
is divergence-free, irrotational, stationary, and two dimensional in water of infinite
depth. We follow Crapper's convention and let the y-direction point *downward*; to
obtain this from a rotation of the usual xy-system in the plane, the x-direction is
assumed to point to the left. The wave is brought to rest by superimposing it on a
stream with constant speed c in the positive x-direction. *Gravity is neglected from
now on, and thus the hydrostatic pressure is zero.* The standard complex variables
and functions are used,

$$z = x + iy,$$
$$w = u - iv,$$
$$\textcircled{} = \phi + i\psi, \tag{9.181}$$

where the complex potential $\textcircled{}$ obeys the Cauchy–Riemann equations, $\phi_x = \psi_y$ and
$\phi_y = -\psi_x$. The ϕ- and ψ-directions are aligned with the x- and y-directions, respectively. As described in Sect. 9.6, $\textcircled{}$ is used as the independent variable, and $z(\textcircled{})$ is
the unknown function. In Chap. 5 on jets, wakes, and cavities, the function $\ln w(\textcircled{})$
was of central importance. Quite similarly, with $q = \sqrt{u^2 + v^2}$, consider now

$$Q(\phi, \psi) = \ln q, \tag{9.182}$$
$$P(\phi, \psi) = \mathrm{atan}(v/u). \tag{9.183}$$

Using $\cos P = u/q$ and $\sin P = v/q$, one has

$$w = u - iv = q\,(\cos P - i \sin P) = qe^{-iP} = e^{Q-iP}. \tag{9.184}$$

With $w = \mathrm{d}\textcircled{}/\mathrm{d}z$, it follows that

$$\ln\left(\frac{\mathrm{d}\textcircled{}}{\mathrm{d}z}\right) = Q - iP. \tag{9.185}$$

This is an analytic function, and hence Q and P obey the Cauchy–Riemann equations,

$$\frac{\partial Q}{\partial \phi} = -\frac{\partial P}{\partial \psi}, \qquad\qquad \frac{\partial Q}{\partial \psi} = \frac{\partial P}{\partial \phi}. \tag{9.186}$$

This gives as the Laplace equation for Q and P,

$$\boxed{Q_{\phi\phi} + Q_{\psi\psi} = P_{\phi\phi} + P_{\psi\psi} = 0} \tag{9.187}$$

Fig. 9.17 Radius of
curvature of a trajectory

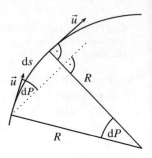

The central new physical ingredient is surface tension T. This has units of an energy
density and establishes a new pressure

$$p = \frac{T}{R}, \tag{9.188}$$

where R is the radius of curvature of the fluid surface (the sign is important).

From Fig. 9.17 one reads off $dP = ds/R$, or

$$\frac{1}{R} = \frac{dP}{ds}, \tag{9.189}$$

where ds is the arclength along a streamline. By definition of the velocity potential,

$$q = \frac{d\phi}{ds}. \tag{9.190}$$

Along a streamline, $\psi = $ const, and (9.189) becomes

$$\frac{1}{R} = q \left. \frac{\partial P}{\partial \phi} \right|_{\psi=\text{const}}. \tag{9.191}$$

This suffices for the following calculation, but we also include the formal definition
of surface tension. The force per volume $\rho \vec{g}$ from a force stress tensor T is, see (2.57),

$$\rho \vec{g} = \nabla \cdot \mathsf{T}. \tag{9.192}$$

Let $\hat{n}$ be the unit normal vector to a free fluid surface. Then the force from surface
tension is, by definition,

$$\mathsf{T} \cdot \hat{n} = \frac{T}{R} \hat{n}, \tag{9.193}$$

where T is the coefficient of surface tension. Since $\hat{n}$ is arbitrary, one has

$$\mathsf{T} = \frac{T}{R} \mathbb{1}, \tag{9.194}$$

i.e., surface tension corresponds to an isotropic pressure. Further details on this can be found in Wehausen and Laitone (1960), p. 452. Let then p be the pressure due to surface tension. This pressure p vanishes for $y \to \infty$, where streamlines become straight. There is no wave motion at great depth, thus $q = c$. Since the Bernoulli constant is a global constant for irrotational flows, the Bernoulli equation at the surface is

$$\frac{p}{\rho} + \frac{q^2}{2} = \frac{c^2}{2} \qquad \text{for} \qquad \psi = 0. \tag{9.195}$$

The equation of motion and the boundary conditions are (cf. 9.158),

$$\phi_{xx} + \phi_{yy} = 0 \qquad \text{for} \qquad 0 < \psi < \infty, \tag{9.196}$$

$$\frac{T}{\rho R} + \frac{q^2}{2} = \frac{c^2}{2} \qquad \text{for} \qquad \psi = 0, \tag{9.197}$$

$$\phi_x = c, \phi_y = 0 \qquad \text{for} \qquad \psi \to \infty. \tag{9.198}$$

These equations are rewritten now in terms of the functions $Q(\phi, \psi)$ and $P(\phi, \psi)$. First, inserting (9.191) for R in (9.197) gives, for $\psi = 0$,

$$\frac{Tq\, P_\phi}{\rho} + \frac{q^2}{2} = \frac{c^2}{2}. \tag{9.199}$$

From now on, c is used as unit of speed (i.e., set to unity), and $T/\rho c^2$ as unit of length. Dividing (9.199) by c^2 gives, for $\psi = 0$,

$$e^Q P_\phi + \frac{e^{2Q}}{2} = \frac{1}{2}. \tag{9.200}$$

Using $P_\phi = Q_\psi$ gives the remarkable condition at the water surface,

$$\boxed{Q_\psi = - \sinh Q \qquad \text{for} \qquad \psi = 0} \tag{9.201}$$

As $y \to \infty$, the speed becomes horizontal and approaches unity, therefore

$$P = Q = 0 \qquad \text{for} \qquad y \to \infty. \tag{9.202}$$

Altogether, the equations for Q are

$$Q_{\phi\phi} + Q_{\psi\psi} = 0 \qquad \text{for} \qquad 0 < \psi < \infty, \tag{9.203}$$

$$Q_\psi = - \sinh Q \qquad \text{for} \qquad \psi = 0, \tag{9.204}$$

$$Q = 0 \qquad \text{for} \qquad \psi \to \infty. \tag{9.205}$$

Similar equations hold for P, with (9.204) replaced by $P_\phi = - \sinh Q$. Equation (9.204) at the free surface is a Neumann boundary condition, and (9.205) at the ground is a Dirichlet boundary condition. Boundary conditions as $\phi \to \pm\infty$ are trivial and not further considered.

Quite clearly, (9.204) is the most relevant equation to be solved in the following. Crapper's central trick (see his Eq. 17) is to look for solutions of the Laplace equation (9.203) that satisfy the equation

$$Q_\psi = -f(\psi) \, \sinh \, Q \quad \text{for} \quad 0 \le \psi < \infty \tag{9.206}$$

for some function $f(\psi)$ with $f(0) = 1$.

> Physically, the import of this assumption is that *any* streamline in the moving fluid has potentially all the properties required of a free boundary. If the fluid on one side of a streamline were suddenly removed, the fluid on the other side could continue its flow undisturbed. (Kinnersley 1976, p. 231)

Equation (9.206) can be written as

$$\int \frac{\mathrm{d}Q}{\sinh Q} = -\int \mathrm{d}\psi \, f(\psi), \tag{9.207}$$

which gives

$$\ln \, \tanh\!\left(\tfrac{1}{2} Q(\phi, \psi)\right) = F(\psi) + G(\phi), \tag{9.208}$$

where $G(\phi)$ is an arbitrary function and $F(\psi)$ is the negative antiderivative of $f(\psi)$,

$$\frac{\mathrm{d}F(\psi)}{\mathrm{d}\psi} = -f(\psi). \tag{9.209}$$

Inversion of (9.208) gives

$$e^Q = \frac{1 + e^{F+G}}{1 - e^{F+G}}, \tag{9.210}$$

or

$$Q = \ln \frac{e^{-F(\psi)} + e^{G(\phi)}}{e^{-F(\psi)} - e^{G(\phi)}}. \tag{9.211}$$

Inserting Q into (9.203) gives equations for F and G. A quite straightforward calculation (see Eqs. 24 to 42 in Crapper 1957) gives the expressions known from standard wave solutions of the Laplace equation (see Eq. 9.9 and subsequent equations),

$$e^{-F(\psi)} = \sinh(k\psi + a),$$
$$e^{G(\phi)} = i \sin(k\phi + b), \tag{9.212}$$

with integration constants a, b and wavenumber k, to be further specified. A second sign $\pm k$ was dropped as it is trivial. Also amplitudes are not needed in (9.212), since they cancel anyway in (9.211). That Q from (9.211) with F and G from (9.212) indeed satisfies the Laplace equation (9.203) is shown below. The second relevant function, P, is obtained from $Q_\psi = P_\phi$, i.e., a differentiation followed by an integration. Using $\sinh(ix) = i \sin x$, $\cosh(ix) = \cos x$, and the addition theorems for sinh and cosh together with $\tanh x = \sinh x / \cosh x$ and $\tanh' x = \cosh^{-2} x$, one finds from (9.211) and (9.212),

$$Q = \ln\left(\frac{\tanh S}{\tanh T}\right), \tag{9.213}$$

$$P = -i \, \ln(\tanh S \, \tanh T), \tag{9.214}$$

where

$$S = \tfrac{1}{2}(k\psi + a + ik\phi + ib),$$
$$T = \tfrac{1}{2}(k\psi + a - ik\phi - ib). \tag{9.215}$$

Here Q in (9.213) is directly obtained, and for P in (9.214) one has indeed

$$Q_\psi = \frac{k}{2}\left(\frac{1}{\sinh S \, \cosh S} - \frac{1}{\sinh T \, \cosh T}\right) = P_\phi. \tag{9.216}$$

Furthermore,

$$Q_\phi = \frac{ik}{2}\left(\frac{1}{\sinh S \, \cosh S} + \frac{1}{\sinh T \, \cosh T}\right) \tag{9.217}$$

and

$$Q_{\psi\psi} = \frac{k^2}{4}\left(-\frac{1}{\cosh^2 S} - \frac{1}{\sinh^2 S} + \frac{1}{\cosh^2 T} + \frac{1}{\sinh^2 T}\right),$$
$$Q_{\phi\phi} = \frac{k^2}{4}\left(\frac{1}{\cosh^2 S} + \frac{1}{\sinh^2 S} - \frac{1}{\cosh^2 T} - \frac{1}{\sinh^2 T}\right), \tag{9.218}$$

so that Q indeed fulfills (9.203), $Q_{\phi\phi} + Q_{\psi\psi} = 0$, and similarly for P. It remains to fix the constants a and b. An arbitrary constant can be added to ϕ, and thus b in G is arbitrary; let

$$b = \frac{\pi}{2}. \tag{9.219}$$

Regarding a, one has from (9.209),

$$f(\psi) = k \coth(k\psi + a). \tag{9.220}$$

The boundary condition $f(0) = 1$ gives then

$$a = \ln\left(i\sqrt{\frac{k+1}{k-1}}\right). \tag{9.221}$$

Furthermore, the trivial boundary condition (9.205) is then also fulfilled, since $Q = \ln 1 = 0$ for $\psi \to \infty$. Inserting Q and P from (9.213) and (9.214) in (9.185), one finally arrives at

$$\frac{dz}{d\Phi} = e^{-Q}\, e^{iP} = \frac{\tanh T}{\tanh S}\, \tanh S \tanh T = \frac{\sinh^2 T}{\cosh^2 T}$$

$$= \left(\frac{e^{k\psi+a-ik\phi-i\pi/2} - 1}{e^{k\psi+a-ik\phi-i\pi/2} + 1}\right)^2 = \left(\frac{1 - \sqrt{\frac{k-1}{k+1}}\, e^{ik\Phi}}{1 + \sqrt{\frac{k-1}{k+1}}\, e^{ik\Phi}}\right)^2, \tag{9.222}$$

see Eq. (56) in Crapper (1957). Integration gives the solution

$$z = \Phi - \frac{4i}{k}\, \frac{1}{1 + \sqrt{\frac{k-1}{k+1}}\, e^{ik\Phi}} + d, \tag{9.223}$$

with a constant d. From (9.222), $w = d\Phi/dz$ has period $\Delta\Phi = 2\pi/k$, and from (9.223), this corresponds to $\Delta z = \Delta\Phi = 2\pi/k$. Therefore, k is indeed the wavenumber, and $\lambda = 2\pi/k$ the wavelength. To find the surface form of the wave, use

$$\psi = 0, \qquad \sigma = \phi/\lambda \tag{9.224}$$

in $\Phi = \phi + i\psi$, where σ acts as the curve variable. Then

$$\boxed{\frac{z}{\lambda} = \sigma - \frac{2i}{\pi}\, \frac{1}{1 + \sqrt{\frac{k-1}{k+1}}\, e^{2\pi i\sigma}} + \frac{2i}{\pi}} \tag{9.225}$$

The constant d is chosen so that $\Phi = z$ or $w = 1$ when $k = 1$. The waveform $y(x)$ is easily found from $z = x + iy$ and is shown in Fig. 9.18 for $k = 2$. The broad crests in Fig. 9.18 are almost semicircles, which corresponds to constant surface tension. The most striking feature in the figure is that *the wave surface becomes vertically tangent to itself toward the troughs.* Furthermore, the solution continues below this point, where a cavity forms. This region of deepest wave depression encloses a submerged, fluid-free bubble of air or water vapor. Thus, wind-induced capillary waves may cause 'foaminess' of the water.

Longuet-Higgins (1988) gives a significant simplification of Crapper's wave. Putting $k = 2$ in (9.222) gives

$$\frac{dz}{d\Phi} = \left(\frac{1 - 3^{-1/2}\, e^{2i\Phi}}{1 + 3^{-1/2}\, e^{2i\Phi}}\right)^2. \tag{9.226}$$

The factor $3^{-1/2}$ can be taken into ψ, thus shifting the origin of ψ,

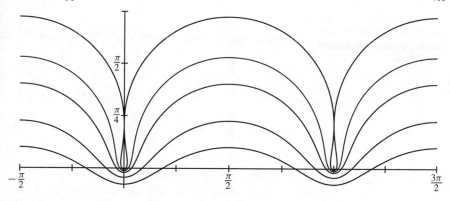

Fig. 9.18 Capillary waves of highest elevation (Crapper 1957). For $k = 2$, this wave has $\psi = 0.394$ (Longuet-Higgins 1988). The other curves are for $\psi = 0.5, 0.6, 0.8,$ and 1

$$3^{-1/2} e^{-2\psi} = e^{-2(\psi + \frac{1}{4} \ln 3)} = e^{-2\hat{\psi}}, \tag{9.227}$$

which gives, where now $\Phi = \phi + i\hat{\psi}$,

$$\frac{dz}{d\Phi} = \left(\frac{e^{i\Phi} - e^{-i\Phi}}{e^{i\Phi} + e^{-i\Phi}} \right)^2 = \left(\frac{i \sin \Phi}{\cos \Phi} \right)^2, \tag{9.228}$$

or

$$\frac{dz}{d\Phi} = -\tan^2 \Phi. \tag{9.229}$$

This gives Crapper's complex speed potential in implicit form,

$$\boxed{z = \Phi - \tan \Phi} \tag{9.230}$$

Taking real and imaginary parts, one obtains from this

$$x = \varphi - \frac{(1 - \tanh^2 \psi) \tan \varphi}{1 + \tan^2 \varphi \, \tanh^2 \psi}, \qquad y = \psi - \frac{(1 + \tan^2 \varphi) \tanh \psi}{1 + \tan^2 \varphi \, \tanh^2 \psi}, \tag{9.231}$$

and these curves are shown in Fig. 9.18.

Kinnersley (1976) generalizes the analysis performed by Crapper (1957) to fluids of finite depth. The resulting wave profile is shown in Fig. 2 of his paper. Vanden-Broeck and Keller (1980) extend Crapper's unique bubble solution to a one-parameter family of solutions with submerged bubbles of varying size and internal pressure. Gravity, which was neglected in the present derivation, is included together with surface tension in a paper by Schwartz and Vanden-Broeck (1979). Akers et al. (2014), also including gravity, find the capillary–gravity wave with maximum size of the attached bubble, when neighboring bubbles come into contact,

Fig. 9.19 Capillary–gravity
wave with maximum bubble
size. Reproduced from Akers
et al. (2014)

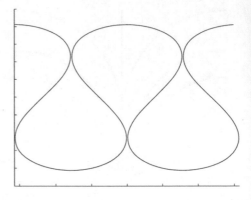

see Fig. 9.19. Somewhat awkward, gravity points upward here, i.e., away from the
fluid of infinite depth. This may be considered as a rough approximation to surface
underpressure caused by horizontal winds. The resulting wave is a standing wave
and 'consists of a string of bubbles and droplets at the fluid interface' (Akers et al.
2014, p. 3).

References

Ahlfors, L.V. 1966. *Complex analysis*, 2nd ed. New York: McGraw-Hill.
Akers, B.F., D.M. Ambrose, and J.D. Wright. 2014. Gravity perturbed Crapper waves. *Proceedings
of the Royal Society A* 470: 20130526.
Amick, C.J., L.E. Fraenkel, and J.F. Toland. 1982. On the Stokes conjecture for the wave of extreme
form. *Acta Mathematica* 148: 193.
Crapper, G.D. 1957. An exact solution for progressive capillary waves of arbitrary amplitude.
Journal of Fluid Mechanics 2: 532.
Crapper, G.D. 1984. *Introduction to water waves*. Hemel: Ellis Horwood Ltd.
Crew, S.C., and P.H. Trinh. 2016. New singularities for Stokes waves. *Journal of Fluid Mechanics*
798: 256.
Duncan, J.H. 1981. An experimental investigation of breaking waves produced by a towed hydrofoil.
Proceedings of the Royal Society of London A 377: 331.
Dyachenko, S.A., P.M. Lushnikov, and A.O. Korotkevich. 2013. Complex singularity of a Stokes
wave. *JETP Letters* 98: 675.
Fraenkel, L.E. 2007. A constructive existence proof for the extreme Stokes wave. *Archive for
Rational Mechanics and Analysis* 183: 187.
Grant, M.A. 1973a. The singularity at the crest of a finite amplitude progressive Stokes wave.
Journal of Fluid Mechanics 59: 257.
Grant, M.A. 1973b. Standing Stokes waves of maximum height. *Journal of Fluid Mechanics* 60:
593.
Jeffreys, H. 1924. On water waves near the shore. *Philosophical Magazine Series 6*, 48: 44.
Keady, G., and J. Norbury. 1978. On the existence theory for irrotational water waves. *Mathematical
Proceedings of the Cambridge Philosophical Society* 83: 137.
Kinnersley, W. 1976. Exact large amplitude capillary waves on sheets of fluid. *Journal of Fluid
Mechanics* 77: 229.
Kinsman, B. 1965. *Wind waves*. Englewood Cliffs: Prentice-Hall.

Kolscher, M. 1940. Unstetige Strömungen mit endlichem Totwasser. *Luftfahrtforschung* (Berlin-Adlershof) vol. 17, Lfg. 5, München: Oldenbourg.

Kotschin, N.J., I.A. Kibel, and N.W. Rose. 1954. *Theoretische Hydromechanik*, vol. I. Berlin: Akademie-Verlag.

Krasovskii, Yu.P. 1961. On the theory of steady-state waves of finite amplitude. *U.S.S.R. Computational Mathematics and Mathematical Physics* 1: 996.

Lamb, H. 1932. *Hydrodynamics*. Cambridge: Cambridge University Press; New York: Dover (1945).

Lighthill, J. 1978. *Waves in fluids*. Cambridge: Cambridge University Press.

Liu, X. 2016. A laboratory study of spilling breakers in the presence of light-wind and surfactants. *Journal of Geophysical Research: Oceans* 121: 1846.

Longuet-Higgins, M.S. 1956. The refraction of sea waves in shallow water. *Journal of Fluid Mechanics* 1: 163.

Longuet-Higgins, M.S. 1963. The generation of capillary waves by steep gravity waves. *Journal of Fluid Mechanics* 16: 138.

Longuet-Higgins, M.S. 1973. On the form of the highest progressive and standing waves in deep water. *Proceedings of the Royal Society of London A* 331: 445.

Longuet-Higgins, M.S. 1979. The almost-highest wave: A simple approximation. *Journal of Fluid Mechanics* 94: 269.

Longuet-Higgins, M.S. 1988. Limiting forms for capillary-gravity waves. *Journal of Fluid Mechanics* 194: 351.

Longuet-Higgins, M.S., and M.J.H. Fox. 1978. Theory of the almost-highest wave. Part 2. Matching and analytic extension. *Journal of Fluid Mechanics 85*: 769.

Longuet-Higgins, M.S., and R.W. Stewart. 1960. Changes in the form of short gravity waves on long waves and tidal currents. *Journal of Fluid Mechanics* 8: 565.

Longuet-Higgins, M.S., and R.W. Stewart. 1961. The changes in amplitude of short gravity waves on steady non-uniform currents. *Journal of Fluid Mechanics* 10: 529.

Longuet-Higgins, M.S., and J.S. Turner. 1974. An 'entraining plume' model of a spilling breaker. *Journal of Fluid Mechanics* 63: 1.

Lushnikov, P.M. 2016. Structure and location of branch point singularities for Stokes waves on deep water. *Journal of Fluid Mechanics* 800: 557.

Michell, J.H. 1893. The highest waves in water. *Philosophical Magazine Series 5*, 36: 430.

Moore, D.W. 1979. The spontaneous appearance of a singularity in the shape of an evolving vortex sheet. *Proceedings of the Royal Society of London A* 365: 105.

Phillips, O.M. 1963. On the attenuation of long gravity waves by short breaking waves. *Journal of Fluid Mechanics* 16: 321.

Rankine, W.J. 1865. Supplement to a paper on stream-lines. *Philosophical Magazine Series 4*, 29: 25.

Schwartz, L.W. 1972. *Analytic continuation of Stokes' expansion for gravity waves*. Ph.D. dissertation, Stanford University.

Schwartz, L.W. 1974. Computer extension and analytical continuation of Stokes' expansion for gravity waves. *Journal of Fluid Mechanics* 62: 553.

Schwartz, L.W., and J.D. Fenton. 1982. Strongly nonlinear waves. *Annual Reviews in Fluid Mechanics* 14: 39.

Schwartz, L.W., and J.-M. Vanden-Broeck. 1979. Numerical solution of the exact equations for capillary-gravity waves. *Journal of Fluid Mechanics* 95: 119.

Sommerfeld, A. 1957. *Mechanik der deformierbaren Medien*, 4th ed. Leipzig: Geest & Portig.

Stoker, J.J. 1957. *Water waves. The mathematical theory with applications*. New York: Interscience Publishers.

Stokes, G.G. 1847 and 1880. On the theory of oscillatory waves. *Transactions of the Cambridge Philosophical Society* 8: 441 (1849); in *Mathematical and physical papers by G.G. Stokes*, vol. 1, 197 and 314 (Supplement). Cambridge: Cambridge University Press (1880).

Stoneley, R. 1935. The refraction of a wave group. *Mathematical Proceedings of the Cambridge Philosophical Society* 31: 360.

Tanveer, S. 1991. Singularities in water waves and Rayleigh-Taylor instability. *Proceedings of the Royal Society of London A* 435: 137.

Unna, P.J.H. 1942. Waves and tidal streams. *Nature* 149: 219.

Vanden-Broeck, J.-M., and J.B. Keller. 1980. A new family of capillary waves. *Journal of Fluid Mechanics* 98: 161.

Wehausen, J.V., and E.V. Laitone 1960. Surface waves. In *Handbuch der Physik = Encyclopedia of Physics*, vol. 3/9, Strömungsmechanik 3, ed. S. Flügge and C. Truesdell, 446. Berlin: Springer.

Whitham, G.B. 1962. Mass, momentum and energy flux in water waves. *Journal of Fluid Mechanics* 12: 135.

Williams, J.M. 1981. Limiting gravity waves in water of finite depth. *Philosophical Transactions of the Royal Society of London A* 302: 139.

Williams, J.M. 1985. Near-limiting gravity waves in water of finite depth. *Philosophical Transactions of the Royal Society of London A* 314: 353.

Wilton, J.R. 1915. On ripples. *Philosophical Magazine Series 6*, 29: 688.

Chapter 10
Existence Proof for Weakly Nonlinear Water Waves

10.1 Introduction

The availability of closed analytic solutions for nonlinear water waves is the rare exception; some examples are the Crapper capillary wave from the last section, as well as cnoidal waves and the Gerstner wave for free surface waves.

Great efforts are therefore made to prove at least the existence of nonlinear solutions of the wave equations with the appropriate boundary conditions. The first modern existence and uniqueness proof for solutions of ordinary differential equations is the Picard–Lindelöf theorem from ca. 1890. The task is obviously much harder for partial differential equations. Here, it was noted early on (Nekrasov 1921) that a (re-)formulation of the persistent boundary value problem as an integral equation may be advantageous. An example of such an equation is the Urysohn integral equation

$$f(s) = \int_a^b \mathrm{d}t \; K(s, t, f(t)) + g(s), \tag{10.1}$$

where s and t are real variables, $[a, b]$ is a finite real interval, $g(s)$ is a known continuous function, $f(s)$ is the unknown function, and K is a continuous kernel function of three arguments. The values of f, g, and K can be either real or complex. The functions f and g belong to some Banach space (a normed linear vector space), in which the integral over K in (10.1) acts as an operator.

There exists a highly developed theory of nonlinear functional analysis addressing such operator equations and their solutions. This material cannot be covered here but is subject of specialized texts. According to Hutson and Pym (1980, p. 110), 'the three volumes of Krasnoselskii (1964a, b, 1972) are probably the nearest to a complete account,' but this refers to the period up to 1970 only.

However, it is possible here to address the question of existence of *weakly* nonlinear solutions, formerly termed nonlinear 'im kleinen' (in the small), as contrasted to fully nonlinear solutions 'im grossen' (in the large). The techniques required for this are only moderately challenging and are covered in the present chapter; they also lie at the foundation of more modern approaches: Krasnoselskii (1964b) on p. 1 cites

© Springer Nature Switzerland AG 2019

A. Feldmeier, *Theoretical Fluid Dynamics*, Theoretical and Mathematical Physics,
https://doi.org/10.1007/978-3-030-31022-6_10

the paper by Schmidt (1908) and the book by Lichtenstein (1931), both of which are extensively discussed in the following, as among the first sources of nonlinear functional analysis.

Existence proofs were long considered a realm of pure mathematics rather than physics. This restrictive viewpoint has widened over the last few decades for mainly two reasons. First, existence proofs often employ an iteration scheme for the solution of an operator equation. The mathematical aspect of the existence theorem consists in proving convergence of the scheme; the physical aspect is that the scheme can be implemented numerically, to actually obtain the solution, often to any desired accuracy. Second, the so-called *qualitative* theory of differential equations does not deal with a specific ('quantitative') solution to given boundary conditions, but rather with general topological properties of a whole class of solutions (say, subsonic versus transonic solutions). With general topological considerations becoming more important in physics (e.g., in elementary particle physics, general relativity, or indeed nonlinear dynamics), so do existence proofs.

Interestingly, one of the first major applications of modern nonlinear functional analysis is the proof by Krasovskii (1961) of the existence of fully nonlinear free surface waves 'in the large,' up to but not including the Stokes angle of 120° introduced in Sect. 9.5. The proof of the existence of fully nonlinear water waves can be found in the books by Hutson and Pym (1980) and Buffoni and Toland (2003), and in the review article by Toland (1996).

Similarly, one of the first applications of the nonlinear theory 'in the small' was also for surface water waves, by Nekrasov (1921). Other, independent existence proofs for the existence of weakly nonlinear wave solutions are by Levi-Civita (1925) using a power series approach, and by Lichtenstein (1931) using the theory of nonlinear integral equations developed by Schmidt (1908). This latter proof is presented in the following; it is also developed in detail in Wainberg and Trenogin (1973). Yet another, alternative proof by Littman and Nirenberg (see Littman 1957) uses an iteration scheme and can be found in a very clear and concise form in Stoker's book (1957).

10.2 Boundary Condition

An inviscid fluid with constant density and stationary, solenoidal, irrotational, 2-D planar velocity field is assumed. The waveform $y(x)$ is represented by a complex variable $z = x + iy$, where y points *downward*. This simplifies the signs in the final eigenvalue analysis in Sect. 10.6. The complex speed potential is $\Phi = \phi + i\psi$, with $\psi = 0$ at the wave surface and $\psi > 0$ is the interior of the fluid. As before, the complex speed potential Φ is treated as independent variable, and the wave profile z as function of Φ. A wave with speed c and constant shape is superimposed on a stream with speed $-c$ and is thus stationary. A new complex variable ω is introduced,

$$\boxed{\omega = \theta + i\tau} \tag{10.2}$$

such that for the complex speed w,

$$w = \frac{d\mathcal{O}}{dz} = u - iv = ce^{-i\omega} = ce^{\tau} e^{-i\theta}. \tag{10.3}$$

Thus,

$$u = ce^{\tau} \cos\theta, \qquad\qquad v = ce^{\tau} \sin\theta. \tag{10.4}$$

The kinematic boundary condition at the surface is that the fluid velocity $\vec{u} = (u, v)$ is tangent to the waveform,

$$\frac{dy}{dx} = \frac{v}{u}. \tag{10.5}$$

Thus the angle θ is the (azimuthal) angle of the wave surface with the x-axis (horizontal line). The Stokes angle at the crest corresponds to $\theta = \pm30°$ as the steepest possible angle at which the smoothness of the solution ceases to exist. The quantity τ in the above formula is of secondary importance in the following, since it measures the absolute flow speed.

In the remainder of this section, the boundary condition is derived that results from the Bernoulli equation at the wave surface, for the functions θ and τ in terms of the independent variables ϕ and ψ. At the upper water surface, with pressure $p = $ const,

$$\tfrac{1}{2} |w|^2 - gy = C, \tag{10.6}$$

where C is a global constant for irrotational flow and the sign of $-gy$ results from the choice of direction of the y-axis. One has then

$$\frac{dz}{d\mathcal{O}} = \frac{1}{d\mathcal{O}/dz} = \frac{1}{u - iv} = \frac{u + iv}{u^2 + v^2} = \frac{\partial z}{\partial \phi} = \frac{\partial x}{\partial \phi} + i\frac{\partial y}{\partial \phi}, \tag{10.7}$$

and therefore

$$\frac{\partial y}{\partial \phi} = \frac{v}{|w|^2}. \tag{10.8}$$

Differentiating (10.6) with respect to ϕ gives

$$|w|\frac{\partial}{\partial \phi}|w| - g\frac{\partial y}{\partial \phi} = 0. \tag{10.9}$$

Inserting here $|w| = ce^{\tau}$ together with (10.8) and $v = ce^{\tau} \sin\theta$ gives

$$\frac{\partial(ce^{\tau})}{\partial \phi} = \frac{g \sin\theta}{c^2 e^{2\tau}}, \tag{10.10}$$

or

$$\frac{\partial \tau}{\partial \phi} = \frac{g}{c^3} e^{-3\tau} \sin \theta, \tag{10.11}$$

see Levi-Civita (1925), Eq. I, p. 264. Using the Cauchy–Riemann equation $\theta_\psi = -\tau_\phi$, this becomes

$$\boxed{\frac{\partial \theta}{\partial \psi} = -\lambda e^{-3\tau} \sin \theta \qquad \text{for} \qquad \psi = 0} \tag{10.12}$$

with a constant

$$\lambda = \frac{g}{c^3}. \tag{10.13}$$

Equation (10.12) is the nonlinear boundary condition for two-dimensional free surface waves of inviscid, incompressible fluids; it lies at the heart of the following considerations. The next three sections introduce the theory of linear integral equations and elements of the theory of nonlinear integral equations. In Sect. 10.6, we return to the problem of nonlinear water waves.

10.3   Linear Integral Equations

This section states general existence theorems for linear integral equations. Wave problems can be written as both differential and integral equations, but regarding existence theorems, integral equations are often of advantage. The reason is that the existence of a solution depends crucially on the boundary conditions. For differential equations, boundary conditions appear as extra equations, while for integral equations, they are inherent to the equation.

First, some fundamental theorems on linear integral equations are proved, following Smirnow (1988). A standard, if somewhat outdated reference, is Hellinger and Toeplitz (1927), and the classic paper on the theory is Fredholm (1903). The subject was then re-formulated as theory of compact operators in the Banach spaces. Starting point is Fredholm's integral equation of the second kind,

$$f(s) = g(s) + \lambda \int_a^b \mathrm{d}t \ K(s,t) \ f(t), \tag{10.14}$$

where g is a known and f an unknown function, K is called integral *kernel*, and $[a, b]$ is a closed and bounded real-valued interval. All numbers and functions (i.e., $a, b, s, t, \lambda, f, g, K$) are assumed to be real; a generalization to complex numbers is straightforward but somewhat tedious: instead of the transposed τ of an operator the *adjoint* † has to be used then, i.e., the transposed and complex-conjugated. Equation (10.14) is written in operator notation as

$$f = g + \lambda \mathbf{K} f, \tag{10.15}$$

where the boldface notation for operators follows Krasnoselskii (1964a, b). Note that for $K(s, t) = K(s - t)$, the integral in (10.14) becomes a convolution. In this case, the Fourier transformation of (10.14) becomes relevant. Furthermore, the following theory is easily generalized to multiple integrals.

The central condition on g and K is that they are continuous on $[a, b]$ and $[a, b] \times [a, b]$, respectively. It follows then that f is also continuous on $[a, b]$ (see Smirnow 1988, p. 19).

The expression $\mathbf{K}f \equiv \int_a^b dt\ K(s, t)\ f(t)$ can be understood as a *linear* operator $\mathbf{K}$ acting on a Banach space vector f, since

$$\mathbf{K}(af + bg) = a\mathbf{K}f + b\mathbf{K}g, \tag{10.16}$$

with continuous functions f, g and real numbers a, b.

A real-valued scalar product $(.\ ,\ .)$ is introduced by

$$(f, g) \equiv \int_a^b dt\ f(t)\ g(t), \tag{10.17}$$

which is symmetric, $(f, g) = (g, f)$.

The *transposed* equation to (10.14) is given by

$$h(s) = g(s) + \lambda \int_a^b dt\ K(t, s)\ h(t), \tag{10.18}$$

or, with the transposed operator $\mathbf{K}^\tau$,

$$h = g + \lambda \mathbf{K}^\tau h. \tag{10.19}$$

Generally, f in (10.14) and h in (10.18) are different functions for the same λ. The following relation holds for arbitrary continuous f, h, and $\mathbf{K}$:

$$(h, \mathbf{K}f) = (\mathbf{K}^\tau h, f) \tag{10.20}$$

since

$$\int_a^b ds\ h(s) \left[\int_a^b dt\ K(s, t)\ f(t) \right] = \int_a^b dt \left[\int_a^b ds\ K(s, t)\ h(s) \right] f(t). \tag{10.21}$$

Let the function $f \neq 0$ be a solution of the *homogeneous* integral equation,

$$f = \lambda \mathbf{K}f. \tag{10.22}$$

Then f is called *eigenfunction* (or eigenvector) and λ *characteristic value* of $\mathbf{K}$ (or $K(s, t)$). By convention, the trivial solution $f = 0$ has no characteristic value. If f_i are eigenfunctions to one single characteristic value λ, then obviously $\sum_i a_i f_i$ is also eigenfunction.

A central result of the theory of integral equations is that if the integration interval $[a, b]$ is finite, then (10.22) has solutions $f \neq 0$ only for a *discrete* spectrum of λ-values.

There are three different conventions, to either write

$$f = \lambda \mathbf{K} f \qquad \text{or} \qquad \lambda f = \mathbf{K} f \qquad \text{or} \qquad f + \mathbf{K} f = 0. \qquad (10.23)$$

In the following, the first alternative is used, since in applications the number λ is often part of the kernel K.

An operator product $\mathbf{KL}$ is introduced by

$$\mathbf{KL}(s, t) = \int_a^b du \, K(s, u) \, L(u, t). \qquad (10.24)$$

One has then

$$
\begin{aligned}
(\mathbf{KL}) f &\equiv \int_a^b dt \left[\int_a^b du \, K(s, u) \, L(u, t) \right] f(t) \\
&= \int_a^b du \, K(s, u) \left[\int_a^b dt \, L(u, t) f(t) \right] \equiv \mathbf{K}(\mathbf{L}f),
\end{aligned}
\qquad (10.25)
$$

which means that the brackets can be dropped. Since all functions in this chapter are assumed to be continuous, integrals are interchangeable. *Powers* of the kernel $K(s, t)$ and the operator $\mathbf{K}$ are defined by, with $n \in \mathbb{N}$,

$$
\begin{aligned}
K^1(s, t) &= K(s, t), \\
K^n(s, t) &= \int_a^b du \, K^{n-1}(s, u) K(u, t).
\end{aligned}
\qquad (10.26)
$$

A central operator in the following is the *resolvent* $\mathbf{R}(\lambda; \mathbf{K})$ of the operator $\mathbf{K}$ to the value λ, defined by the infinite *Neumann series*,

$$\boxed{\mathbf{R}(\lambda; \mathbf{K}) = \mathbf{K} + \lambda \mathbf{K}^2 + \lambda^2 \mathbf{K}^3 + \cdots} \qquad (10.27)$$

see Hellinger and Toeplitz (1927), p. 1383. The symbol $\mathbf{R}(\lambda; \mathbf{K})$ for the resolvent is that of Dunford and Schwartz (1958), p. 566. To solve the operator equation

$$f = g + \lambda \mathbf{K} f, \qquad (10.28)$$

one expands f in an infinite series,

$$f = f_0 + \lambda f_1 + \lambda^2 f_2 + \cdots \qquad (10.29)$$

Inserting this in (10.28) and equating powers of λ gives

$$f_0 = g,$$
$$f_1 = \mathbf{K} f_0,$$
$$f_2 = \mathbf{K} f_1,$$
$$\cdots$$
$$f_n = \mathbf{K} f_{n-1}. \tag{10.30}$$

This *iteration* method goes back to Liouville. Setting $\mathbf{K}^0 = 1$ (as a number or an operator), one has then

$$f_n = \mathbf{K} f_{n-1} = \mathbf{K}^2 f_{n-2} = \cdots = \mathbf{K}^n f_0 = \mathbf{K}^n g. \tag{10.31}$$

Inserting this in (10.29) and using (10.27) for $\mathbf{R}(\lambda; \mathbf{K})$ gives

$$f = \sum_{n=0}^{\infty} \lambda^n f_n$$
$$= \sum_{n=0}^{\infty} \lambda^n \mathbf{K}^n g$$
$$= g + \lambda \sum_{n=0}^{\infty} \lambda^n \mathbf{K}^{n+1} g$$
$$= g + \lambda \mathbf{R}(\lambda; \mathbf{K}) g. \tag{10.32}$$

Therefore, the operator equation

$$f = g + \lambda \mathbf{K} f \tag{10.33}$$

has the solution

$$\boxed{f = g + \lambda \mathbf{R}(\lambda; \mathbf{K}) g} \tag{10.34}$$

where the right side of the latter equation is assumed to be known. The convergence properties of the Neumann series are discussed in Smirnow (1988), p. 22.

The resolvent $\mathbf{R}(\lambda; \mathbf{K})$ obeys two fundamental equations. Multiplying (10.27) with $\lambda \mathbf{K}$ from either the left or the right gives

$$\lambda \mathbf{K} \mathbf{R}(\lambda; \mathbf{K}) = \lambda \mathbf{R}(\lambda; \mathbf{K}) \mathbf{K} = \sum_{n=0}^{\infty} \lambda^{n+1} \mathbf{K}^{n+2}$$
$$= \sum_{n=0}^{\infty} \lambda^n \mathbf{K}^{n+1} - \mathbf{K}$$
$$= \mathbf{R}(\lambda; \mathbf{K}) - \mathbf{K}, \tag{10.35}$$

where the sum in the second line is expanded by a new first term, $+\mathbf{K}$, which is then subtracted off again. Therefore,

$$\boxed{\mathbf{R}(\lambda; \mathbf{K}) = \mathbf{K} + \lambda \mathbf{K} \mathbf{R}(\lambda; \mathbf{K}) = \mathbf{K} + \lambda \mathbf{R}(\lambda; \mathbf{K})\mathbf{K}} \qquad (10.36)$$

One can from now on avoid any explicit reference to the series expansions (10.27) and (10.29), since all relevant information is contained in (10.36). This leads to the following theorem.

Theorem 1 *For given λ, let the resolvent $\mathbf{R}(\lambda; \mathbf{K})$ of the operator $\mathbf{K}$ exist, be continuous, and obey (10.36). Then*

$$f = g + \lambda \mathbf{K} f \qquad \leftrightarrow \qquad f = g + \lambda \mathbf{R}(\lambda; \mathbf{K})g. \qquad (10.37)$$

Proof '$\rightarrow$'. Abbreviate $\mathbf{R} = \mathbf{R}(\lambda; \mathbf{K})$, and let f be the solution of $f = g + \lambda \mathbf{K} f$. Using (10.36), one has

$$\lambda \mathbf{R} f = \lambda \mathbf{R} g + \lambda^2 \mathbf{R} \mathbf{K} f = \lambda \mathbf{R} g + \lambda (\mathbf{R} - \mathbf{K}) f, \qquad (10.38)$$

or $\lambda \mathbf{R} g = \lambda \mathbf{K} f$. Inserting this in $f = g + \lambda \mathbf{K} f$ gives $f = g + \lambda \mathbf{R} g$.
'$\leftarrow$'. Let $f = g + \lambda \mathbf{R} g$. According to (10.36),

$$\lambda \mathbf{R} g = \lambda (\mathbf{K} + \lambda \mathbf{K} \mathbf{R})g, \qquad (10.39)$$

or

$$g + \lambda \mathbf{R} g = g + \lambda \mathbf{K}(g + \lambda \mathbf{R} g), \qquad (10.40)$$

or, by the definition of f,

$$f = g + \lambda \mathbf{K} f, \qquad (10.41)$$

thus f obeys the operator equation. $\square$

The resolvent $\mathbf{R}$ is given by an infinite operator series (10.27), and thus criteria have to be established as to when $\mathbf{R}$ exists. To find these, the kernel integrals can be written as the Riemann sums, which turn the integral equation into a matrix equation. For $n \in \mathbb{N}$, let $x_0 = a, x_1 \ldots, x_{n-1}, x_n = b$ be an equidistant partition of $[a, b]$. Let the complex function $D(\lambda)$ be given by the power series

$$D(\lambda) = 1 + \sum_{n=1}^{\infty} (-1)^n \frac{\lambda^n}{n!} \int_a^b dx_1 \ldots \int_a^b dx_n \begin{vmatrix} K(x_1, x_1) & \ldots & K(x_1, x_n) \\ \vdots & & \vdots \\ K(x_n, x_1) & \ldots & K(x_n, x_n) \end{vmatrix}, \qquad (10.42)$$

where the vertical bars $|\ldots|$ indicate a determinant, and let the complex function $P(s, t, \lambda)$ be given by

$$P(s, t, \lambda) =$$

$$= 1 + \sum_{n=1}^{\infty} (-1)^n \frac{\lambda^n}{n!} \int_a^b dx_1 \dots \int_a^b dx_n \begin{vmatrix} K(s, t) & K(s, x_1) \dots & K(s, x_n) \\ K(x_1, t) & K(x_1, x_1) \dots & K(x_1, x_n) \\ \vdots & \vdots & \vdots \\ K(x_n, t) & K(x_n, x_1) \dots & K(x_n, x_n) \end{vmatrix}. \quad (10.43)$$

Both D and P are *entire* functions in the complex λ-plane. (An entire function is analytic on all of $\mathbb{C}$, except at infinity.)

Theorem 2 *Let g be an arbitrary continuous function. Then the equation $f = g + \lambda K f$ has the solution $f = g + \lambda R(\lambda; K)g$ if $D(\lambda) \neq 0$, where $R(\lambda; K)$ is the operator corresponding to the kernel*

$$R(s, t) = \frac{P(s, t, \lambda)}{D(\lambda)}. \quad (10.44)$$

For the proof, see Smirnow (1988), pp. 25–31. The theorem applies in the complex λ-plane with exception of poles of $R(s, t)$. One can show that these poles of R are exactly the zeros of D.

Let $D_\tau(\lambda)$ and $P_\tau(s, t, \lambda)$ be D and P for the transposed equation. The determinant of a matrix equals the determinant of the transposed matrix. From the definitions (10.42) and (10.43), one obtains then directly

$$D_\tau(\lambda) = D(\lambda),$$
$$P_\tau(s, t, \lambda) = P(t, s, \lambda). \quad (10.45)$$

Note the exchange of s and t in the latter equation. The determinant D is called the *Fredholm* determinant. The theory of infinite determinants goes back to von Koch, see Hellinger and Toeplitz (1927), p. 1417.

The next theorem shows that the homogeneous equation $f = \lambda K f$ (no function g) has a nontrivial solution if and only if the Fredholm determinant vanishes. This is in direct correspondence to the existence of a nontrivial solution of a finite system of homogeneous, linear equations.

Theorem 3 $D(\lambda_0) = 0 \leftrightarrow$ *the homogeneous equation $f = \lambda_0 K f$ has solution $f \neq 0$, i.e., λ_0 is characteristic value of K.*

Proof (Smirnow 1988, pp. 31–33) '$\rightarrow$'. Assume that a denominator $D = 0$ gives all the poles of $R(s, t)$ (corresponding to $R(\lambda; K)$) in (10.44). Let λ_0 be such a pole of order r. Then the Laurent series of $R(s, t)$ about λ_0 is

$$R(s, t) = \sum_{j=-r}^{\infty} (\lambda - \lambda_0)^j a_j(s, t), \quad (10.46)$$

with real-valued functions $a_j(s, t)$. It is readily shown (but skipped here) that $a_{-r}(s, t)$ is continuous in s and t. Therefore, integrals can be performed with R from (10.46). Equation (10.36) is

$$R(s, t) = K(s, t) + \lambda \int_a^b du\, K(s, u)\, R(u, t). \tag{10.47}$$

Inserting (10.46) gives

$$\sum_{j=-r}^{\infty} (\lambda - \lambda_0)^j\, a_j(s, t) = K(s, t) + \lambda \int_a^b du\, K(s, u) \sum_{j=-r}^{\infty} (\lambda - \lambda_0)^j\, a_j(u, t). \tag{10.48}$$

Multiplying both sides by $(\lambda - \lambda_0)^r$ and setting $\lambda = \lambda_0$, one obtains

$$a_{-r}(s, t) = \lambda_0 \int_a^b du\, K(s, u)\, a_{-r}(u, t). \tag{10.49}$$

For any number t, define the function $\tilde{a}_{-r}(s) = a_{-r}(s, t)$. Then $\tilde{a}_{-r} \neq 0$ is a solution of the homogeneous equation $\tilde{a}_{-r} = \lambda_0 \mathbf{K}\tilde{a}_{-r}$, and thus λ_0 is a characteristic value of $\mathbf{K}$.

'$\leftarrow$'. Assume $D(\lambda) \neq 0$, λ is *not* a zero of D. Then from Theorem 2, $f = g + \lambda \mathbf{K}f$ has solution $f = g + \lambda \mathbf{R}(\lambda; \mathbf{K})g$. For the homogeneous integral equation, $g = 0$ and therefore $f = 0$, and the homogeneous equation has *no* solution $f \neq 0$ and λ is *not* a characteristic value of $\mathbf{K}$. $\square$

Let λ be a characteristic value of $\mathbf{K}$, with $\phi_i = \lambda \mathbf{K}\phi_i$. Then the number of linearly independent eigenfunctions ϕ_i is *finite*. For the proof of this statement using Bessel's inequality, see Smirnow (1988), pp. 35–36.

The next theorem shows how to build an operator $\mathbf{L}$ with *no* eigenvalues and eigenfunctions, from an operator $\mathbf{K}$ that has p eigenvalues and eigenfunctions. This theorem will be important in the existence proof for water waves.

Theorem 4 *For operators $\mathbf{K}$ and $\mathbf{K}^\tau$ let ϕ_i and ψ_j, respectively, be nonvanishing, linearly independent, and normalized eigenfunctions,*

$$(\phi_i, \phi_j) = (\psi_i, \psi_j) = \delta_{ij} \tag{10.50}$$

(this can be achieved by Schmidt orthogonalization), with characteristic value λ,

$$\begin{aligned} \phi_i &= \lambda \mathbf{K}\phi_i, & i &= 1, \ldots, p, \\ \psi_j &= \lambda \mathbf{K}^\tau \psi_j, & j &= 1, \ldots, q, \end{aligned} \tag{10.51}$$

where p and q is the maximum number of independent ϕ_i and ψ_j.

(i) Then $p = q$.
  Define the kernel $L(s, t)$ (with corresponding operator $\mathbf{L}$) by

$$\boxed{\lambda L(s, t) = \lambda K(s, t) - \sum_{j=1}^{p} \psi_j(s) \phi_j(t)} \qquad (10.52)$$

(ii) Then the equations

$$f = \lambda \mathbf{L} f, \qquad (10.53)$$
$$h = \lambda \mathbf{L}^{\tau} h \qquad (10.54)$$

have only trivial solutions $f = h = 0$.

Proof (Smirnow 1988, pp. 36–38; Lichtenstein 1931, p. 22)
We assume $q \geq p$ and prove part (ii) first. Let f be a solution of (10.53),

$$f(s) = \lambda \int_a^b dt \, L(s, t) \, f(t) = \lambda \int_a^b dt \, K(s, t) \, f(t) - \sum_{j=1}^{p} \psi_j(s) \int_a^b dt \, \phi_j(t) \, f(t),$$
$$(10.55)$$

or, as operator equation,

$$f = \lambda \mathbf{L} f = \lambda \mathbf{K} f - \sum_{j=1}^{p} \psi_j(\phi_j, f). \qquad (10.56)$$

Performing a scalar product with ψ_i from the left, where $i = 1, \ldots, p$ (not q), gives

$$(\psi_i, f) = (\psi_i, \lambda \mathbf{K} f) - \sum_{j=1}^{p} (\psi_i, \psi_j)(\phi_j, f)$$

$$= (\lambda \mathbf{K}^{\tau} \psi_i, f) - \sum_{j=1}^{p} \delta_{ij} \, (\phi_j, f)$$

$$= (\psi_i, f) - (\phi_i, f), \qquad (10.57)$$

where (10.20) was used in the second line. Therefore,

$$(\phi_i, f) = 0, \qquad\qquad i = 1, \ldots, p. \qquad (10.58)$$

Using this in (10.56) gives

$$f = \lambda \mathbf{K} f. \qquad (10.59)$$

Thus f is eigenfunction of $\mathbf{K}$ to characteristic value λ and can be written as

$$f = \sum_{j=1}^{p} c_j \phi_j,\tag{10.60}$$

with real numbers c_j. Performing a scalar product with ϕ_i for $i = 1$ to p from the left and using again (10.58) gives

$$0 = (\phi_i, f) = \sum_{j=1}^{p} c_j (\phi_i, \phi_j) = c_i, \qquad\qquad i = 1, \ldots, p.\tag{10.61}$$

Thus indeed $f = 0$. The case $p \geq q$ is treated correspondingly, and similarly one shows $h = 0$ for the transposed equation (10.54). This finishes the proof of part (ii).

Next (i) can be proved by contradiction: assume $q > p$, and we show that then (10.54) has a nontrivial solution, which cannot be. Equation (10.54) is

$$h(s) = \lambda \int_a^b dt\, L(t, s)\, h(t) = \lambda \int_a^b dt\, K(t, s)\, h(t) - \sum_{j=1}^{p} \int_a^b dt\, \psi_j(t)\, h(t)\, \phi_j(s),\tag{10.62}$$

or, as operator equation,

$$h = \lambda \mathbf{L}^\tau h = \lambda \mathbf{K}^\tau h - \sum_{j=1}^{p} (\psi_j, h) \phi_j.\tag{10.63}$$

Inserting here $h = \psi_{p+k}$ with $k = 1, \ldots, q - p$ gives, using (10.51),

$$\lambda \mathbf{L}^\tau \psi_{p+k} = \lambda \mathbf{K}^\tau \psi_{p+k} - \sum_{j=1}^{p} (\psi_j, \psi_{p+k}) \phi_j$$

$$= \lambda \mathbf{K}^\tau \psi_{p+k} = \psi_{p+k}.\tag{10.64}$$

Thus $\psi_{p+k} \neq 0$ is a solution of $h = \lambda \mathbf{L}^\tau h$, in contradiction to part (ii). The case $q < p$ is treated correspondingly. $\square$

We have obtained so far:

– if $D(\lambda) = 0$ then $f = \lambda \mathbf{K} f$ has solutions $f \neq 0$,
– if $D(\lambda) \neq 0$ then $f = \lambda \mathbf{K} f$ has as only solution $f = 0$,
– if $D(\lambda) \neq 0$ then $f = g + \lambda \mathbf{K} f$ has solution $f = g + \lambda \mathbf{R}(\lambda; \mathbf{K}) g$,

and turn now to the missing and most interesting case:

– if $D(\lambda) = 0$, find solutions of $f = g + \lambda \mathbf{K} f$.

Since $D(\lambda) = 0$, the first case, together with Theorem 4, establishes the existence of functions $h \neq 0$ such that $h = \lambda \mathbf{K}^\tau h$. Then

$$(h, f) = (h, g) + (h, \lambda \mathbf{K} f) = (h, g) + (\lambda \mathbf{K}^\tau h, f) = (h, g) + (h, f), \quad (10.65)$$

and therefore

$$\boxed{(h, g) = 0} \tag{10.66}$$

This gives the following theorem.

Theorem 5 *For $D(\lambda) = 0$, let $\psi_i \neq 0$ with $i = 1$ to p be eigenfunctions of $\mathbf{K}^\tau$ to characteristic value λ, $\psi_i = \lambda \mathbf{K}^\tau \psi_i$. Then for $f = g + \lambda \mathbf{K} f$ to have a solution $f \neq 0$, $(h, g) = 0$ must hold. Therefore, $f = g + \lambda \mathbf{K} f$ does not generally have a solution if $D(\lambda) = 0$.*

According to Theorem 4, the homogeneous equation $f = \lambda \mathbf{K} f$ has a solution $f \neq 0$ if and only if $h = \lambda \mathbf{K}^\tau h$ has a solution $h \neq 0$. The results so far can then be summarized in the *Fredholm alternative* (see Smirnow 1988, p. 35):

(1) Either $f_0 = \lambda \mathbf{K} f_0$ has only the trivial solution $f_0 = 0$,
    and $f = g + \lambda \mathbf{K} f$ has a solution $f \neq 0$ for any continuous g
(2) or $f_0 = \lambda \mathbf{K} f_0$ has nontrivial solutions $f_0 \neq 0$,
    and $f = g + \lambda \mathbf{K} f$ has a solution only if $(h_0, g) = 0$,
    where h_0 is any solution of $h_0 = \lambda \mathbf{K}^\tau h_0$.

Note that there is no reference to the infinite determinant D in statements (1) and (2); however, statement (1) corresponds to $D \neq 0$, where the homogeneous equation has no nontrivial solution, and the inhomogeneous equation has always a solution determined by the resolvent of $\mathbf{K}$; and statement (2) corresponds to $D = 0$, where the homogeneous equation has nontrivial solution(s) but the inhomogeneous equation only under the condition $(h_0, g) = 0$. All this is in agreement with corresponding statements for finite systems of linear equations. The alternative (2), $D = 0$, is more relevant to physics, where indeed nontrivial solutions exist in the homogeneous limit of the integral equation.

Finally, for this interesting case (2), one can give an explicit formula for the solution f of $f = g + \lambda \mathbf{K} f$ with $D(\lambda) = 0$. The trick is to use the operator $\mathbf{L}$ introduced in Theorem 4.

Theorem 6 *Let $D(\lambda) = 0$, and let $\phi_i = \lambda \mathbf{K} \phi_i$ and $\psi_i = \lambda \mathbf{K}^\tau \psi_i$ for $i = 1, \ldots, p$. Then $(\psi_i, g) = 0$ for $i = 1$ to p is sufficient for the solvability of the equation $f = g + \lambda \mathbf{K} f$, and a solution is given by*

$$f = g + \lambda \mathbf{R}(\lambda; \mathbf{L}) g, \tag{10.67}$$

where $\mathbf{R}(\lambda; \mathbf{L})$ *is the resolvent of the operator* $\mathbf{L}$ *corresponding to the kernel* $L(s, t)$ *introduced in (10.52).*

Proof The operator $\mathbf{L}$ has no characteristic value and no eigenfunction $\neq 0$ (Theorem 4). Thus the Fredholm determinant of $\mathbf{L}$ does not vanish (Theorem 3), which implies that the resolvent $\mathbf{R}(\lambda; \mathbf{L})$ exists (Theorem 2). Therefore, according to Theorem 1, for any g the solution of the inhomogeneous equation

$$f = g + \lambda \mathbf{L} f \tag{10.68}$$

is

$$f = g + \lambda \mathbf{R}(\lambda; \mathbf{L}) g. \tag{10.69}$$

To see that this f solves $f = g + \lambda \mathbf{K} f$, write out $\lambda \mathbf{L}$ in (10.68), giving

$$f = g + \lambda \mathbf{K} f - \sum_{j=1}^{p} \psi_j (\phi_j, f). \tag{10.70}$$

Therefore, with $i = 1$ to p, and using $(\psi_i, g) = 0$ and (10.21), one obtains

$$(\psi_i, f) = (\psi_i, g) + (\psi_i, \lambda \mathbf{K} f) - \sum_{j=1}^{p} (\psi_i, \psi_j)(\phi_j, f)$$

$$= (\lambda \mathbf{K}^{\tau} \psi_i, f) - \sum_{j=1}^{p} \delta_{ij} (\phi_j, f)$$

$$= (\psi_i, f) - (\phi_i, f), \tag{10.71}$$

and thus, for $i = 1$ to p,

$$(\phi_i, f) = 0. \tag{10.72}$$

Using this in (10.70) gives

$$f = g + \lambda \mathbf{K} f, \tag{10.73}$$

and f in (10.69) is indeed a solution of this equation. $\square$

Corollary *With the assumptions of the theorem, the general solution of*

$$f = g + \lambda \mathbf{K} f \tag{10.74}$$

is

$$\boxed{f = g + \lambda \mathbf{R}(\lambda; \mathbf{L}) g + \sum_{j=1}^{p} c_j \phi_j} \tag{10.75}$$

with **L** *from above and arbitrary real numbers* c_j.

The proof is by simple insertion into the integral equation.

Theorem 6 is due to Schmidt (1907). He also shows that for kernels of the form

$$K(s, t) = u_1(s)\, v_1(t) + \cdots + u_n(s)\, v_n(t), \tag{10.76}$$

the Fredholm determinants are not required. For Theorems 1 to 6, see also Hellinger and Toeplitz (1927), pp. 1370–1374.

In modern texts, the theory of this section is formulated in terms of *linear, continuous*, and *compact* operators (the kernel integrals) acting on the Banach spaces of continuous functions f, see, e.g., Dieudonné (1969), Chap. 11, and Hutson and Pym (1980). Since this requires familiarity with *relative compactness*, we have followed here the traditional approach using the Fredholm integral equations.

10.4 Schmidt's Nonlinear Integral Equation

In the existence proof for weakly nonlinear water waves, a fundamental theorem by Schmidt (1908) on the existence of solutions of nonlinear integral equations is used. This theorem is proved here, following Chap. 1 of Lichtenstein (1931) with some simplifications. In modern accounts, the existence of solutions of nonlinear operator equations is proved instead using Schauder's fixed-point theorem (or variants thereof), discussed in most texts on nonlinear functional analysis (Dunford and Schwartz 1958, p. 456; Hutson and Pym 1980, p. 207; Krasnoselskii 1964a, p. 67, 1964b, p. 124; Schwartz 1969, p. 96). The proof of Schauder's theorem, however, is more intricate than that of Schmidt's theorem, which suffices for the present purposes.

Let f and g be continuous functions of one real argument, where g is a known function and f is unknown. The theorem of Schmidt (1908) applies to *nonlinear* functions U_{mn} of f and g of the form

$$U_{mn}(s) = \int_a^b dt\, K_{mn}(s, t)\, f^m(t)\, g^n(t), \tag{10.77}$$

with a continuous kernel K_{mn} and $[a, b]$ again a closed and bounded real interval (note that f and g in (10.77) have the same argument t). Expressions of the form (10.77) are typically obtained from the Taylor expansions of given functions specific to the considered application of the theory.

The constant term U_{00} is of no further interest. The linear term U_{10} containing $f^1 g^0$ belongs to a linear integral equation, as was considered in the foregoing section and will be taken up again in the next section; here, this term is dropped. The inhomogeneous linear term U_{01} containing $f^0 g^1$ will be considered in the present section, but separate from the nonlinear terms. Thus, a sum over the U_{mn} in (10.77) is performed with $m + n > 1$ (up to infinity in both m and n), containing fg, f^2, g^2, and higher products of f and g. Note that fg is linear in f, but of second order in products of f and g, and therefore relevant here.

A norm on functions is introduced by

$$\| f \| = \max_s |f(s)|, \tag{10.78}$$

and similarly for U_{mn}. Since $[a, b]$ is closed, the maximum is attained in this interval (some books use instead the supremum in 10.78). A norm on operators is defined by

$$\| K_{mn} \| = \max_s \int_a^b dt \, |K_{mn}(s, t)|. \tag{10.79}$$

Let $A > 0$ and $B > 0$ be two real numbers such that the infinite double series

$$\sum_{m+n>1} \| K_{mn} \| \, A^m B^n \tag{10.80}$$

converges. Let

$$\| f \| \le A, \quad \| g \| \le B. \tag{10.81}$$

Then for any s,

$$\sum_{m+n>1} |U_{mn}(f, g)(s)| \le \sum_{m+n>1} \int_a^b dt \, |K_{mn}(s, t)| \, |f^m(t)| \, |g^n(t)|$$

$$\le \sum_{m+n>1} \| K_{mn} \| \, A^m B^n = W(A, B). \tag{10.82}$$

Here W is a differentiable function of A and B, with

$$W(0, 0) = W_A(0, 0) = W_B(0, 0) = 0, \tag{10.83}$$

where $W_A = \partial W/\partial A$, etc. Using Taylor's theorem up to second-order terms and the mean-value theorem of calculus, one has

$$W(A, B) = \tfrac{1}{2} A^2 W_{AA}(qA, qB) + ABW_{AB}(qA, qB) + \tfrac{1}{2} B^2 W_{BB}(qA, qB) \tag{10.84}$$

for some $0 < q < 1$. Let

$$\alpha = \max\!\left(\tfrac{1}{2} |W_{AA}(qA, qB)|, |W_{AB}(qA, qB)|, \tfrac{1}{2} |W_{BB}(qA, qB)|\right). \tag{10.85}$$

Note that this depends on the kernels K_{mn} and the numbers A and B, but not on g or even the unknown f. Then

$$\boxed{\sum_{m+n>1} |U_{mn}(f, g)(s)| \leq W(A, B) \leq \alpha(A^2 + AB + B^2)}$$ (10.86)

Let $\hat{f}$ be another function with

$$\| \hat{f} \| \leq A.$$ (10.87)

Then (10.86) holds also with f replaced by $\hat{f}$. Furthermore, for all s,

$$\sum_{m+n>1} |U_{mn}(f, g)(s) - U_{mn}(\hat{f}, g)(s)|$$

$$\leq \sum_{m+n>1} \int_a^b dt \ |K_{mn}(s, t)| \ |f^m(t) - \hat{f}^m(t)| \ |g^n(t)|$$

$$\stackrel{(a)}{=} \sum_{m+n>1} \int_a^b dt \ |K_{mn}(s, t)| \ |f(t) - \hat{f}(t)| \times$$

$$\times |f^{m-1}(t) + f^{m-2}(t)\hat{f}(t) + \cdots + f(t)\hat{f}^{m-2}(t) + \hat{f}^{m-1}(t)| \ |g^n(t)|$$

$$\stackrel{(b)}{\leq} \sum_{m+n>1} \| K_{mn} \| \ m A^{m-1} B^n \ \| f - \hat{f} \|$$

$$\stackrel{(c)}{=} W_A(A, B) \ \| f - \hat{f} \| .$$ (10.88)

The factorization in (a) may not seem obvious but is trivially checked to be correct. Regarding (b), the expression $Q = |f^{m-1} + f^{m-2}\hat{f} + \cdots + f \hat{f}^{m-2} + \hat{f}^{m-1}|$ contains m terms, each of total power $m - 1$ in powers of f and $\hat{f}$. According to (10.81) and (10.87), thus $Q \leq m A^{m-1}$. Also according to (10.81), $\| g \| \leq B$, and one obtains the estimate in (b). Finally, step (c) is a direct consequence of the definition of the function $W(A, B)$ in (10.82). Again after the Taylor expansion and using the mean-value theorem,

$$W_A(A, B) = A W_{AA}(\bar{q} A, \bar{q} B) + B W_{AB}(\bar{q} A, \bar{q} B)$$ (10.89)

for some $0 < \bar{q} < 1$. Let

$$\beta = \max(|W_{AA}(\bar{q} A, \bar{q} B)|, |W_{AB}(\bar{q} A, \bar{q} B)|).$$ (10.90)

Then

$$\boxed{\sum_{m+n>1} |U_{mn}(f, g) - U_{mn}(\hat{f}, g)| \leq \beta(A + B) \| f - \hat{f} \|}$$ (10.91)

For these considerations, see Lichtenstein (1931), pp. 1–4 and Wainberg and Trenogin (1973), pp. 82–84. The latter authors discuss that because of (10.86) and (10.91), U_{mn} is a contracting operator (their Theorems 7.2 and 7.3), which also lies at the heart of the proof by Littman (1957). The nonlinear integral equation and a convergent iteration scheme to obtain its solution can be formulated now.

Theorem 7 (Schmidt 1908; Lichtenstein 1931) *Consider the nonlinear integral equation*

$$\boxed{f(s) = \int_a^b dt\ K_{01}(s,t)\ g(t) + \sum_{m+n>1} \int_a^b dt\ K_{mn}(s,t)\ f^m(t)\ g^n(t)} \qquad (10.92)$$

with known continuous functions g *and* $K_{01}, K_{20}, K_{11}, K_{02}, \dots$ *(no* K_{10}*). Define a sequence* f_k *of successive approximations by*

$$f_0(s) = U_{01}(s),$$

$$f_k(s) = U_{01}(s) + \sum_{m+n>1} \int_a^b dt\ K_{mn}(s,t)\ f_{k-1}^m(t)\ g^n(t) \qquad (10.93)$$

for $k \geq 1$. *Let* $A > 0$ *and* $B > 0$ *be such that the series*

$$\sum_{m+n>1} \| K_{mn} \|\ A^m B^n \qquad (10.94)$$

converges. Then for sufficiently small $\| g \|$*, the following statements hold:*

(i) *the* f_k *converges uniformly to a solution* f *of (10.92),*
(ii) *this solution* f *is unique,*
(iii) $\| f \| \to 0$ *if* $\| g \| \to 0$.

Comments

1. A linear term $\int_a^b dt\ K_{10}(s,t)\ f(t)$ is only added to the integral equation (10.92) in the next section.
2. The theorem holds for sufficiently small $\| g \|$. This is why it can be used to prove the existence of weakly nonlinear waves only.

Proof (i) With g and K_{01} specified in the statement of the theorem, let

$$\bar{B} = \max\ (\| g \|,\ \| U_{01} \|). \qquad (10.95)$$

From the kernels K_{mn} and numbers A and B given in (10.94), determine a number $\alpha > 0$ according to (10.85). Define $\bar{A}$ by the quadratic equation

$$\boxed{\bar{A} = \bar{B} + \alpha(\bar{A}^2 + \bar{A}\bar{B} + \bar{B}^2)} \qquad (10.96)$$

The solutions of (10.96) are

$$\bar{A}_\pm = \frac{1}{2\alpha} - \frac{\bar{B}}{2} \pm \left[\left(\frac{1}{2\alpha} - \frac{\bar{B}}{2} \right)^2 - \bar{B} \left(\frac{1}{\alpha} + \bar{B} \right) \right]^{1/2}. \qquad (10.97)$$

Choose the smaller one, and assume $\alpha\bar{B} \ll 1$, i.e., small $\bar{B}$. The Taylor expansion to second order gives (writing $\bar{A}$ instead of $\bar{A}_-$)

$$\bar{A} = \bar{B} + 3\alpha\bar{B}^2 + \cdots. \tag{10.98}$$

Thus

$$\bar{A} > \bar{B} \tag{10.99}$$

and

$$\bar{A} \to 0 \qquad \text{if} \qquad \bar{B} \to 0. \tag{10.100}$$

U_{01} depends linearly on g, and thus its norm becomes arbitrarily small with g. Assume that $\|g\|$ is sufficiently small, so that

$$\bar{B} \le B,$$
$$\bar{A} \le A. \tag{10.101}$$

Then (10.86) can be applied to the iteration scheme (10.93), giving

$$f_0(s) \le \bar{B} < \bar{A},$$
$$f_1(s) \le \bar{B} + \alpha(\bar{A}^2 + \bar{A}\bar{B} + \bar{B}^2) = \bar{A},$$
$$f_2(s) \le \bar{B} + \alpha(\bar{A}^2 + \bar{A}\bar{B} + \bar{B}^2) = \bar{A}, \tag{10.102}$$

and so forth; f_{k-1} is bounded by $\bar{A}$, so f_k is bounded by $\bar{A}$, as is f itself. Furthermore, for $k \ge 2$ and for arbitrary s, one has according to (10.91),

$$|f_k(s) - f_{k-1}(s)| = \left| \sum_{m+n\ge2} \left(U_{mn}(f_{k-1}, g)(s) - U_{mn}(f_{k-2}, g)(s) \right) \right|$$
$$\le \beta(\bar{A} + \bar{B}) \, \|f_{k-1} - f_{k-2}\| \,. \tag{10.103}$$

Choose $\|g\|$ and thus $\bar{A}$ and $\bar{B}$ sufficiently small so that

$$|f_k(s) - f_{k-1}(s)| \le \eta \, \|f_{k-1} - f_{k-2}\| \tag{10.104}$$

with

$$\eta < 1. \tag{10.105}$$

Since this holds for all s, one has

$$\| f_k - f_{k-1} \| \le \eta \, \| f_{k-1} - f_{k-2} \| \,. \tag{10.106}$$

Therefore, the infinite series

$$f = f_0 + (f_1 - f_0) + (f_2 - f_1) + \cdots \tag{10.107}$$

converges uniformly (independent of s), and

$$f_k \to f. \tag{10.108}$$

Taking this limit in the scheme (10.93) yields

$$f(s) = g(s) + \sum_{m+n \geq 2} \int_a^b dt\, K_{mn}(s, t)\, f^m(t)\, g^n(t), \tag{10.109}$$

thus $f(s)$ is solution of the integral equation (10.92). From (10.102),

$$\| f \| \leq \bar{A}. \tag{10.110}$$

(ii) Let $\hat{f}$ be another solution of (10.92). For $\hat{f}$, repeat the steps that lead to (10.106). Consider the equations (the first is the integral equation, the second the iteration scheme 10.93)

$$\hat{f}(s) = U_{01}(s) + \sum_{m+n \geq 2} \int_a^b dt\, K_{mn}(s, t)\, \hat{f}^m(t)\, g^n(t),$$

$$f_k(s) = U_{01}(s) + \sum_{m+n \geq 2} \int_a^b dt\, K_{mn}(s, t)\, f_{k-1}^m(t)\, g^n(t). \tag{10.111}$$

Subtracting the equations and applying (10.91) with $\beta(\bar{A} + \bar{B}) = \eta < 1$ yields

$$|\hat{f}(s) - f_1(s)| \leq \eta |\hat{f}(s) - f_0(s)|,$$
$$|\hat{f}(s) - f_2(s)| \leq \eta |\hat{f}(s) - f_1(s)| \leq \eta^2 |\hat{f}(s) - f_0(s)|,$$
$$\cdots$$
$$|\hat{f}(s) - f_k(s)| \leq \eta^k |\hat{f}(s) - f_0(s)|. \tag{10.112}$$

Thus

$$f_k \to \hat{f}. \tag{10.113}$$

Since from (10.108) also $f_k \to f$, it follows that $f = \hat{f}$.
(iii) From (10.95) and since U_{01} is linear in g,

$$\bar{B} \to 0 \qquad \text{if} \qquad \| g \| \to 0. \tag{10.114}$$

Therefore by (10.100) and (10.110),

$$\| f \| \to 0 \quad \text{if} \quad \bar{A} \to 0 \quad \text{if} \quad \bar{B} \to 0 \quad \text{if} \quad \| g \| \to 0, \quad (10.115)$$

which completes the proof.   □

This proof is due to Lichtenstein (1931, §2) and can also be found in Wainberg and Trenogin (1973), pp. 85–87. The original proof by Schmidt (1908), pp. 381–384, contains an interesting convergence argument but is rather technical. Instead of successive approximations, Schmidt uses a series expansion for f.

10.5   General Nonlinear Integral Equations

Section 10.3 dealt with linear integral equations, Sect. 10.4 with a purely nonlinear integral equation, and thus the next step is to combine both, i.e., to include a linear term $\lambda \int \mathrm{d}t\, K_{10}\, f$ in the nonlinear equation (10.92), giving

$$f(s) = \lambda \int_a^b \mathrm{d}t\, K_{10}(s, t)\, f(t) + \int_a^b \mathrm{d}t\, K_{01}(s, t)\, g(t)$$

$$+ \sum_{m+n>1} \int_a^b \mathrm{d}t\, K_{mn}(s, t)\, f^m(t)\, g^n(t). \quad (10.116)$$

Here, λ is included (somewhat arbitrarily) only in the linear term. As in (10.77), let

$$U_{mn}(s) = \int_a^b \mathrm{d}t\, K_{mn}(s, t)\, f^m(t)\, g^n(t), \quad (10.117)$$

and in the following, the nonlinear function $\sum_{m+n>1} U_{mn}(s)$ of argument s is abbreviated ΣU_{mn}. The basic idea is then to read the operator equation

$$f = \lambda \mathbf{K}_{10} f + \mathbf{K}_{01} g + \Sigma U_{mn} \quad (10.118)$$

as

$$f = \lambda \mathbf{K}_{10} f + \hat{g}, \quad (10.119)$$

with $\hat{g} = \mathbf{K}_{01} g + \Sigma U_{mn}$, and to apply the theory of linear integral equations to (10.119).

The two theorems of the present section appear somewhat unwieldy, but they are rather obvious combinations of the results obtained for linear and purely nonlinear integral equations. Since the integral equation is now necessarily inhomogeneous due to the presence of nonlinear terms, only two cases need to be considered: first, that a resolvent of the linear part of the integral equation exists and allows to directly obtain an implicit solution of the nonlinear integral equation, similar to Theorem 1 of Sect. 10.3 (see Theorem 8). Second, that no such resolvent exists; then one proceeds as in Theorem 6 of Sect. 10.3, introducing the operator $\mathbf{L}$ (see Theorem 9).

Theorem 8 *Let* λ *not be a characteristic value of* $\mathbf{K}_{10}$ *and let* $\mathbf{R}(\lambda, \mathbf{K}_{10})$ *be the resolvent of* $\mathbf{K}_{10}$. *Then the equation*

$$f = \lambda \mathbf{K}_{10} f + \mathbf{K}_{01} g + \Sigma U_{mn} \tag{10.120}$$

has, for sufficiently small $\|\mathbf{K}_{01}g\|$ *and* $\|\Sigma U_{mn}\|$, *the implicit solution*

$$f = \left[\mathbf{1} + \lambda \mathbf{R}(\lambda, \mathbf{K}_{10}) \right] \left(\mathbf{K}_{01} g + \Sigma U_{mn} \right), \tag{10.121}$$

where $\mathbf{1}$ *is the identity operator,* $\mathbf{1} f = f$ *and* $\mathbf{1K} = \mathbf{K}$. *Equation (10.121) is an implicit solution, since* f *also appears in* ΣU_{mn} *on the right side;* f *can be obtained from (10.121) by successive approximations.*

Proof (Lichtenstein 1931, p. 15). With $\hat{g} = \mathbf{K}_{01} g + \Sigma U_{mn}$, the solution of $f = \hat{g} + \lambda \mathbf{K}_{10} f$ is $f = \hat{g} + \lambda \mathbf{R}(\lambda, \mathbf{K}_{10}) \hat{g}$. Inserting $\hat{g}$ gives (10.121). □

Theorem 9 *For given characteristic value* λ *of* $\mathbf{K}_{10}$, *let* $f = \lambda \mathbf{K}_{10} f$ *have solutions (eigenfunctions)* ϕ_1 *to* ϕ_p, *and* $h = \lambda \mathbf{K}_{10}^{\tau} h$ *have solutions* ψ_1 *to* ψ_p. *Then for*

$$f = \lambda \mathbf{K}_{10} f + \mathbf{K}_{01} g + \Sigma U_{mn} \tag{10.122}$$

to have a solution, the relations

$$(\psi_i, \mathbf{K}_{01} g + \Sigma U_{mn}) = 0 \tag{10.123}$$

must hold for $i = 1$ *to* p. *For sufficiently small* $\|\mathbf{K}_{01}g\|$ *and* $\|\Sigma U_{mn}\|$, *the implicit solution of (10.122) is*

$$f = \left[\mathbf{1} + \lambda \mathbf{R}(\lambda, \mathbf{L}_{10}) \right] \left(\mathbf{K}_{01} g + \Sigma U_{mn} + \sum_{j=1}^{p} \psi_j (\phi_j, f) \right), \tag{10.124}$$

where $\mathbf{R}(\lambda; \mathbf{L}_{10})$ *is the resolvent of the operator* $\mathbf{L}_{10}$ *defined in (10.52).*

Proof For $f = \hat{g} + \lambda \mathbf{K}_{10} f$ with $\hat{g} = \mathbf{K}_{01} g + \Sigma U_{mn}$ to have a solution, (10.66) must hold, $(\psi_i | \hat{g}) = 0$ for $i = 1$ to p, which is (10.123). To obtain a solution of (10.122), write

$$
\begin{aligned}
f &= \mathbf{K}_{01} g + \Sigma U_{mn} + \lambda \mathbf{K}_{10} f \\
&= \mathbf{K}_{01} g + \Sigma U_{mn} + \lambda \mathbf{L}_{10} f + \sum_{j=1}^{p} \psi_j (\phi_j, f) \\
&= \tilde{g} + \lambda \mathbf{L}_{10} f, \tag{10.125}
\end{aligned}
$$

with

$$\tilde{g} = \mathbf{K}_{01} g + \Sigma U_{mn} + \sum_{j=1}^{p} \psi_j(\phi_j, f). \tag{10.126}$$

The solution of (10.125) is $f = \tilde{g} + \lambda \mathbf{R}(\lambda; \mathbf{L}_{10}) \, \tilde{g}$, with resolvent $\mathbf{R}$ of $\mathbf{L}_{10}$. Inserting $\tilde{g}$ gives (10.124). $\qquad \square$

10.6 Integral Equations for Nonlinear Waves

We turn to the existence proof for weakly nonlinear waves. The boundary condition at the free surface is used to transform Green's integral theorem into an integral equation, to which the above existence theorems apply. This section follows Chap. 2 in the book by Lichtenstein (1931).

The complex variables z, $\mathbb{O}$, $\omega = \theta + i\tau$, and $w = ce^{-i\omega}$ have the same meaning as in Sect. 10.2. The free surface of the wave is determined by the two functions θ, the angle of the horizontal plane with the tangent to the waveform, and τ, the logarithm of the absolute speed.

Boundary Condition Along Unit Circle

The wave is periodic, with wavelength λ, so that $\mathbb{O}(z + \lambda) = \mathbb{O}(z)$. This suggests to introduce an angle variable with periodicity 2π. More specifically (see Levi-Civita 1925, Sect. 9), the fluid body in the *upper* half-plane $\psi > 0$ of the complex $\mathbb{O}$-plane is conformally mapped to the interior of the unit circle centered at the origin of a new, complex ζ-plane. The polar representation of ζ is

$$\zeta = re^{i\varphi}. \tag{10.127}$$

Note that the polar coordinates r and φ do not refer to the z-plane of the waveform $y(x)$, but to the ζ-plane. To avoid proportionality constants, it is assumed that the wavelength is normalized to $\lambda = 2\pi$, and the wave speed to $c = 1$. Then the mapping from $\mathbb{O}$ to ζ is simply given by (see Eq. 10.3 in Levi-Civita 1925),

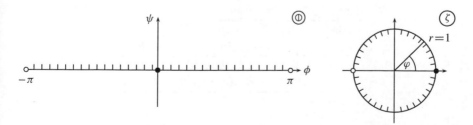

Fig. 10.1 Conformal mapping $\zeta = e^{i\mathbb{O}}$ (with polar representation $\zeta = re^{i\varphi}$). The upper half-plane of the $\mathbb{O}$-plane is mapped conformally to the interior of the unit circle in the ζ-plane

Table 10.1 Complex variables for weakly nonlinear waves

$z = x + iy$		Waveform $y(x)$
$\Phi = \phi + i\psi$		Speed potential
$w = u - iv$	$w = d\Phi/dz$	Complex speed
$\omega = \theta + i\tau$	$w = ce^{-i\omega}$	Angle of surface; Speed magnitude
$\zeta = re^{i\varphi}$	$\zeta = e^{i\Phi}$	Interior of unit circle

$$\zeta = e^{i\Phi} = e^{-\psi}\, e^{i\phi}, \tag{10.128}$$

and this change in independent variables from Φ to ζ is shown in Fig. 10.1. The relevant function is then $\omega(\zeta)$, or its real and imaginary parts

$$\theta(r, \varphi), \qquad\qquad \tau(r, \varphi). \tag{10.129}$$

All complex variables used in the following are listed in Table 10.1.

From (10.127), the circle $|\zeta| = r = 1$ corresponds to the fluid surface with $\psi = 0$, and furthermore

$$\varphi = \phi, \qquad\qquad \frac{\partial}{\partial \varphi} = \frac{\partial}{\partial \phi}, \tag{10.130}$$

so that the nonlinear boundary condition (10.11) in Sect. 10.3 becomes

$$\frac{\partial \tau}{\partial \varphi} = \lambda\, e^{-3\tau}\, \sin\theta, \tag{10.131}$$

with $\lambda = g/c^3$. To replace the left side by the more relevant derivative $\partial\theta/\partial r$, one has to take some care with signs in the Cauchy–Riemann equations. With

$$\ln(w/c) = \tau - i\theta,$$
$$\ln\zeta = \ln r + i\varphi, \tag{10.132}$$

and assuming that $w(\zeta)$ is analytic, one has on the unit circle $r = 1$,

$$\begin{aligned}
\frac{d\ln(w/c)}{d\ln\zeta} &= \frac{d\tau - id\theta}{r^{-1}\,dr + id\varphi} \\
&= \frac{\tau_r dr - i\tau_\varphi id\varphi - i\theta_r dr - i\theta_\varphi d\varphi}{dr + id\varphi} \\
&= \frac{(\tau_r - i\theta_r)dr + i(-\theta_\varphi - i\tau_\varphi)d\varphi}{dr + id\varphi}.
\end{aligned} \tag{10.133}$$

(In the second line, $r = 1$ on the unit circle was used.) The two brackets in the last equation have to be equal, in order that $dr + id\varphi$ cancels in the numerator and

denominator, making the differential quotient independent of the direction in the
ζ-plane; thus the Cauchy–Riemann equations on the unit circle $r = 1$ are

$$\frac{\partial \tau}{\partial r} = -\frac{\partial \theta}{\partial \varphi},$$

$$\frac{\partial \tau}{\partial \varphi} = \frac{\partial \theta}{\partial r}, \tag{10.134}$$

which means that due to the minus sign in $\ln(w/c)\tau - i\theta$, the plus and minus signs in
the Cauchy–Riemann equations exchange their usual places. The differential quotient
of the tangent angle θ of the wave with respect to the direction r normal to the circle
$r = 1$ in the ζ-plane has thus to obey the Neumann boundary condition

$$\left.\frac{\partial \theta(r, \varphi)}{\partial r}\right|_{r=1} = \lambda e^{-3\tau(1,\varphi)} \sin(\theta(1, \varphi)). \tag{10.135}$$

(Lichtenstein, p. 48 uses here the *inward* normal to the circle instead of r.) In the
following, the constant λ will become the characteristic value in an integral equation.

The wave profile is assumed to be symmetric, $z(-\textcircled{0}) = z(\textcircled{0})$. This implies that
the slope θ is an odd function, and the logarithm of the speed τ is an even function
of ϕ (in the $\textcircled{0}$-plane) and of φ (in the ζ-plane),

$$\theta(r, -\varphi) = -\theta(r, \varphi),$$

$$\tau(r, -\varphi) = \tau(r, \varphi), \tag{10.136}$$

hence $\theta(r, 0) = 0$ (no surface slope at the maximum elevation of the wave).

The Integral Equation

With the boundary condition (10.135), Green's integral theorem for harmonic functions along the boundary $r = 1$ of the unit circle becomes an integral equation for
$\theta(1, \varphi)$. We repeat the steps from (3.37) to (3.39) in Sect. 3.6 for the Green's function
G of the 2-D Laplace equation in the Euclidean plane with position vector $\vec{r}$,

$$\Delta G(\vec{r}) = -\delta(\vec{r}),$$

$$G(\vec{r}) = -\frac{1}{2\pi} \ln |\vec{r}|. \tag{10.137}$$

The point $\vec{r}$ shall lie on the unit circle, i.e., the domain boundary over which the line
integral is performed in (3.37), which leads to a singularity at $\vec{r}' = \vec{r}$. This singularity
poses no problems if some care is applied.

Let $B_\varepsilon(\vec{r})$ be a disk with infinitesimal radius ε and center $\vec{r}$. In the domain $D \backslash B_\varepsilon$
with boundary $C \cup \bar{C}$ (see Fig. 10.2), both θ and G are harmonic, the latter since
$\vec{r} - \vec{r}' \neq 0$, where $\vec{r}' \in C \cup \bar{C}$. Thus the left side of Green's integral theorem (3.18)
in Sect. 3.2 vanishes, and one has, with polar coordinates (r, φ) and (r', φ') of $\vec{r}$ and
$\vec{r}'$, respectively, and dividing by 2π for convenience,

$$0 = \frac{1}{2\pi} \oint_{C \cup \bar{C}} dl' \left[\theta(1, \varphi') \frac{\partial \ln |\vec{r}' - \vec{r}|}{\partial r'} - \frac{\partial \theta(r', \varphi')}{\partial r'} \ln |\vec{r}' - \vec{r}| \right]. \qquad (10.138)$$

The integration over $\bar{C}$ can be extended to the semicircle limited by the dashed vertical line in Fig. 10.2, since the length difference between the curve $\bar{C}$ and this semicircle is of second order ('curvature'): to first, linear order, the part of the unit circle crossing through $B_\varepsilon(\vec{r})$ is replaced by a straight diameter of length 2ε and $\bar{C}$ becomes indeed a semicircle. As in Eq. (3.38) in Sect. 3.6, the integral over $\bar{C}$ in (10.138) becomes then

$$\frac{1}{2\pi} \int_{-\pi/2}^{-3\pi/2} \varepsilon \, d\varphi \left[\frac{\theta(1, \varphi)}{\varepsilon} - \frac{\partial \theta(r', \varphi)}{\partial r'} \Big|_{r'=\varepsilon} \ln \varepsilon \right] = -\frac{1}{2} \theta(1, \varphi), \qquad (10.139)$$

where for the continuous function θ the value on the boundary $\bar{C}$ was replaced by $\theta(1, \varphi)$ at $\vec{r}$. The second term in the bracket on the left side of (10.139) vanishes for ε approaching zero, since $\lim_{\varepsilon \to 0} \varepsilon \ln \varepsilon = 0$. Thus in (10.138)

$$\frac{1}{2} \theta(1, \varphi) = \frac{1}{2\pi} \int_C dl' \left[\theta(1, \varphi') \frac{\partial \ln |\vec{r}' - \vec{r}|}{\partial r'} - \frac{\partial \theta(r', \varphi')}{\partial r'} \ln |\vec{r}' - \vec{r}| \right]_{r,r'=1}. \qquad (10.140)$$

The first integral on the right side is a classical expression from potential theory (e.g., Sternberg 1925; Neumann 1877, Chap. 4) and corresponds to the potential from a dipole layer in the plane. This integral has a discontinuity, when the point $\vec{r}$ crosses the unit circle, see the theorem below. We are interested here in the case that $\vec{r}$ lies *on* the circle. Then, using $dl' = d\varphi'$ on the unit circle, and without any limit from the interior of the disk to its boundary circle,

$$\frac{1}{2\pi} \int_C d\varphi' \, \theta(1, \varphi') \frac{\partial \ln |\vec{r}' - \vec{r}|}{\partial r'} \Big|_{r,r'=1} = \frac{1}{2\pi} \int_C d\varphi' \, \theta(1, \varphi') \left[\hat{r}' \cdot \nabla' \ln |\vec{r}' - \vec{r}| \right]_{r,r'=1}$$

$$= \frac{1}{2\pi} \int_C d\varphi' \, \theta(1, \varphi') \left[\hat{r}' \cdot \frac{\vec{r}' - \vec{r}}{|\vec{r}' - \vec{r}|^2} \right]_{r,r'=1}$$

$$= \frac{1}{2\pi} \int_C d\varphi' \, \theta(1, \varphi') \frac{\cos \vartheta}{|\vec{r}' - \vec{r}|} \Big|_{r,r'=1}, \qquad (10.141)$$

Fig. 10.2 Boundary segments C and $\bar{C}$ of $D \backslash B_\epsilon(\vec{r})$ for $\vec{r}$ on ∂D

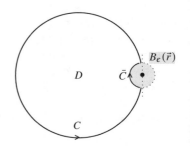

where ϑ is the angle between $\vec{r}'$ and $\vec{r}' - \vec{r}$, see Fig. 10.3. Also from this figure, $\cos \vartheta = |\vec{r}' - \vec{r}|/2$ since $r = r' = 1$ (and only then, not for the case $r < 1$), or

$$\left. \frac{\cos \vartheta}{|\vec{r}' - \vec{r}|} \right|_{r,r'=1} = \frac{1}{2}. \tag{10.142}$$

Note again that the point $\vec{r}$ is excluded from C, and thus a zero does not occur in the denominator of this expression. Since $\cos \vartheta / |\vec{r}' - \vec{r}|$ is a finite constant, the removal of $\vec{r}$ does not correspond to the removal of a singularity, which is denoted by a Cauchy principal value, but to an unproblematic removal of an infinitesimal number from the integral. Replacing the integral over the open curve C therefore by a full integral over 2π, one has

$$\frac{1}{2\pi} \oint_C d\varphi' \, \theta(1, \varphi') \left. \frac{\partial \ln |\vec{r}' - \vec{r}|}{\partial r'} \right|_{r,r'=1} = \frac{1}{4\pi} \int_0^{2\pi} d\varphi' \, \theta(1, \varphi') = 0, \tag{10.143}$$

since by assumption $\theta(r, \varphi)$ is an odd function of φ. Green's integral theorem in the form (10.140) becomes therefore

$$\frac{1}{2\pi} \int_0^{2\pi} d\varphi' \left[\frac{\partial \theta(r', \varphi')}{\partial r'} \ln |\vec{r}' - \vec{r}| \right]_{r,r'=1} = -\frac{1}{2} \theta(1, \varphi) \tag{10.144}$$

The above derivation was tailored to the present purpose, where $\vec{r}$ lies on the unit circle. To put this in a somewhat broader context of potential theory, a theorem is proved now on the jump in the potential when crossing a dipole layer. This, however, is not needed in the following treatment on water waves, and the reader can jump directly to (10.154).

Theorem *Let D be an arbitrary, simply connected, open domain D in the plane, with (piecewise) smooth boundary curve ∂D with normal vector $\hat{n}'$ pointing away from D. Let $\vec{r}' \in \partial D$ be arbitrary and $D_c = \mathbb{R}^2 \backslash (D \cup \partial D)$. Then*

$$\frac{1}{2\pi} \oint_{\partial D} dl' \, \frac{\partial \ln |\vec{r}' - \vec{r}|}{\partial n'} = \begin{cases} 1 & \text{if } \vec{r} \in D \\ \frac{1}{2} & \text{if } \vec{r} \in \partial D \\ 0 & \text{if } \vec{r} \in D_c. \end{cases} \tag{10.145}$$

Fig. 10.3 Definition of ϑ

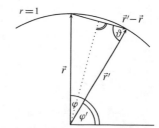

Comment Note that (10.141) together with (10.142) also gives $1/2$ for this integral, when D is the unit disk and $\vec{r} \in \partial D$, and if the function θ is put to unity.

Proof (Levandosky 2003; Schmidt 1910) Equation (10.137) is used throughout, $G(\vec{r}) = -(2\pi)^{-1} \ln |\vec{r}|$. Assume first that $\vec{r} \in D_c$. Then

$$\oint_{\partial D} dl' \, \frac{\partial}{\partial n'} \, G(\vec{r}' - \vec{r}) = \oint_{\partial D} dl' \, \hat{n}' \cdot \nabla' G(\vec{r}' - \vec{r})$$

$$= \int_D da' \, \Delta' G(\vec{r}' - \vec{r}) = 0, \tag{10.146}$$

using Gauss' theorem in the plane. In the last line, $\vec{r}' \in D$ and $\vec{r} \in D_c$, thus $|\vec{r}' - \vec{r}| > 0$ and $\Delta' G = 0$.

Assume next $\vec{r} \in D$, thus $|\vec{r}' - \vec{r}| = 0$ is possible. Let $B_\varepsilon(\vec{r})$ be a disk with radius ε centered at $\vec{r}$, which lies fully within D. On $D \backslash B_\varepsilon(\vec{r})$ one has $|\vec{r}' - \vec{r}| > 0$ and $\Delta G = 0$, giving

$$0 = \int_{D \backslash B_\varepsilon(\vec{r})} da' \, \Delta' G(\vec{r}' - \vec{r})$$

$$= \oint_{\partial(D \backslash B_\varepsilon(\vec{r}))} dl' \, \hat{n}' \cdot \nabla' G(\vec{r}' - \vec{r})$$

$$= \oint_{\partial D} dl' \, \hat{n}' \cdot \nabla' G(\vec{r}' - \vec{r}) + \oint_{\partial B_\varepsilon(\vec{r})} dl' \, \hat{n}' \cdot \nabla' G(\vec{r}' - \vec{r}). \tag{10.147}$$

One has

$$\nabla' G(\vec{r}' - \vec{r}) = -\frac{1}{2\pi} \, \nabla' \ln |\vec{r}' - \vec{r}| = -\frac{1}{2\pi} \, \frac{\vec{r}' - \vec{r}}{|\vec{r}' - \vec{r}|^2}. \tag{10.148}$$

Let $\vec{r}'$ lie on the boundary of the disk $B_\varepsilon(\vec{r})$. The unit normal of $\partial B_\varepsilon(\vec{r})$ pointing away from $D \backslash B_\varepsilon(\vec{r})$ (Fig. 10.4) is

$$\hat{n}' = -\frac{\vec{r}' - \vec{r}}{|\vec{r}' - \vec{r}|}. \tag{10.149}$$

Therefore,

$$\oint_{\partial D} dl' \, \hat{n}' \cdot \nabla' G(\vec{r}' - \vec{r}) = -\frac{1}{2\pi} \oint_{\partial B_\varepsilon(\vec{r})} dl' \, \frac{\hat{n}' \cdot \hat{n}'}{|\vec{r}' - \vec{r}|}$$

$$= -\frac{1}{2\pi} \oint_{\partial B_\varepsilon(\vec{r})} \frac{dl'}{|\vec{r}' - \vec{r}|} = -\frac{1}{2\pi \varepsilon} \oint_{\partial B_\varepsilon(\vec{r})} dl' = -1. \tag{10.150}$$

Fig. 10.4 Outer unit normal
to the boundary of the disk
$B_\varepsilon(\vec{r})$

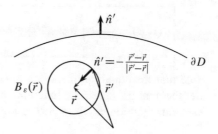

Finally, let both $\vec{r} \in \partial D$ and $\vec{r}' \in \partial D$. Let $B_\varepsilon(\vec{r})$ be a disk with radius ε and center $\vec{r}$.
Let C and $\bar{C}$ be the boundary segments of $D \backslash B_\varepsilon(\vec{r})$ as shown in Fig. 10.2 (the unit
circle replaced by any smooth boundary ∂D now). Then (10.147) is modified to

$$
0 = \int_{D \backslash B_\varepsilon(\vec{r})} da' \, \Delta' G(\vec{r}' - \vec{r})
$$
$$
= \int_C dl' \, \hat{n}' \cdot \nabla' G(\vec{r}' - \vec{r}) + \int_{\bar{C}} dl' \, \hat{n}' \cdot \nabla' G(\vec{r}' - \vec{r}). \tag{10.151}
$$

This gives, similar to (10.150),

$$
\int_C dl' \, \hat{n}' \cdot \nabla' G(\vec{r}' - \vec{r}) = -\frac{1}{2\pi} \int_{\bar{C}} \frac{dl'}{|\vec{r}' - \vec{r}|} = -\frac{1}{2\pi\varepsilon} \int_{\bar{C}} dl' = -\frac{1}{2} + O(\varepsilon),
$$
$$
\tag{10.152}
$$
where the integral over $\bar{C}$ was replaced, to first order in ε, by an integral over a
semicircle, see Fig. 10.2. For $\varepsilon \to 0$, the left side of (10.152) converges to the full
line integral along the closed curve ∂D, giving

$$
\oint_{\partial D} dl' \, \hat{n}' \cdot \nabla' G(\vec{r}' - \vec{r}) = -\frac{1}{2}, \tag{10.153}
$$

and the theorem is proved. $\qquad\qquad\qquad\qquad\qquad\qquad\qquad\qquad\qquad\qquad\square$

We return to the two central equations derived before this theorem, the Bernoulli
law (10.135) at the free fluid boundary and Green's integral theorem (10.144) along
the unit circle,

$$
\left.\frac{\partial \theta(r, \varphi)}{\partial r}\right|_{r=1} = \lambda e^{-3\tau(1, \varphi)} \sin(\theta(1, \varphi)), \tag{10.154}
$$
$$
\frac{1}{2\pi} \int_0^{2\pi} d\varphi' \left[\frac{\partial \theta(r', \varphi')}{\partial r'} \ln |\vec{r}' - \vec{r}| \right]_{r, r'=1} = -\frac{1}{2} \theta(1, \varphi). \tag{10.155}
$$

Inserting (10.154) in (10.155) gives a nonlinear integral equation for $\theta(1, \varphi)$ along
the waveform, i.e., the unit circle in the ζ-plane (replacing the line $\psi = 0$ in the
Φ-plane),

$$\theta(1, \varphi) = -\frac{\lambda}{\pi} \int\limits_0^{2\pi} d\varphi' \, e^{-3\tau(1,\varphi')} \, \sin(\theta(1, \varphi')) \, \ln|\vec{r} - \vec{r}'|_{r,r'=1} \qquad (10.156)$$

Note that the equation of motion is inherent in this equation, since Green's integral theorem for *harmonic* functions was used in its derivation. Next, the Cauchy–Riemann equation $\partial\theta/\partial r = \partial\tau/\partial\varphi$ applied to the Bernoulli equation (10.154) gives

$$\tau(1, \varphi) - \tau(1, 0) = \lambda \int\limits_0^{\varphi} d\varphi' \, e^{-3\tau(1,\varphi')} \, \sin(\theta(1, \varphi')). \qquad (10.157)$$

Equations (10.156) and (10.157) are Eq. (6) on p. 48 in Lichtenstein (1931), and express the full problem of nonlinear water waves. One has therefore to deal with two coupled, nonlinear integral equations. To reduce the (rather obvious) algebra, we assume that $\tau(1, \varphi)$ from (10.157) is already known, and consider only the single Eq. (10.156). For a treatment of the full system (10.156) and (10.157), see Lichtenstein (1931).

Splitting off the Homogeneous Case

The next steps are to separate off the linear part of the integral equation (10.157), to find a series expansion of the kernel $\ln|\vec{r} - \vec{r}'|_{r,r'=1}$ in terms of the argument $\varphi - \varphi'$, and to obtain the characteristic values and eigenfunctions of this kernel. To start with, one writes $e^{-3\tau} \sin\theta$ in (10.156) as

$$e^{-3\tau} \sin\theta = \theta + (e^{-3\tau} \, \sin\theta - \theta). \qquad (10.158)$$

The first term on the right will give the linear part of the integral equation, and the term in brackets the nonlinear part. Here

$$e^{-3\tau} \sin\theta - \theta = (1 - 3\tau \pm \ldots)(\theta \pm \ldots) - \theta = -3\tau\theta + O(3) \qquad (10.159)$$

is second-order small in products of τ and θ, and the same was assumed in the nonlinear integral equation (10.92) with $m + n > 1$, and was used in the convergence proof for the iteration scheme (10.93). When writing $e^{-3\tau}$ and $\sin\theta$ in the following, infinite series expansions in terms of τ and θ are understood. Taking the sum over powers of $\tau(1, \varphi)$ and $\theta(1, \varphi)$ in (10.156) out of the integral over φ gives a Schmidt series as in (10.92). This termwise integration is allowed because of uniform convergence of the series expansions. With (10.158), Eq. (10.156) becomes

$$\theta(1, \varphi) + \frac{\lambda}{\pi} \int_0^{2\pi} d\varphi' \, \ln |\vec{r} - \vec{r}'|_{r,r'=1} \, \theta(1, \varphi') =$$

$$= -\frac{\lambda}{\pi} \int_0^{2\pi} d\varphi' \, \ln |\vec{r} - \vec{r}'|_{r,r'=1} \left(e^{-3\tau} \sin\theta - \theta\right). \tag{10.160}$$

The linear and homogeneous part of (10.160) is

$$\boxed{\theta(1, \varphi) + \frac{n}{\pi} \int_0^{2\pi} d\varphi' \, \ln |\vec{r} - \vec{r}'|_{r,r'=1} \, \theta(1, \varphi') = 0} \tag{10.161}$$

where λ was replaced by n, since it will be shown below that (10.161) has nontrivial solutions only for $n \in \mathbb{N}$. Equation (10.161) is a classic integral equation, solved by Picard (1910), p. 96. Replacing $\vec{r}$ and $\vec{r}'$ in the real plane by two complex numbers ζ and ζ' in the complex ζ-plane, one has (the limit $r = 1$ is taken below)

$$\zeta = re^{i\varphi}, \qquad\qquad \zeta' = e^{i\varphi'}, \tag{10.162}$$

giving (since ln is an analytic function, thus $\ln(\bar{z}) = \overline{\ln(z)}$ from a power series expansion),

$$\ln |\vec{r}' - \vec{r}| = \ln \sqrt{(\zeta' - \zeta)(\bar{\zeta}' - \bar{\zeta})} = \frac{1}{2} \left[\ln(\zeta' - \zeta) + \ln(\bar{\zeta}' - \bar{\zeta})\right]$$

$$= \frac{1}{2} \left[\ln(\zeta' - \zeta) + \overline{\ln(\zeta' - \zeta)}\right], \tag{10.163}$$

and thus

$$\ln |\vec{r}' - \vec{r}| = \text{Re}(\ln(\zeta' - \zeta)). \tag{10.164}$$

Using a well-known power series expansion for the logarithm, one obtains

$$\ln(\zeta' - \zeta) = \ln \left(\zeta' \left(1 - \frac{\zeta}{\zeta'}\right)\right) = \ln \zeta' + \ln \left(1 - \frac{\zeta}{\zeta'}\right)$$

$$= \ln \zeta' - \frac{\zeta}{\zeta'} - \frac{1}{2} \left(\frac{\zeta}{\zeta'}\right)^2 - \frac{1}{3} \left(\frac{\zeta}{\zeta'}\right)^3 - \frac{1}{4} \left(\frac{\zeta}{\zeta'}\right)^4 - \cdots . \tag{10.165}$$

Here, for $r' = 1$,

$$\text{Re} \left(\frac{\zeta}{\zeta'}\right)^k = r^k \cos(k(\varphi - \varphi')). \tag{10.166}$$

This gives, with $\text{Re}(\ln \zeta') = \text{Re}\left(\ln e^{i\varphi'}\right) = \text{Re}(i\varphi') = 0$,

$$\ln |\vec{r} - \vec{r}'|_{r'=1} = -\sum_{k=1}^{\infty} \frac{r^k}{k} \cos(k(\varphi - \varphi')),$$ (10.167)

or

$$\ln |\vec{r} - \vec{r}'|_{r'=1} = -\sum_{k=1}^{\infty} \frac{r^k}{k} \left(\cos(k\varphi) \; \cos(k\varphi') + \sin(k\varphi) \; \sin(k\varphi') \right)$$ (10.168)

which for $r < 1$ is a convergent geometric series. The integration over cosines and sines suggests a Fourier expansion of the function $\theta(1, \varphi)$. This function has period 2π, thus

$$\theta(1, \varphi) = \frac{a_0}{2} + \sum_{l=1}^{\infty} \left(a_l \cos(l\varphi) + b_l \sin(l\varphi) \right).$$ (10.169)

Inserting (10.168) and (10.169) in (10.161) and putting $r = 1$ gives

$$\frac{a_0}{2} + \sum_{l=1}^{\infty} \left(a_l \cos(l\varphi) + b_l \sin(l\varphi) \right) =$$

$$= \frac{n}{\pi} \int_0^{2\pi} d\varphi' \sum_{k=1}^{\infty} \frac{1}{k} \left(\cos(k\varphi) \; \cos(k\varphi') + \sin(k\varphi) \; \sin(k\varphi') \right) \times$$

$$\times \left(\frac{a_0}{2} + \sum_{l=1}^{\infty} \left(a_l \cos(l\varphi') + b_l \sin(l\varphi') \right) \right).$$ (10.170)

Using the orthogonality relations $\int_0^{2\pi} dx \; \sin(kx) \; \cos(lx) = 0$ and

$$\int_0^{2\pi} dx \; \sin(kx) \; \sin(lx) = \int_0^{2\pi} dx \; \cos(kx) \; \cos(lx) = \pi \, \delta_{kl},$$ (10.171)

one obtains

$$\frac{a_0}{2} + \sum_{l=1}^{\infty} \left(a_l \cos(l\varphi) + b_l \sin(l\varphi) \right) = n \sum_{l=1}^{\infty} \frac{1}{l} \left(a_l \cos(l\varphi) + b_l \sin(l\varphi) \right).$$ (10.172)

This has as solution $a_0 = 0$ and

$$a_l = b_l = \delta_{ln}, \qquad\qquad n \in \mathbb{N}$$ (10.173)

The kernel

$$K(\varphi, \varphi') = -\frac{1}{\pi} \ln |\vec{r} - \vec{r}'|_{r,r'=1}$$ (10.174)

(the sign adheres to the convention $f = \lambda \mathbf{K} f$) in (10.161) has therefore characteristic values $n \in \mathbb{N}$, and the two normalized (as required in Theorem 4 in Sect. 10.4) eigenfunctions to the characteristic value n are

$$\phi_1(\varphi) = \pi^{-1/2} \cos(n\varphi), \qquad\qquad \phi_2(\varphi) = \pi^{-1/2} \sin(n\varphi). \qquad (10.175)$$

Since $K(\varphi, \varphi') = K(\varphi', \varphi)$, one has $\mathbf{K} = \mathbf{K}^\tau$, giving

$$\psi_1 = \phi_1, \qquad\qquad \psi_2 = \phi_2. \qquad (10.176)$$

Next, the operator L from (10.52) is introduced,

$$nL(s, t) = nK(s, t) - \phi_1(s)\phi_1(t) - \phi_2(s)\phi_2(t), \qquad (10.177)$$

giving, for ϕ_1 and ϕ_2 from (10.175),

$$\begin{aligned} nL(\varphi, \varphi') &= \frac{n}{\pi} \sum_{k=1}^{\infty} \frac{1}{k} \cos(k(\varphi - \varphi')) \quad - \quad \frac{1}{\pi} \cos(n(\varphi - \varphi')) \\ &= \frac{n}{\pi} \sum_{k \neq n} \frac{1}{k} \cos(k(\varphi - \varphi')). \end{aligned} \qquad (10.178)$$

For all symmetric and real kernels K, the formula

$$\lambda \mathbf{L}\phi_k = \lambda \mathbf{K}\phi_k - \sum_j \psi_j(\phi_j, \phi_k) = \phi_k - \psi_k = 0 \qquad (10.179)$$

holds. Indeed, for the present case,

$$\int_0^{2\pi} d\varphi' \, nL(\varphi, \varphi') \, \phi_1(\varphi') = \frac{n}{\pi^{3/2}} \sum_{k \neq n} \frac{1}{k} \int_0^{2\pi} d\varphi' \, \cos(k(\varphi - \varphi')) \, \cos(n\varphi') = 0,$$

and similarly for $\phi_2(\varphi)$. Therefore, $n \in \mathbb{N}$ is *not* a characteristic value of L, and its resolvent exists, as is intended by the definition of L. Since $\phi_1(\varphi)$ is even and $\theta(1, \varphi)$ is odd, one has $(\phi_1, \theta) = \pi^{-1/2} \int_0^{2\pi} d\varphi' \, \phi_1(\varphi') \, \theta(1, \varphi') = 0$.

Two small real numbers $|\varepsilon_1| \ll 1$ and $|\varepsilon_2| \ll 1$ are introduced now. First, the presence of small nonlinear terms in the full integral equation (10.160) will shift the characteristic value slightly, from $\lambda = n$ with $n \in \mathbb{N}$ to

$$\lambda = n - \varepsilon_1. \qquad (10.180)$$

To guarantee a smooth transition to infinitesimally small, linear waves, corresponding to the linear and homogeneous integral equation (10.161), ε_1 must vanish together with the nonlinear terms in (10.160). Second, the number ε_2 is introduced by

$$\varepsilon_2 = \phi_2(\phi_2, \theta) = \frac{1}{\pi} \int_0^{2\pi} d\varphi' \, \sin(n\varphi') \, \theta(1, \varphi'), \tag{10.181}$$

and $|\varepsilon_2| \ll 1$ since θ is small for weakly nonlinear waves. With this, Eq. (10.160) becomes, using (10.174) and (10.177), the slightly awkward expression

$$\theta(1, \varphi) - n \int_0^{2\pi} d\varphi' \, L(\varphi, \varphi') \, \theta(1, \varphi') =$$

$$= \varepsilon_2 \, \sin(n\varphi) + \frac{\varepsilon_1}{\pi} \int_0^{2\pi} d\varphi' \, \ln |\vec{r} - \vec{r}'|_{r,r'=1} \, \theta(1, \varphi')$$

$$- \frac{n - \varepsilon_1}{\pi} \int_0^{2\pi} d\varphi' \, \ln |\vec{r} - \vec{r}'|_{r,r'=1} \left(e^{-3\tau(1,\varphi')} \, \sin \theta(1, \varphi') - \theta(1, \varphi') \right), \tag{10.182}$$

which is Eq. (16) on p. 51 of Lichtenstein (1931). Theorem 8 applies, which extends the central existence Theorem 7 of Schmidt. Therefore, (10.182) *has a unique solution, if the norm of its right side is sufficiently small*. This requires that $|\varepsilon_1|$ and $|\varepsilon_2|$ are sufficiently small. Then $\|\theta\|$ (for given small $\|\tau\|$) is small as solution of (10.156). The functions θ and τ can be obtained by successive approximations. This finishes the proof of the existence of weakly nonlinear solutions to the surface wave equations.

For simplicity, it was assumed in this chapter that the functions f and g and the kernel functions $K_{m,n}(s, t)$ are continuous. But the kernel $\ln |\vec{r} - \vec{r}'|_{r,r'=1}$ in (10.182) is not bounded and hence not continuous at $\varphi' = \varphi$. Actually, Schmidt's theorem and Theorems 8 and 9 only require that the integrals $\int_a^b dt \, K_{mn}(s, t) \, f^m(t) \, g^n(t)$ (including now $m + n = 1$) exist and give continuous functions, and that the infinite sum $\sum_{m,n} \int_a^b dt \, K_{mn}(s, t) \, f^m(t) \, g^n(t)$ converges, see especially (10.80). As was seen in the calculation leading from (10.169) to (10.172), the integration over the singular logarithm poses no problem. We give now an upper bound for the integrals in (10.182). First, from Fig. 10.3,

$$|\vec{r} - \vec{r}'|_{r,r'=1} = 2 \, \sin \frac{|\varphi - \varphi'|}{2}. \tag{10.183}$$

To show that the integrals are finite despite the singularity of the function $\ln[2 \sin(|\varphi - \varphi'|/2)]$ at $\varphi - \varphi' = 0$, one can put $\varphi = 0$ without loss of generality. Then, for arbitrary m and n, and writing φ for φ',

$$\left| \int_0^{2\pi} d\varphi \, \ln \left(2 \, \sin \frac{|\varphi|}{2} \right) \theta^m(1, \varphi) \, \tau^n(1, \varphi) \right|$$

$$\overset{(a)}{\leq} 2 \left| \int_0^{\pi} d\varphi \, \ln \left(2 \sin \frac{\varphi}{2} \right) \theta^m(1, \varphi) \, \tau^n(1, \varphi) \right|$$

$$\overset{(b)}{\leq} 2 \left\{ \int_0^\pi d\varphi \ \left(\ln \left(2 \sin \frac{\varphi}{2} \right) \right)^2 \right\}^{1/2} \left\{ \int_0^\pi d\varphi \ \theta^{2m}(1, \varphi) \ \tau^{2n}(1, \varphi) \right\}^{1/2}$$

$$\overset{(c)}{\leq} 2 \left\{ \int_0^\pi d\varphi \ (\ln \varphi)^2 \right\}^{1/2} \left\{ \int_0^\pi d\varphi \ \theta^{2m}(1, \varphi) \ \tau^{2n}(1, \varphi) \right\}^{1/2}$$

$$\overset{(d)}{\leq} 2\pi^{1/2} \left\{ \int_0^\pi d\varphi \ (\ln \varphi)^2 \right\}^{1/2} \ \|\theta\|^m \ \|\tau\|^n$$

$$\overset{(e)}{=} 2\pi^{1/2} \left\{ \left[\varphi(\ln \varphi)^2 - 2\varphi \ln \varphi + 2\varphi \right]_0^\pi \right\}^{1/2} \ \|\theta\|^m \ \|\tau\|^n$$

$$\overset{(f)}{=} 2\pi^{1/2} \left[\pi (\ln \pi)^2 - 2\pi \ln \pi + 2\pi \right]^{1/2} \ \|\theta\|^m \ \|\tau\|^n . \tag{10.184}$$

The steps in this calculation are

(a) $\theta(1, -\varphi) = -\theta(1, \varphi)$ and $\tau(1, -\varphi) = \tau(1, \varphi)$ according to (10.136).
(b) Cauchy's inequality, $\left(\int dx \ f(x) \ g(x) \right)^2 \leq \int dx \ (f(x))^2 \int dx \ (g(x))^2$.
(c) $\sin x \leq x$ for $x \geq 0$.
(d) From (10.78), $\|\theta\| = \max_\varphi |\theta(\varphi)|$, and similarly for $\|\tau\|$. Then

$$\left\{ \int_0^\pi d\varphi \ \theta^{2m} \tau^{2n} \right\}^{1/2} \leq \left\{ \|\theta\|^{2m} \ \|\tau\|^{2n} \int_0^\pi d\varphi \right\}^{1/2} = \pi^{1/2} \ \|\theta\|^m \ \|\tau\|^n . \tag{10.185}$$

(e) $\int d\varphi \ (\ln \varphi)^2$ has the given closed solution.
(f) The expression in square brackets in (e) is finite, since

$$\lim_{x \to 0} x \ln x = \lim_{x \to 0} x(\ln x)^2 = 0, \tag{10.186}$$

as is easily seen using de l'Hospital's rule:

$$\lim_{x \to 0} x \ln x = \lim_{x \to 0} \frac{\ln x}{x^{-1}} = \lim_{x \to 0} \frac{x^{-1}}{-x^{-2}} = \lim_{x \to 0} (-x) = 0, \tag{10.187}$$

and

$$\lim_{x \to 0} x(\ln x)^2 = \lim_{x \to 0} \frac{(\ln x)^2}{x^{-1}} = \lim_{x \to 0} \frac{2x^{-1} \ln x}{-x^{-2}} = \lim_{x \to 0} (-2x \ln x) = 0. \tag{10.188}$$

This shows that the integrals in (10.182) are finite and the sum in (10.80) converges for sufficiently small $\|\theta\|$ and $\|\tau\|$. Therefore, the theorems on the existence of solutions of nonlinear integral equations are applicable. A few extra considerations are required to ensure that θ and τ are differentiable, using a theorem of Weierstrass from complex function theory (Schmidt's theorem ensures only the existence of a continuous solution). For the differentiability of the solution, see Wainberg and Trenogin (1973), p. 85, Lichtenstein (1931), p. 53, and Stoker (1957), p. 536.

The efforts taken in this chapter to demonstrate the existence of weakly nonlinear surface wave solutions are an indicator of the depth and complexity of the mathematical theorems required to prove the existence of fully nonlinear surface water waves up to (and beyond?) the Stokes angle of 120°. With a temporal distance of half a century, it seems fair to say that two of the most groundbreaking results in the mathematical theory of classical mechanics in the twentieth century are the almost contemporaneous KAM theorem (Kolmogorov 1954; Arnold 1963; Moser 1962) on the stability of phase-space tori of Hamiltonian systems under small but finite perturbations, and the existence theorem by Krasovskii (1961) for fully nonlinear water waves using Krasnoselskii's theorems from nonlinear functional analysis (see, e.g., Krasnoselskii 1964a).

References

Arnold, V.I. 1963. Proof of a theorem by A. N. Kolmogorov on the invariance of quasi-periodic motions under small perturbations of the Hamiltonian. *Uspekhi Matematicheskikh Nauk* 18: 13. *Russian Mathematical Surveys* 18: 9.

Buffoni, B., and J. Toland. 2003. *Analytic theory of global bifurcation*. Princeton: Princeton University Press.

Dieudonné, J. 1969. *Foundations of modern analysis*. New York: Academic Press.

Dunford, N., and J.T. Schwartz. 1958. *Linear operators. Part I: General theory*. New York: Interscience Publishers.

Fredholm, I. 1903. Sur une classe d'équations fonctionnelles. *Acta Mathematica* 27: 365.

Hellinger, E., and O. Toeplitz. 1927. Integralgleichungen und Gleichungen mit unendlich vielen Unbekannten. In *Encyklopädie der mathematischen Wissenschaften*, vol. 2-3-2, 1335. Leipzig: Teubner (digital at GDZ Göttingen).

Hutson, V., and J.S. Pym. 1980. *Applications of functional analysis and operator theory*. London: Academic Press.

Kolmogorov, A.N. 1954. The general theory of dynamical systems and classical mechanics. In *Proceedings of the international congress of mathematicians, Amsterdam*, vol. 1, 315. North Holland, Amsterdam, 1957 (in Russian). English translation as Appendix in R.H. Abraham and J.E. Marsden, *Foundations of mechanics*, 2nd ed. Benjamin/Cummings, 1978.

Krasnoselskii, M.A. 1964a. *Positive solutions of operator equations*. Groningen: P. Noordhoff Ltd.

Krasnoselskii, M.A. 1964b. *Topological methods in the theory of nonlinear integral equations*. New York: The Macmillan Company; Oxford: Pergamon Press.

Krasnoselskii, M.A., et al. 1972. *Approximate solutions of operator equations*. Groningen: Wolters-Noordhoff Publishing.

Krasovskii, Yu.P. 1961. On the theory of steady-state waves of finite amplitude. *U.S.S.R. Computational Mathematics and Mathematical Physics* 1: 996.

Levandosky, J. 2003. *Partial differential equations of applied mathematics*. Course Notes for Math 220B. http://web.stanford.edu/class/math220b/handouts/potential.pdf.

Levi-Civita, T. 1925. Détermination rigoureuse des ondes permanentes d'ampleur finie. *Mathematische Annalen* 93: 264.

Lichtenstein, L. 1931. *Vorlesungen über einige Klassen nichtlinearer Integralgleichungen und Integro-Differentialgleichungen*. Berlin: Springer.

Littman, W. 1957. On the existence of periodic waves near critical speed. *Communications on Pure and Applied Mathematics* 10: 241.

Moser, J. 1962. On invariant curves of area-preserving mappings of an annulus. *Nachrichten der Akademie der Wissenschaften in Göttingen, Mathematisch-Physikalische Klasse II* 1: 1.

Nekrasov, A.I. 1921. On steady waves. *Izv. Ivanovo-Voznesenskogo politekhn.* in-ta 3. Translated by D.V. Thampuran: *The exact theory of steady state waves on the surface of a heavy liquid*. 1967, Technical Summary Report No. 813, Mathematics Research Center, United States Army,

University of Wisconsin, ed. C.W. Cryer. Available at Hathi Trust Digital Library. https://www. hathitrust.org/.

Neumann, C. 1877. *Untersuchungen über das logarithmische und Newtonsche Potential*. Leipzig: Teubner.

Picard, É. 1910. Sur un théorème général relatif aux équations intégrales de première espèce et sur quelques problèmes de physique mathématique. *Rendiconti del Circolo Matematico di Palermo* 29: 79.

Schmidt, E. 1907. Zur Theorie der linearen und nichtlinearen Integralgleichungen. *Mathematische Annalen* 63: 433 (part 1) and 64: 161 (part 2).

Schmidt, E. 1908. Zur Theorie der linearen und nichtlinearen Integralgleichungen. *Mathematische Annalen* 65: 370 (part 3).

Schmidt, E. 1910. Bemerkung zur Potentialtheorie. *Mathematische Annalen* 68: 107.

Schwartz, J.T. 1969. *Nonlinear functional analysis*. New York: Gordon and Breach.

Smirnow, W.I. 1988. *Lehrgang der höheren Mathematik, part IV/1*. Berlin: Deutscher Verlag der Wissenschaften.

Sternberg, W. 1925. *Potentialtheorie. I. Die Elemente der Potentialtheorie*. Berlin: Walter de Gruyter & Co.

Stoker, J.J. 1957. *Water waves. The mathematical theory with applications*. New York: Interscience Publishers.

Toland, J.F. 1996. Stokes waves. *Topological Methods in Nonlinear Analysis* 7: 1.

Wainberg, M.M., and W.A. Trenogin. 1973. *Theorie der Lösungsverzweigung bei nichtlinearen Gleichungen*. Berlin: Akademie-Verlag.

Chapter 11
Sound and Internal Gravity Waves

In the last three chapters, shallow and free surface water waves were treated, with gravity and capillarity as the driving forces. The subject of the present, brief chapter are two new wave types with (partially) different driving forces: first, sound waves due to perturbations of the thermal pressure, and, second, internal gravity waves due to buoyancy perturbations. Both these pressure and buoyancy fluctuations are directly linked to local variations in the fluid density in stratified fluids, due to a depth-dependence of, e.g., salinity or temperature. Thus in this and also in the next chapter, the fluid density is no longer assumed to be constant. Sound waves give rise to the vast field of acoustics, and internal gravity waves have found ample consideration in geophysical fluid dynamics. We consider here only some elementary examples of internal waves; for further discussions, the reader is referred to dedicated books on fluid dynamics of the oceans and the atmosphere, like the one by Vallis (2017).

11.1 Wave Equation

Sound is the fast propagation of high-frequency variations in thermal pressure and density in an elastic medium like a gas, plasma, liquid, or solid. The corresponding considerations from elasticity theory and thermodynamics are of no concern here, and give as general formula for the sound speed (see Huang 1963),

$$a = \sqrt{\frac{\partial p}{\partial \rho}}. \tag{11.1}$$

For an incompressible fluid, $\delta \rho \to 0$, and $a \to \infty$. Since liquids are less compressible than gases, the sound speed in water is much larger than in air.

© Springer Nature Switzerland AG 2019
A. Feldmeier, *Theoretical Fluid Dynamics*, Theoretical and Mathematical Physics,
https://doi.org/10.1007/978-3-030-31022-6_11

One has to specify in (11.1), which thermodynamic state variable like entropy or temperature is (held) constant while density and pressure change. The compression and expansion of fluid regions by pressure changes is usually much faster than the heat exchange by heat conduction (or emission and absorption of radiation) between parts of the fluid; therefore, sound is to a good approximation an *adiabatic* process. According to Landau and Lifshitz (1987), p. 3, an *ideal fluid* (to be distinguished from an ideal gas) has by definition negligible heat exchange and viscosity, which implies that 'the motion is adiabatic throughout the fluid.'

In an adiabatic process of an ideal gas, $p \sim \rho^{\gamma}$ holds, with adiabatic exponent $\gamma > 1$. In the kinetic theory of gases, the remarkable formula

$$\gamma = 1 + \frac{2}{f} \tag{11.2}$$

is derived, where f is the number of degrees of freedoms of a molecule due to translation of its center of mass, rotation, vibration, electronic excitation, dissociation, and ionization (the latter two change the number of particles). For air at room temperature, which consists mainly of N_2 and O_2 molecules, only the three translational ('thermal motion') and two rotational degrees of freedom are excited; the latter are excited starting at temperatures between 2 and 100 K for different gases (see Zel'dovich and Raizer 1966, p. 178, for measured values), and thus quantum effects are negligible at room temperature and rotation can be treated classically. In this case, a linear arrangement of point masses, and especially a diatomic gas, has two degrees of freedom for rotation, viz., about two independent axes normal to the line connecting the mass centers. For rotation about the latter axis (if one does not exclude this on pure symmetry grounds), the rotation energy vanishes and there is thus no corresponding degree of freedom, since the moment of inertia J about this axes vanishes: to have $J > 0$, masses must orbit the rotation axis at normal distances $r_{\perp} > 0$, which is not the case for a diatomic or linear molecule. Thus, 'if all the atoms are collinear (and in particular for a diatomic molecule) there are only two rotational degrees of freedom' (Landau and Lifshitz 1980, p. 130). Furthermore, molecular vibrations are only excited far above room temperature, at temperatures of the order of 1.000 K (Zel'dovich and Raizer 1966, p. 178), and the other degrees of freedom are excited at even higher temperatures. Thus $f = 5$ for air at room temperature, and $\gamma = 7/5$. The adiabatic sound speed is then given by

$$a = \sqrt{\left.\frac{\partial p}{\partial \rho}\right|_{S}} = \sqrt{\frac{\gamma p}{\rho}}, \tag{11.3}$$

with constant specific entropy S. In an isothermal process, on the other hand, $p \sim \rho$ for an ideal gas (the Boyle–Mariotte law). The isothermal sound speed is then

$$A = \sqrt{\left.\frac{\partial p}{\partial \rho}\right|_{T}} = \sqrt{\frac{p}{\rho}}, \tag{11.4}$$

giving

$$a^2 = \gamma A^2. \tag{11.5}$$

For liquids, the adiabatic exponent γ is usually close to unity, and thus the difference between isothermal and adiabatic sound speed is rather small.

We consider one-dimensional, plane sound waves, for which the continuity and Euler equations are

$$\frac{\partial \rho}{\partial t} + \frac{\partial}{\partial z}(\rho w) = 0, \tag{11.6}$$

$$\frac{\partial w}{\partial t} + w \frac{\partial w}{\partial z} = -\frac{1}{\rho} \frac{\partial p}{\partial z}, \tag{11.7}$$

with w the speed component in the z-direction. A closed system of equations for the unknown fields ρ, w, and p is obtained by adding the thermodynamic relation $p \sim \rho^\gamma$, which yields, when moving with a fluid parcel,

$$\frac{\mathrm{d}p}{\mathrm{d}t} = a^2 \frac{\mathrm{d}\rho}{\mathrm{d}t}. \tag{11.8}$$

Let

$$\rho_0 = \text{const}, \qquad p_0 = \text{const}, \qquad w_0 = 0 \tag{11.9}$$

be a static solution of (11.6) and (11.7), and let

$$\rho_1 \ll \rho_0, \qquad p_1 \ll p_0, \qquad w_1 \ll a \tag{11.10}$$

be small perturbations (note especially the last condition on w_1). A constant sound speed is assumed in all the following. Writing $\rho = \rho_0 + \rho_1$, etc., in (11.6) to (11.8) and keeping only terms with at most one subscript 1 (corresponding to a linear, first-order analysis) gives

$$\dot{\rho}_1 + \rho_0 w_1' = 0, \tag{11.11}$$

$$\dot{w}_1 + \frac{p_1'}{\rho_0} = 0, \tag{11.12}$$

$$\dot{p}_1 - a^2 \dot{\rho}_1 = 0, \tag{11.13}$$

where a prime indicates differentiation with respect to z. From (11.11) to (11.13) one obtains by differentiation with respect to t and z, followed by cross-elimination,

$$\ddot{w}_1 - a^2 w_1'' = 0, \tag{11.14}$$

$$\ddot{p}_1 - a^2 p_1'' = 0. \tag{11.15}$$

$$\ddot{\rho}_1 - p_1'' = 0. \tag{11.16}$$

Equations (11.14) and (11.15) are wave equations, with solutions

$$w_1 = w_-(z - at) + w_+(z + at),$$
$$p_1 = p_-(z - at) + p_+(z + at), \tag{11.17}$$

with $w_\pm$ and $p_\pm$ arbitrary smooth functions of one real argument, see (8.42). Inserting (11.17) in the Euler equation (11.7) gives

$$\frac{w'_-}{a} = \frac{p'_-}{a^2 \rho_0}, \qquad \frac{w'_+}{a} = -\frac{p'_+}{a^2 \rho_0}, \tag{11.18}$$

where $w'_\pm = dw_\pm/d\zeta_\pm$, with $\zeta_\pm = z \pm at$ and assuming that w_- and w_+ are independent functions, as are p_- and p_+. For harmonic perturbations,

$$w_\pm = \tilde{w}_\pm \, e^{i(z \pm at)},$$
$$p_\pm = \tilde{p}_\pm \, e^{i(z \pm at)}. \tag{11.19}$$

Equation (11.18) gives then the amplitude relations

$$\frac{\tilde{w}_\pm}{a} = \mp \frac{\tilde{p}_\pm}{a^2 \rho_0}. \tag{11.20}$$

Thus, the speed and pressure perturbations have phase difference 0 for the wave $w_-(z - at)$ propagating to larger z, and phase difference π for the wave $w_+(z + at)$ propagating to smaller z. The phase and group speeds of sound waves agree with one another,

$$\frac{\omega}{k} = \frac{d\omega}{dk} = a. \tag{11.21}$$

11.2 Acoustic Cutoff

We assume now that a sound wave propagates vertically upward (in z-direction) through a barometric density stratification. The unperturbed, static atmosphere (subscript 0) is described by the Euler equation for hydrostatic equilibrium,

$$0 = -\frac{p'_0}{\rho_0} - g. \tag{11.22}$$

For simplicity, the atmosphere is assumed to be isothermal. According to the ideal gas law, the isothermal sound speed A (i.e., for $\gamma = 1$) is proportional to the square root of the temperature, and thus the atmospheric temperature can be characterized by A. In the following, A is a parameter without direct dynamical significance. The square of the isothermal sound speed is given by

$$A^2 = \frac{p_0}{\rho_0}. \tag{11.23}$$

The *scale height* H of the atmosphere is defined by

$$H = \frac{A^2}{g}. \tag{11.24}$$

Close to the Earth surface, $H \approx 8\,\text{km}$. For constant A, (11.22) becomes with (11.23),

$$\frac{\rho_0'}{\rho_0} = \frac{\mathrm{d}}{\mathrm{d}z}(\ln \rho_0) = -\frac{1}{H}, \tag{11.25}$$

which, for constant g and H, has as solution the barometric law,

$$\rho_0(z) = \rho_0(0)\, e^{-z/H}. \tag{11.26}$$

Although the atmosphere is isothermal, sound shall propagate at the adiabatic sound speed. (Similarly, sound propagates at adiabatic speed through a room of given constant temperature.) The vertical speed component in the static atmosphere is $w_0 = 0$, and the linearized continuity, Euler, and energy equations become

$$\dot{\rho}_1 + \rho_0 w_1' - \frac{\rho_0 w_1}{H} = 0, \tag{11.27}$$

$$\dot{w}_1 + \frac{p_1'}{\rho_0} + \frac{g\rho_1}{\rho_0} = 0, \tag{11.28}$$

$$\dot{p}_1 - a^2 \dot{\rho}_1 + (\gamma - 1)g\rho_0 w_1 = 0, \tag{11.29}$$

where the subscript 1 refers to small perturbations. Differentiating these equations with respect to time and space gives, after rearrangement,

$$\ddot{w}_1 - a^2 w_1'' + \gamma g w_1' = 0, \tag{11.30}$$

$$\ddot{p}_1 - a^2 p_1'' - \gamma g p_1' + g\left(p_1' - a^2 \rho_1'\right) - (\gamma - 1)g^2 \rho_1 = 0, \tag{11.31}$$

$$\ddot{\rho}_1 - p_1'' - g\rho_1' = 0, \tag{11.32}$$

see also (11.14)–(11.16). Harmonic waves

$$w_1 = \tilde{w}_1\, e^{i(\omega t - Kz)}, \tag{11.33}$$

$$\rho_1 = \tilde{\rho}_1\, e^{i(\omega t - Kz)}, \tag{11.34}$$

$$p_1 = \tilde{p}_1\, e^{i(\omega t - Kz)} \tag{11.35}$$

are assumed, with constants $\tilde{w}_1$, $\tilde{\rho}_1$, and $\tilde{p}_1$. Inserting (11.33) in (11.30) for the speed w_1 gives the complex equation

$$\omega^2 - a^2 K^2 + i\gamma g K = 0, \tag{11.36}$$

with $K \in \mathbb{C}$. Setting the imaginary part of (11.36) to zero gives

$$\text{Im } K = \frac{\gamma g}{2a^2} = \frac{1}{2H}. \tag{11.37}$$

Setting the real part of (11.36) to zero and using (11.37) yields, with $k = \text{Re } K$,

$$\boxed{\omega^2 - a^2 k^2 - \omega_a^2 = 0, \qquad \omega_a = \frac{a}{2H}}. \tag{11.38}$$

This is the dispersion relation for sound propagating vertically upward through an isothermal atmosphere. The angular frequency ω_a in (11.38) is the acoustic cutoff frequency introduced by Lamb (1909). For the Earth atmosphere, the corresponding acoustic cutoff period is $2\pi/\omega_a \approx 7$ min.

With (11.37) in (11.33), one obtains $w_1 \sim \exp(z/(2H))$, and thus the amplitude of the speed w_1 *grows* exponentially with height. This is not related to an instability but expresses conservation of kinetic wave energy, $\frac{1}{2}\rho_0 w_1^2$.

Inserting, on the other hand, (11.34) and (11.35) for density and pressure perturbations, respectively, into the wave equations (11.31) and (11.32), yields

$$\begin{pmatrix} \omega^2 - a^2 K^2 - i(\gamma - 1)gK & (\gamma - 1)g^2 - iga^2 K \\ -K^2 & \omega^2 - igK \end{pmatrix} \cdot \begin{pmatrix} \tilde{p}_1 \\ \tilde{\rho}_1 \end{pmatrix} = \begin{pmatrix} 0 \\ 0 \end{pmatrix}. \tag{11.39}$$

This homogeneous system has nontrivial solutions $(\tilde{p}_1, \tilde{\rho}_1) \neq (0, 0)$ only if the determinant of the system vanishes. After substantial cancellations, this gives

$$\omega^2 - a^2 K^2 - i\gamma g K = 0. \tag{11.40}$$

Thus now (compare with (11.37))

$$\text{Im } K = -\frac{1}{2H}, \tag{11.41}$$

and both density and pressure perturbations *drop* exponentially with height, according to $\rho_1, p_1 \sim \exp(-z/(2H))$. This means that density and pressure perturbations *grow* relative to the static atmosphere, for which $\rho_0, p_0 \sim \exp(-z/H)$. The real part of (11.40) gives again the dispersion relation (11.38).

Due to the exponential growth of w_1/a and ρ_1/ρ_0 with height (see (11.33) and (11.34)), sound waves propagating upward through an atmosphere cannot stay linear and will eventually steepen into shock fronts (see the next chapter). The dissipation of wave energy in shocks can contribute to upper atmospheric heating. Furthermore, shocks transport momentum through the atmosphere, and the corresponding ram pressure, i.e., the pressure due to bulk motion of the fluid, adds to the thermal gas pressure. Most notably then, the scale height H can increase in the presence of shocks. This and related phenomena are termed *shock levitation*.

Cutoff Frequency

We turn now to the meaning of the acoustic cutoff frequency ω_a in (11.38). The latter dispersion relation gives for the phase speed,

$$v_p = \frac{\omega}{k} = a \left(1 + \frac{1}{4H^2 k^2} \right)^{1/2}, \tag{11.42}$$

and for the group speed,

$$v_g = \frac{d\omega}{dk} = a \left(1 + \frac{1}{4H^2 k^2} \right)^{-1/2}. \tag{11.43}$$

Thus, sound propagating vertically in an atmosphere is dispersive, and

$$\boxed{v_p \, v_g = a^2} \tag{11.44}$$

From (11.42) and (11.43), waves with $k = 0$ propagate with $v_p \to \infty$ and $v_g = 0$, where the latter means that they do not transport energy. Most notably, an infinite wavelength already occurs at finite frequency, since putting $\omega = \omega_a$ in (11.38) gives $k = 0$ or $\lambda \to \infty$. For $\omega < \omega_a$, the dispersion relation (11.38) gives purely imaginary k. The wave is then a standing wave, since there is no spatial progression for imaginary k. Due to its exponential drop in amplitude, the wave is *evanescent*. Evanescent waves have thus the following properties:

– they are standing waves,
– their phase speed is infinite, and
– their group speed and energy flux are zero.

Note especially that sound can propagate vertically upward through an atmosphere only for $\omega > \omega_a$. The phase difference between ρ_1 and w_1 is obtained from (11.27) with the use of (11.33) and (11.34) as

$$\frac{\tilde{\rho}_1/\rho_0}{\tilde{w}_1/(\omega/k)} = 1 - \frac{i}{2Hk}. \tag{11.45}$$

At the acoustic cutoff, $k \to 0$, the second term on the right dominates, and the phase shift between ρ_1 and w_1 is $\pi/2$. This expresses again that no energy is transported by the wave.

Sound waves close to the acoustic cutoff affect the atmosphere over substantial parts of its height and can be observed after volcano eruptions and during solar eclipses. In the latter case, Moon's shadow causes significant changes in the thermal balance of the atmosphere. The fast-propagating shadow triggers atmospheric relaxation oscillations near the acoustic cutoff. The motion of Moon's shadow relative to the Earth during an eclipse is seen in the two frames of Fig. 11.1.

Fig. 11.1 'NASA's Earth Polychromatic Imaging Camera (EPIC) captured 12 natural color images of the Moon's shadow crossing over North America on Aug. 21, 2017.' The red + sign is located at the Gulf of California in both images. YouTube video at https://www.youtube.com/watch?v=pm7tfLvHmXA

11.3   The Schwarzschild Criterion

The shallow water and free surface waves given in Chaps. 8 and 9 were caused by the weight difference of water columns of different heights: a higher, heavier column exerts net hydrostatic pressure on neighboring columns. In this and the next section, *internal* gravity oscillations and waves are discussed, which occur in oceans and the atmosphere; their driving force is *buoyancy* caused by density perturbations in the fluid. As is the case for sound waves, the fluid must therefore be assumed to be compressible. Alternatively, a stratified, incompressible fluid can be assumed, but this is not treated here.

Since sound waves and internal gravity waves have density perturbations as their common origin, a mixed wave type occurs, called *gravo-acoustic* waves. As was found above, sound can propagate vertically through an atmosphere only at frequencies above the acoustic cutoff, $\omega > \omega_a$. Internal gravity waves, by contrast, can propagate *horizontally* only at frequencies *below* a certain cutoff, $\omega < \omega_b$, called the Brunt–Väisälä angular frequency. This frequency ω_b is close to but not identical to the acoustic cutoff ω_a. Only the simplest example of gravo-acoustic waves in an isothermal atmosphere is discussed in the following.

All hydrodynamic fields shall depend on the height coordinate z only, pointing vertically upward, and the atmosphere is assumed to be in hydrostatic equilibrium, $p_0' = -\rho_0 g$. Let $H = A^2/g$ again be the scale height, with isothermal sound speed A. A fluid parcel shall experience a perturbation that carries it upward from z to $z + \zeta$, with small ζ. The atmospheric density ρ_i of the adjacent gas at $z + \zeta$ is

$$\rho_i = \rho_0 + \rho_0'\zeta = \rho_0 - \frac{\rho_0 g \zeta}{A^2}, \tag{11.46}$$

and the atmospheric pressure at $z + \zeta$ is

$$p_0 + p_0'\zeta = p_0 - \rho_0 g \zeta. \tag{11.47}$$

The fluid parcel therefore experiences a pressure change of $-\rho_0 g \zeta$. Again, the re-sponse of a fluid parcel to the forces that cause wave motion shall be fast, i.e., on a dynamical timescale, so that heat exchange between the parcel and its surroundings plays no role (adiabatic process). Then the pressure and density perturbations of the parcel are related by

$$\delta p = a^2 \, \delta \rho, \tag{11.48}$$

with adiabatic sound speed a. Thus the new parcel density ρ_j at $z + \zeta$ is, from (11.47),

$$\rho_j = \rho_0 - \frac{\rho_0 g \zeta}{a^2}. \tag{11.49}$$

Due to the difference between the density of the parcel, ρ_j, and that of the surrounding gas, ρ_i, the parcel experiences a buoyancy force $\rho_0 \, g_b$ per volume, which is given by

$$\rho_0 \, g_b = g(\rho_i - \rho_j). \tag{11.50}$$

The sign is obvious: if the parcel is denser than its surroundings, it sinks. From (11.46) and (11.49), one has

$$\rho_0 \, g_b = -g^2 \rho_0 \zeta \left(\frac{1}{A^2} - \frac{1}{a^2} \right). \tag{11.51}$$

Assuming a purely mechanical model of a point mass moving according to Newton's second law, the equation of motion for the parcel is

$$\ddot{\zeta} = g_b = -\omega_b^2 \zeta, \tag{11.52}$$

with the Brunt–Väisälä frequency

$$\omega_b = g \left(\frac{1}{A^2} - \frac{1}{a^2} \right)^{1/2}. \tag{11.53}$$

If $\omega_b \in \mathbb{R}$, buoyancy opposes the displacement ζ of the fluid parcel. For adiabatic buoyancy oscillations in an isothermal atmosphere, one has

$$\boxed{\omega_b = \sqrt{\gamma - 1} \, \frac{g}{a}} \tag{11.54}$$

which is indeed real since $\gamma \geq 1$. For comparison, the cutoff frequency ω_a in (11.38) can be written as $\omega_a = \gamma g / (2a)$. Assuming $\gamma = 7/5$ for air gives $\omega_a = 0.70 \, g/a$ and $\omega_b = 0.63 \, g/a$. Generally,

$$\frac{\omega_a}{\omega_b} = \frac{\gamma}{2\sqrt{\gamma - 1}} > 1. \tag{11.55}$$

We have thus found that an isothermal atmosphere is stable to adiabatic perturbations. If, on the other hand, $\omega_b \in i\mathbb{R}$ in an atmosphere, a buoyancy instability occurs with increasing perturbation amplitudes. The condition for stability is therefore

$$\frac{1}{A^2} - \frac{1}{a^2} > 0. \tag{11.56}$$

This criterion was given for the first time by Schwarzschild (1906) in a slightly different form. He considers the solar atmosphere, assumed to be in radiative equilibrium, i.e., the temperature and density stratifications are determined by absorption and emission of radiation (Schwarzschild applies the Kirchhoff and the Stefan law). To keep the stability condition general, the atmospheric stratification is indicated by 'atm,' and compared to adiabatic changes ('adi') of a fluid parcel experiencing an accidental uplift (or a depression) in the atmosphere. Multiplying (11.56) with $\rho_0 g$ and using (11.46) and (11.49), one obtains

$$\frac{\rho_0\, g}{A^2} - \frac{\rho_0\, g}{a^2} = \frac{\rho_0 - \rho_i}{\zeta} - \frac{\rho_0 - \rho_j}{\zeta} > 0 \tag{11.57}$$

as condition for a stable atmosphere. Here $(\rho_0 - \rho_i)/\zeta > 0$ is the absolute density gradient of the atmosphere, and $(\rho_0 - \rho_j)/\zeta > 0$ is the absolute density gradient that a fluid parcel experiences in its adiabatic uplift. Therefore, the atmosphere is stable if

$$\left|\frac{\mathrm{d}\rho}{\mathrm{d}z}\right|_{\mathrm{atm}} > \left|\frac{\mathrm{d}\rho}{\mathrm{d}z}\right|_{\mathrm{adi}}. \tag{11.58}$$

Assuming that the density and temperature behave inversely, this becomes

$$\boxed{\left|\frac{\mathrm{d}T}{\mathrm{d}z}\right|_{\mathrm{atm}} < \left|\frac{\mathrm{d}T}{\mathrm{d}z}\right|_{\mathrm{adi}}} \tag{11.59}$$

which is the Schwarzschild criterion for atmospheric stability:

Ist nämlich der Temperaturgradient kleiner als bei adiabatischem Gleichgewicht, so gerät eine aufsteigende Luftmasse in Schichten, welche wärmer und dünner sind, als sie selbst ankommt, sie erfährt daher einen Druck nach unten [...] Ein Gleichgewicht mit kleinerem Temperaturgradienten, als das adiabatische, ist daher stabil, umgekehrt eines mit größerem Temperaturgradienten instabil.[1] (Schwarzschild 1906, p. 47)

---

[1] For if the temperature gradient is smaller than in adiabatic equilibrium, a rising mass of air approaches layers that are warmer and thinner than the mass itself upon arrival, and therefore it experiences a downward pressure [force] [...] An equilibrium with smaller temperature gradient than the adiabatic one is therefore stable; conversely, one with a larger temperature gradient is unstable.

This hydrodynamic instability for temperature gradients that are steeper than the adiabatic temperature gradient and lead to unbalanced buoyancy forces is termed *convection*. In everyday experience, convection is observed in fluids that are heated from below (a pot of water on a stove; air above a radiator): the heated fluid expands and rises through the heavier fluid atop of it (in a rather complicated process, not considered in this book; see Chandrasekhar 1961).

11.4 Gravo-Acoustic Waves

We turn from this mechanical model to a fluid description and consider two-dimensional plane waves in Cartesian xz-coordinates, the latter pointing vertically upward. The unperturbed atmosphere is static and isothermal, with $\partial \rho_0 / \partial x = 0$. Let ρ, u, w, p be the perturbed density, speed, and pressure (no subscript 1 now). Then the linearized continuity, Euler, and adiabatic energy equations are

$$\rho_t + \rho_0 \, u_x + \rho_0 \, w_z - \frac{\rho_0 \, w}{H} = 0, \tag{11.60}$$

$$u_t + \frac{p_x}{\rho_0} = 0, \tag{11.61}$$

$$w_t + \frac{p_z}{\rho_0} + \frac{g\rho}{\rho_0} = 0, \tag{11.62}$$

$$p_t - a^2 \rho_t + (\gamma - 1)g\rho_0 \, w = 0. \tag{11.63}$$

Equations (11.62) and (11.63) are identical to (11.28) and (11.29) for vertical sound. Differentiate (11.62) with respect to t and (11.63) with respect to z. Cross-eliminate p_{zt} from the two equations, and eliminate ρ_t and ρ_{tz} by inserting (11.60) and its z-derivative. With this, (11.62) becomes

$$w_{tt} - a^2 w_{zz} + \gamma g w_z + (\gamma - 1)g u_x - a^2 u_{xz} = 0. \tag{11.64}$$

The corresponding procedure applied to the t-derivative of (11.61) gives

$$u_{tt} - a^2 u_{xx} + g w_x - a^2 w_{xz} = 0. \tag{11.65}$$

Equations (11.64) and (11.65) are two coupled wave equations for the perturbations u and w. Inserting plane waves,

$$u = \tilde{u} \, e^{i(\omega t - Kz - lx)},$$
$$w = \tilde{w} \, e^{i(\omega t - Kz - lx)}, \tag{11.66}$$

into (11.64) and (11.65) and setting the determinant of the emerging 2×2 matrix equation to zero, one obtains the dispersion relation

$$\omega^4 - \omega^2(a^2 l^2 + a^2 K^2 - i\gamma g K) + (\gamma - 1)g^2 l^2 = 0. \tag{11.67}$$

From (11.54),

$$(\gamma - 1)g^2 l^2 = a^2 \omega_b^2 l^2. \tag{11.68}$$

For complex K, and setting the imaginary part of (11.67) to zero, one finds again

$$\operatorname{Im} K = \frac{1}{2H}, \tag{11.69}$$

as for pure sound waves. Setting the real part of (11.67) to zero gives

$$\boxed{\omega^4 - a^2\omega^2(k^2 + l^2) - \omega_a^2\omega^2 + a^2\omega_b^2 l^2 = 0} \tag{11.70}$$

This is the simplest dispersion relation for combined gravo-acoustic waves (sound plus internal gravity waves) in an isothermal atmosphere. Pure sound waves are obtained for $\omega \gg \omega_a$ and $\omega \gg \omega_b$, with

$$\omega^2 = a^2(k^2 + l^2). \tag{11.71}$$

To suppress sound, assume zero compressibility or $a \to \infty$, which gives

$$\omega^2 = \omega_b^2 \, \frac{l^2}{k^2 + l^2}, \tag{11.72}$$

which is the dispersion relation for pure internal gravity waves. The Brunt–Väisälä frequency is thus an *upper* cutoff frequency,

$$\omega \le \omega_b. \tag{11.73}$$

For real-valued ω_b and $\omega \le \omega_b$, one has therefore stable buoyancy-driven waves. If instead the temperature stratification is such that ω_b is imaginary, buoyancy forces lead to convective instability. We have therefore arrived at the important conclusion that

Internal gravity waves and convection may be considered as stable and unstable manifestations of the same mode. (Stein and Leibacher 1974)

Next, let α be the angle of the wave vector (l, k) with the x-axis,

$$\tan \alpha = k/l. \tag{11.74}$$

Then (11.72) becomes

$$\cos \alpha = \omega/\omega_b, \tag{11.75}$$

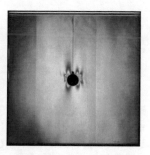

Fig. 11.2 Internal gravity waves excited by vertical oscillations of a metal rod immersed in stratified saltwater, photographed with the Moiré technique. The period is 8, 5, and 3.8 s, from left to right. The Brunt–Väisälä period is 3.9 s, thus in the frame at the right no internal gravity wave can occur. Photograph: Satoshi Sakai, at https://www.gfd-dennou.org/library/gfd_exp/exp_e/index0.htm

and thus a given $\omega \leq \omega_b$ fixes the propagation direction of the internal gravity wave! This is demonstrated in Fig. 11.2, for vertical oscillations of a metal cylinder in stratified saltwater.

Turning to the full dispersion relation (11.70), the latter can be solved for k^2, giving

$$k^2 = \frac{\omega^2(\omega^2 - \omega_a^2) + a^2 l^2(\omega_b^2 - \omega^2)}{\omega^2 a^2}. \tag{11.76}$$

From (11.55), $\omega_b < \omega_a$. For real ω, the numerator is negative in the interval $\omega_b < \omega < \omega_a$, whereas the denominator is positive. Thus, there is no progressive wave (neither sound nor internal gravity) in any direction for $\omega_b < \omega < \omega_a$. Here k is purely imaginary, i.e., the wave decays exponentially, and one has standing, evanescent waves in this frequency interval.

Figure 11.3 shows the *diagnostic diagram* $\omega = \omega(l)$ for $k = $ const. At all wavenumbers l, there is a gap between the acoustic and the internal gravity branch. Let α be as defined in (11.74), and then $\cos^2 \alpha = l^2/(k^2 + l^2)$. Equation (11.70) can be rewritten as

Fig. 11.3 Diagnostic diagram for gravo-acoustic waves in an isothermal atmosphere. Plotted is the dispersion relation (11.70) for $a = 1$, $\omega_b = 1$, and $k^2 + \omega_a^2 = 2$, or $\omega^4 - \omega^2(2 + l^2) + l^2 = 0$

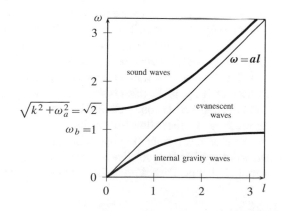

$$\omega^4 - \omega^2 \left(a^2(k^2 + l^2) + \omega_a^2 \right) + a^2 \omega_b^2 (k^2 + l^2) \cos^2 \alpha = 0, \tag{11.77}$$

or

$$\frac{\omega^2}{a^2(k^2 + l^2)} = \frac{\omega^2 - \omega_b^2 \cos^2 \alpha}{\omega^2 - \omega_a^2}. \tag{11.78}$$

For sound waves, $\omega > \omega_a$. The numerator and denominator on the right side of (11.78) are then both positive for all α. This means that sound can propagate in any direction. For internal gravity waves, on the other hand, $\omega < \omega_b < \omega_a$. The denominator on the right side of (11.78) is then < 0, and hence the numerator must also be < 0. Thus, plane internal gravity waves can propagate within an angle of

$$|\alpha| \leq \mathrm{acos}\,\frac{\omega}{\omega_b} \tag{11.79}$$

with the horizontal direction; for $\omega = \omega_b$ especially, internal gravity waves can only propagate horizontally. Figure 11.4 shows contour lines of ω from (11.70) in the lk-plane. For sound, the contour lines are ellipses; for internal gravity waves, they are hyperbolas. The phase propagation of a wave is along the wave vector (l, k), whereas the group propagation is along the vector $(\partial \omega / \partial l, \partial \omega / \partial k)$. This is the gradient $\nabla \omega$ of $\omega(l, k)$ in the lk-plane. The gradient of ω is normal to the contour lines of ω. For sound (ellipses), the vector (l, k) and the gradient $\nabla \omega$ are nearly aligned, whereas for internal gravity waves, the vector (l, k) and the gradient $\nabla \omega$ can become almost normal.

More information on internal gravity waves can be found in the textbook by Mihalas and Weibel-Mihalas (1984), Sect. 5.2, from which the material in this section is drawn. We close with some comments on wave trapping and helioseismology. In different regions of the solar interior, either radiation, convection, turbulence, or magnetic fields can dominate the local energy balance and transport, making the local temperature in the sun a complicated function of radius. One can then analyze the influence of temperature *jumps* on wave propagation. A wave that can travel freely through one region may be evanescent in another. If the evanescent region has finite extent, the wave is partially transmitted through it. A *wave cavity* is a region through which a wave can propagate freely. Wave cavities are enclosed by regions in which the wave is evanescent, and thus the wave is reflected at both the upper and lower cavity boundaries. The wave interferes with itself and a standing wave forms.

Solar wave cavities occur in so-called transition regions. The solar *five minute oscillations* were discovered by Leighton et al. (1962) and interpreted by Ulrich (1970) to be 'acoustic waves trapped [in a wave cavity] below the solar photosphere' (Ulrich 1970, p. 995); this results in a *discrete* wave spectrum. The cavity extends from the convection zone (relatively far out) to the deep solar interior. The observed acoustic oscillations allow to draw information on the Sun's interior, most notably on the rotation rate of inner regions. The surprising results are that from roughly 0.2 to 0.7 solar radii, in the radiative interior, the solar plasma rotates at a constant angular speed Ω, that is, like a rigid body. Outside 0.7 solar radii, in the convection zone, the

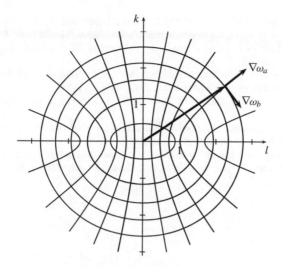

Fig. 11.4 Dispersion relation (11.70) for gravo-acoustic waves in the lk-plane for $a = \omega_b = 1$ and $\omega_a = 1.1$. *Ellipses*: sound waves; *hyperbolas*: internal gravity waves. The ω-contours are at 0.2, 0.4, 0.6, 0.8, 0.95 for internal gravity waves, and at 1.2, 1.6, 2.0, 2.4, 2.8 for sound waves. The long bold vector is (l, k), the short bold vectors are the gradients $\nabla \omega_a$ and $\nabla \omega_b$ for sound and internal gravity waves, respectively

angular speed Ω is still roughly independent of the radius but depends strongly on latitude, with large Ω at small latitudes, and vice versa. For a review on the internal rotation of the Sun, see Thompson et al. (2003). Facts on early helioseismology can be found in the review by Stein and Leibacher (1974), and a recent textbook is that by Lang (2016).

References

Chandrasekhar, S. 1961. *Hydrodynamic and hydromagnetic stability.* Oxford: Oxford University Press; New York: Dover (1981).

Huang, K. 1963. *Introduction to statistical physics.* New York: Wiley.

Lamb, H. 1909. On the theory of waves propagated vertically in the atmosphere. *Proceedings of the London Mathematical Society, Series 2*, 7: 122.

Landau, L.D., and E.M. Lifshitz. 1980. *Statistical physics, Course of theoretical physics*, vol. 5, 3rd ed. Amsterdam: Elsevier Butterworth-Heinemann.

Landau, L.D., and E.M. Lifshitz. 1987. *Fluid mechanics, Course of theoretical physics*, vol. 6, 2nd ed. Amsterdam: Elsevier Butterworth-Heinemann.

Lang, K. 2016. *The Sun from space*, 2nd ed. Berlin: Springer.

Leighton, R.B., R.W. Noyes, and G.W. Simon. 1962. Velocity fields in the solar atmosphere. I. Preliminary report. *Astrophysical Journal* 135: 474.

Mihalas, D., and B. Weibel-Mihalas. 1984. *Foundations of radiation hydrodynamics.* Oxford: Oxford University Press; Mineola: Dover (1999).

Schwarzschild, K. 1906. Ueber das Gleichgewicht der Sonnenatmosphäre. *Nachrichten von der Gesellschaft der Wissenschaften zu Göttingen, Mathematisch-Physikalische Klasse* 1906: 41.

Stein, R.F., and J. Leibacher. 1974. Waves in the solar atmosphere. *Annual Review of Astronomy and Astrophysics* 12: 407.

Thompson, M.J., J. Christensen-Dalsgaard, M.S. Miesch, and J. Toomre. 2003. The internal rotation of the Sun. *Annual Review of Astronomy and Astrophysics* 41: 599.

Ulrich, R.K. 1970. The five-minute oscillations on the solar surface. *Astrophysical Journal* 162: 993.

Vallis, G.K. 2017. *Atmospheric and oceanic fluid dynamics: Fundamentals and large-scale circulation*, 2nd ed. Cambridge: Cambridge University Press.

Zel'dovich, Ya.B., and Yu.P. Raizer. 1966. *Physics of shock waves and high-temperature hydrodynamic phenomena*, vol. 1. New York: Academic Press; Mineola: Dover (2002).

Chapter 12
Supersonic Flow and Shocks

12.1 Shock Kinematics and Entropy

The present chapter treats supersonic flows and shock fronts (also termed shock waves or simply shocks). A shock is an enduring and smooth surface in a fluid through which the fluid passes; in this passage, the fluid density, pressure, velocity, and temperature all undergo abrupt changes ('jumps').

Supersonic flows can be generated in the laboratory in a Laval (or de Laval) nozzle, invented in the late nineteenth century and used since the 1940s for jet propulsion. Figure 12.1 shows a Laval nozzle setup at the University of Sydney, from which the Schlieren photographs of the Mach cones in Fig. 12.3 were obtained.

Shocks can be generated in the laboratory by explosives, supersonic projectiles and in shock tubes. Figure 12.2 shows the '*dome* thrown up by the shock wave' (Kolsky et al. 1949, p. 400) from an underwater explosion. This shock manifests itself first in the whitening of the water surface seen in the left photograph:

> High-speed photography showed a new phenomenon during the first few milliseconds after detonation, which was provisionally termed the "crack". It is seen as a whitening below the surface of the water, which grows at a rate closely corresponding to the arrival at the surface of the shock wave. [...] The water in the dome [...] is so completely broken up that the incident light is scattered equally in all directions, giving it a white appearance. (Kolsky et al. 1949, p. 379, 389)

Figure 12.3 shows the Mach cones of a projectile in a supersonic flow. Here air experiences a sharp pressure increase when hit by the cone. In a shock tube, finally, two gases at significantly different pressures are separated by a diaphragm, which is an aluminum foil in Fig. 12.4. After the diaphragm bursts in a controlled manner, the high-pressure gas expands into the volume filled by the low-pressure gas. For appropriate initial conditions, this expansion is supersonic, and a shock front propagates at the head of the high-pressure gas region into the low-pressure gas and compresses it. Since supersonic flows and shocks are mostly studied in gases and plasmas, we will often use in the following the more specific word 'gas' instead of the more general 'fluid.'

© Springer Nature Switzerland AG 2019
A. Feldmeier, *Theoretical Fluid Dynamics*, Theoretical and Mathematical Physics,
https://doi.org/10.1007/978-3-030-31022-6_12

Fig. 12.1 A Laval nozzle used in supersonic wind tunnel experiments. 'The nozzle is 2-D with constant width of 1", not axisymmetric. The side walls consist of two 1/4" thick treated glass plates securely clamped to the top and bottom by the steel plates and bolts with hand tensioned nuts (8 of which can be seen in the picture.) The glass walls allow a Schlieren system's parallel light beams to go through the test section. The background in the picture is a felt board located outside and behind the section; when in operation this is removed. Models are inserted from down stream on long rods. The one in the picture is a blunt nose bullet shape.' Photograph and text (priv. comm.): D. Auld. See http://www.aerodynamics4students.com

Fig. 12.2 Water dome after an underwater explosion in a pond. The left image is taken 0.03 s after the explosion, the right image 0.77 s. Reproduced from Kolsky et al. (1949)

Another way to perform controlled shock experiments is to drive a piston at supersonic speed into a straight cylindrical tube filled with gas at rest. This will drive a shock front into the gas ahead of the piston. Furthermore, much of modern shock research takes place in astrophysics, where due to large relative speeds, say of the ejecta of an exploding star penetrating into the ambient interstellar gas, shock heating up to very high temperatures can occur, leading to directly observable ultraviolet, optical, and infrared emission of the gas behind the shock.

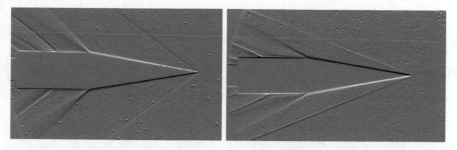

Fig. 12.3 The Mach cones from a projectile at rest in a supersonic air stream. Left: Ma = 1.7, right: Ma = 3. The pictures are digital enhancements of the original Schlieren polaroids using a graphics gradient enhancing tool kit. Photograph: D. Auld. Reproduced from http://www.aerodynamics4 students.com

Fig. 12.4 'Aluminium foil used as a diaphragm between shock tube pipe segments' in a shock tube at University of Ottawa. *Source* Wikipedia entry 'Shock tube.' Photograph: Achim Hering. https://commons.wikimedia. org/w/index.php? curid=58529132

Turning then to the theory of shock fronts, certain conceptual issues were left unclear until around 1905. Among some of the historic difficulties with the shock concept, the following may be mentioned:

(i) for inviscid fluids, shocks correspond to true mathematical discontinuities in the fluid fields ρ, p, u, T, violating the assumed smoothness (differentiability) of these fields;

(ii) even allowing for nonvanishing viscosity, shocks are unresolved on laboratory flow lengthscales, their widths corresponding to thermal, microscopic lengthscales, roughly three mean free paths of the atomic constituents of the flow;

(iii) in front of the shock and behind it, the fluid is often isentropic (uniform entropy), but there is a finite entropy jump within the shock. In shock fronts, parcels are heated and their entropy increases.

Items (i) and (ii) lead to the concept of a front, i.e., a sharp transition layer from a state before the shock to one behind it. Shocks are *transitions*, and gas passes from one side to the other. If shocks are treated as discontinuities, this transition is instantaneous.

According to (iii), shocks transform a certain fraction of the macroscopic kinetic energy of the gas (corresponding to the supersonic entrance speed relative to the front) into heat or microscopic kinetic energy. Shock transitions cannot be isentropic, i.e., entropy conserving, as is detailed in the following quote.

> That the entropy should increase across a shock was first pointed out by G. Zemplén (1905). Zemplén's remark, which today seems obvious enough, is actually far from a triviality. At the time when his note was published, both Kelvin and Rayleigh were of the opinion that the entropy jump is zero at a shock surface. They therefore had serious doubts as to the validity of the shock wave hypothesis as a model for the behavior of real gases, since the equations for mass, momentum, and energy conservation are incompatible with the presumed truth of zero entropy jump. (Serrin 1959, p. 219)

Some of this confusion can still be felt on p. 486 in Lamb's Hydrodynamics (1932), where he discusses entropy production in shocks (with some apparent unease).

Since a shock transforms kinetic energy into heat, gas is always slowed down relative to the front. The transformation of macroscopic to microscopic kinetic energy is performed by viscosity. Newtonian friction is proportional to the velocity gradient, and thus viscous forces are large if large velocity changes occur over short lengthscales. This explains the occurrence of a sharp front.

Gas enters a shock always at supersonic speed, and leaves it at subsonic speed. Due to the discontinuity in the gas speed, shocks have interesting kinematics. Gas flows from the so-called preshock to the postshock side. Let v_0, v_1, v_{sh} be the speed before and after the shock, and the shock speed itself, respectively. To obtain properties and concepts that hold for all inertial systems, only speed differences $u_0 = v_0 - v_{sh}$, $u_1 = v_1 - v_{sh}$, and $u_{ju} = v_0 - v_1$ enter the following discussion.

For the shock to be a transition, as opposed to a sink or source, u_0 and u_1 must have the same sign in any frame of reference; only then there is a mass flux through the shock. The shock speed v_{sh} is therefore either larger than the maximum of v_0 and v_1 or smaller than the minimum of v_0 and v_1. The shock speed cannot lie in between the adjacent gas speeds.

To avoid case distinctions, all speeds v are assumed to be positive or zero in the following. Then in a *forward* shock, $v_{sh} > \max(v_0, v_1)$ and thus $u_0, u_1 < 0$ relative to the shock front. The shock overtakes the gas ahead of it and leaves it behind. The kinematics is shown in the xv-diagram in Fig. 12.5, where the slow gas passes through the shock from the right to the left and is accelerated in the shock transition, in the direction of shock propagation. The postshock speed is closer to the shock speed than the preshock speed was, and thus the kinetic energy of the gas *relative to the shock front* decreases in the transition. Forward shocks typically occur in

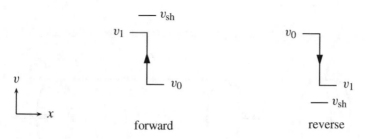

Fig. 12.5 Kinematics of forward and reverse shocks. Speeds v are indicated by horizontal bars, and shocks correspond to the vertical jumps with arrows

explosions, where gas (air) at rest is overtaken by the explosion front and has a finite speed (relative to the ground) thereafter.

In a *reverse* shock, by contrast, $v_{sh} < \min(v_0, v_1)$ and thus $u_0, u_1 > 0$. Here fast gas overtakes the front and is decelerated in it (see Fig. 12.5); again, the gas-kinetic energy relative to the front decreases in the transition. The simplest example of a reverse shock is a supersonic stream of gas hitting an obstacle, for example, a wall: the presence of the wall makes itself manifest by a shock front that propagates upstream through the gas and (partially) stops the newly incoming material. The pileup of cars on a freeway in heavy fog is a discrete approximation to a reverse shock.

Two classic books on shock physics are Courant and Friedrichs (1948), mostly employing the theory of characteristics (see p. 128 for the distinction between forward and reverse shocks), and the two volumes by Zel'dovich and Raizer (1966, 1967) on more physical aspects, for example, the internal layering of shocks. The classic monograph by Whitham (1974) treats the kinematics of shocks, and there is a three-volume handbook on shocks by Ben-Dor et al. (2000) as well as an encyclopedic book by Krehl (2009) containing 700 pages of chronology.

12.2   Jump Conditions at Shocks

A shock is a narrow transition zone, which in the limit of vanishing viscosity becomes a discontinuity ('jump') in fluid properties. Vanishing viscosity is assumed in the following to demonstrate that no mathematical problems result from the non-continuity and non-differentiability of the fluid fields $\rho, \vec{u}, p, T$; the jump conditions at the shock are obtained from the conservation law for mass and the balance laws for momentum and energy. Note that within the shock, viscosity plays a key role in transforming kinetic energy into heat. These microscopic processes, however, need *not* be considered when applying the conservation and balance laws to the pre- and postshock gas, which to a very good approximation are inviscid.

Since the shock is very narrow, ideally a two-dimensional surface, and since the shock transition is almost instantaneous, the jump conditions at any position of

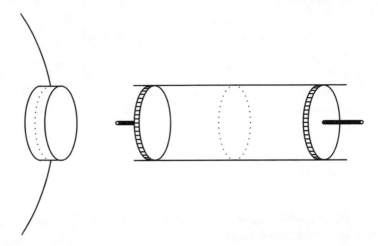

Fig. 12.6 Left: spherical shock front and planar pillbox. Right: tube with two pistons

the shock front are found by considering constant, planar flow through the tangential plane of this front: to obtain the inflow–outflow balance at a shock, the time interval δt of shock passage can be assumed to be arbitrarily small due to the vanishing thickness of the shock, and the extent $\delta \vec{r} = \vec{u}\,\delta t$ of the gas regions entering and leaving the shock becomes arbitrary small, and thus the functions $\rho, \vec{u}, p, T$, which are smooth before and after the shock, can be approximated by constants. Expressed differently, the shock adapts instantaneously to any (continuous) changes in the incoming gas.

To simplify matters further, it is assumed that the gas speed $\vec{u}$ has only a normal component u to the shock front both on the pre- and postshock sides, which defines a *perpendicular* shock. The case of *oblique* shocks, like the Mach cone of a supersonic jet plane, is not treated here.

The transition through the shock can then be analyzed by considering an infinitesimal pillbox that is cut into two halves by the shock, the two volumes containing the gas before and after the transition, see Fig. 12.6. Scaling this to macroscopic dimensions, the flow in the pillbox can be treated as a one-dimensional flow through a straight cylindrical tube with no velocity component normal to the cylinder axis and with constant states ahead of the shock and behind it. The latter statement includes the assumption that the gas before and behind the shock is isentropic, i.e., has constant specific entropy (though with different values in the two domains). Following Zel'dovich and Raizer (1966, p. 47), the tube flow is enclosed by two pistons, where one piston is driven into the gas at some pressure p_0 that enforces a shock transition, and the postshock gas drives the second piston at pressure p_1 through the tube.

Mass Conservation

Let the cross-sectional area of the tube be unity. The speed relative to the shock front is again denoted by u. Then the rate at which mass flows into the shock is $\rho_0 u_0$, where a subscript zero refers to the preshock site, and the rate at which mass leaves the shock is $\rho_1 u_1$. Thus mass conservation implies

$$\boxed{\rho_0 u_0 = \rho_1 u_1} \tag{12.1}$$

Momentum Balance

The gas momentum density in the frame of the shock front is ρu. Gas momentum enters the shock (of unit cross-sectional area) at a rate $\rho_0 u_0^2$ and leaves it at a rate $\rho_1 u_1^2$. The gas is pushed into the shock by a pressure force p_0 per area, and the gas behind the shock pushes the piston ahead of it with a pressure force p_1. According to Newton's second law, the force–momentum balance equation is then

$$\rho_1 u_1^2 - \rho_0 u_0^2 = p_0 - p_1, \tag{12.2}$$

(p_0 has a positive sign since it drives the gas), or

$$\boxed{\rho_0 u_0^2 + p_0 = \rho_1 u_1^2 + p_1} \tag{12.3}$$

Energy Balance

The total gas energy consists of kinetic and thermal energy. Any changes in this total energy are ruled by the energy law of mechanics and the first law of thermodynamics, according to the heat supply and to the work done in the shock transition. Any heat exchange (e.g., by heat conduction) is assumed to be very slow, and thus negligible in the shock. Furthermore, gravitational work can be neglected since it scales linearly with the distance crossed against the gravitational field, and the shock has negligible width. Energy changes can then only result from the pressure work $-p\,\delta V$ performed when compressing the gas.

The pressure work per area done on the gas by the piston in Fig. 12.6 (corresponding to energy gained by the gas) when driving the gas into the shock is $p_0 u_0$, and the work done by the gas in driving the second piston (energy lost to the gas) is $-p_1 u_1$; therefore, the net work done per unit surface of the gas is $p_0 u_0 - p_1 u_1$ (see also Courant and Friedrichs 1948, p. 122).

Let dE be the sum of the kinetic and thermal energy of a fluid parcel with volume dV, and $e = dE/dV$ the energy density of the fluid; then e is a scalar field defined at any location and time, similar to the mass density ρ. Let

$$e = \frac{1}{2}\rho u^2 + \epsilon, \tag{12.4}$$

with thermal energy density ϵ of the gas. The rate at which gas energy enters the shock is $e_0 u_0$, and gas energy leaves the shock at a rate $e_1 u_1$. The difference between the two is due to the pressure work done in the shock transition, and therefore the energy balance reads

$$\left[\frac{1}{2}\rho u_1^2 + \epsilon_1\right] u_1 - \left[\frac{1}{2}\rho u_0^2 + \epsilon_0\right] u_0 = p_0 u_0 - p_1 u_1 \tag{12.5}$$

(since p_0 drives the gas *into* the shock, it must appear with a plus sign). It remains to express the thermal energy density ϵ by the known scalar fields ρ, p, T of the fluid. The gas on both sides of the shock is in thermal equilibrium, and thus the Maxwell–Boltzmann equipartition theorem from kinetic gas theory applies, which states that the mean thermal energy η of a single molecule with f degrees of freedom is

$$\eta = \frac{f}{2} k_B T, \tag{12.6}$$

with the Boltzmann constant k_B. Replacing f by the adiabatic exponent γ through the relation $\gamma = 1 + 2/f$ gives

$$\eta = \frac{k_B T}{\gamma - 1}. \tag{12.7}$$

The particle density dN/dV (number of molecules per infinitesimal volume) is, for a mean particle mass $\mu = dM/dN$ (with total mass dM of dN particles),

$$\frac{dN}{dV} = \frac{dM}{dV}\frac{dN}{dM} = \frac{\rho}{\mu}. \tag{12.8}$$

Therefore,

$$\epsilon = \frac{dN}{dV}\,\eta = \frac{\rho\eta}{\mu} = \frac{\rho k_B T}{\mu(\gamma - 1)}. \tag{12.9}$$

From the ideal gas law,

$$\frac{p}{\rho} = \frac{k_B T}{\mu}, \tag{12.10}$$

which leads to the remarkably simple expression

$$\epsilon = \frac{p}{\gamma - 1} \tag{12.11}$$

for the thermal energy density of an ideal gas. Thus the energy balance at a shock becomes

$$\left[\frac{1}{2}\rho_1 u_1^2 + \frac{p_1}{\gamma - 1}\right] u_1 - \left[\frac{1}{2}\rho_0 u_0^2 + \frac{p_0}{\gamma - 1}\right] u_0 = p_0 u_0 - p_1 u_1, \tag{12.12}$$

or

$$\boxed{\frac{1}{2}\rho_0 u_0^3 + \frac{\gamma}{\gamma - 1}\,p_0 u_0 = \frac{1}{2}\rho_1 u_1^3 + \frac{\gamma}{\gamma - 1}\,p_1 u_1} \tag{12.13}$$

If the shock is sufficiently strong to dissociate molecules into atoms, then γ takes different values on the left and right sides of this equation, i.e., one has to distinguish also between γ_0 and γ_1.

Equations (12.1), (12.3), and (12.13) are the jump conditions at a shock and allow to derive the ratios ρ_1/ρ_0, u_1/u_0, and p_1/p_0. To simplify matters, one assumes first a *strong* shock defined by

$$p_1/p_0 \to \infty \quad \text{or} \quad p_0 = 0. \tag{12.14}$$

This postulates that the gas pressure must grow in a shock, and thus allows to distinguish the pre- and postshock sides and gives a unique direction in which the shock is crossed. This (entropy) condition is discussed further in Sect. 12.4. From (12.3),

$$\rho_1 u_1^2 + p_1 = \rho_0 u_0^2. \tag{12.15}$$

With (12.1) this becomes

$$\rho_0 u_0 u_1 + p_1 = \rho_0 u_0^2, \tag{12.16}$$

and thus

$$u_1 = u_0 - \frac{p_1}{\rho_0 u_0}. \tag{12.17}$$

Note again that u_0 and u_1 are measured relative to the shock. Equation (12.13) gives

$$\frac{1}{2}\rho_1 u_1^3 + \frac{\gamma}{\gamma - 1} p_1 u_1 = \frac{1}{2}\rho_0 u_0^3, \tag{12.18}$$

or, after division by $\rho_1 u_1^2$ and using (12.1) again,

$$\frac{u_1}{2} + \frac{\gamma}{\gamma - 1}\frac{p_1}{\rho_0 u_0} = \frac{u_0^2}{2u_1}, \tag{12.19}$$

or

$$\frac{p_1}{\rho_0 u_0} = \frac{\gamma - 1}{2\gamma}\left(\frac{u_0^2}{u_1} - u_1\right). \tag{12.20}$$

Using (12.17) yields

$$u_1 = u_0 - \frac{\gamma - 1}{2\gamma}\left(\frac{u_0^2}{u_1} - u_1\right). \tag{12.21}$$

This is a quadratic equation in both u_0 and u_1,

$$(\gamma + 1)u_1^2 - 2\gamma u_1 u_0 + (\gamma - 1)u_0^2 = 0, \tag{12.22}$$

and has the trivial solution $u_0 = u_1$ (no shock) and the nontrivial solution

$$\frac{u_1}{u_0} = \frac{\gamma - 1}{\gamma + 1}. \tag{12.23}$$

From (12.1) follows then

$$\boxed{\frac{\rho_1}{\rho_0} = \frac{\gamma + 1}{\gamma - 1}} \tag{12.24}$$

The postshock pressure is, using (12.20),

$$p_1 = \frac{2}{\gamma + 1} \rho_0 u_0^2. \tag{12.25}$$

For a monatomic gas, $\gamma = 5/3$, and

$$\frac{\rho_1}{\rho_0} = \frac{u_0}{u_1} = 4, \qquad p_1 = \frac{3}{4} \rho_0 u_0^2. \tag{12.26}$$

A useful quantity is the velocity jump u_{ju} at the shock,

$$u_{\mathrm{ju}} = u_0 - u_1 = \frac{2}{\gamma + 1} u_0. \tag{12.27}$$

For $\gamma = 5/3$, this gives $u_1 = u_{\mathrm{ju}}/3$.

We also give the jump conditions at an arbitrary (not strong) shock. Let

$$\mathrm{Ma} = u_0/a \tag{12.28}$$

be the Mach number of a shock, u_0 again the preshock speed relative to the front, and a the sound speed in the gas ahead of the shock. Then (see Landau and Lifshitz 1987, p. 335 for details of the derivation)

$$\frac{\rho_1}{\rho_0} = \frac{u_0}{u_1} = \frac{(\gamma + 1)\,\mathrm{Ma}^2}{(\gamma - 1)\,\mathrm{Ma}^2 + 2},$$
$$\frac{p_1}{p_0} = \frac{2\gamma\,\mathrm{Ma}^2 + 1 - \gamma}{\gamma + 1}. \tag{12.29}$$

Finally, as a somewhat simpler case, we consider arbitrary *isothermal* shocks with $\gamma = 1$. The assumption $T = \mathrm{const}$ (which may be realized when energy from shock heating is immediately radiated away) solves the energy equation, which can therefore be dropped. For a gas at constant temperature, $p = a^2 \rho$, and the jump conditions become

$$\rho_1 u_1 = \rho_0 u_0,$$
$$\rho_1(u_1^2 + a^2) = \rho_0(u_0^2 + a^2). \tag{12.30}$$

Division of the two equations gives

$$\frac{u_1^2 + a^2}{u_1} = \frac{u_0^2 + a^2}{u_0}, \tag{12.31}$$

or

$$u_1^2 u_0 + a^2 u_0 = u_0^2 u_1 + a^2 u_1. \tag{12.32}$$

From inspection, this has the solution

$$a^2 = u_0 u_1, \tag{12.33}$$

or, with the Mach number $\mathrm{Ma} = u_0/a$,

$$u_1 = \mathrm{Ma}^{-2} u_0, \qquad \rho_1 = \mathrm{Ma}^2 \rho_0. \tag{12.34}$$

For a strong shock, $p_0 = 0$, therefore $a = 0$ and $\mathrm{Ma} \to \infty$.

To illustrate the jump conditions, we give two simple examples for shocks in flows with $\gamma = 5/3$. First, let a supersonic gas in a straight tube hit a solid wall terminating the tube, see Fig. 12.7. The incoming gas speed relative to the terminating wall shall be $v_0 = 1$. One can patch together two solutions with $\rho = \text{const}$ and $v = \text{const}$ left and right of the shock. For these it must hold that

$$u_0 = v_0 - v_{\text{sh}} = 4(v_1 - v_{\text{sh}}) = 4u_1. \tag{12.35}$$

For $v_0 = 1$ and $v_1 = 0$ (the gas is stopped by the wall), this gives a shock speed $v_{\text{sh}} = -\frac{1}{3}$. The gas accumulates between the wall and the shock; the finite postshock pressure pushes the shock through the tube, so that it meets the gas at relative speed $u_0 = \frac{4}{3}$.

Fig. 12.7 Kinematics of a wall shock

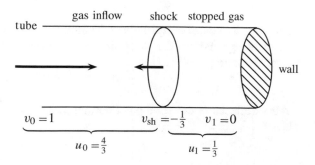

Fig. 12.8 Double shock
consisting of a reverse shock
to the left, moving at speed
$v_{\rm sh} = 2/3$, and a forward
shock to the right, moving at
$v_{\rm sh} = 4/3$

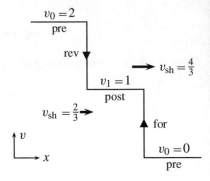

As a second example, consider an infinitely long, straight tube. At $t = 0$, the gas
at $x > 0$ shall be at rest, $v_0 = 0$. At $t > 0$, this gas is hit by a gas stream coming from
$x < 0$ with speed $v_0 = 2$. We look for a solution that features a *pair* of an equally
strong reverse and forward shock. The postshock gas in the region between the two
shocks has then $v_1 = 1$ by symmetry. To obey the 4:1 jumps in relative speeds for
$\gamma = \frac{5}{3}$, the reverse shock at small x has $v_{\rm sh} = \frac{2}{3}$, and the forward shock at large x
has $v_{\rm sh} = \frac{4}{3}$, see Fig. 12.8. Thus, the forward shock *overtakes* gas and accelerates
it, and the reverse shock *slows* gas *down*. In between the two shocks is a region
of trapped postshock gas. Obviously, one can assume quite arbitrary ratios for the
relative strengths of the reverse and forward shocks, which is further discussed in
Sturrock and Spreiter (1965).

12.3   Shock Speed

Next, we can derive a general formula for the shock speed, following Lax (1973),
p. 7. Let ρ be any density field (like mass, momentum, or energy density), and $\vec{j}(\rho)$ the
corresponding flux. Given an arbitrary but fixed volume V, the general conservation
law is

$$\frac{{\rm d}M}{{\rm d}t} = \frac{{\rm d}}{{\rm d}t} \int_V {\rm d}V\, \rho = - \int_{\partial V} {\rm d}\vec{a} \cdot \vec{j}, \qquad (12.36)$$

where M is the conserved quantity with volume density ρ (and may be mass, momentum, or energy). For fixed V, the time derivative can be taken inside the integral, and
Gauss' theorem gives

$$\dot{\rho} + \nabla \cdot \vec{j} = 0. \qquad (12.37)$$

Let the Cartesian x-direction point normal to the shock front, and let $j' = \partial j_x / \partial x$.
Then

$$\dot{\rho} + j' = 0 \qquad (12.38)$$

holds in the 1-D planar approximation at the shock. A point on the shock front has a trajectory $X(t)$. Let ρ_0 and ρ_1 denote again the pre- and postshock densities, respectively, and the jump across $X(t)$ is written as $[\rho] = \rho_1 - \rho_0$. To be specific, let the preshock (postshock) side be left (right) of X. At time t, choose x_- and x_+ to be two points far away from $X(t)$, with $x_- \ll X \ll x_+$; moreover, let A be an arbitrary, constant area normal to x (planar flow). From the first equality in (12.36), one has

$$
\begin{aligned}
\frac{1}{A} \frac{dM}{dt} &= \frac{d}{dt} \int_{x_-}^{x_+} dx\ \rho(t, x) \\
&= \frac{d}{dt} \left(\int_{x_-}^{X(t)} dx\ \rho + \int_{X(t)}^{x_+} dx\ \rho \right) \\
&= \int_{x_-}^{X(t)} dx\ \dot{\rho} + \dot{X} \rho_0 + \int_{X(t)}^{x_+} dx\ \dot{\rho} - \dot{X} \rho_1 \\
&= - \int_{x_-}^{X(t)} dx\ j' - \int_{X(t)}^{x_+} dx\ j' - \dot{X}[\rho] \\
&= j(t, x_-) - j(t, x_+) + [j] - \dot{X}[\rho].
\end{aligned}
\tag{12.39}
$$

Alternatively, from the second equality in (12.36), one has

$$
\frac{1}{A} \frac{dM}{dt} = j(t, x_-) - j(t, x_+).
\tag{12.40}
$$

Equations (12.39) and (12.40) give for the shock speed,

$$
\boxed{\dot{X} = \frac{[j]}{[\rho]}}
\tag{12.41}
$$

Specifically in the shock frame where $\dot{X} = 0$, one obtains

$$
\boxed{[j] = 0}
\tag{12.42}
$$

which corresponds to the jump conditions (12.1)–(12.13). Equation (12.42) is called the Rankine–Hugoniot jump condition at a shock.

12.4 Shock Entropy and Supersonic Inflow

It was mentioned above that gas can pass through a shock in one direction only. The reason is that the gas entropy increases when gas passes the shock in this allowed direction, but would decrease in the opposite direction—which is forbidden according to the second law of thermodynamics.

In all possible shock transitions, the gas pressure, temperature, and entropy increase (constant temperature in an isothermal shock is a somewhat artificial assumption). Furthermore, shocks in polytropic gases (where pressure scales as some power of density, see Sect. 2.2) are always compressive, as was shown by Jouguet (1901) and Zemplén (1905). Thus, in polytropic gases p, T, and also ρ grow in a shock transition. This is also true for most non-polytropic gases, but the proof is harder, and is given in Courant and Friedrichs (1948), p. 144, following Weyl (1944). Details can also be found in Landau and Lifshitz (1987), p. 329. For a modern mathematical version of the proof, see Lax (1973), pp. 9–17. The proof requires that a convexity condition holds along adiabatic curves (for which entropy S is constant),

$$\left.\frac{\partial^2 \rho^{-1}}{\partial p^2}\right|_S > 0. \tag{12.43}$$

From this one derives that at shocks,

$$\frac{\partial^2 \rho^{-1}}{\partial p^2} > 0, \tag{12.44}$$

which in turn is used to show that density increases in a shock. However, $[\partial^2 \rho^{-1}/\partial p^2]_S$ can actually change sign, as happens in *phase transitions*. In this case, then, *rarefaction shocks* (jumps to lower densities) are possible. They are further treated in Zel'dovich and Raizer (1966, 1967), in sections I §19 and XI §20.

Let us return to the standard case of a compression shock, with a jump to higher densities. Here, kinetic flow energy is consumed in slowing the gas down relative to the front. For $\rho_1/\rho_0 > 0$ and $p_1/p_0 > 0$, Eq. (12.29) imply that Ma > 1 at the shock. Thus gas is supersonic ahead and supersonic behind the shock. We have thus obtained a result on shock kinematics from an entropy condition. This is awkward since kinematics is basic, but entropy is an involved concept; indeed, a simpler argument (counting unknowns and equations) can be given for Ma > 1 ahead of any shock, see Landau and Lifshitz (1987), §88 and Zel'dovich and Raizer (1966), p. 61.

12.5   The Laval Nozzle and Solar Wind

After these elementary considerations on shocks, we study two supersonic flow types that can harbor shocks, first flow through a Laval nozzle, and then the solar wind.

The Laval Nozzle

The standard technological method to generate steady supersonic flows is to expand high-pressure gas through a Laval nozzle, see Fig. 12.9. The cross section of a Laval nozzle is large both at the inlet and outlet, and has a single minimum in between. The gas enters the nozzle at high pressure and leaves it supersonically.

For simplicity, a stationary flow under adiabatic conditions is assumed through the nozzle. The flow shall be 1-D in the axis direction of the nozzle, and there is no flow component normal to the nozzle axis. Let $q(x)$ be the cross-sectional area of

Fig. 12.9 The Laval nozzle: longitudinal section of an RD-107 rocket engine in the Tsiolkovsky State Museum of the History of Cosmonautics. Image source: Wikipedia entry 'de Laval nozzle', Photograph: Albina-belenkaya

the nozzle as function of x. In steady state, the continuity and Euler equations are then, with a prime indicating an x-derivative,

$$(\rho u q)' = 0,$$

$$u u' = -a^2 \, \frac{\rho'}{\rho}, \tag{12.45}$$

with adiabatic sound speed a (gravity is neglected). Dividing the continuity equation by $\rho u q$ gives

$$\frac{\rho'}{\rho} + \frac{u'}{u} + \frac{q'}{q} = 0. \tag{12.46}$$

Inserting ρ'/ρ from this equation in the Euler equation yields

$$\left(u - \frac{a^2}{u}\right) u' = a^2 \frac{q'}{q},$$
(12.47)

or

$$u' = \frac{a^2 q'/q}{u - a^2/u}.$$
(12.48)

To have always finite u', the numerator in (12.48) must vanish if the denominator vanishes,

$$u = a \quad \rightarrow \quad q' = 0.$$
(12.49)

In a nozzle with a profile that first narrows and then widens again, the flow *may* become supersonic exactly at the location of smallest cross section. Note, however, that $u < a$ and $u' = 0$ at $q' = 0$ is another possible solution. Equation (12.48) gives the basic working principle of the Laval nozzle: a subsonic flow *accelerates* up to the location where $q' = 0$, since both $u - a^2/u < 0$ and $a^2 q'/q < 0$, thus $u' > 0$. If $u = a$ at the location where $q' = 0$, then the flow *accelerates* further beyond the constriction, in the widening part of the nozzle, since $u - a^2/u > 0$ and $a^2 q'/q > 0$, thus again $u' > 0$. If, however, the flow has $u < a$ at $q' = 0$, this subsonic flow decelerates again in the widening part of the nozzle, $u - a^2/u < 0$ and $a^2 q'/q > 0$, thus $u' < 0$. Whether the flow stays subsonic throughout the nozzle or becomes supersonic at the constriction is controlled by the pressure ratio between inlet and outlet.

   To understand better why supersonic flow, in contrast to subsonic flow, accelerates when it expands, we rewrite the Euler equation in the form

$$\frac{d\rho}{\rho} = -\frac{u\,du}{a^2} = -\frac{u^2}{a^2}\frac{du}{u},$$
(12.50)

or, with the Mach number $\mathrm{Ma} = u/a$,

$$\frac{d\rho}{\rho} = -\mathrm{Ma}^2 \frac{du}{u}.$$
(12.51)

The continuity equation is then

$$-\frac{dq}{q} = \frac{d(\rho u)}{\rho u} = \frac{d\rho}{\rho} + \frac{du}{u} = \left(1 - \mathrm{Ma}^2\right)\frac{du}{u}.$$
(12.52)

Equation (12.51) shows that for $\mathrm{Ma} < 1$, changes in speed outweigh changes in density (with incompressible flow for $\mathrm{Ma} \to 0$), and that the opposite is true for $\mathrm{Ma} > 1$. Since ρ scales with the thermal pressure p, one can say in the latter case that supersonic flow is dominated by thermal properties. For accelerating supersonic flow, the relative drop $d\rho/\rho$ in density (and pressure) is larger in magnitude than the relative increase du/u in speed, thus the mass flux ρu drops according to (12.52), and the accelerating flow needs 'more space' at this low pressure, $dq/q > 0$. For this discussion and more details on the Laval nozzle, see Zierep (1991), p. 53.

Solar Wind

Quite surprisingly, the flow through a Laval nozzle and the solar wind have certain similarities. For an interesting historical account of the intricate plasma research undertaken that lead eventually to this rather simple, revolutionary insight, see Parker (1999). The outer layers of the solar atmosphere are heated to $\approx 10^6$ K, and this *corona* extends from one to approximately 1.5 solar radii. The complicated mechanisms that are responsible for heating the corona are not addressed here. Parker (1958) suggested that the coronal gas expands, and at some distance from the Sun becomes supersonic, with speeds around 500 km/s. This largely radial outflow from the Sun is termed *solar wind*.

To model this outflow, we assume a stationary, 1-D, spherically symmetric flow in the radial direction, with r the distance to the center of the Sun. Due to efficient heat conduction, the flow shall have constant temperature; this is one of Parker's central assumptions. The hydrodynamic equations are then, with a prime for an r-derivative now, G the gravitational constant, and M the solar mass,

$$(\rho u r^2)' = 0,$$

$$uu' = -a^2 \frac{\rho'}{\rho} - \frac{GM}{r^2}. \tag{12.53}$$

Dividing the continuity equation by $\rho u r^2$ yields

$$\frac{\rho'}{\rho} + \frac{u'}{u} + \frac{2}{r} = 0, \tag{12.54}$$

and inserting ρ'/ρ from (12.54) into the Euler equation (12.53) gives

$$\left(u - \frac{a^2}{u}\right) u' = \frac{2a^2}{r} - \frac{GM}{r^2}, \tag{12.55}$$

or

$$u' = \frac{2a^2/r - GM/r^2}{u - a^2/u}. \tag{12.56}$$

To have always finite u', the numerator must again vanish if the denominator vanishes. This gives for the location r_a of the sonic point at which $u(r_a) = a$,

$$r_a = \frac{GM}{2a^2}. \tag{12.57}$$

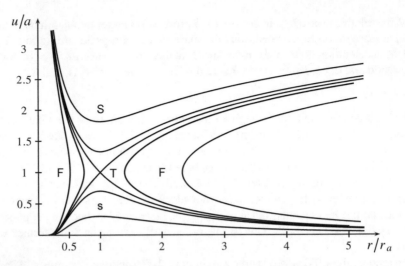

Fig. 12.10 The Parker solar wind solutions $u(r)$ from (12.59), for two transonic solutions T passing through $(r, u) = (r_a, a)$ (with integration constant $C = -3$), fully subsonic and supersonic solutions, s and S, respectively (with $C > -3$), and failed solutions F (with $C < -3$), for which $u(r)$ is imaginary in a finite r-interval

Inserting typical numbers for the Sun, $r_a \approx 1.5$ solar radii. The expression $u - a^2/u = 0$ in the denominator is called a singularity condition, and $2a^2/r - GM/r^2 = 0$ in the numerator is called a regularity condition: in order that a singularity can occur, a regularity condition must be obeyed. Equation (12.55) can be written as

$$u \, du - a^2 \frac{du/a}{u/a} = 2a^2 \frac{dr/r_a}{r/r_a} - GM \frac{dr}{r^2}. \tag{12.58}$$

This is readily integrated to give an implicit equation for $u(r)$,

$$\frac{u^2}{a^2} - \ln \frac{u^2}{a^2} = \frac{4r_a}{r} + 4 \ln \frac{r}{r_a} + C, \tag{12.59}$$

with an integration constant C. Inserting $u = a$ at $r = r_a$ gives $C = -3$ for the transonic Parker solution. This has $u < a$ below and $u > a$ above the sonic radius r_a.

Both the Laval nozzle and the solar wind show an interesting X-type solution topology in an ru-diagram. The full set of solutions for the solar wind velocity law $u(r)$ is obtained by varying the constant C in (12.59) and is shown in Fig. 12.10. Depending on the value of C, it consists of three families of solution curves: purely subsonic flow (marked s in the figure); purely supersonic flow (S); and failed wind (F); plus two unique curves of transonic flow (T). The failed solutions do not exist in a finite interval around the sonic point, but only at smaller and larger radii; their mass flux ρu is too large to be accelerated at all radii.

We do not go here into the extensive literature on the solar wind, but briefly discuss two serious challenges raised against this idea, both of them met excellently by the theory developed by Parker (1958, 1960a, b).

First, the solar wind model was disputed by Chamberlain (1960), who proposed subsonic *breeze* solutions instead of a transonic flow. Chamberlain envisages that the corona *evaporates* into space, and does not expand as a hydrodynamic flow that becomes eventually supersonic. He suggests that particles from the tail of the Maxwellian velocity distribution, with speeds in excess of the local escape speed from Sun, escape to infinity and are lost to the thermal pool. They are replenished and on the assumption that this process is steady a stationary subsonic outflow results. The expansion velocity for thermal evaporation has a maximum at a certain distance r, as do the hydrodynamic solar wind *breezes*, marked s in Fig. 12.10. Thus, the solar wind should be an overall subsonic evaporation outflow, and Chamberlain surmises that Parker picked the wrong flow solution.

The main argument of Parker (1960a) against Chamberlain is that evaporation is only important in very tenuous gases, when collisions play almost no role (large mean free paths). By contrast, the (expanding) solar corona is so dense that collisions are important; hence, a hydrodynamic flow is set up, not gas-kinetic evaporation. Second, Parker shows that breeze solutions have larger pressure at infinity than a supersonic flow (eventually, by orders of magnitude), which could not meet the very low pressure of the interplanetary or interstellar gas. He concludes that 'if there is nothing to contain this enormous pressure, it is obvious that the expansion will go more rapidly than we have computed, and [the speed] will increase' (Parker 1960a, p. 180). The ultimate decision between the transonic wind and breeze evaporation came in the years 1962–64, when direct measurements of particle densities and velocities became possible in space and highly supersonic speeds were found.

Cannon and Thomas (1977) challenge Parker's solar wind model by lessening the role of the corona in driving the solar wind. They suggest that

> the rates of mass loss [...] are determined by conditions imposed on the flow at or below photospheric levels and that the warm chromosphere and hot corona of a solar-like star are simply consequences of this imposed photospheric flow and dissipation in the resultant stellar wind. (Wolfson and Holzer 1982, p. 610)

In their own words,

> We reformulate the wind-tunnel [Laval nozzle] analogy to stellar winds, suggesting that stars satisfy an "imperfect," rather than "perfect," such model, i.e., transonic shocks occur before the throat [the location of $q' = 0$ in the Laval nozzle], corresponding to an imposed outward velocity in the [...] subatmosphere. (Cannon and Thomas 1977, p. 910)

This argument builds upon the failed solar wind solutions (F in Fig. 12.10). Cannon and Thomas (1977) suggest that all stars show sub-photospheric outward motions with a prespecified mass flux. Usually, this mass flux will be too large to leave the star, and stalls in a shock jump that starts on a failed (F) solution and ends on a subsonic (s) solution in Fig. 12.10. This shock heats the chromosphere and corona, and a part of the original flow is accelerated to supersonic speeds as solar wind.

Fig. 12.11 Entropy for different thermal wind solutions. The Cannon and Thomas (1977) imperfect nozzle corresponds to the forbidden shock at A. Abscissa: distance from the Sun in units of the distance of the sonic point. Ordinate: the Mach number squared. Reproduced from Wolfson and Holzer (1982)

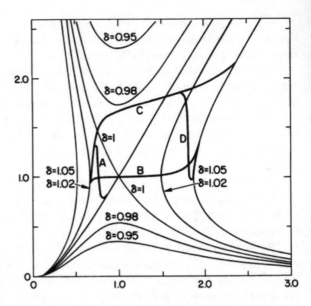

Parker (1981) rules out the Thomas model by showing that the sub-photospheric inflow speed into the solar wind is of no relevance for the wind dynamics, but that the photosphere only replenishes the mass lost in the wind. He finds that 'blocking the flow at the photosphere would produce no detectable effect in the wind velocity, nor any decline in the density detectable within the millennium' (Parker 1981, p. 270). Thus, the photospheric inflow speed at the inner boundary of the wind is, contrary to the assertion by Cannon and Thomas (1977), not the 'prime mover' of the wind.

Wolfson and Holzer (1982) rule out the Cannon and Thomas (1977) model by an ingenious argument. They show that the different solutions (transonic, breeze, and failed) of a polytropic wind have constant entropy at distinct values. The δ-values in Fig. 12.11 scale monotonically with entropy. The scenario suggested by Thomas would correspond to the shock A, which is a reverse shock decelerating fast, inner flow, and heating the corona. At this shock, however, the flow undergoes a transition from high to low entropy, which is ruled out by the second law of thermodynamics. By contrast, shocks like D in Fig. 12.11, located above the sonic point, are allowed, since the entropy markers indicate a jump from a state of low to one of large entropy. (Wolfson and Holzer 1982 'integrate the solar wind equations right through the shocks, rather than matching separate inviscid solutions across a discontinuity representing the shocks.' R. Wolfson, priv. comm.)

12.6   Supersonic Spots

The Laval nozzle was treated above in a simplified 1-D approximation. We include now the sideward flow component in a 2-D transonic flow, following a paper by Tomotika and Tamada (1950). The flow is assumed to be stationary in the xy-plane.

The xy-cuts through the nozzle are stacked upon each other in the z-direction, and thus the nozzle is slab-like, not axially symmetric. The flow is assumed to be inviscid and irrotational, $\nabla \times \vec{u} = 0$. The latter implies the existence of a potential ϕ with

$$\phi_x = u = q \ \cos\varphi,$$
$$\phi_y = v = q \ \sin\varphi, \tag{12.60}$$

where q is the magnitude of the speed and φ is the angle of the velocity vector with the x-axis. Let

$$\rho\vec{u} = \nabla \times (\psi\hat{z}), \tag{12.61}$$

with generalized streamfunction ψ (note that ρ is included), or

$$\psi_x = -\rho v = -\rho q \ \sin\varphi,$$
$$\psi_y = \rho u = \rho q \ \cos\varphi. \tag{12.62}$$

The total differentials of the potentials ϕ and ψ are

$$d\phi = \phi_x \ dx + \phi_y \ dy,$$
$$d\psi = \psi_x \ dx + \psi_y \ dy. \tag{12.63}$$

Inserting here (12.60) and (12.62) gives

$$\rho \ d\phi = \rho q \ \cos\varphi \ dx + \rho q \ \sin\varphi \ dy, \tag{12.64}$$
$$d\psi = -\rho q \ \sin\varphi \ dx + \rho q \ \cos\varphi \ dy. \tag{12.65}$$

Next, multiply (12.64) by $\cos\varphi$ and (12.65) by $\sin\varphi$ and subtract the resulting equations. Then multiply (12.64) by $\sin\varphi$ and (12.65) by $\cos\varphi$ and add the resulting equations. This gives

$$\rho \cos\varphi \ d\phi - \sin\varphi \ d\psi = \rho q \ dx,$$
$$\rho \sin\varphi \ d\phi + \cos\varphi \ d\psi = \rho q \ dy. \tag{12.66}$$

As was done before, x and y are considered as functions of the independent variables ϕ and ψ. Then

$$dx = x_\phi \ d\phi + x_\psi \ d\psi,$$
$$dy = y_\phi \ d\phi + y_\psi \ d\psi. \tag{12.67}$$

Comparing the coefficients in (12.67) with the corresponding ones in (12.66) gives

$$x_\phi = \frac{\cos\varphi}{q}, \qquad x_\psi = -\frac{\sin\varphi}{\rho q}, \tag{12.68}$$

$$y_\phi = \frac{\sin\varphi}{q}, \qquad y_\psi = \frac{\cos\varphi}{\rho q}. \qquad (12.69)$$

Eliminating x and y by differentiation using $x_{\phi\psi} = x_{\psi\phi}$ and $y_{\phi\psi} = y_{\psi\phi}$ yields

$$\left(\frac{\cos\varphi}{q}\right)_\psi = -\left(\frac{\sin\varphi}{\rho q}\right)_\phi,$$

$$\left(\frac{\sin\varphi}{q}\right)_\psi = \left(\frac{\cos\varphi}{\rho q}\right)_\phi. \qquad (12.70)$$

Performing the ϕ- and ψ-differentiations, one finds

$$\rho^2 q \,\sin\varphi\,\varphi_\psi + \rho^2\,\cos\varphi\,q_\psi = \rho q\,\cos\varphi\,\varphi_\phi - \sin\varphi\,(\rho q)_\phi, \qquad (12.71)$$

$$-\rho^2 q\,\cos\varphi\,\varphi_\psi + \rho^2\,\sin\varphi\,q_\psi = \rho q\,\sin\varphi\,\varphi_\phi + \cos\varphi\,(\rho q)_\phi. \qquad (12.72)$$

For stationary flow, the Euler equation along a streamline is

$$q\,\mathrm{d}q = -\frac{\mathrm{d}p}{\rho}. \qquad (12.73)$$

Assume here that p is a function of ρ alone, and let

$$\frac{\mathrm{d}p}{\mathrm{d}\rho} = a^2. \qquad (12.74)$$

To simplify the algebra somewhat, constant flow temperature is assumed, with sound speed $a = \mathrm{const} = 1$. The case of adiabatic conditions is treated in Tomotika and Tamada (1950). Then from (12.73),

$$\frac{\mathrm{d}q}{\mathrm{d}\rho} = -\frac{1}{\rho q}\frac{\mathrm{d}p}{\mathrm{d}\rho}, \qquad (12.75)$$

or

$$\frac{\mathrm{d}\rho}{\mathrm{d}q} = -\rho q. \qquad (12.76)$$

Therefore, assuming a density $\rho = 1$ at the sonic point where $q = a = 1$,

$$\rho(q) = \exp[(1 - q^2)/2]. \qquad (12.77)$$

There are two unknowns, q and φ. Multiplying (12.71) by $\cos\varphi$ and (12.72) by $\sin\varphi$ and adding the resulting equations leads to

$$\varphi_\phi = \frac{\rho}{q}\,q_\psi. \qquad (12.78)$$

Similarly, from [(12.71) times $\sin \varphi$], [(12.72) times $\cos \varphi$], and subtracting the resulting expressions, one obtains

$$\varphi_\psi = -\frac{1}{\rho^2 q} (\rho q)_\phi = -\frac{1}{\rho^2 q} \left(q \frac{d\rho}{dq} + \rho \right) q_\phi, \qquad (12.79)$$

or, using (12.76),

$$\varphi_\psi = \frac{q}{\rho} \left(1 - \frac{1}{q^2} \right) q_\phi. \qquad (12.80)$$

It seems straightforward to cross-differentiate (12.78) and (12.80) now, which should give a Laplace-type equation for the unknowns φ and q in terms of the variables ϕ and ψ. The problem is, however, that ρ and q are not constant (as they would be on the free streamline for a wake or cavity, see Sect. 5.2). Tomotika and Tamada proceed by introducing a function (with dummy integration variable $\tilde{q}$)

$$w(q) = \int_1^q d\tilde{q} \, \frac{\rho(\tilde{q})}{\tilde{q}}, \qquad (12.81)$$

where $\rho = \rho(q)$ from (12.76). Note that $w > 0$, if the flow is supersonic, i.e., $q > 1$, and else $w < 0$. Then $dw/dq = \rho/q$, and using $q = q(\phi, \psi)$,

$$w_\phi = \frac{dw}{dq} q_\phi = \frac{\rho}{q} q_\phi,$$
$$w_\psi = \frac{dw}{dq} q_\psi = \frac{\rho}{q} q_\psi. \qquad (12.82)$$

With this, (12.78) and (12.80) become

$$\varphi_\phi = w_\psi, \qquad (12.83)$$
$$\varphi_\psi = X w_\phi, \qquad (12.84)$$

where

$$X = (q^2 - 1)/\rho^2. \qquad (12.85)$$

So far, ρ and q are eliminated from (12.83), but not yet from (12.84). To proceed, one has to make an approximation for X. First, from (12.77) and (12.85),

$$X(q) = (q^2 - 1) \, \exp(q^2 - 1). \qquad (12.86)$$

Introducing $w(q)$ from (12.81), the function $X(q)$ can be understood as a function $\tilde{X}(w)$. The tilde is omitted in the following since the variables will indicate which of the functions X and $\tilde{X}$ is meant. The Taylor expansion at the sonic point $q = 1$ and $w = 0$ gives

$$X(w) = X(w = 0) + \left.\frac{dX}{dq}\right|_{q=1} \left.\frac{dq}{dw}\right|_{w=0} w + \cdots, \tag{12.87}$$

and a simple calculation yields

$$X(w = 0) = 0,$$

$$\left.\frac{dX}{dq}\right|_{q=1} = 2, \tag{12.88}$$

$$\left.\frac{dq}{dw}\right|_{w=0} = 1.$$

Thus, to *first order* in w,

$$X(w) \approx 2w. \tag{12.89}$$

This approximation for small $w \approx 0$ is valid near the sonic point $q = 1$. Equations (12.83) and (12.84) become now

$$\varphi_\phi = w_\psi, \tag{12.90}$$

$$\varphi_\psi = (w^2)_\phi. \tag{12.91}$$

Eliminating φ by cross-differentiation gives the remarkable equation

$$\boxed{w_{\psi\psi} = (w^2)_{\phi\phi}} \tag{12.92}$$

Next, Tomotika and Tamada (1950) assume w to be of the form

$$\boxed{w = Z(\phi + \psi^2) + 2\psi^2} \tag{12.93}$$

where $Z(\phi + \psi^2)$ is a new function of the variable $\phi + \psi^2$. Abbreviate the latter as

$$s = \phi + \psi^2. \tag{12.94}$$

Inserting (12.93) in (12.92) gives, after a brief calculation, the ordinary differential equation

$$\frac{d}{ds}\left(Z\frac{dZ}{ds}\right) - \frac{dZ}{ds} - 2 = 0, \tag{12.95}$$

and from trivial integration, one obtains

$$Z\frac{dZ}{ds} - Z - 2s = 0. \tag{12.96}$$

The constant of integration is absorbed into s; this corresponds to shifting the coordinate origin of the s-axis. We try to solve (12.96) by a polynomial P of the form

$$P(Z, s) = Z^3 + aZ^2s + bZs^2 + cs^3 = C, \tag{12.97}$$

where C is a constant. Then

$$0 = dP = P_Z \, dZ + P_s \, ds, \tag{12.98}$$

or (implicit differentiation)

$$\frac{dZ}{ds} = -\frac{P_s}{P_Z} = -\frac{aZ^2 + 2bZs + 3cs^2}{3Z^2 + 2aZs + bs^2}. \tag{12.99}$$

Inserting this in (12.96) gives $a = -3$, $b = 0$, and $c = 4$. A solution of (12.96) is thus given (implicitly) by $Z^3 - 3Z^2s + 4s^3 = C$, or

$$(Z - 2s)^2 (Z + s) = C. \tag{12.100}$$

This function $Z(s)$ is shown in Fig. 12.12. Special significance has $C = 0$, thus

$$Z = -s \qquad \text{or} \qquad Z = 2s, \tag{12.101}$$

which gives the two straight lines in the figure.

Fig. 12.12 Solutions of (12.100) for $C = 0$ (straight lines), $C = -0.002$ (solid curves), and $C = 0.002$ (dashed)

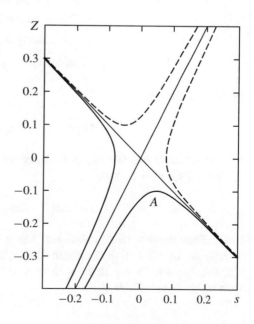

It remains to determine φ; the relevant equations are

$$w = Z + 2\psi^2,$$
$$s = \phi + \psi^2,$$
$$ZZ_s = Z + 2s,$$

according to (12.93), (12.94), and (12.96), and where $Z_s = dZ/ds$. One has $Z_\phi = Z_s s_\phi = Z_s$ and

$$Z_\psi = Z_s s_\psi = 2\psi Z_s = 2\psi Z_\phi. \tag{12.102}$$

Then from (12.90),

$$\varphi_\phi = w_\psi = Z_\psi + 4\psi = 2\psi Z_\phi + 4\psi. \tag{12.103}$$

This can be integrated immediately,

$$\varphi = 4\psi\phi + 2\psi Z + F(\psi), \tag{12.104}$$

where F is an arbitrary function of ψ. From (12.91),

$$\begin{aligned}
\varphi_\psi &= 2w w_\phi \\
&= 2(Z + 2\psi^2)Z_\phi \\
&= 2ZZ_\phi + 2\psi Z_\psi \\
&= 2ZZ_s + 2\psi Z_\psi \\
&= 2(Z + 2s) + 2\psi Z_\psi \\
&= 2Z + 4\phi + 4\psi^2 + 2\psi Z_\psi \\
&= 4\phi + 4\psi^2 + 2(\psi Z)_\psi. \tag{12.105}
\end{aligned}$$

Integrating this gives

$$\varphi = 4\psi\phi + 4\psi^3/3 + 2\psi Z + G(\phi), \tag{12.106}$$

with an arbitrary function $G(\phi)$. Comparison with (12.104) yields finally $F(\psi) = 4\psi^3/3$ and $G(\phi) = 0$; thus,

$$\varphi = 2\psi(Z + 2\phi + 2\psi^2/3). \tag{12.107}$$

Streamlines are now readily obtained. Along an individual streamline, ϕ runs from $-\infty$ to ∞ with $\phi = 0$ at the nozzle throat, and varying ψ gives the family of all streamlines, with $\psi = 0$ in the midplane of the nozzle. The relevant formulas are, with integration variable $\tilde{q}$,

$$s = \phi + \psi^2,$$

$$Z \quad \text{from} \quad (Z - 2s)^2 (Z + s) = C,$$

$$w = Z + 2\psi^2,$$

$$\varphi = 2\psi \left(Z + 2\phi + 2\psi^2/3 \right),$$

$$q \quad \text{from} \quad w = \int_1^q \frac{d\tilde{q}}{\tilde{q}} \exp\left[\tfrac{1}{2}(1 - \tilde{q}^2) \right], \tag{12.108}$$

and the streamlines are then calculated from

$$x = \int_0^\phi d\tilde{\phi} \, \frac{\cos \varphi}{q}, \tag{12.109}$$

$$y = \int_0^\phi d\tilde{\phi} \, \frac{\sin \varphi}{q} + y_0, \tag{12.110}$$

$$y_0 = \int_0^\psi d\tilde{\psi} \, \frac{\exp\left[-\tfrac{1}{2}(1 - q_0(\tilde{\psi})^2) \right]}{q_0(\tilde{\psi})}, \tag{12.111}$$

with dummy integration variables $\tilde{\phi}$ and $\tilde{\psi}$. Equations (12.109) and (12.110) are (12.68) and (12.69), respectively, for fixed ψ along a streamline, and y_0 in (12.110) and (12.111) is the distance of a streamline from the mid-axis at the throat, obtained from (12.69) and (12.77), where $q_0(\psi)$ is the speed magnitude at the nozzle throat, $x = \phi = 0$, on a streamline.

Integration from $-\infty < \phi < \infty$ is possible only along the curve marked A in Fig. 12.12, and similar curves for other constants C. For the center streamline, one has $\psi = 0$ and thus $w = Z$. Since $Z < 0$ along the whole curve A, also $w < 0$ on the center streamline, along A, the flow on the center streamline is everywhere subsonic.

However, above some threshold value for ψ, i.e., beyond a certain distance from the midplane of the nozzle, the function $w = Z + 2\psi^2$ can become positive along the curve A, and thus the flow becomes supersonic. From Fig. 12.12, this will be the case in a small interval of s-values that includes the point where $Z(s)$ has its maximum. Some feeling for the locations of these supersonic 'pockets' in the nozzle can be obtained from $w = Z + 2\psi^2 > 0$ and using for $Z(s)$ the two straight-line solutions (12.101), $Z = 2s$ for $s < 0$ and $Z = -s$ for $s > 0$. (Note that these actually have a sonic point on the center streamline, $\psi = s = Z = w = 0$.) This gives the conditions

$$w = 2s + 2\psi^2 > 0 \quad \text{for} \quad s < 0,$$

$$w = -s + 2\psi^2 > 0 \quad \text{for} \quad s > 0, \tag{12.112}$$

or, using that $s = \phi + \psi^2$,

$$\psi^2 > -\tfrac{1}{2}\phi \quad \text{and} \quad \psi^2 < -\phi, \tag{12.113}$$

$$\psi^2 > \phi \quad \text{and} \quad \psi^2 > -\phi. \tag{12.114}$$

Fig. 12.13 Supersonic
regions in the $\phi\psi^2$-plane

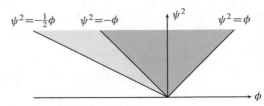

Fig. 12.14 The Laval nozzle
flow with supersonic wall
pockets after Taylor (1930)

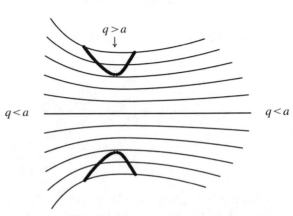

The first pair of conditions (12.113) is fulfilled in the yellow region in Fig. 12.13, and
the second pair (12.114) holds in the pink region. This shape of the domain $w > 0$
is in qualitative agreement with the shape of supersonic pockets in the xy-plane in
Fig. 12.14 (the mapping $(\phi, \psi) \to (x, y)$ is monotonic), obtained from a numerical
calculation for $C = -0.002$, see curve A in Fig. 12.12. The streamlines in Fig. 12.14
undergo sonic transitions along the bold curves, i.e., the flow is supersonic within the
V-shaped regions near the throat walls. This type of nozzle solution with subsonic
streamlines near the mid-axis and supersonic wall pockets was first found by Taylor
(1930).

A two-dimensional flow through a Laval nozzle that is transonic for all $\psi \geq 0$ was
first calculated in the dissertation of Th. Meyer (1908). Unfortunately, this solution
cannot be derived with $Z(s)$ from (12.100), as there is no curve $Z(s)$ in Fig. 12.12
that corresponds to a flow that is always subsonic before and always supersonic
behind a certain location (recall $w = Z + 2\psi^2$). Still, a limiting case of a flow that
is transonic at all ψ can be found, by integrating along the steep straight line $Z = 2s$
for $C = 0$. Figure 12.15 shows the flow solution for this limiting case. In the solution
for the Laval nozzle found by Meyer (1908), the flow becomes supersonic near the
wall *before* the nozzle throat, and on the mid-axis only *behind* the throat.

Fig. 12.15 Transonic flow
through a Laval nozzle,
obtained from $Z = 2s$ in
(12.101), i.e., the steep
straight line in Fig. 12.12

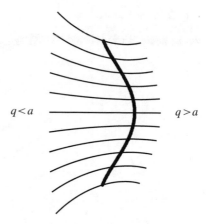

$q < a$

$q > a$

A number of papers discuss the transition between these flow types after Taylor
(1930) and Meyer (1908). This transition can be mediated by shock waves that are
created in the supersonic Taylor pockets, and propagate then downstream through the
nozzle. Shocks are dissipative, and viscosity must be included to study this transition
between solution types. Sichel (1966) generalizes (12.95) to include (longitudinal)
viscosity; (12.95) can be written as

$$ZZ_{ss} + (Z_s - 2)(Z_s + 1) = 0, \tag{12.115}$$

and, by including viscosity, Sichel (1966) amends this to (see his Eq. 12),

$$-\frac{Z_{sss}}{2} + ZZ_{ss} + (Z_s - 2)(Z_s + 1) = 0. \tag{12.116}$$

Division by Z_{ss} gives

$$\frac{Z_{sss}}{Z_{ss}} = \frac{2(Z_s - 2)(Z_s + 1)}{Z_{ss}} + 2Z. \tag{12.117}$$

Let

$$p = Z_s, \qquad q = Z_{ss} \tag{12.118}$$

(this is not the q from before). Then

$$\frac{dp/ds}{dq/ds} = \frac{dp}{dq} = \frac{q}{2(p-2)(p+1) + 2Zq}, \tag{12.119}$$

where s is the common variable of the functions p and q. Figure 12.16 shows the
integral curves of the ordinary differential equation (12.119) for the choice $Z = -1$,
i.e., the figure gives a 2-D cut through the three-dimensional ZZ_sZ_{ss}-space.

Fig. 12.16 Saddle and focus
in a two-dimensional Laval
nozzle flow of a viscous fluid

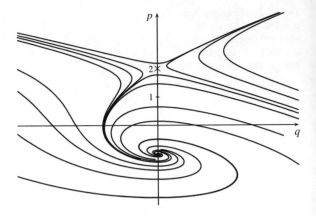

One finds a saddle point at $(q, p) = (0, 2)$ and a focus at $(q, p) = (0, -1)$. For Z different from -1, one finds a node instead of this focus (Sichel 1966). We do not go further into the physics of these flows, but notice (again) the rich solution topology of flows of viscous fluids—which may seem a little counterintuitive at first, if one thinks of viscosity as only a decelerating agent that 'lessens' the dynamics, see Sect. 4.3.

12.7 Solar Wind Exhibiting a Shock Pair

Besides the flow through a Laval nozzle, the second example in Sect. 12.5 of a transonic flow was the solar wind expanding from the hot solar corona. The present section treats a simplified model for the formation of a *shock pair* in the solar wind, as could occur after a solar flare (a sudden energy release in the solar atmosphere caused by magnetic reconnection). This example brings together some of the central ideas of this and earlier chapters: shocks, supersonic flows, and similarity variables. The model was first considered by Simon and Axford (1966), and to a certain extent is a direct continuation of the final pages of the book by Courant and Friedrichs (1948).

The solar wind is treated as a one-dimensional, spherically symmetric flow of an inviscid ideal gas under adiabatic conditions. The shock structure is treated locally only, and thus gravity can be neglected. This is mandatory for a *similarity solution* to be available, since without gravity a physical scale besides those of time and space is absent. The method of similarity variables was introduced in Sects. 8.8 and 8.9.

The continuity, Euler, and energy equations outside shock regions read in the radial direction of spherical coordinates

$$\frac{\partial \rho}{\partial t} + \frac{1}{r^2} \frac{\partial}{\partial r} (r^2 \rho v) = 0,$$

$$\frac{\partial v}{\partial t} + v \frac{\partial v}{\partial r} = -\frac{1}{\rho} \frac{\partial p}{\partial r},$$

$$\frac{\partial}{\partial t} (p\rho^{-\gamma}) + v \frac{\partial}{\partial r} (p\rho^{-\gamma}) = 0. \tag{12.120}$$

Here, v refers to absolute speeds, and u will again be used for speeds in a shock frame. The energy equation expresses conservation of the specific entropy $S = p\rho^{-\gamma}$ of fluid parcels during the motion: $\mathrm{d}S/\mathrm{d}t = 0$ (see Landau and Lifshitz 1987, p. 4).

The dimensionless *similarity variable* is chosen as

$$\eta = \frac{r}{\alpha t}, \tag{12.121}$$

with a constant speed α that will be fixed in course of the calculation. Introducing η, the equations (12.120) can be written as a set of ordinary differential equations. The differentials then become

$$\frac{\partial}{\partial r} = \frac{\partial \eta}{\partial r} \frac{\mathrm{d}}{\mathrm{d}\eta} = \frac{\eta}{r} \frac{\mathrm{d}}{\mathrm{d}\eta},$$

$$\frac{\partial}{\partial t} = \frac{\partial \eta}{\partial t} \frac{\mathrm{d}}{\mathrm{d}\eta} = -\frac{\eta}{t} \frac{\mathrm{d}}{\mathrm{d}\eta}. \tag{12.122}$$

The η-derivative acts on dimensionless functions $V(\eta), \sigma(\eta)$, and $\Pi(\eta)$ that are defined by

$$v(t, r) = \frac{r \, V(\eta)}{t},$$

$$\rho(t, r) = \frac{\rho_0 \, \sigma(\eta)}{r^2},$$

$$p(t, r) = \frac{\rho_0 \, \Pi(\eta)}{t^2}, \tag{12.123}$$

and a dimensionless sound speed A is introduced by

$$A^2 = \frac{\gamma \Pi}{\sigma} = \frac{a^2 \, t^2}{r^2}. \tag{12.124}$$

Inserting the above definitions in (12.120) and performing elementary rearrangements, one obtains

$$(1 - V)\, \eta\, \frac{d\sigma}{d\eta} = \sigma V + \sigma \eta\, \frac{dV}{d\eta},$$

$$(1 - V)\, \eta\, \frac{dV}{d\eta} = -V(1 - V) + \frac{\eta}{\sigma}\, \frac{d\Pi}{d\eta},$$

$$(1 - V)\, \eta\, \frac{d\Pi}{d\eta} = 2(\gamma V - 1)\, \Pi + (1 - V)\, \frac{\gamma \Pi}{\sigma}\, \eta\, \frac{d\sigma}{d\eta}. \tag{12.125}$$

Eliminating the differentials on the right sides by using consecutively the left sides gives

$$[(1 - V)^2 - A^2]\, \eta\, \frac{d\sigma}{d\eta} = -2A^2\sigma\, \frac{1 - \gamma V}{\gamma(1 - V)}, \tag{12.126}$$

$$[(1 - V)^2 - A^2]\, \eta\, \frac{dV}{d\eta} = -V(1 - V)^2 + A^2(3V - 2/\gamma), \tag{12.127}$$

$$[(1 - V)^2 - A^2]\, \eta\, \frac{dA}{d\eta} = -\frac{A(1 - \gamma V)[(1 - V)^2 - A^2/\gamma]}{1 - V}. \tag{12.128}$$

In the last line, $d\Pi/d\eta$ was replaced by $dA/d\eta$, using

$$\frac{dA}{d\eta} = \frac{1}{2A\sigma}\left(\gamma\, \frac{d\Pi}{d\eta} - A^2\, \frac{d\sigma}{d\eta}\right). \tag{12.129}$$

Finally, dividing (12.128) by (12.127), one finds

$$\boxed{\frac{dA}{dV} = \frac{A(1 - \gamma V)[(1 - V)^2 - A^2/\gamma]}{(1 - V)[V(1 - V)^2 - A^2(3V - 2/\gamma)]}} \tag{12.130}$$

Note that σ does not enter here, but only V and A. For (12.126)–(12.130), see Eqs. (160.05)–(160.07) in Courant and Friedrichs (1948), and their rather impressive figure VI.11 (p. 426) for the solution manifold of (12.130) in the (V, A)-plane.

We are interested in solar wind solutions with a dense shell enclosed by two shocks at its inner and outer boundaries: a fast gas flow from small r (the solar flare) is decelerated in a reverse shock, and ahead of the reverse shock, a forward shock accelerates the outer gas. The two postshock regions are separated by a so-called *contact discontinuity*. At the latter, the density ρ shows a jump, but the pressure p and the speed v are continuous. There is thus no dynamics across a contact discontinuity, where simply two gas regions meet at different densities.

Figure 12.17 shows this intended solution of (12.130) in the (V, A)-plane by assuming $\gamma = 5/3$ and a strong forward shock. Let v_0 and v_1 be the pre- and postshock speed at this forward shock, respectively. The shock overtakes a 'quiet' solar wind gas, and one can assume $v_0 = 0$ for simplicity. Then (12.23) gives for the speeds relative to the forward shock (that has speed v_f),

$$v_f - v_1 = \frac{v_f - v_0}{4} = \frac{v_f}{4}, \tag{12.131}$$

or $v_1 = 3v_f/4$. The postshock sound speed is calculated from (12.26),

$$a_1^2 = \frac{\gamma p_1}{\rho_1} = \frac{5}{3} \cdot \frac{3}{4} \frac{\rho_0 v_f^2}{4\rho_0} = \frac{5}{16} v_f^2. \tag{12.132}$$

We choose now as similarity parameter $\alpha = v_f$, the speed (relative to the Sun) of the forward shock. Then $\eta_f = 1$, and

$$V_f = \frac{t v_f}{r_f} = \frac{t\alpha}{r_f} = \frac{1}{\eta_f} = 1. \tag{12.133}$$

Therefore, one has on the postshock side of the forward shock that

$$V_1 = \frac{3}{4}, \qquad A_1 = \frac{\sqrt{5}}{4}. \tag{12.134}$$

The numerical integration of the ordinary differential equation (12.130) shown in Fig. 12.17 starts at the point $f = (V_1, A_1)$ and stops when the contact discontinuity at $(1, 0)$ is reached. With $A = 0$, the flow has cooled there back to its preshock temperature. (Recall that for strong shocks, the preshock pressure and temperature are zero.) This cooling is due to adiabatic expansion of the gas in the spherically expanding flow. The pressure in the postshock region is roughly constant, since sound waves flatten out any pressure gradients in this subsonic region. One has thus finite pressure p, but $a = 0$ at the contact discontinuity. Since $\sigma = \gamma \Pi/A^2$ according to (12.124), this implies $\sigma \to \infty$ at the contact discontinuity.

Next, a strong reverse shock r is assumed to occur at $V > 1$. At this reverse shock, (12.132) becomes, with $v_r < v_0$ and $v_r < v_1$,

$$a_1 = \frac{\sqrt{5}}{4} (v_0 - v_r) = \sqrt{5} (v_1 - v_r). \tag{12.135}$$

This gives, with $a_1 = r_r A_1/t$ and $v_1 = r_r V_1/t$,

$$A_1 = \sqrt{5} \left(V_1 - \frac{t v_r}{r_r} \right) = \sqrt{5} (V_1 - 1), \tag{12.136}$$

and this determines the immediate postshock region of the reverse shock. Equation (12.136) gives the straight, dashed line in Fig. 12.17. One can start integration of (12.130) on any point of this straight line. Specifically, Figs. 12.17 and 12.18 assume $V_1 = 1.6$ at the reverse shock. Integration proceeds again until the contact discontinuity at $(1, 0)$ is reached.

With $A(V)$ known between the two shocks, η can be calculated from (12.127). One starts at $\eta_f = 1$, and thus obtains η_c of the contact discontinuity. The value for η_r is guessed and changed until the integration from r to c gives the same value η_c.

The gas density σ can be freely normalized; we choose a postshock density $\sigma_1 = 1$ at the forward shock. This determines the pressure at the contact discontinuity.

Fig. 12.17 Double shock
solution of (12.130).
Immediate postshock values
at the forward shock f are
from (12.134), and for the
reverse shock r they lie on
the straight dashed line given
by (12.136). The contact
discontinuity is at $(1, 0)$

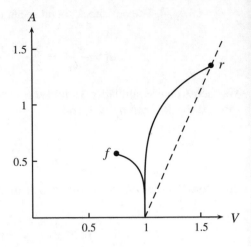

Fig. 12.18 Density $\sigma(\eta)$
from (12.126)

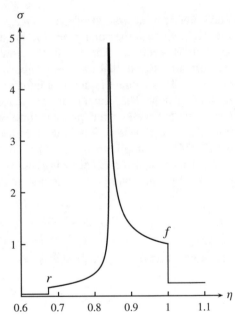

The postshock density at the reverse shock is again found by trial and error from
the condition that the pressure p_c at the contact discontinuity is the same both for
integration from r to c and for integration from f to c. Figure 12.18 shows the
resulting curve for $\sigma(\eta)$.

The somewhat surprising result is that the planar double shock solution shown
in Fig. 12.8 allows for finite constant speed, density, pressure, and temperature in
the postshock regions between the reverse and forward shocks (there is no contact
discontinuity in this example), whereas in the present case, the density in Fig. 12.18

becomes infinite at the contact discontinuity. The physical reason for this is quite clear: the spherically expanding, shock-heated gas cools adiabatically, and settles on the contact discontinuity in an infinitesimally thin layer of infinite density, since $\rho \to \infty$ for $T \to 0$ at a finite postshock pressure p. (Here, $T \to 0$ is not an absolute statement referring to zero Kelvin, but only an approximate expression for low temperatures compared to postshock temperatures.)

12.8 The Riemann Sheets for the Burgers Equation

The analysis by Moore (1979) in Sect. 6.8 showed that a curvature singularity can form in a vortex sheet within a finite critical time from smooth initial conditions. A core question regarding shocks in the limit of an inviscid fluid is therefore, whether finite shock jumps, i.e., discontinuities, can develop from smooth initial values within finite time? In the simple kinematic examples of shock tubes and wall shocks at the beginning of this chapter, jumps were already present in the initial conditions (two gas regions at different pressures or speeds), and the result was merely that these discontinuities are maintained at all times, and propagate through the system. The question whether discontinuities can *form* in a smooth flow is more subtle, and we give in the following an example where a shock wave does indeed develop from smooth initial values within finite time. However, the example also shows that the discontinuity is present at *all* times, residing at imaginary locations at early times, as was already the case for the singularity formation in a vortex sheet discussed in Sect. 6.9.

One of the simplest nonlinear equations that allows to study shock formation is the inviscid, one-dimensional, planar *Burgers* equation, which is a model equation for pure advection,

$$\frac{\partial u}{\partial t} + u \frac{\partial u}{\partial x} = 0. \tag{12.137}$$

For this equation, the velocity field $u(t, x)$ can become multivalued after a finite time interval. Shortly before u becomes multivalued, strong gradients occur, viscosity becomes important, and shocks will form. Given the appearance of a multivalued solution, Bessis and Fournier (1984) study the Riemann surface of solutions of the Burgers equation (12.137), and find how poles for the case of a viscosity coefficient $\nu > 0$ are related to shocks for the inviscid case $\nu = 0$.

> The lack of understanding of the delicate analytic structure of the Riemann surface in this [zero viscosity] limit, is the reason for paradoxes [...] concerning the high wave number behaviour of the velocity. (Bessis and Fournier 1990, p. 253)

Of special interest is here the motion of branch points in the *complex* x-plane. The initial conditions for u are chosen to be an odd function of x,

$$u(0, x) = -ax + bx^3 + cx^5 + dx^7 + \cdots. \tag{12.138}$$

For coefficients $a, b > 0$, a negative gradient $u' = \partial u / \partial x < 0$ at $x = 0$ steepens in the course of time, $u' \to -\infty$, and turns over to make $u' > 0$ at some finite t_* with u becoming multivalued. To see this, consider the Burgers equation $\dot{u}(t, 0) = -u(t, 0)u'(t, x)|_{x=0}$ at $x = 0$. For the initial value $u(0, 0) = 0$, this has the solution $u(t, 0) = 0$, thus the fluid parcel at $x = 0$ remains at rest at all times. Next, differentiate the Burgers equation $\dot{u} + uu' = 0$ with respect to x, which yields

$$\dot{u}' + u'^2 + uu'' = 0. \tag{12.139}$$

At $x = 0$, this becomes, with $u(t, 0) = 0$,

$$\left[\dot{u}'(t, x) + u'^2(t, x)\right]_{x=0} = 0. \tag{12.140}$$

This can be integrated once (note that $u' = \partial u / \partial x$),

$$\int_{u'(0,0)}^{\infty} \frac{du'(t, 0)}{u'^2(t, 0)} = -\int_{0}^{t_*} dt, \tag{12.141}$$

where t_* is the time it takes the initial gradient $u'(0, 0)$ to become infinite,

$$t_* = -\frac{1}{u'(0, 0)}. \tag{12.142}$$

For (12.138), this time is given by $t_* = a^{-1}$. Bessis and Fournier (1984) put $c = d = \cdots = 0$ in (12.138), and use as initial conditions,

$$\boxed{u(0, x) = -\frac{x}{t_*} + 4x^3} \tag{12.143}$$

where $b = 4$ is chosen for later convenience. Remarkably, the nonlinear equation (12.137) can be solved analytically for the initial conditions (12.143). The total differential of $u(t, x)$ is given by

$$du(t, x) = \dot{u}\, dt + u'\, dx. \tag{12.144}$$

Multiplying the Burgers equation (12.137) by dt gives

$$\dot{u}\, dt + uu'\, dt = 0. \tag{12.145}$$

Comparing the last two equations shows that $du = 0$ along those curves $x(t)$ in a kinematic xt-plane for which $dx = u\, dt$, or

$$\frac{dx}{dt} = u. \tag{12.146}$$

Fig. 12.19 Kinematics of the Burgers equation for initial conditions $u(0, x) = -x$

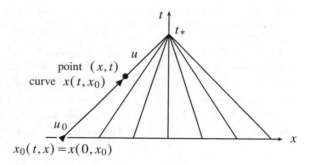

These are the so-called *characteristics* in the xt-plane. Note that, since the speed u does not change along a characteristic, $du = 0$, the characteristics are straight lines here. We follow Caflisch et al. (1993) and introduce two functions,

$$x(t, x_0) \qquad \text{and} \qquad x_0(t, x). \tag{12.147}$$

Here $x(., x_0)$ is a one-parameter family of curves, defined by the relation

$$\frac{dx}{dt}(t, x_0) = u(t, x) \tag{12.148}$$

for arbitrary times t, see Fig. 12.19, which shows the simplified case of initial conditions $u(0, x) = -x$, whereas Fig. 12.20 shows characteristics for the full initial conditions (12.143) with $t_* = \frac{1}{4}$. Thus $x(t, x_0)$ are the integral curves through the slopes $u(t, x)$ in the xt-plane. The curve parameter x_0 is the value at which the curve starts at $t = 0$ on the x-axis,

$$x(0, x_0) = x_0. \tag{12.149}$$

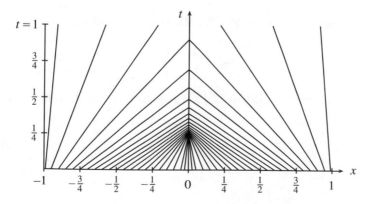

Fig. 12.20 Kinematics of the Burgers equation with initial conditions $u(0, x) = -4x + 4x^3$, i.e., $t_* = \frac{1}{4}$ according to (12.143)

The value x_0 is the initial position of the straight line characteristic with speed $u_0(x_0)$ on which $u = u_0(x_0)$ is constant. (Caflisch et al. (1993), p. 458, with symbols adapted)

For any arbitrary point (x, t) of the xt-plane with $x \neq 0, t > 0$ there is a unique straight line $x(t, x_0)$ (with $xx_0 > 0$) that crosses through (x, t) at time t, i.e., $x(t, x_0) = x$ (we do not distinguish here between points (x, t) and functions $x(t)$). The mapping $x_0 = x_0(t, x)$, the second function in (12.147), is then unique, and gives for any (x, t) the point x_0 on the x-axis at which the integral curve through (x, t) starts. The points $(x_0, 0)$ and (x, t) are connected by a straight line with inverse slope

$$u(0, x_0) = \frac{x - x_0(t, x)}{t}. \tag{12.150}$$

Inserting the initial conditions (12.143) gives

$$-\frac{x_0(t, x)}{t_*} + 4x_0^3(t, x) = \frac{x}{t} - \frac{x_0(t, x)}{t}, \tag{12.151}$$

or

$$4x_0^3(t, x) + \left(\frac{1}{t} - \frac{1}{t_*}\right) x_0(t, x) - \frac{x}{t} = 0. \tag{12.152}$$

Thus x_0 is determined by the zeros of a third-order polynomial. Equations (12.150) and (12.152) solve Burgers' equation for the initial condition (12.143).

We treat t now as a parameter, and x and x_0 as complex variables. The highest orders in (12.152) read $x_0 \sim x^{1/3}$, and thus the three complex Riemann x-sheets are required to fully cover one complex x_0-sheet: the x_0-values lie in one single complex plane, and the x-values in the three different Riemann sheets. At each time t, the branch points in x are given by $dx_0/dx \to \infty$. Writing (12.152) as $P(x_0, x) = 0$, one has

$$0 = \frac{dx}{dx_0} = -\frac{\partial P/\partial x_0}{\partial P/\partial x} \tag{12.153}$$

at the branch points. According to (12.152), $\partial P/\partial x_0$ vanishes for

$$12x_0^2 + \frac{1}{t} - \frac{1}{t_*} = 0. \tag{12.154}$$

Inserting x_0 from this equation in (12.152) gives as x-values of the branch points,

$$x_\pm = \pm\frac{i}{3} \sqrt{\frac{(t_* - t)^3}{3tt_*^3}}. \tag{12.155}$$

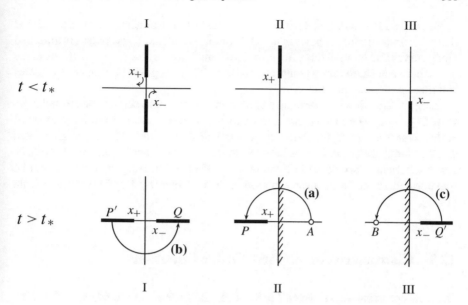

Fig. 12.21 The Riemann sheets I, II, III for the complex coordinate x in (12.152) at times $t < t_*$ (single-valued solution) and $t > t_*$ (multivalued or shock solution). Bold lines are cuts from the branch points $x_\pm$ to the branch point at infinity. Small arrows indicate the motion of the points $x_\pm$ in time. For the lower three panels, the two physical half-planes are indicated by hashes: the full physical plane (including the full real x-axis) is distributed over the two Riemann sheets

Each of these two branch points is of first order, i.e., connects two sheets. A further branch point lies at infinity and connects all three sheets. Furthermore, for $t < t_*$ the branch points $x_\pm$ are purely imaginary, whereas for $t > t_*$, both branch points $x_\pm$ lie on the real x-axis. Figure 12.21 shows the geometry of this Riemann surface.

The upper three panels show the three sheets I, II, III at $t < t_*$. Sheet I is the physical plane, and there is no multivaluedness with respect to the real x-axis. The lower three panels show the sheets at times $t > t_*$, the physical plane is made up of the half-plane $\text{Re } x > 0$ on sheet II and the half-plane $\text{Re } x < 0$ on sheet III.

The path (a) + (b) + (c) in Fig. 12.21 leads from point A on the positive real x-axis to point B on the negative real x-axis. It switches at point P from sheet II to P' on plane I, and at point Q on sheet I to Q' on sheet III. This path thus leads from $x > 0$ to $x < 0$, both members of the physical plane, through three unphysical regions, 1. $\text{Re } x < 0$ on sheet II; 2. sheet I; and 3. $\text{Re } x > 0$ on sheet III.

For the lower three panels, $u = dx/dt$ becomes multivalued at points x_+ and x_- on the real x-axis. In reality, this multivalued solution is replaced by a shock, where the shock jump connects the three allowed values for u.

A corresponding entropy condition, which fixes the shock location by the requirement that the areas of the three-valued curve $u(t, x)$ to the left and right sides of the vertical shock jump are equal, is derived in Landau and Lifshitz (1987), §101.

Bessis and Fournier (1984) also treat the case of nonvanishing viscosity and find then an infinite number of poles in u in the complex x-plane. These poles are arranged along curves that move in time. In the limit of vanishing viscosity, $\nu \to 0$, the distance $d \sim \nu$ between neighboring poles tends to zero. This merging and vanishing of poles is termed *pole condensation*.

Burgers' equation can serve as a simple model for supersonic shock turbulence. Ordinary turbulence in subsonic flows shows a vortex cascade from large to small scales, whereas shock turbulence in supersonic flows shows shock merging from small to large scales, an *inverse* cascade. A classic paper on the statistical mechanics of supersonic turbulence is Tatsumi and Kida (1972). Burgers' equation can be transformed to the linear heat equation, as was shown by Hopf (1950) and Cole (1951).

12.9   Characteristics for First-Order Equations

On a few occasions so far (Sects. 7.5, 8.7, 8.8, 12.8), we encountered *characteristics*, i.e., curves in a kinematic xt-diagram (or xt-plane for simplicity) along which certain fluid properties are constant. The present section gives a brief introduction to the method of characteristics. This method is strongest for time-dependent flow problems in one spatial dimension, i.e., with two independent variables t and x (or r), and with two unknown fluid fields $u(t, x)$ and $\rho(t, x)$. The case of three and four independent variables is treated in Courant and Friedrichs (1948), p. 75.

The fluid shall be inviscid, the flow one-dimensional and there shall be no heat exchange among fluid parcels (adiabatic conditions). The fluid fields $u(t, x)$ and $\rho(t, x)$ are given at time t_0, and shall be found at $t > t_0$. (The pressure p can be obtained from ρ.)

One looks for fluid fields $L_+(t, x)$ and $L_-(t, x)$ that are functions of u and ρ, such that the contour lines $x_+(t)$ and $x_-(t)$ along which $L_+ = C_+ =$ const and $L_- = C_- =$ const form two one-parameter families (with parameters $C_\pm$) of curves foliating the kinematic xt-plane, and both $x_+(t)$ and $x_-(t)$ shall have physically relevant kinematic properties. More specifically, it will be found that $\dot{x}_\pm = u \pm a$, with sound speed a.

Such curves $x_\pm(t)$ are called *characteristics*, and the $L_\pm$ are the *Riemann invariants*. The changes of L_+ along the x_--characteristics and of L_- along the x_+-characteristics are given by two ordinary differential equations that replace the original partial differential equations for u and ρ and to which standard solution procedures can be applied. Finally, one obtains $u(x, t)$ and $\rho(x, t)$ by (algebraic) inversion from $L_+(x_+, x_-)$ and $L_-(x_+, x_-)$.

This construction is the basis of both some existence proofs and some methods of numerical computation in gas dynamics. (Chorin and Marsden 1990, p. 104)

The theory of characteristics was initiated by G. Monge in 1770–73, and some references to his work are given in Krehl (2009), p. 255 and in Courant and Hilbert

(1937). The modern mathematical theory of characteristics goes back to Riemann (1860).

Consider first a simple advection example,

$$u_t + a u_x = 0, \tag{12.156}$$

with constant speed $a =$ (see Lax 1973, p. 5 and Chorin and Marsden 1990, p. 106). For the initial conditions $u(0, x) = u_0(x)$, the solution of (12.156) is

$$u(t, x) = u_0(x - at). \tag{12.157}$$

This expresses that the function u_0 is shifted as a whole with speed a in the x-direction. Therefore, $u = u_0$ is constant along straight characteristics $x(t)$ with constant slope $\dot{x} = a$. One says that the values $u_0(x)$ are propagated or transported along the characteristics $x(t)$. Note that there is only one family of characteristics in this example, as there is only one fluid field $u(t, x)$.

Consider then an arbitrary curve in the xt-plane that has slope less than a everywhere. It is possible to specify initial values on this curve, instead of specifying them on the coordinate line $t = 0$. From any point on the curve, a characteristic 'carries' a certain value u_0 to larger t, and from this the whole field $u(t, x)$ can be constructed.

Next, (12.156) is generalized to

$$u_t + a(u) u_x = 0. \tag{12.158}$$

The characteristics are still straight lines, but now with varying slopes $dx/dt = a(u)$, depending on their initial value at $t = 0$. Equation (12.157) takes the form

$$u(t, x) = u_0(x - t a(u(t, x))). \tag{12.159}$$

Differentiation with respect to t gives

$$u_t = -[a(u) + t a' u_t] u_0', \tag{12.160}$$

where a prime indicates differentiation with respect to the single variable of a function. Therefore,

$$u_t = -\frac{u_0' a}{1 + u_0' a' t}. \tag{12.161}$$

Similarly, differentiating (12.159) with respect to x gives

$$u_x = \frac{u_0'}{1 + u_0' a' t}. \tag{12.162}$$

If $u_0' < 0$ at x_0 and $a' > 0$, then both u_t and u_x become infinite at time

$$t_* = -\frac{1}{u_0'(x_0)a'} \tag{12.163}$$

(see the example in the foregoing section). At t_*, two converging straight characteristics cross. This causes multivaluedness, which in reality is prevented by shock formation.

We turn from these pure advection examples to gas dynamics, and consider a compressible, one-dimensional planar gas flow along a straight tube with length coordinate x. Let $p\rho^{-\gamma} = \text{const}$, with the adiabatic sound speed a given by

$$a^2 = \frac{dp}{d\rho} = \frac{\gamma p}{\rho}. \tag{12.164}$$

The Euler and continuity equations are

$$0 = \dot{u} + uu' + a^2 \frac{\rho'}{\rho}, \tag{12.165}$$

$$0 = \dot{\rho} + u\rho' + \rho u', \tag{12.166}$$

with unknown fields $u(t, x)$ and $\rho(t, x)$. Let $t = t(\sigma)$, $x = x(\sigma)$ be the parametric representation of a curve in the xt-plane with arclength σ. (For the following derivations, see Courant and Friedrichs 1948, p. 40.) Differentiation along the curve gives

$$\frac{\partial u}{\partial \sigma} = \frac{\partial u}{\partial t}\frac{\partial t}{\partial \sigma} + \frac{\partial u}{\partial x}\frac{\partial x}{\partial \sigma}. \tag{12.167}$$

In an xt-diagram, let the curve have inverse slope

$$u = \frac{dx}{dt} = \frac{x_\sigma}{t_\sigma} = \frac{B}{A}, \tag{12.168}$$

with some functions A and B. Let $x_\sigma = BC$ and $t_\sigma = AC$, with another function C. Then

$$\frac{\partial u}{\partial \sigma} = C\left(A\frac{\partial u}{\partial t} + B\frac{\partial u}{\partial x}\right). \tag{12.169}$$

The scaling C is irrelevant in the following, and one only needs that the expression

$$A\frac{\partial}{\partial t} + B\frac{\partial}{\partial x} \tag{12.170}$$

corresponds to differentiation in a direction

$$\frac{dx}{dt} = \frac{B}{A}. \tag{12.171}$$

The fluid equations (12.165) and (12.166) have the general form

$$H_1 = A_1 \dot{u} + B_1 u' + C_1 \dot{\rho} + D_1 \rho' + E_1 = 0,$$
$$H_2 = A_2 \dot{u} + B_2 u' + C_2 \dot{\rho} + D_2 \rho' + E_2 = 0, \qquad (12.172)$$

where A_1 to E_2 are functions of u and ρ (and of t and x), but there appear no differentials of u and ρ as arguments in A_1 to E_2, so that the equations are linear in all differentials (so-called *quasi-linear* differential equations; see Courant and Hilbert 1937, p. 2.)

A central question, first posed by Riemann, is as follows: are there any linear combinations $\lambda_1 H_1 + \lambda_2 H_2$, i.e.,

$$(\lambda_1 A_1 + \lambda_2 A_2)\dot{u} + (\lambda_1 B_1 + \lambda_2 B_2)u' +$$
$$+ (\lambda_1 C_1 + \lambda_2 C_2)\dot{\rho} + (\lambda_1 D_1 + \lambda_2 D_2)\rho' = -\lambda_1 E_1 - \lambda_2 E_2, \qquad (12.173)$$

in which both u and ρ are differentiated in the same direction in the xt-plane? If this is the case, Eq. (12.172) takes the simple form

$$F(u, \rho)\, \frac{\partial u}{\partial \sigma} + G(u, \rho)\, \frac{\partial \rho}{\partial \sigma} = -\lambda_1 E_1 - \lambda_2 E_2, \qquad (12.174)$$

with new functions F and G determined by the coefficients A_1 to D_2, and where the variable σ indicates differentiation in this common direction, and could be the arclength along a curve pointing in this direction. According to (12.170) and (12.171), the inverse slope dx/dt in this common direction of differentiation of u and ρ is given by

$$\boxed{\frac{dx}{dt} = \frac{\lambda_1 B_1 + \lambda_2 B_2}{\lambda_1 A_1 + \lambda_2 A_2} = \frac{\lambda_1 D_1 + \lambda_2 D_2}{\lambda_1 C_1 + \lambda_2 C_2}} \qquad (12.175)$$

To fully explicate this important step, note that the numerators in the latter two fractions in (12.175) are the coefficients of the x-derivatives of u and ρ in (12.173), while the denominators are the coefficients of the t-derivatives of u and ρ in (12.173), in complete correspondence with Eqs. (12.170) and (12.171). (On physical grounds, all speeds $dx/dt < \infty$, and we can assume that no denominator vanishes.)

Assume that $\lambda_2 \neq 0$ (else the problem is trivial). Dividing then (12.175) by λ_2 gives a quadratic equation for the unknown λ_1/λ_2, with two solutions λ_1/λ_2, distinguished in the following by subscripts $+$ and $-$, that determine the common directions of differentiation of u and ρ by (12.175). These two solutions exist in each point (x, t) of the xt-plane and form a net of (continuously differentiable) curves. This curve net with associated coordinates σ_+ and σ_- along the curves, e.g., their arclengths, establishes a non-orthogonal curvilinear coordinate system in the xt-plane. The σ_+ and σ_- are the 'natural' coordinates of the problem at hand, in the sense that Eqs. (12.174) are two coupled *ordinary* differential equations in independent variables σ_+ and σ_-.

Inserting A, B, C, D from (12.165) and (12.166) in (12.175), one obtains

$$\frac{dx}{dt} = \frac{\lambda_1 u + \lambda_2 \rho}{\lambda_1} = \frac{\lambda_1 a^2/\rho + \lambda_2 u}{\lambda_2}. \tag{12.176}$$

The second equation gives $\lambda_1^2/\lambda_2^2 = \rho^2/a^2$, and thus

$$\frac{\lambda_1}{\lambda_2} = \pm\frac{\rho}{a}. \tag{12.177}$$

The first equation then becomes

$$\boxed{\frac{dx_\pm}{dt} = u \pm a} \tag{12.178}$$

Next, multiply the continuity equation with $\lambda_2/\lambda_1 = \pm a/\rho$. Adding this to the Euler equation and rearranging gives the two equations

$$\dot{u} + uu' \pm au' \pm \frac{a}{\rho}\left(\dot{\rho} + u\rho' \pm a\rho'\right) = 0. \tag{12.179}$$

This can be simplified by introducing a new differential D (not to be confused with the function D), defined by

$$D_\pm = \partial_t + (u \pm a)\, \partial_x, \tag{12.180}$$

which means differentiation along the characteristics $x_\pm(t)$ with $dx_\pm/dt = u \pm a$. With this, (12.179) becomes

$$D_\pm u \pm \frac{a}{\rho} D_\pm \rho = 0. \tag{12.181}$$

The chain rule applies to D, thus

$$\frac{D\rho}{\rho} = \frac{D(\rho/\rho_0)}{\rho/\rho_0} = D\ln(\rho/\rho_0), \tag{12.182}$$

where for the reference density $\rho_0 = 1$ is assumed. For flow with constant temperature, $a = $ const, and one has then

$$\boxed{D_\pm L_\pm = 0, \qquad L_\pm = u \pm a \ln \rho \qquad (\gamma = 1)} \tag{12.183}$$

Therefore, L_+ is constant along curves of the family x_+, and correspondingly for L_- on x_-. The L_+ and L_- in (12.183) are the Riemann invariants for gas flows at constant temperature. For flows under adiabatic conditions, $p\rho^{-\gamma} = $ const with $\gamma > 1$, $a^2 = \gamma p/\rho$ implies $a^2 \rho^{1-\gamma} = $ const, or $\rho \sim a^{\frac{2}{\gamma-1}}$. This gives

$$\frac{D\rho}{\rho} = \frac{2}{\gamma - 1} \frac{Da}{a}. \tag{12.184}$$

Therefore, from (12.181),

$$\boxed{D_{\pm}L_{\pm} = 0, \qquad L_{\pm} = u \pm 2a/(\gamma - 1) \qquad (\gamma > 1)} \tag{12.185}$$

Note that ρ no longer appears here in L, in contrast to (12.183).

In the absence of shocks, characteristics are the curves of fastest signal propagation (i.e., at sound speed) in the gas flow. Characteristics are thus connected with the (causal) evolution of the flow. To work this out, we return to the general case (12.172) of two coupled partial differential equations. To avoid unwieldy sub-subscripts $+$ and $-$, the variables σ_{+} and σ_{-} are renamed σ and τ. For the following derivation, see Courant and Hilbert (1937), p. 303. Let $\tau(t, x) = C = $ const determine a one-parameter family (parameter C) of curves in the xt-plane, and let σ be the arclength variable along these curves. A change of variables is performed from x and t to σ and τ. One has

$$\dot{u} = \dot{\tau}\, u_{\tau} + \dot{\sigma}\, u_{\sigma},$$
$$u' = \tau' u_{\tau} + \sigma' u_{\sigma}, \tag{12.186}$$

and

$$\dot{\rho} = \dot{\tau}\, \rho_{\tau} + \dot{\sigma}\, \rho_{\sigma},$$
$$\rho' = \tau' \rho_{\tau} + \sigma' \rho_{\sigma}. \tag{12.187}$$

Equations (12.172) become then

$$(A_1\dot{\tau} + B_1\tau')u_{\tau} + (C_1\dot{\tau} + D_1\tau')\rho_{\tau} = -(A_1\dot{\sigma} + B_1\sigma')u_{\sigma} + (C_1\dot{\sigma} + D_1\sigma')\rho_{\sigma} - E_1,$$
$$(A_2\dot{\tau} + B_2\tau')u_{\tau} + (C_2\dot{\tau} + D_2\tau')\rho_{\tau} = -(A_2\dot{\sigma} + B_2\sigma')u_{\sigma} + (C_2\dot{\sigma} + D_2\sigma')\rho_{\sigma} - E_2. \tag{12.188}$$

Initial values for u and ρ shall be given along the curve $\tau = 0$ as functions of σ. Then u_{σ} and ρ_{σ}, and thus the right sides in (12.188) are known. However, the derivatives u_{τ} and ρ_{τ} that point away from the curve $\tau = 0$ are *not* uniquely determined, *if* the determinant of the system (12.188) vanishes,

$$\begin{vmatrix} A_1\dot{\tau} + B_1\tau' & C_1\dot{\tau} + D_1\tau' \\ A_2\dot{\tau} + B_2\tau' & C_2\dot{\tau} + D_2\tau' \end{vmatrix} = 0. \tag{12.189}$$

The curve $\tau = 0$ has an inverse slope

$$\frac{\mathrm{d}x}{\mathrm{d}t} = -\frac{\dot{\tau}}{\tau'}. \tag{12.190}$$

Inserting this in (12.189) gives

$$\left(-A_1 \frac{dx}{dt} + B_1\right)\left(-C_2 \frac{dx}{dt} + D_2\right) = \left(-A_2 \frac{dx}{dt} + B_2\right)\left(-C_1 \frac{dx}{dt} + D_1\right).$$
(12.191)

In the first bracket on both sides, put

$$\frac{dx}{dt} = \frac{\lambda_1 B_1 + \lambda_2 B_2}{\lambda_1 A_1 + \lambda_2 A_2},$$
(12.192)

and in the second bracket on both sides, put

$$\frac{dx}{dt} = \frac{\lambda_1 D_1 + \lambda_2 D_2}{\lambda_1 C_1 + \lambda_2 C_2}.$$
(12.193)

Then (12.191) and thus (12.189) are fulfilled identically, which means that the derivatives u_τ and ρ_τ pointing away from the curve $\tau = 0$ are not uniquely known. Since (12.192) and (12.193) are the characteristic slope dx/dt from (12.175), this is exactly the case if $\tau = 0$ is a characteristic.

We have thus the important result that *initial conditions cannot be specified on a characteristic* $\tau = 0$, since then the differential equations cannot be integrated forward in time, due to the non-uniqueness of the differentials u_τ and ρ_τ pointing away from a characteristic.

However, if an arbitrary curve has a slope that is everywhere smaller than the local characteristic slope, then this curve can serve to specify initial conditions on it, since time integration away from this curve is possible. This corresponds to *spacelike* curves in the theory of relativity.

Let two characteristics $x_+(t)$ and $x_-(t)$ cross at point P in the xt-plane. The flow shall be subsonic, $|u| < a$, and the characteristics have slopes $u \pm a$. The lines marked i and d in Fig. 12.22 are nowhere as steep as characteristics.

The triangular domain I in Fig. 12.22 is called P's *region (or range) of influence*, and the line segment d is P's *domain of dependence*. Initial values can be posed

Fig. 12.22 Domain of dependence d and region of influence I of point P

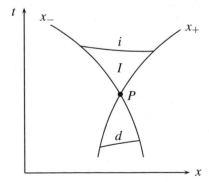

on curves like d. For $u < a$, no signal from P can leave I by crossing x_+ or x_-, which are the lines of fastest signal propagation. Similarly, u and ρ at P cannot depend on initial values beyond the segment d. For the strict theorems addressing these intuitively clear facts, see Lewy (1928).

For $|u| > a$, on the other hand, shock discontinuities can occur that propagate faster than sound. Besides characteristics, the shock trajectories are then crucial in the xt-plane to obtain the full kinematics of the flow. This supersonic case is covered in Courant and Friedrichs (1948).

12.10 Characteristics for Second-Order Equations

This section gives a brief account of characteristics for second-order partial differential equations of hyperbolic type. Only the simplest case is considered of a single differential equation for an unknown fluid field $u(x, y)$ (say, a wave amplitude) that is a function of two independent variables x and y,

$$au_{xx} + 2bu_{xy} + cu_{yy} + d = 0, \qquad (12.194)$$

with coefficients

$$a = a(x, y, u(x, y), u_x(x, y), u_y(x, y)), \qquad (12.195)$$

and similarly for b, c, d. The functions u, u_x, and u_y and the coefficients a, b, c, d are assumed to be once continuously differentiable (abbreviated in the following as C_1) in the considered domain of the xy-plane. Instead of x and y, the variables could be x and t; the former are chosen here for reasons of symmetry only. Note that the differential equation (12.194) is linear in second derivatives (quasi-linear).

The basic idea of introducing characteristic curves for this single equation is similar to that of the foregoing section, to bring Eq. (12.194) into a simpler form that is more suitable to solution. In the last section, this was achieved by considering derivatives of *both* u and ρ in *one* direction in the xt-plane, thereby obtaining two ordinary differential equations of first order instead of two partial differential equations of first order. Here now, one asks for a variable transformation from x and y to new variables σ and τ,

$$\sigma = \sigma(x, y), \qquad \tau = \tau(x, y), \qquad (12.196)$$

that simplifies (12.194). There are obviously different choices for a simplified form of (12.194), and we consider here the case that in the new variables, (12.194) becomes

$$2\beta u_{\sigma\tau} + \delta = 0 \qquad (12.197)$$

(as before, no new symbol is introduced for the function u of the new variables). This has the form of a wave equation, see (8.115).

The further calculations are straightforward: one has for the first-order derivatives of u,

$$u_x = u_\sigma \sigma_x + u_\tau \tau_x,$$
$$u_y = u_\sigma \sigma_y + u_\tau \tau_y. \tag{12.198}$$

Similarly, for the second-order derivatives of u,

$$u_{xx} = u_{\sigma\sigma} \sigma_x^2 + 2u_{\sigma\tau} \sigma_x \tau_x + u_{\tau\tau} \tau_x^2 + u_\sigma \sigma_{xx} + u_\tau \tau_{xx},$$
$$u_{xy} = u_{\sigma\sigma} \sigma_x \sigma_y + u_{\sigma\tau}(\sigma_x \tau_y + \sigma_y \tau_x) + u_{\tau\tau} \tau_x \tau_y + u_\sigma \sigma_{xy} + u_\tau \tau_{xy},$$
$$u_{yy} = u_{\sigma\sigma} \sigma_y^2 + 2u_{\sigma\tau} \sigma_y \tau_y + u_{\tau\tau} \tau_y^2 + u_\sigma \sigma_{yy} + u_\tau \tau_{yy}. \tag{12.199}$$

The last two terms of each equation, e.g., $u_\sigma \sigma_{xx} + u_\tau \tau_{xx}$ in the first line, contain only first-order derivatives of u, and thus can all be added to the coefficient d, giving a new coefficient δ. Inserting the remaining second-order derivatives $u_{\sigma\sigma}, u_{\sigma\tau}, u_{\tau\tau}$ from (12.199) in (12.194) and sorting for these derivatives, one obtains the equation

$$\alpha u_{\sigma\sigma} + 2\beta u_{\sigma\tau} + \gamma u_{\tau\tau} + \delta = 0, \tag{12.200}$$

where

$$\alpha = a\sigma_x^2 + 2b\sigma_x \sigma_y + c\sigma_y^2,$$
$$\beta = a\sigma_x \tau_x + b(\sigma_x \tau_y + \sigma_y \tau_x) + c\sigma_y \tau_y,$$
$$\gamma = a\tau_x^2 + 2b\tau_x \tau_y + c\tau_y^2, \tag{12.201}$$

and

$$\delta = d + u_\sigma(a\sigma_{xx} + 2b\sigma_{xy} + c\sigma_{yy}) + u_\tau(a\tau_{xx} + 2b\tau_{xy} + c\tau_{yy}). \tag{12.202}$$

Demanding that $\alpha = \gamma = 0$ leads to the quadratic equations

$$a\sigma_x^2 + 2b\sigma_x \sigma_y + c\sigma_y^2 = 0,$$
$$a\tau_x^2 + 2b\tau_x \tau_y + c\tau_y^2 = 0, \tag{12.203}$$

which have the same solutions for the quotients σ_x/σ_y and τ_x/τ_y, namely,

$$\frac{dy}{dx} = -\frac{\sigma_x}{\sigma_y} = -\frac{\tau_x}{\tau_y} = \frac{b \pm \sqrt{b^2 - ac}}{a}. \tag{12.204}$$

Thus, if the differential equation is of so-called *hyperbolic* type, which is defined by the condition

$$b^2 - ac > 0 \tag{12.205}$$

for the coefficients a, b, c, for all x, y in the considered domain, then there are indeed two different, real-valued curves $y_+(x)$ and $y_-(x)$ with slopes

$$\frac{dy_\pm}{dx} = \frac{b \pm \sqrt{b^2 - ac}}{a} \tag{12.206}$$

through each point (x, y) of the domain, so that for the arclength coordinates σ and τ along these curves, the partial differential equation (12.194) takes the simplified form (12.197). For the arguments of this section, see Courant and Hilbert (1937), pp. 123–126.

12.11 Derivatives on Characteristics

A *weak* discontinuity is a jump in a first or higher *derivative* of a fluid field like u, ρ, p. We prove in this section that weak discontinuities can only occur across characteristics. Since the speed on characteristics is the wave speed, weak discontinuities (e.g., cusps) propagate the same way linear waves do. Bores and shocks, by contrast, run faster than water waves and sound waves, respectively. Such *strong* discontinuities therefore *cross* characteristics.

Theorem *Consider the second-order partial differential equation*

$$au_{xx} + 2bu_{xy} + cu_{yy} + d = 0 \tag{12.207}$$

for an unknown function $u(x, y)$, with coefficients a, b, c, d that are C_1-functions of x, y, u, u_x, u_y, with C_1-functions u, u_x, u_y of x, y. Equation (12.207) shall be hyperbolic, i.e., $b^2 - ac > 0$ at each point of the considered xy-domain. Let C be an arbitrary smooth curve in the xy-plane, along which the functions u, u_x, u_y are continuously differentiable.

  Then if C is not a characteristic of (12.207), all second and higher derivatives of u are unique along C. If C is a characteristic, however, second derivatives of u are not unique along C.

Proof (Courant and Hilbert 1937, pp. 291–293) Let $\tau(x, y) = 0$ define a curve C in the xy-plane with arclength parameter σ. The functions $x(\sigma), y(\sigma)$ are differentiable since C is smooth. Considering the first derivatives u_x, u_y of u as functions of σ and differentiating them with respect to σ (this is possible according to the assumptions) gives, with $x_\sigma = \partial x / \partial \sigma$, etc.,

$$(u_x)_\sigma = u_{xx} \, x_\sigma + u_{xy} \, y_\sigma,$$
$$(u_y)_\sigma = u_{xy} \, x_\sigma + u_{yy} \, y_\sigma. \tag{12.208}$$

Together with the differential equation (12.207), this gives three equations for the 'unknowns' u_{xx}, u_{xy}, u_{yy},

$$
\begin{aligned}
au_{xx} + 2bu_{xy} + cu_{yy} &= -d, \\
x_\sigma u_{xx} + y_\sigma u_{xy} &= (u_x)_\sigma, \\
x_\sigma u_{xy} + y_\sigma u_{yy} &= (u_y)_\sigma,
\end{aligned}
\tag{12.209}
$$

where the right sides are known (and continuous). The determinant D of this system is

$$
D = \begin{vmatrix} a & 2b & c \\ x_\sigma & y_\sigma & 0 \\ 0 & x_\sigma & y_\sigma \end{vmatrix} = ay_\sigma^2 - 2bx_\sigma y_\sigma + cx_\sigma^2.
\tag{12.210}
$$

If $D \neq 0$, then u_{xx}, u_{xy}, u_{yy} are uniquely determined along the curve $\tau = 0$, by multiplying the right side of (12.209) with the inverse matrix of the left side. All higher derivatives along the curve $\tau = 0$ are then also uniquely determined. To see this, differentiate (12.209) with respect to x, which gives

$$
\begin{aligned}
au_{xxx} + 2bu_{xyx} + cu_{yyx} &= -d_x - a_x u_{xx} - 2b_x u_{xy} - c_x u_{yy}, \\
x_\sigma u_{xxx} + y_\sigma u_{xyx} &= (u_{xx})_\sigma, \\
x_\sigma u_{xyx} + y_\sigma u_{yyx} &= (u_{yx})_\sigma.
\end{aligned}
\tag{12.211}
$$

Here, the right sides are known since $D \neq 0$, and the inhomogeneous system (12.211) has unique solution for $u_{xxx}, u_{xxy}, u_{xyy}$, since its determinant is $D \neq 0$. Differentiating (12.209) with respect to y gives the missing derivative u_{yyy}.

In the opposite case that $D = 0$, the homogeneous system corresponding to (12.209) has nontrivial solutions. The inhomogeneous system has then no unique solution, since any linear combination of solutions of the homogeneous system can be added to such a solution.

To see that indeed $D = 0$ along characteristics, note that along the curve C given by $\tau(x(\sigma), y(\sigma)) = 0$ one has

$$
0 = \frac{d\tau}{d\sigma} = \tau_x x_\sigma + \tau_y y_\sigma,
\tag{12.212}
$$

or

$$
\frac{y_\sigma}{x_\sigma} = -\frac{\tau_x}{\tau_y}.
\tag{12.213}
$$

One can assume that the denominators are zero; else consider the inverse of these fractions. Inserting (12.213) in (12.210) gives

$$
a\tau_x^2 + 2b\tau_x \tau_y + c\tau_y^2 = 0.
\tag{12.214}
$$

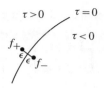

Fig. 12.23 Jump$[f] = \lim_{\epsilon \to 0} (f_+ - f_-)$ in f across a curve $\tau = 0$

According to (12.203), this means that the curve $\tau = 0$ is a characteristic, as is stated in the theorem. □

The following is the central theorem concerning the characteristics of the above second-order partial differential equation. It shows that characteristics are wavefronts, across which derivatives of second order may have jumps. Let $[f] = f_+ - f_-$ denote the jump of a function f across a curve (at any position along the curve) given by $\tau = 0$, where f_+ and f_- are the values of f on the sides where $\tau > 0$ and $\tau < 0$, respectively, with distance $\epsilon \to 0$ from the curve in the normal direction (Fig. 12.23).

Theorem *Consider again the hyperbolic equation*

$$au_{xx} + 2bu_{xy} + cu_{yy} + d = 0 \qquad (12.215)$$

with C_1-functions a, b, c, d as in the last theorem. Along a curve C given by $\tau = 0$ and with arclength σ, the functions u, u_x, u_y shall be continuously differentiable, and these derivatives shall have no jumps across C,

$$[u_\sigma] = [(u_x)_\sigma] = [(u_y)_\sigma] = 0. \qquad (12.216)$$

Then any jumps in u_{xx}, u_{xy}, u_{yy} can only occur across characteristics of equation (12.215).

Proof (Courant and Hilbert 1937, pp. 297–298) Along C one has as in (12.212) and (12.213),

$$\frac{y_\sigma}{x_\sigma} = -\frac{\tau_x}{\tau_y}, \qquad (12.217)$$

and the denominators are again assumed to be $\neq 0$. Let

$$y_\sigma = -\kappa \tau_x, \qquad x_\sigma = \kappa \tau_y, \qquad (12.218)$$

with a common scaling factor $\kappa \neq 0$ that cancels in (12.217). The derivatives of u_x, u_y along C are

$$(u_x)_\sigma = u_{xx}x_\sigma + u_{xy}y_\sigma = \kappa u_{xx}\tau_y - \kappa u_{xy}\tau_x,$$
$$(u_y)_\sigma = u_{yx}x_\sigma + u_{yy}y_\sigma = \kappa u_{yx}\tau_y - \kappa u_{yy}\tau_x. \qquad (12.219)$$

From the assumptions of the theorem, and dividing again through κ,

$$0 = [(u_x)_\sigma] = [u_{xx}]\tau_y - [u_{xy}]\tau_x \tag{12.220}$$

$$0 = [(u_y)_\sigma] = [u_{yx}]\tau_y - [u_{yy}]\tau_x, \tag{12.221}$$

where τ_x and τ_y have no jumps and can be taken out of [.]. Furthermore, from the differential equation (12.215) one has

$$a\,[u_{xx}] + 2b\,[u_{xy}] + c\,[u_{yy}] = 0, \tag{12.222}$$

in which a, b, c, d are C_1-functions by assumption, thus continuous. Note that the inhomogeneous term d has vanished. Equations (12.222), (12.220), and (12.221) are three equations for three unknowns $[u_{xx}]$, $[u_{xy}]$, $[u_{yy}]$,

$$\begin{pmatrix} a & 2b & c \\ \tau_y & -\tau_x & 0 \\ 0 & \tau_y & -\tau_x \end{pmatrix} \cdot \begin{pmatrix} [u_{xx}] \\ [u_{xy}] \\ [u_{yy}] \end{pmatrix} = \begin{pmatrix} 0 \\ 0 \\ 0 \end{pmatrix}. \tag{12.223}$$

The determinant D of this homogeneous system is

$$D = a\tau_x^2 + 2b\tau_x\tau_y + c\tau_y^2. \tag{12.224}$$

For (12.223) to have nontrivial solutions $([u_{xx}], [u_{xy}], [u_{yy}]) \neq (0, 0, 0)$, one must have $D = 0$, which makes (12.224) the equation of a characteristic. Thus, discontinuities in second derivatives can occur across characteristics. $\qquad\square$

This result can be compared with the corresponding result in Sect. 7.5. The material of the present section is expanded to full generality in Courant and Hilbert (1937): higher order differential equations in two variables x, y and systems of differential equations are treated in their Chap. 5, and the case of n independent variables in Chap. 6.

12.12 Simple Waves

We end our considerations with a final geometric construction. The two homogeneous fluid equations (12.179) have the constant-state solutions

$$u(t, x) = C_1, \qquad \rho(t, x) = C_2, \tag{12.225}$$

with global constants C_1 and C_2. In a constant state, both the Riemann invariants $L_+(u, \rho)$ and $L_-(u, \rho)$ are globally constant, whereas so far L_+ (resp. L_-) had constant value only along an x_+ (resp. x_-) characteristic.

Fig. 12.24 The region S adjacent to a constant state K is a simple wave, here even a centered simple wave

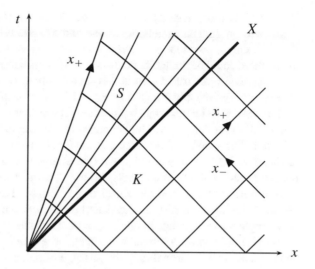

Let K be a region in the xt-plane in which both $u = C_1$ and $\rho = C_2$ are constant, see Fig. 12.24. Then both families of characteristics are straight lines in K, with inverse slopes $dx_\pm/dt = C_1 \pm a(C_2)$. The lines $x_+(t)$ are parallel among themselves, and so are the lines $x_-(t)$. Let S be a region adjacent to K, where S is not a constant state. If in the transition from K to S some *derivative* of a fluid field undergoes a jump, the boundary line between K and S must be a characteristic according to the last section. Courant and Friedrichs (1948, p. 55) and Courant and Hilbert (1937, p. 309) give arguments that even without such a discontinuity, the border line between a constant state and a nonconstant state must be a characteristic.

Consider then the situation in Fig. 12.24, where K and S meet along the x_+-characteristic X (drawn fat). The following arguments hold also if this border is instead an x_--characteristic. Since $a \neq 0$, the x_+- and x_--characteristics have different slopes, thus the x_--characteristics from K must *cross* the line X. These x_--characteristics from K that cross X define a subset of S; we continue to use the symbol S for this. On each x_--characteristic in S, the Riemann invariant L_- is constant. However, L_- has one single, global value on X, since K is a constant state; thus L_- has one single, global value in S too. Since L_+ is by definition constant on each x_+-characteristic in S, one arrives at the equation for the x_+-characteristics,

$$\frac{dx_+}{dt}(u(L_+, L_-), \rho(L_+, L_-)) = \text{const}, \qquad (12.226)$$

where this constant will change with L_+ from one x_+-characteristic to the other. Thus the x_+-characteristics in S are straight but not parallel lines.

If the invariant L_- is constant in the wave region, the x_+ characteristic lines, $L_+ = \text{constant}$, are straight. (Courant and Friedrichs 1948, p. 93; with our symbols)

For a slightly more abstract argument, see Courant and Friedrichs (1948), p. 60. A region with straight characteristics of one of the two families is termed a *simple wave*. We have thus shown that *adjacent to a constant state lies a simple wave*. (Alternatively, there could occur another constant state, separated by a contact discontinuity.)

Simple waves are true waves, and as such do *not* move at the speed of fluid parcels: as is clear from the construction in Fig. 12.24, two straight characteristics bound a simple wave at its left and right boundaries. These characteristics are termed the *head* and *tail* of the simple wave. Fluid parcels enter a simple wave at its head and leave it at its tail. The speed of fluid parcels at the head and tail is u, whereas the speeds of the head and tail are $u + a$ and $u - a$ (the sign is not relevant here). Therefore, the relative speed between fluid parcels and both the head and tail of a simple wave is a.

A simple wave is called *forward* facing if the x_+-characteristics are straight, and *backward* facing if the x_--characteristics are straight. If all the straight characteristics of a simple wave originate in one single point of the xt-plane, the wave is called a *centered* simple wave. A simple wave is called a *rarefaction wave* if both the parcel density and pressure drop during the passage of parcels through the wave.

As an example of a centered rarefaction wave, consider an infinitely long straight tube filled with gas in the domain $x > 0$. Starting at $x = 0$ and $t = 0$, a piston is withdrawn from the tube at constant speed $-U < 0$. Not only the gas at $x < 0$ is affected by this, but the piston motion triggers a sound wave that propagates at speed a_0 with a front $x_+(t) = a_0 t$ into the undisturbed gas. This front is the head of a centered rarefaction wave S, with straight x_+-characteristics orginating at $(x, t) = (0, 0)$. Ahead of this front, the gas is still at rest in a constant state K_0. The parcels are accelerated within the rarefaction wave and leave the latter at its tail at some location $x > 0$, which is determined in (12.234). Beyond this location, i.e., at smaller x, the gas is again in a constant state K_1 with speed $u = -U$ equal to the piston speed (Figs. 12.26 and 12.27). For this example, see Courant and Friedrichs (1948, p. 94) and Landau and Lifshitz (1987, p. 369).

In the rarefaction wave, L_- and dx_+/dt are constants. More specifically, for a flow under adiabatic conditions, with adiabatic exponent $\gamma > 1$, one has from (12.178) and (12.185) that

$$L_- = u - \frac{2a}{\gamma - 1} = \text{const}, \tag{12.227}$$

$$\frac{dx_+}{dt} = u + a = \text{const}. \tag{12.228}$$

With $x_+(0) = 0$ as origin of the x_+-characteristics in the rarefaction wave,

$$\frac{x_+}{t} = u + a. \tag{12.229}$$

From (12.227),

$$u - u_0 = \frac{2}{\gamma - 1}(a - a_0). \tag{12.230}$$

Fig. 12.25 Speed and
density in a centered
rarefaction wave S for a
piston moving at constant
speed $-U$ toward $x < 0$.
The parcel motion is from
right to left

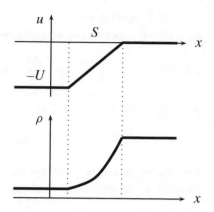

With $u_0 = 0$ in K_0,

$$a = a_0 + \frac{\gamma - 1}{2}\, u. \tag{12.231}$$

Inserting this in (12.229) and dropping the subscript $+$ gives

$$u(t, x) = \frac{2}{1 + \gamma}\left(\frac{x}{t} - a_0\right). \tag{12.232}$$

With $p \sim \rho^\gamma$, the fluid density scales as $\rho \sim a^{2/(\gamma - 1)}$, thus

$$\frac{\rho}{\rho_0} = \left(1 + \frac{\gamma - 1}{2}\frac{u}{a_0}\right)^{\frac{2}{\gamma - 1}}. \tag{12.233}$$

For $\gamma \to 1$, this gives an exponential drop-off, $\rho = \rho_0\,\exp(u/a_0)$. Thus, at any given instance $t > 0$, the speed $u(., x)$ is a linear function, and the gas density drops as a power law in a rarefaction wave, see Fig. 12.25. Note that the gas speed and density have jumps in their derivatives (weak discontinuities) at the head and tail of the rarefaction wave.

The gas behind the centered rarefaction wave, after leaving its tail, moves at the constant speed $u = -U$ of the piston. From (12.231),

$$u + a = a_0 - \frac{\gamma + 1}{2}\, U \quad \text{and} \quad u - a = -a_0 - \frac{3 - \gamma}{2}\, U \tag{12.234}$$

are the slopes of the characteristics in this second constant state K_1.

The shape of the curved x_--characteristics in the rarefaction wave can be derived as follows. To keep the algebra simple, let $\gamma = \frac{5}{3}$ for a monatomic gas. Write the characteristics as $x_+ = x_+(\sigma, \tau)$ and $x_- = x_-(\sigma, \tau)$, where σ is a curve variable (not necessarily the arclength) along x_-, and is also the parameter that labels the x_+-curves: each value $\sigma = $ const corresponds to one full x_+-curve. The role of τ is just interchanged to that of σ, but τ does not appear in the following. *Along* a given x_--characteristic one has, dropping the curve parameter τ,

$$\frac{dx_-}{dt} = \frac{x_{-\sigma}}{t_\sigma} = u - a, \tag{12.235}$$

with $x_{-\sigma} = dx_-/d\sigma$. Therefore,

$$x_{-\sigma} = (u - a)\, t_\sigma. \tag{12.236}$$

The next argument is novel, see Courant and Friedrichs (1948), p. 97 and p. 104: consider the parameterization of the x_+-characteristics with σ,

$$x_+(\sigma) = (u(\sigma) + a(\sigma))\, t, \tag{12.237}$$

where the curve variable τ is suppressed, and along any x_+-characteristic, $u + a$ is constant, and does not depend on time. The idea is to express the characteristic curve $(x_-(\sigma), t(\sigma))$ within the rarefaction wave by the 'coordinate' x_+ in this domain using (12.237),

$$x_-(\sigma) = (u(\sigma) + a(\sigma))\, t(\sigma) \tag{12.238}$$

(note the appearance of $t(\sigma)$ here). Differentiation with respect to σ yields

$$x_{-\sigma} = (u_\sigma + a_\sigma)\, t + (u + a)\, t_\sigma. \tag{12.239}$$

Inserting here (12.236) for the x_--characteristic on the left side gives

$$(u - a)\, t_\sigma = (u_\sigma + a_\sigma)\, t + (u + a)\, t_\sigma, \tag{12.240}$$

or

$$\frac{t_\sigma}{t} = -\frac{u_\sigma + a_\sigma}{2a}, \tag{12.241}$$

or, de-parameterizing,

$$\frac{dt}{t} = -\frac{du + da}{2a}. \tag{12.242}$$

From (12.230) on the x_--characteristic, for $\gamma = \frac{5}{3}$ and $u_0 = 0$,

$$u = 3(a - a_0) \quad \text{and} \quad du = 3\, da, \tag{12.243}$$

and (12.242) becomes

$$\frac{dt}{t} = -2\, \frac{da}{a}, \tag{12.244}$$

with solution

$$\frac{t}{t_0} = \left(\frac{a_0}{a}\right)^2, \tag{12.245}$$

where $a = a_0$ at $t = t_0$. Equation (12.238) can be written as, using the first equation in (12.243),

$$x_-(t) = (4a - 3a_0)\, t, \tag{12.246}$$

which becomes with (12.245),

$$x_-(t) = 4a_0\sqrt{tt_0} - 3a_0 t. \tag{12.247}$$

For given sound speed a_0 in the gas region at rest, this determines the characteristic curves $x_-(t)$. Here t_0 is the curve parameter: $x_-(t; t_0)$ starts on the characteristic $x_+ = a_0\, t$ at time t_0. An xt-diagram of this example with all the x_+- and x_--characteristics is shown in Fig. 12.26, which can be compared to Fig. 20, p. 105 in Courant and Friedrichs (1948). For (12.247), see their Eq. (46.07).

Furthermore, since $a \neq 0$, parcel trajectories (inverse slope u) cannot coincide with characteristics (inverse slopes $u \pm a$), and the above parameterization of a curve in the rarefaction-wave domain using the x_+-characteristics with

$$x_+(\sigma) = (u(\sigma) + a(\sigma))\, t \tag{12.248}$$

can also be applied to the parcel trajectory $(x(\sigma), t(\sigma))$, defined by

$$\frac{dx}{dt} = \frac{x_\sigma}{t_\sigma} = u. \tag{12.249}$$

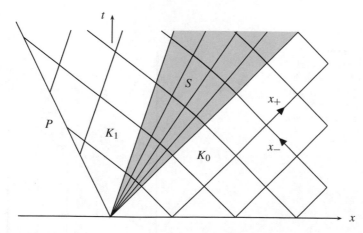

Fig. 12.26 Characteristics x_+ and x_- for piston flow in a long straight tube. A piston P is withdrawn toward $x < 0$ from a tube filled with gas. A rarefaction wave S (in gray) centered at $(x, t) = (0, 0)$ separates two constant states K_0, the still unaffected gas at rest, and K_1, the rarefied gas moving at piston speed

Differentiating (12.248) with respect to σ, including again $t = t(\sigma)$, and equating this to $x_\sigma = u\,t_\sigma$, one has

$$u t_\sigma = (u_\sigma + a_\sigma)\,t + (u + a)\,t_\sigma, \qquad (12.250)$$

which yields

$$\frac{\mathrm{d}t}{t} = -\frac{\mathrm{d}u + \mathrm{d}a}{a} = -4\,\frac{\mathrm{d}a}{a}, \qquad (12.251)$$

with solution

$$\frac{t}{t_0} = \left(\frac{a_0}{a}\right)^4. \qquad (12.252)$$

Equation (12.246) also applies here, if x_- of the characteristic is replaced now by x for the parcel trajectory,

$$x(t) = (4a - 3a_0)\,t. \qquad (12.253)$$

Inserting (12.252), one finds (see Eq. 46.06 in Courant and Friedrichs 1948)

$$x(t) = 4a_0\,t_0^{1/4} t^{3/4} - 3a_0\,t \qquad (12.254)$$

as equation for the parcel trajectories in the centered rarefaction wave. These curves and the straight trajectories ahead and behind the rarefaction wave are shown in Fig. 12.27.

We have reached a point where a fundamental question first raised by Riemann becomes sensible: what are the possible gas motions in a straight tube that result if a barrier is released at time $t = 0$ at location $x = 0$ between two constant, different gas states at $x < 0$ and $x > 0$ ('the Riemann problem')?

Fig. 12.27 Parcel trajectories $x(t)$ from the state at rest K_0 through the rarefaction wave S (in gray) to the constant state K_1 moving at the same speed as the piston P

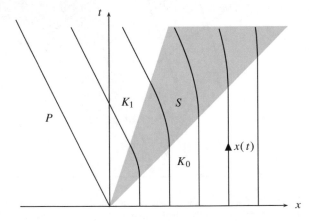

Riemann's answer is: there are four possible types of subsequent flow, inasmuch as in both directions from the origin a shock or a centered rarefaction wave may proceed depending on inequalities prevailing for the initial conditions. (Courant and Friedrichs 1948, p. 181)

At early times, the motion in the tube consists of individual shocks and/or rarefaction waves, which may interact nonlinearly at later times. Riemann's theorem lays at the foundation of an important method in time-dependent computational fluid dynamics, the so-called *Riemann solver*. In a one-dimensional calculation, e.g., the gas state is discretized into a sequence of narrow constant states ('cells'), and the subsequent evolution of discontinuities at the cell interfaces is obtained from the propagation of shocks and rarefaction waves corresponding to the initial value problem at each interface. This evolution is followed over a short interval of time ('time step'), before interactions between shocks and waves from adjacent cells occur. The resulting functions for u, ρ, and p are discretized into a new sequence of constant states, which pose the initial value problem for the next time step, and so forth.

In the last few sections, we have found that shocks, i.e., (strong) discontinuities in the fluid fields ρ, $\vec{u}$, p, T, move through a (preshock) gas at supersonic speeds, whereas (weak) discontinuities in the first, second, or higher derivatives of these fluid fields propagate along characteristics, i.e., at the speed of sound. This is also plausible on physical grounds, since a discontinuity in a derivative can be introduced in a smooth flow solution by making arbitrarily small changes to this solution, and small perturbations propagate through a gas at the speed of sound. The next issue then is to analyze the *interactions* of strong and weak discontinuities (e.g., shocks overtaking shocks or rarefaction waves), since the principle of superposition does not hold for nonlinear waves, and new physical phenomena can occur. This is a major topic in the book by Courant and Friedrichs (1948), see, e.g., their §78–§83, and is not followed upon here.

References

Ben-Dor, G., O. Igra, and T. Elperin. 2000. *Handbook of shock waves*, 3 volumes. New York: Academic Press.

Bessis, D., and J.D. Fournier. 1984. Pole condensation and the Riemann surface associated with a shock in Burgers equation. *Journal de Physique Lettres* 45: 833.

Bessis, D., and J.D. Fournier. 1990. Complex singularities and the Riemann surface for the Burgers equation. In *Nonlinear physics*, vol. 252, ed. G. Chaohao, L. Yishen, T. Guizhang, and Z. Yunbo. Berlin: Springer.

Caflisch, R.E., N. Ercolani, T.Y. Hou, and Y. Landis. 1993. Multi-valued solutions and branch point singularities for nonlinear hyperbolic or elliptic systems. *Communications on Pure and Applied Mathematics* 46: 453.

Cannon, C.J., and R.N. Thomas. 1977. The origin of stellar winds: Subatmospheric nonthermal storage modes versus radiation pressure. *Astrophysical Journal* 211: 910.

Chamberlain, J.W. 1960. Interplanetary gas. II. Expansion of a model solar corona. *Astrophysical Journal* 131: 47.

Chorin, A.J., and J.E. Marsden. 1990. *A mathematical introduction to fluid mechanics*. New York: Springer.

Cole, J.D. 1951. On a quasi-linear parabolic equation occurring in aerodynamics. *Quarterly of Applied Mathematics* 9: 225.

Courant, R., and K.O. Friedrichs. 1948. Supersonic flow and shock waves. New York: Interscience; New York: Springer (1976).

Courant, R., and D. Hilbert. 1937. *Methoden der mathematischen Physik*, zweiter Band. Berlin: Springer (2nd ed., 1968).

Hopf, E. 1950. The partial differential equation $u_t + uu_x = \mu u_{xx}$. *Communications on Pure and Applied Mathematics* 3: 201.

Jouguet, E. 1901. Sur la propagation des discontinuités dans les fluides. *Comptes rendus de l'Académie des sciences Paris* 132: 673.

Kolsky, H., J.P. Lewis, M.T. Sampson, A.C. Shearman, and C.I. Snow. 1949. Splashes from underwater explosions. *Proceedings of the Royal Society of London A* 196: 379.

Krehl, P.O. 2009. *History of shock waves, explosions and impact*. Berlin: Springer.

Lamb, H. 1932. *Hydrodynamics*. Cambridge: Cambridge University Press; New York: Dover (1945).

Landau, L.D., and E.M. Lifshitz. 1987. *Fluid mechanics, Course of theoretical physics*, vol. 6, 2nd ed. Amsterdam: Elsevier; Oxford: Butterworth-Heinemann.

Lax, P.D. 1973. *Hyperbolic systems of conservation laws and the mathematical theory of shock waves*. Philadelphia: Society for Industrial and Applied Mathematics.

Lewy, H. 1928. Über das Anfangswertproblem einer hyperbolischen nichtlinearen partiellen Differentialgleichung zweiter Ordnung mit zwei unabhängigen Veränderlichen. *Mathematische Annalen* 98: 179.

Meyer, Th. 1908. *Ueber zweidimensionale Bewegungsvorgänge in einem Gas, das mit Ueberschallgeschwindigkeit strömt*, Dissertation Universität Göttingen, Sonderabdruck in den Mitteilungen über Forschungsarbeiten des Vereins deutscher Ingenieure, Berlin.

Moore, D.W. 1979. The spontaneous appearance of a singularity in the shape of an evolving vortex sheet. *Proceedings of the Royal Society of London A* 365: 105.

Parker, E.N. 1958. Dynamics of the interplanetary gas and magnetic fields. *Astrophysical Journal* 128: 664.

Parker, E.N. 1960a. The hydrodynamic treatment of the expanding solar corona. *Astrophysical Journal* 132: 175.

Parker, E.N. 1960b. The hydrodynamic theory of solar corpuscular radiation and stellar winds. *Astrophysical Journal* 132: 821.

Parker, E.N. 1981. Photospheric flow and stellar winds. *Astrophysical Journal* 251: 266.

Parker, E.N. 1999. Space physics before the space age. *Astrophysical Journal* 525 (Centennial Issue): 792.

Riemann, B. 1860. Ueber die Fortpflanzung ebener Luftwellen von endlicher Schwingungsweite. *Abhandlungen der Königlichen Gesellschaft der Wissenschaften in Göttingen* 8: 43.

Serrin, J. 1959. Mathematical principles of classical fluid mechanics. In *Handbuch der Physik = Encyclopedia of Physics*, vol. 3/8-1, Strömungsmechanik 1, ed. S. Flügge, and C.A. Truesdell, 125. Berlin: Springer.

Sichel, M. 1966. The effect of longitudinal viscosity on the flow at a nozzle throat. *Journal of Fluid Mechanics* 25: 769.

Simon, M., and W.I. Axford. 1966. Shock waves in the interplanetary medium. *Planetary and Space Science* 14: 901.

Sturrock, P.A., and J.R. Spreiter. 1965. Shock waves in the solar wind and geomagnetic storms. *Journal of Geophysical Research* 70: 5345.

Tatsumi, T., and S. Kida. 1972. Statistical mechanics of the Burgers model of turbulence. *Journal of Fluid Mechanics* 55: 659.

Taylor, G.I. 1930. *The flow of air at high speeds past curved surfaces*, Aeronautical Research Committee, R. & M. 1381, 213, and in *G.I. Taylor, Scientific Papers*, vol. 3, ed. G. K. Batchelor, 128. London: Cambridge University Press (1963).

Tomotika, S., and K. Tamada. 1950. Studies of two-dimensional transonic flows of compressible fluid. I. *Quarterly of Applied Mathematics* 7: 381.

Weyl, H. 1944, *Shock waves in arbitrary fluids*, National Defense Research Committee, Applied Mathematics Panel Note No. 12, and Applied Mathematics Group—New York University No. 46.

Whitham, G.B. 1974. *Linear and nonlinear waves*. New York: Wiley.

Wolfson, R.L.T., and T.E. Holzer. 1982. Intrinsic stellar mass flux and steady stellar winds. *Astrophysical Journal* 255: 610.

Zel'dovich, Ya.B., and Yu.P. Raizer. 1966. *Physics of shock waves and high-temperature hydrodynamic phenomena*, vol. 1. New York: Academic Press; Mineola: Dover (2002).

Zel'dovich, Ya.B., and Yu.P. Raizer. 1967. *Physics of shock waves and high-temperature hydrodynamic phenomena*, vol. 2. New York: Academic Press; Mineola: Dover (2002).

Zemplén, G.V. 1905. Sur l'impossibilité d'ondes de choc négatives dans les gaz. *Comptes rendus de l'Académie des sciences Paris* 141: 710.

Zierep, J. 1991. *Theoretische Gasdynamik*. Karlsruhe: G. Braun.

Correction to: Theoretical Fluid Dynamics

Correction to:
A. Feldmeier, *Theoretical Fluid Dynamics*,
Theoretical and Mathematical Physics,
https://doi.org/10.1007/978-3-030-31022-6

In the original version of this book, the following belated corrections have been incorporated:

Page 18, Eq. (2.54): replace $d\vec{F}$ by $d\vec{F}(d\vec{a})$.
Page 18: delete full sentence after boxed Eq. (2.55).
Page 34, item (d): add "and" before "using Gauss' theorem."
Page 213, Eqs. (6.21) and (6.23): replace $\oint$ by $\int$.
Page 214, Eqs. (6.25) and (6.27): replace $\oint$ by $\int$.
Page 215, Eq. (6.29): replace $\oint$ by $\int$.
The erratum book has been updated with the changes.

The updated version of these chapters can be found at
https://doi.org/10.1007/978-3-030-31022-6_2
https://doi.org/10.1007/978-3-030-31022-6_6

Appendix A
Analytic and Meromorphic Functions

We list here some standard theorems for analytic and meromorphic functions from complex function theory without proofs. All functions are assumed to be single-valued. Curves (along which integrals are taken) are piecewise differentiable, and domains are connected (no disjunct pieces) and often even simply connected (no holes; see Sect. 3.10 for details).

Exchanging sum and integral

A series $\sum_{j=0}^{\infty} f_j$ with $f_j : D \to \mathbb{C}$ converges *uniformly* in $B \subset D$ if

$$\forall \epsilon > 0 \quad \exists N(\epsilon) \in \mathbb{N} \quad \forall n \geq N \quad \forall z \in B : \quad \left| \sum_{j=n}^{\infty} f_j(z) \right| < \epsilon. \qquad (A.1)$$

Note that N does not depend on z. Let C be a curve and $\sum_{j=0}^{\infty} f_j$ a function series. Let $\sum f_j$ converge uniformly on C to a function f defined on C. (We identify the curve C—a map—with its set of image points.) Then

$$\sum_j \int_C dz \, f_j(z) = \int_C dz \sum_j f_j(z) = \int_C dz \, f(z). \qquad (A.2)$$

Proof: Remmert (1992), p. 142, Bieberbach (1921), p. 111.

The Goursat lemma

Let the function f be analytic in the domain D. Consider a triangle whose boundary and interior lie completely in D. Let C be the boundary of this triangle in D. Then

$$\oint_C dz \, f(z) = 0. \qquad (A.3)$$

© Springer Nature Switzerland AG 2019
A. Feldmeier, *Theoretical Fluid Dynamics*, Theoretical and Mathematical Physics,
https://doi.org/10.1007/978-3-030-31022-6

Lemma and proof are classic, see Remmert, p. 148, Bieberbach, p. 118.

The Cauchy theorem

A domain D is *star-shaped* if there exists an interior point a such that for any point $b \in D$, the straight line from a to b lies in D. Let f be analytic in the star-shaped domain D. Then for each closed curve C in D (not touching its boundary),

$$\oint_C dz\, f(z) = 0. \tag{A.4}$$

Thus the line integral from one point to another depends only on these points, not on the path taken. The theorem also holds if D is not star-shaped, but simply connected. For the proof, see Remmert p. 151, Bieberbach, p. 115, and Ahlfors (1966), pp. 139–143. C is called *homologous to 0* if C does not enclose points outside D. This means, C does not go round holes in D. For an exact definition using *winding numbers*, see (A.22). Cauchy's theorem holds if f is analytic in a (multiply connected) domain D, and if C is a cycle that is homologous to 0 in D. Proof: Ahlfors (1966), pp. 144–145, Dixon (1971).

The Cauchy integral formula

Let f be analytic in a domain D. Let B be the set of points within a circle ∂B in the complex plane. (In $\mathbb{R}^3$, B is a ball; in the complex plane, B is a disk.) Let $B \cup \partial B \subset D$. Then for all points z in B (not on ∂B),

$$\boxed{f(z) = \frac{1}{2\pi i} \oint_{\partial B} d\zeta\, \frac{f(\zeta)}{\zeta - z}} \tag{A.5}$$

Thus, any f analytic in a disk is fully determined by f on the boundary circle of the disk. For the proof, see Remmert p. 159 or Bieberbach, p. 123 and p. 128. Since Cauchy's formula is so fundamental, we give a simple derivation. (The modern proof is more elegant, using $\frac{f(\zeta)-f(z)}{\zeta-z}$ and $f'(z)$.) Let Γ be a circle with radius r and center z. For points on Γ,

$$\zeta = z + re^{i\varphi},$$
$$d\zeta = \frac{\partial \zeta}{\partial \varphi} d\varphi = ire^{i\varphi}\, d\varphi, \tag{A.6}$$

giving

$$\oint_\Gamma \frac{d\zeta}{\zeta - z} = i \int_0^{2\pi} d\varphi = 2\pi i. \tag{A.7}$$

Since f is continuous, one has for $r \to \epsilon$, with $\epsilon \ll 1$,

$$f(z) \oint_\Gamma \frac{d\zeta}{\zeta - z} = \oint_\Gamma d\zeta\, \frac{f(\zeta)}{\zeta - z} \tag{A.8}$$

(for small ϵ, Γ lies completely in B), or

$$2\pi i f(z) = \oint_\Gamma d\zeta \, \frac{f(\zeta)}{\zeta - z}. \tag{A.9}$$

Let K be an arbitrary curve that starts on ∂B and ends on Γ. Then the curve $\Pi = K - \Gamma - K + \partial B$ defines a closed cycle. The signs refer to the direction in which the segments are passed. (Regarding K and $-K$, one can think of two banks of one curve or of two curves that have infinitesimal distance everywhere.) f is analytic in the simply connected domain inside Π. From Cauchy's theorem, with contributions from K and $-K$ cancelling,

$$\oint_{\partial B} d\zeta \, \frac{f(\zeta)}{\zeta - z} - \oint_\Gamma d\zeta \, \frac{f(\zeta)}{\zeta - z} = 0, \tag{A.10}$$

and using (A.9), Cauchy's formula (A.5) follows.

The Morera theorem

In Cauchy's theorem, 'f is analytic' implies $\oint dz \, f(z) = 0$. Morera's theorem gives conditions for the implication in the opposite direction. Let f be continuous in a domain D. Let $\oint_C dz \, f(z) = 0$ for all closed curves C in D. Then f is analytic in D. Proof: Remmert p. 185, Bieberbach, p. 133.

Normal convergence

A series $\sum_{j=0}^{\infty} f_j$ of functions $f_j : D \to \mathbb{C}$ converges *normally* in D if every point in D has a neighborhood U such that $\sum |f_j|_U < \infty$.

Theorems on normal convergence

If all f_j in $\sum f_j = f$ are continuous and converge normally, f is continuous. Let $\tau : \mathbb{N} \to \mathbb{N}$ be one-to-one. If $\sum f_j$ is normally convergent, then $\sum f_{\tau(j)}$ is. The limit is in both cases the same function f. Proof: Remmert, p. 82.

The Cauchy–Taylor theorem

Let B be the interior of a circle with center c, and f be analytic in B. Then there is a series $\sum_{j=0}^{\infty} a_j (z - c)^j$ that converges normally to f in B. The a_j are given by, with C a circle that lies wholly within B,

$$a_j = \frac{1}{2\pi i} \oint_C d\zeta \, \frac{f(\zeta)}{(\zeta - c)^{j+1}}. \tag{A.11}$$

Furthermore, the Taylor formula holds,

$$a_j = \frac{1}{j!} \frac{d^j}{dz^j} f(z) \bigg|_{z=c}. \tag{A.12}$$

Thus, every analytic function has a Taylor series expansion,

$$f(z) = \sum_{j=0}^{\infty} a_j (z - c)^j, \tag{A.13}$$

where c is an arbitrary but fixed point in the analytic domain. Proof: Remmert, p. 165, Bieberbach, p. 135.

The Riemann continuation theorem

Let D be a domain, and $c \in D$ an arbitrary point. Let f be analytic in $D\backslash c$ (short for $D\backslash\{c\}$) and continuous in D. Then f is analytic in D. Proof: Remmert, p. 167, Bieberbach, p. 145.

Infinite differentiability

Let f be analytic in a domain D. Then f is arbitrarily often complex differentiable in D. Thus every analytic function can be differentiated infinitely often. This is quite clear from the Cauchy–Taylor theorem above. Proof: Remmert p. 169, Bieberbach, p. 132.

Identity theorem. I

Let f and g be analytic in the interior B of a circle. Let $f|U = g|U$ for an arbitrarily small neighborhood $U \subset B$. Then $f = g$ in B. The reason is that the Taylor coefficients can be calculated in U and held in B.

Identity theorem. II

Let f and g be two functions analytic in a domain D. Let the set of points c_j where $f(c_j) = g(c_j)$ have a limit point b in D. Then $f = g$ in D. For the proof, see Remmert, p. 179 or Bieberbach, p. 137. The proof idea is as follows. At b, all derivatives of f and g agree. By the Cauchy–Taylor theorem, f and g are then identical in D.

Double series theorem (Weierstrass)

Let all $f_j(z)$ with $j = 0, 1, 2, \ldots$ be analytic in a domain D. Let $f(z) = \sum_{j=0}^{\infty} f_j(z)$ converge *uniformly*. Then f is analytic, and $f' = \sum f_j'$. Proof: Remmert, p. 196, Bieberbach, p. 153.

The Vitali theorem

This is an extension of the foregoing theorem. Let D be a domain and $C = \{c_j\} \subset D$ a point set with limit point in D. We assume $\exists M \; \forall m \; \forall z \in D : \sum_{j=0}^{m} f_j(z) < M$.

Then $f = \sum_{j=0}^{\infty}$ converges *uniformly* in D, and thus f is analytic. Proof: Bieberbach, p. 166.

Discrete zeros

Let $f \neq 0$ be analytic in a domain D. Then the set of zeros of f is discrete: for any two zeros there exist nonoverlapping neighborhoods. Proof: Remmert, p. 182, Bieberbach, p. 138.

Conformal mapping

$f : D \to \mathbb{C}$ is called *conformal* if it preserves angles locally. If two curves cross at angle α, their image curves also cross at angle α. Let f be analytic in a domain D, and nonconstant in every neighborhood. Then there exists a discrete set $A \subset D$ so that f is conformal in $D \backslash A$. The set A consists of the points where $f' = 0$. At these points, f is not conformal. Proof: Remmert, p. 187, Bieberbach, p. 41.

Liouville's theorem

Let f be analytic and bounded in the whole complex plane. This means there is some $M > 0$ such that

$$|f(z)| \leq M \qquad \text{for all } z \in \mathbb{C}. \tag{A.14}$$

Then f is constant on $\mathbb{C}$. Proof: Remmert, p. 192, Bieberbach, p. 150.

Maximum property or maximum principle

Let f be analytic in D and let $|f|$ have a local maximum in D. Then f is constant in D (compare with Sect. 3.4). Equivalently, if ∂D is the boundary of D, and $\bar{D} = D \cup \partial D$, then

$$|f(z)| \leq |f(\zeta)| \qquad \text{for all } z \in \bar{D} \text{ and } \zeta \in \partial D. \tag{A.15}$$

Proof: Remmert, p. 191 and p. 203, Bieberbach, p. 143.

Open mapping theorem

If f is analytic in D and not constant in D, then $f(G)$ is a domain. Thus, analytic functions map open sets U to open sets $f(U)$. Proof: Remmert, p. 202, Bieberbach, p. 187. The general invariance of domain theorem for two dimensions is as follows: If $U \subset \mathbb{R}^2$ is open and $f : U \to \mathbb{R}^2$ is injective and continuous, then $f(U)$ is open. The proof is by Schoenflies (1899), with simplifications by Osgood (1900) and Bernstein (1900), and can be found in Markushevich (1965), p. 95. The proof of the invariance of domain theorem for general $\mathbb{R}^n$ is by Brouwer (1912).

Lemma of Schwarz

If $f : \mathbb{E} \to \mathbb{E}$, with unit disk $\mathbb{E} = \{z : |z| < 1\}$, and if $f(0) = 0$, then

$$|f(z)| \leq |z| \qquad \text{for all } z \in \mathbb{E}. \tag{A.16}$$

Proof: Remmert, p. 212.

Study theorem

Let both $f : \mathbb{E} \to D$ (with unit disk $\mathbb{E}$) and f^{-1} be analytic. Let D_r be the image under f of a circle with radius $0 < r < 1$. Then if D is convex, so are all D_r. Proof: Remmert, p. 215, using the lemma of Schwarz.

We now allow f to have *isolated* singularities where f is not analytic. *Isolated* means that for any two singularities, disjunct neighborhoods exist. Singularities are not generally isolated, e.g., there exist continuous curves of singularities. For example, for the function defined by the series $\sum\limits_{j=1}^{\infty} z^{2^j}$, each point on the convergence circle $|z| = 1$ is singular. This so-called Weierstrass series is discussed in Bieberbach, p. 213. A function

$$f(z) = \sum_{j=-m}^{\infty} a_j (z - c)^j \qquad (A.17)$$

with $a_{-m} \neq 0$ has a *pole* of order m at c. The series (A.17) with $m \geq 1$ is termed a *Laurent series*. If $m \equiv \infty$, then f is said to have an *essential singularity* at c. If f has a pole at $z = c$, then $1/f$ is bounded in some neighborhood of c. This is *not* the case if c is an essential singularity. At poles, one often considers $1/f$ instead of f.

Poles are isolated

A limit point of poles is an essential singularity. This is a direct consequence of Liouville's theorem. Proof: Bieberbach, p. 150.

Meromorphic and rational functions

A complex function f is *meromorphic* if f is analytic in D except at isolated poles. A *rational* function is the quotient of two polynomials in z.

Poles and rational functions

If f is meromorphic in the whole of $\mathbb{C}$, then f is a rational function. Proof: Bieberbach, p. 151, Neumann (1884), p. 60.

The Casorati–Weierstrass theorem

Let c be an isolated, essential singularity, and U any neighborhood of c. Then $f(U \backslash c)$ is dense in $\mathbb{C}$. More specifically, for any $\epsilon > 0$ and any $b \in \mathbb{C}$, there is an $a \in U$ so that $|f(a) - b| < \epsilon$, i.e., $f(a)$ comes close to b. The case $f(a) = b$ is subject of Picard's theorem. Proof: Remmert, p. 242, Bieberbach, p. 149.

Identity theorem. III

Let f and g be two meromorphic functions in a domain D. The set of points where $f(c_j) = g(c_j) < \infty$ shall have a limit point. Then $f = g$ in D. Proof: Remmert, p. 252, Bieberbach, p. 137.

The Cauchy theorem for rings

A *ring* R is a complex domain with an inner and outer circular boundary. A ring is twofold connected. Let f be analytic in the ring. Let the circles C_1, C_2 belong to the interior of the ring. Then

$$\oint_{C_1} dz \, f(z) = \oint_{C_2} dz \, f(z). \tag{A.18}$$

Proof: Remmert, p. 273, Bieberbach, p. 127 and p. 129.

The Cauchy integral formula for rings

The boundary ∂R of a ring consists of two circles. Integration proceeds in opposite directions along these circles. Let f be analytic in D, and R be a ring so that $R \cup \partial R \subset D$. Then for $z \in R$,

$$f(z) = \frac{1}{2\pi i} \int_{\partial R} d\zeta \, \frac{f(\zeta)}{\zeta - z}. \tag{A.19}$$

Proof: Remmert, p. 274, Bieberbach, p. 139 and p. 145.

The Laurent series

A direct consequence of this formula is the Laurent series theorem. Let f be analytic in the ring R with center c. Then there is a unique Laurent series that converges normally to f,

$$f(z) = \sum_{j=-\infty}^{\infty} a_j (z - c)^j. \tag{A.20}$$

For any circle C inside the ring (not its boundary), the a_j are given by

$$a_j = \frac{1}{2\pi i} \oint_C d\zeta \, \frac{f(\zeta)}{(\zeta - z)^{j+1}}. \tag{A.21}$$

Proof: Remmert, p. 276, Bieberbach, p. 139.

Winding number

Let C be a closed curve and z a point not lying on C. Then

$$\mathrm{ind}_C(z) = \frac{1}{2\pi i} \oint_C \frac{d\zeta}{\zeta - z} \quad \in \mathbb{Z} \tag{A.22}$$

is the *winding number* (number of turns) of C around z. The *interior* of C is the set of all points with ind $\neq 0$. The *exterior* of C is the set of all points with ind $= 0$. A curve is *simple closed*, if ind $= 1$ for all points in its interior.

Null-homolog

A curve C is *null-homolog* in a domain D, if its interior lies in D. Alternatively, C is null-homolog if it does not go round points of the complement of D. That is, a curve is null-homolog if $\mathrm{ind}_C(z) = 0$ for all $z \in \mathbb{C} \backslash D$.

Residue

Let C be a circle with center c and infinitesimal radius. Let f be analytic except at isolated poles or essential singularities. The *residue* $\mathrm{Res}_c f$ of f with respect to c is the integral

$$\mathrm{Res}_c f = \frac{1}{2\pi i} \oint_C dz \, f(z). \tag{A.23}$$

If f is analytic at c, then $\mathrm{Res}_c f = 0$ according to Cauchy's theorem. Let

$$f(z) = \sum_{j=-\infty}^{\infty} a_j (z - c)^j \tag{A.24}$$

for f analytic in a neighborhood U of $D \backslash c$. Then

$$\mathrm{Res}_c f = \frac{1}{2\pi i} \oint_C dz \, f(z) = a_{-1}, \tag{A.25}$$

where the circle C lies in U. For the residue at the point $c \to \infty$, see Bieberbach, p. 172 and p. 174. We have now immediately (see Remmert, p. 304, Bieberbach, p. 197): if c is a pole of f of first order, then

$$\mathrm{Res}_c f = \lim_{z \to c} (z - c) \, f(z). \tag{A.26}$$

If c is a pole of f of order m, then

$$\mathrm{Res}_c f = \frac{1}{(m-1)!} \lim_{z \to c} \frac{d^{m-1}}{dz^{m-1}} \left[(z - c)^m \, f(z) \right]. \tag{A.27}$$

For essential singularities, there is no simple formula for the residue.

Residue theorem

Let C be a closed, null-homolog curve in a domain D. Let $\{c_j\}$ be a finite set of n points, where number of the c_j lies on C. Let f be analytic in $D \backslash \{c_j\}$. Then

$$\frac{1}{2\pi i} \oint_C dz \, f(z) = \sum_{j=1}^{n} \mathrm{ind}_C(c_j) \, \mathrm{Res}_{c_j} f. \tag{A.28}$$

For the proof, see Remmert, p. 306 or Bieberbach, p. 174. In most applications, C is chosen so that $\mathrm{ind}_C(c_j) = 1$ for all c_j: if C is simple closed and null-homolog in D, then

$$\frac{1}{2\pi i} \oint_C dz\, f(z) = \sum_{j=1}^{n} \operatorname{Res}_{c_j} f \qquad (A.29)$$

There are many applications of this theorem; we mention one. Let f be analytic on the whole real axis, except possibly at infinity. Let f be analytic in the upper half-plane except at singularities c_1 to c_n. Finally, f shall vanish more strongly than $1/z$ for large z,

$$\forall \epsilon > 0 \,\, \exists r > 0 \,\, \forall z : \,\, |z| > r \,\, \rightarrow \,\, |z\, f(z)| < \epsilon. \qquad (A.30)$$

Then the integral of f along the real axis converges and is given by

$$\int_{-\infty}^{\infty} dz\, f(z) = 2\pi i \sum_{j=1}^{n} \operatorname{Res}_{c_j} f. \qquad (A.31)$$

Proof: Bieberbach, p. 175.

The Rouche theorem

Let f and g be analytic in D. Let C be a simple closed and null-homolog curve. For all points z on C, let

$$|f(z) - g(z)| < |g(z)|. \qquad (A.32)$$

This also means that $g(z) \neq 0$ on C. Then f and g have the same number of zeros in the interior of C. Proof: Remmert, p. 310, Bieberbach, p. 185.

Monodromy theorem

Let D be an open disk with center z, and $f_0 : D \to \mathbb{C}$ analytic (see Fig. A.1). Let C and C' be two homotopic curves from z to z', and D' an open disk with center z'. If $f, f' : D' \to \mathbb{C}$ are analytic continuations of f_0 along C and C', then $f = f'$. Note that general curves C and C' can only be homotopic if they lie in a simply connected domain. Proof: Jänich (1993), p. 59 and Markushevich (1967), p. 269.

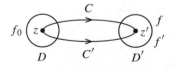

Fig. A.1 Analytic continuation of the function element (f_0, D) centered at z to the function element (f, D') centered at z' along the curve C, and to the function element (f', D') along the homotopic curve C'

References

Ahlfors, L.V. 1966. *Complex analysis*, 2nd ed. New York: McGraw-Hill.

Bernstein, F. 1900. Ueber einen Schönflies'schen Satz der Theorie der stetigen Funktionen zweier reeller Veränderlichen. *Nachrichten von der Gesellschaft der Wissenschaften zu Göttingen, Mathematisch-Physikalische Klasse* 1900: 98.

Bieberbach, L. 1921. *Lehrbuch der Funktionentheorie, Band I, Elemente der Funktionentheorie.* Leipzig: B.G. Teubner. Wiesbaden: Springer Fachmedien.

Brouwer, L.E.J. 1912. Beweis der Invarianz des n-dimensionalen Gebiets. *Mathematische Annalen* 71: 305 and 72: 55.

Dixon, J.D. 1971. A brief proof of Cauchy's integral theorem. *Proceedings of the American Mathematical Society* 29: 625.

Jänich, K. 1993. *Funktionentheorie.* Berlin: Springer.

Markushevich, A.I. 1965. *Theory of functions of a complex variable*, vol. I. Englewood Cliffs: Prentice-Hall.

Markushevich, A.I. 1967. *Theory of functions of a complex variable*, vol. III. Englewood Cliffs: Prentice-Hall.

Neumann, C. 1884. *Vorlesungen über Riemann's Theorie der Abel'schen Integrale*, 2nd ed. Leipzig: Teubner.

Osgood, W.F. 1900. Ueber einen Satz des Herrn Schönflies aus der Theorie der Functionen zweier reeller Veränderlichen. *Nachrichten von der Gesellschaft der Wissenschaften zu Göttingen, Mathematisch-Physikalische Klasse* 1900: 94.

Remmert, R. 1992. *Funktionentheorie 1*, 3rd ed. Berlin: Springer.

Schoenflies, A. 1899. Ueber einen Satz der Analysis Situs, Nachrichten von der Gesellschaft der Wissenschaften zu Göttingen. *Mathematisch-Physikalische Klasse* 1899: 282.

Index

© Springer Nature Switzerland AG 2019
A. Feldmeier, *Theoretical Fluid Dynamics*, Theoretical and Mathematical Physics,
https://doi.org/10.1007/978-3-030-31022-6

Printed in the United States
by Baker & Taylor Publisher Services